PROJECT MASTERY: Bridging Processes, People, and Resources

A Competency-Based Guidebook Focused on Technical Project Management, Leadership, and Business Management

First Edition 2024

ISBN: 979-8-218-37811-0

Centrestar Learning
State College, Pennsylvania, USA
www.centrestar.com

PROJECT MASTERY:

Bridging Processes, People, and Resources

A Competency-Based Guidebook Focused on Technical Project Management, Leadership, and Business Management

Preface

PROJECT MASTERY: Bridging Processes, People, and Resources is a practical guidebook designed for individuals who manage or anticipate managing workplace projects. This audience includes frontline supervisors, managers, engineers, technical specialists, team members, and project stakeholders. The twelve-chapter guidebook is segmented into three parts.

PART I: **Processes** focuses on the technical side of the traditional project management life cycle and covers the tools and techniques most widely used to manage workplace projects.

PART II: **People** focuses on the leadership, management, and people side of project management.

PART III: **Resources** focuses on the business management side of project management, including controlling resources such as time, money, and risk.

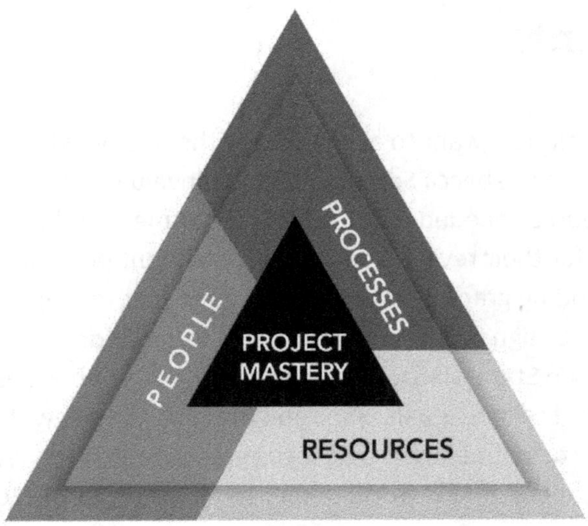

This book grew out of my roles as a project manager, entrepreneur, and business owner. Early in my career as a project engineer, I experienced and observed the impact that project managers and other leaders had on organizations, both positive and negative. As a result, I knew that I needed to hone my skills on the technical or "tools" side of project management as well as the "numbers" side of business to justify projects that I thought we should pursue. However, I also knew that I needed to focus on the "soft skill" or human side to better deal with the many people problems associated with managing projects.

Later, as a business owner, I learned that without customers and loyal employees you have nothing. This gave me a new appreciation for the value and impact of project leadership. Along with this experience came the insight that to retain good employees the golden rule of "treating people the way you want to be treated" is not just a saying, but a rule for us to live by.

Anyone at the operational or senior levels of an organization knows the value and necessity of taking a "bottoms-up" approach to strategic planning. This translates to asking the people doing the work, managing your projects, and those who buy and use your products or services for their input.

No matter what leadership level you achieve, one thing remains true; you either get better or you get worse. This is true for individuals as well as organizations. This book is based on sound research, years of experience in a variety of industries, and the know-how gained in managing projects and teaching others how to be successful leaders. Effective project managers and other leaders achieve success through a combination of learning, planning, and acting.

As famous author Alan Lakein once said, "Planning is bringing the future into the present so that you can do something about it now."

Dr. Wesley E. Donahue
2024

Acknowledgments

We did not create this book alone. I want to acknowledge the people who contributed significantly to the work: Drs. Lisa Donahue and Rebecca Sarnaski for their invaluable research efforts in tracking down articles, books, and other sources needed for content development and identifying examples; Dr. Lisa Turner and Billie Tomlinson for their review and editing of content presented in this book; Valentine Platon and Sommoade for adding graphics that help bring the text to life; Zizi Iryaspraha Subiyarta and Alex Donahue for thoughtful design suggestions; Amit Dey for text formatting; former instructors and professional associates at Penn State Management who shared their business and industry wisdom and years of teaching experience; the individuals who took the time to review the manuscript, and the many thousands of people who participated in surveys, focus groups, and interviews, without whom this workbook would lack the richness of real-world detail. To all these people I offer a sincere and heartfelt thank you!

Contents

Introduction

Audience

PROJECT MASTERY: Bridging Processes, People, and Resources is a practical three-part guidebook designed for individuals who manage or anticipate managing workplace projects. The target audience includes frontline supervisors, managers, engineers, technical specialists, team members, and project stakeholders from any industry.

The twelve-chapter guidebook is segmented into the three parts illustrated in the following table. The overarching competencies covered are broadly described as Technical Project Management, Leadership, and Business Management. However, for each chapter, the three individual competencies most closely associated and aligned with the Centrestar Competency Model are subsequently noted and described.

PROJECT MASTERY: Bridging Processes, People, and Resources		
Part 1 **Processes**	Part 2 **People**	Part 3 **Resources**
• Contemporary Project Management • Defining Project Scope • Project Planning and Scheduling Tools • Monitoring and Controlling Projects	• Enhancing Project Communication • Developing Interpersonal Project Relationships • Building High-Performance Project Teams • Leading and Managing Projects	• Identifying and Mitigating Project Risk • Controlling Project Costs and Budgets • Managing Time and Project Meetings • Completing and Closing Projects

Part 1, **Processes,** focuses on the technical side of the traditional project management life cycle and covers the tools and techniques most widely used to manage workplace projects. This part primarily covers the competency areas of Planning and Evaluation, Resource Management, Problem Solving, Communication, Leadership and Coaching, and Management Controls.

Part 2, **People**, focuses on leadership and the people side of project management. It primarily covers the competency areas of Oral and Written Communication, Active Listening, Interpersonal Skills, Interpersonal Relationship Building, Conflict Management, Influencing and Negotiating, Teamwork and Cooperation, Leadership and Coaching, and Managing a Diverse Workforce.

Book 3, **Resources**, focuses on the business management side of project management. It primarily covers the competency areas of Resource Usage and Management, Understanding Systems, Financial Management and Budgeting, Self-Direction, Decisiveness, Planning and Evaluation, and Management Controls.

All three parts are aligned with the Project Management Institute's (PMI)® Project Management Body of Knowledge (PMBOK® Guide) and are competency-based, focusing on sets of skills, knowledge, attitudes, and behaviors that are observable and measurable.

As a bonus, each of the 12 chapters in the series contains a 20-question knowledge review test with answers for individuals interested in potentially pursuing certification through the Project Management Institute or obtaining Google's Project Management Certificate.

Structured Learning Design

All our books and workbooks align with the research-based **Centrestar Competency Model.** The model, which is rooted in work by the U.S. Department of Labor and others, gives you a framework for structured learning by helping you identify and develop specific competencies.

If you ask people to define competency, you may be amazed at the variety of responses you receive. We define a *competency* as a set of skills, knowledge, attitudes, and behaviors that are observable and measurable. *The emphasis here is on observable and measurable*. It is not enough that you *think* you are competent in an area. You must be able to *behave* in ways that demonstrate your competence to others.

Our framework has **35 competency dimensions** associated with successful performance in leadership and professional roles.

Centrestar Competency Model™

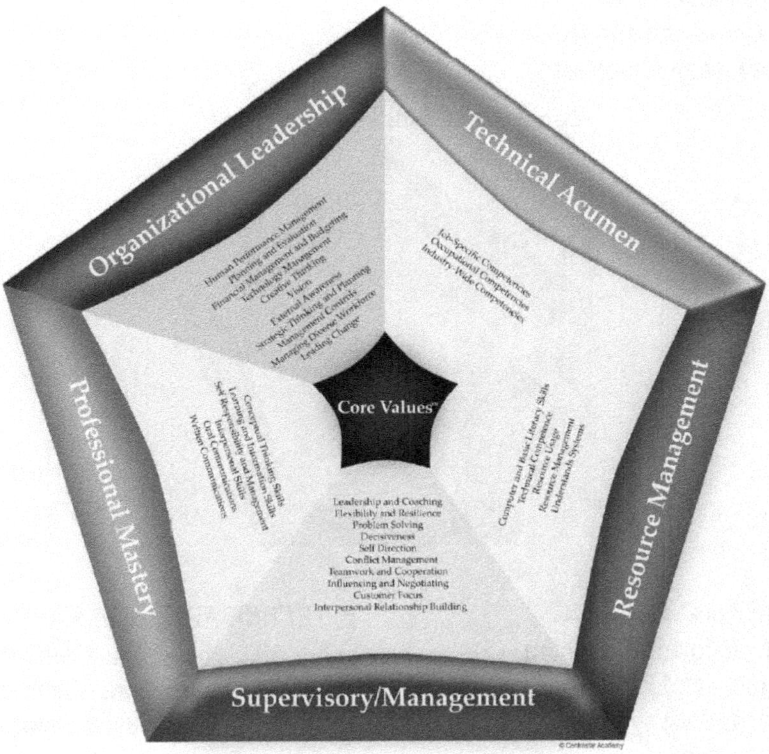

Based on thousands of business participant responses from many industries, we clustered these 35 competencies into five competency domains, which we named and coded as follows:

A. Resource Management
B. Professional Competence
C. Supervisory/Management
D. Organizational Leadership
E. Technical Acumen

Competencies and clusters may overlap. The content of this **PROJECT MASTERY: Bridging Processes, People, and Resources** book are most closely associated with the competency clusters indicated below.

A. RESOURCE MANAGEMENT

_____ 1. Computer and Literacy Skills
_____ 2. Technical Competence
__X__ 3. Resource Usage
__X__ 4. Resource Management
__X__ 5. Understands Systems

B. PROFESSIONAL COMPETENCE

_____ 6. Conceptual Thinking
_____ 7. Learning and Information Skills
__X__ 8. Self-Responsibility and Management
__X__ 9. Interpersonal Skills
__X__ 10. Oral Communication
__X__ 11. Written Communication

D. ORGANIZATIONAL LEADERSHIP

_____ 22. Human Performance Management
__X__ 23. Planning and Evaluation
__X__ 24. Financial Management and Budgeting
_____ 25. Technology Management
_____ 26. Creative Thinking
_____ 27. Vision
_____ 28. External Awareness
_____ 29. Strategic Thinking and Planning
__X__ 30. Management Controls
__X__ 31. Diverse Workforce
_____ 32. Leading Change

Centrestar Competency Model

C. SUPERVISORY MANAGEMENT

__X__ 12. Leadership and Coaching
_____ 13. Flexibility and Resilience
__X__ 14. Problem Solving
__X__ 15. Decisiveness
__X__ 16. Self-Direction
__X__ 17. Conflict Management
__X__ 18. Teamwork and Cooperation
__X__ 19. Influencing and Negotiating
_____ 20. Customer Focus
__X__ 21. Interpersonal Relationship Building

E. TECHNICAL ACUMEN

_____ 33. Job-Specific Competencies
_____ 34. Occupational Competencies
_____ 35. Industry-Wide Competencies

This **PROJECT MASTERY: Bridging Processes, People, and Resources** guidebook primarily focuses on the competency areas of:

- **Resource Usage**. Identifies, organizes, uses, and allocates resources.
- **Resource Management.** Demonstrates awareness of technical resources; knows how to apply resources to achieve desired outcomes.
- **Understand Systems.** Grasps complex interrelationships and interdependencies.
- **Self-Responsibility and Management.** Displays responsibility, self-confidence, emotional self-control, integrity, and honesty.
- **Interpersonal Skills.** Appropriately sociable, interacts effectively with others.
- **Oral Communication.** Makes clear and effective presentations to individuals and groups; listens to others.
- **Written Communication.** Communicates effectively in writing; can critically review and comprehend information written by others.
- **Leadership and Coaching.** Models and encourages high standards of ethical behavior; adapts leadership styles to situations and people; empowers, motivates, guides, and coaches.
- **Problem Solving.** Recognizes and defines problems; analyzes relevant information; encourages alternative solutions, develops plans to solve problems.
- **Decisiveness.** Decides and responds quickly; makes difficult decisions as necessary.
- **Self-Direction.** Realistically assesses strengths and weaknesses; invests in self-development; demonstrates self-confidence; works persistently toward a goal; manages time effectively.
- **Conflict Management.** Anticipates and seeks to resolve disagreements, complaints, and confrontations in a constructive manner.
- **Teamwork and Cooperation.** Demonstrates and fosters cooperation, communication, and consensus among individuals and groups.
- **Influencing and Negotiating**. Keeps groups and individuals informed; appropriately uses negotiation, persuasion, and authority in working with others to achieve goals; builds productive networks.
- **Interpersonal Relationship Building.** Considers and responds appropriately to the needs, feelings, and capabilities of others; seeks feedback and accurately assesses impact on others; provides helpful feedback; builds trust with others.
- **Planning and Evaluation.** Establishes policies, guidelines, plans, and priorities; plans and coordinates with others; aligns required resources; monitors progress and evaluates outcomes; improves organizational efficiency and effectiveness.
- **Financial Management and Budgeting**. Understands budget process; prepares and justifies budgets; monitors expenses; manages profit/loss ratios as appropriate.
- **Management Controls.** Ensures the integrity of the organization's processes; promotes ethical and effective practices.
- **Diverse Workforce**. Recognizes the value of cultural, ethnic, gender, and other individual differences; provides employment and development opportunities for a diverse workforce.

How to Use This Book

This book is structured to be a hands-on guide with options for how you can use the material.

One way to use the book is to read it straight through. Or you can jump to specific chapters depending on your interests and goals. We recommend you start by reviewing the table of contents, so you understand the workbook's organization. Read the overview of each chapter and then complete the introductory assessment by rating your level of agreement with the ten statements. This will help you clarify your current thinking. After that, visually scan the book. Scanning will help you pinpoint areas where you may have an information gap and areas where you feel confident. From there, you can set your learning goals and dive deeper into the material.

The book will help you develop your competence by addressing the ten most vital concepts associated with each of the twelve topical chapters. The content for the ten concepts is structured in a consistent learning format.

Each concept starts with several paragraphs of relevant content and ends with three **What to Do** action suggestions, followed by a **Remember** section which lists important learning points. To **Enhance Your Learning**, we offer links to additional resources. Each concept area concludes with a **Reinforce Your Learning** application activity.

At the end of each topical chapter is a **Summary** and a **Recap Checklist**. The recap lists all 30 of the **What to Do** actions from each concept in the workbook.

As with most things in life, you will get out of this book only what you put into it. To learn and grow, you must engage fully with the material presented. Ask yourself questions as you read the material in the book. Do the activities and answer the questions in each concept. Make notes. Look for ideas new to you and consider how they fit with your current knowledge. Recognize what you already know and can build on, and what might be a new way of looking at something.

This book aims to ensure that your skills are at the highest level they can be. Engaging with the concepts and materials will help you become a more rounded professional who experiences success in your chosen career.

Part I: Processes

Part I: **Processes** focuses on the technical side of the traditional project management life cycle and covers the tools and techniques most widely used to manage workplace projects.

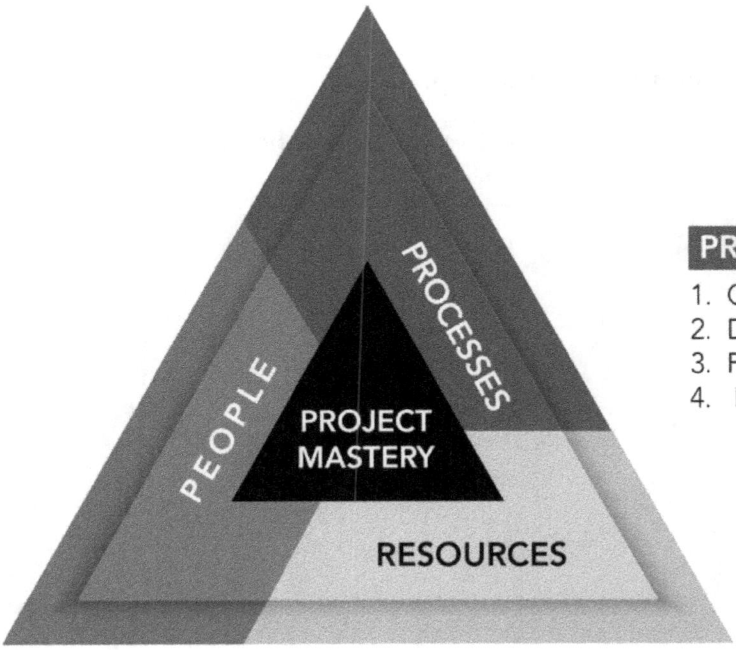

PROCESSES

1. Contemporary Project Management
2. Defining Project Scope
3. Project Planning and Scheduling Tools
4. Monitoring and Controlling Projects

Plans are nothing; planning is everything.

—Dwight D. Eisenhower

Chapter 1

Contemporary Project Management

Overview

"Plans are nothing; planning is everything."
—Dwight D. Eisenhower

A *project* is a temporary work assignment to achieve a specific goal or outcome. It is essential to recognize that the project life cycle includes the phases of beginning, planning, doing the work, and closing, and that people's actions may account for the majority of project problems. These problems include issues related to the management of the project.

This chapter describes the project management process and examines project problems, issues, and challenges with a focus on the human side. Understanding the roles, responsibilities, and attributes of project managers is essential for selecting individuals with the appropriate qualifications.

In this chapter, you will learn how to:

- Distinguish project management from conventional management
- Describe the general phases of a project
- Understand the importance of a formalized project management methodology
- Recognize the attributes of a project manager
- Understand the key roles of project managers and the responsibilities associated with each

The competencies associated with this chapter include:

Planning and Evaluation	Resource Management	Leadership and Coaching

Take Your Temperature for Contemporary Project Management

With yourself and your organization in mind, read each statement carefully. Next to each statement, write the number (from 1 to 10) that indicates your level of agreement with the statement.

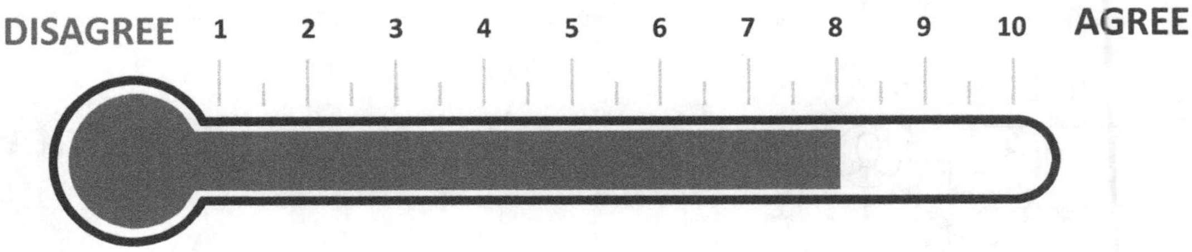

_____ 1. I can explain what a project is and the benefits of project management.

_____ 2. I am aware of Project Management Institute (PMI)® professional certifications.

_____ 3. I know the benefits of using a project management methodology.

_____ 4. I know how project problems, issues, and challenges affect project success.

_____ 5. I can explain the project management life cycle.

_____ 6. I can describe the project management phases.

_____ 7. I am confident in my ability to discuss project manager responsibilities.

_____ 8. I can list important project manager attributes.

_____ 9. I can describe the roles a project manager fulfills.

_____ 10. I know the merits of having a project management office.

_____ **Total**

Take a moment to reflect on the following before you move forward in the chapter.
In your work, what problems have you experienced or observed related to project management?

Take a few minutes to also reflect on your self-assessment. List two or three areas you want to improve.

As you progress through this chapter consider what actions you can take to demonstrate competence in the following three competency areas:

1. **Planning and Evaluation**. Establishes policies, guidelines, plans, and priorities; plans and coordinates with others; aligns required resources; monitors progress and evaluates outcomes; improves organizational efficiency and effectiveness.

2. **Resource Management.** Demonstrates awareness of technical resources; knows how to apply resources to achieve desired outcomes.

3. **Leadership and Coaching.** Models and encourages high standards of ethical behavior; adapts leadership styles to situations and people; empowers, motivates, guides, and coaches.

Contemporary Project Management #1.1

Define Project and Project Management

According to the **Project Management Institute (PMI)®** and ***A Guide to the Project Management Body of Knowledge* (*PMBOK*)®**,[1] *project management* is the application of knowledge, skills, tools, and techniques to project activities to meet project requirements. A *project* is a temporary endeavor undertaken to create a specific product, service, or result.

Project management differs from conventional management in that projects have a finite life with a specific beginning and end.

Examples of Activities Associated with Projects and Project Management	Examples of Activities Associated with Conventional Management
Constructing a building	Maintaining a building
Relocating an office	Managing an office
Installing a piece of equipment	Operating equipment
Developing a strategic plan	Leading an organization
Developing a training program	Managing performance

Here are the typical characteristics of most projects:

- Limited time span
- Led by a project manager
- Completed by a team whose job is in their functional area, so team members report back to their operational areas after the project is over, and they sometimes work on a project part-time
- Constrained by human performance, time, and budget
- Organized to achieve a specific goal (product, service, or result)

Program management is the management of a group of related projects. It may involve sharing resources between projects. The work product of one project may affect associated projects.

Portfolio management typically includes the management of multiple programs. It is the most complex level in the project management hierarchy because of the number of programs and associated projects.

The graphic that follows illustrates the Traditional Project Management Hierarchy, with the broadest or most encompassing level at the top and moving to the most specific level, which is the management of a single project. Notice the "we are here" arrow. It highlights that this workbook is primarily concerned with the management of one project and not with program or portfolio management.

[1] PMI and *PMBOK* are registered marks of the Project Management Institute, Inc.

TRADITIONAL PROJECT MANAGEMENT HIERARCHY

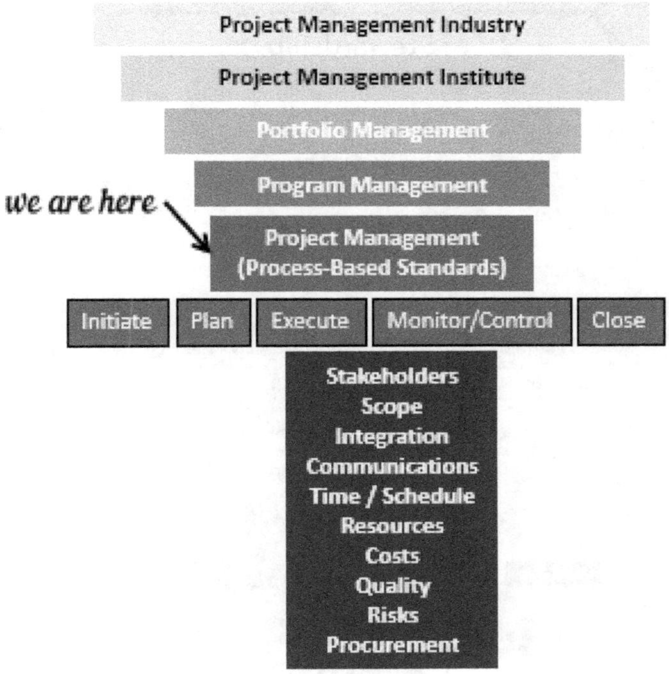

Also notice that project management is divided into major phases: Initiate; Plan; Execute; Monitor & Control; Close. Note that managing a project requires consideration of ten related knowledge or process areas. These knowledge and process areas are stakeholders, project scope, integration, communications, time and schedule, resources, costs, quality, risks, and procurement. These ten areas are the ten *knowledge areas* within the Project Management Body of Knowledge (PMBOK® Guide, 6th Edition). We will discuss these knowledge areas throughout this course.

Recent Developments by the Project Management Institute (PMI)®

The Project Management Institute (PMI) has also recently developed standards (PMBOK® Guide, 7th Edition) by which project management may be viewed from a couple of additional perspectives. In addition to looking at **project management** from a **process-based** standards orientation, project management may be viewed from a **principles-based** orientation as well as a **systems-based** orientation. For the process-based orientation, the focus is on the deliverables; for the principles-based orientation the focus is on outcomes, and for the systems-based orientation, the focus is on linking business capabilities to meet an organization's strategy, value, and business objectives.

These additional orientations are not meant to negate the process-based standards orientation, but rather to compliment them. The additional orientations add a broader global business perspective to the field of project management by permitting a shift in thinking from the traditional ten knowledge areas (illustrated in the previous graphic) towards the additional twelve guiding project management principles and eight project performance domains as depicted in the Contemporary Project Management Hierarchy graphic that follows. The additional orientations may be particularly useful from the perspective of governing larger projects, programs, and portfolios, and linking them to overarching business capabilities, strategies, and objectives.

Contemporary Project Management Environment

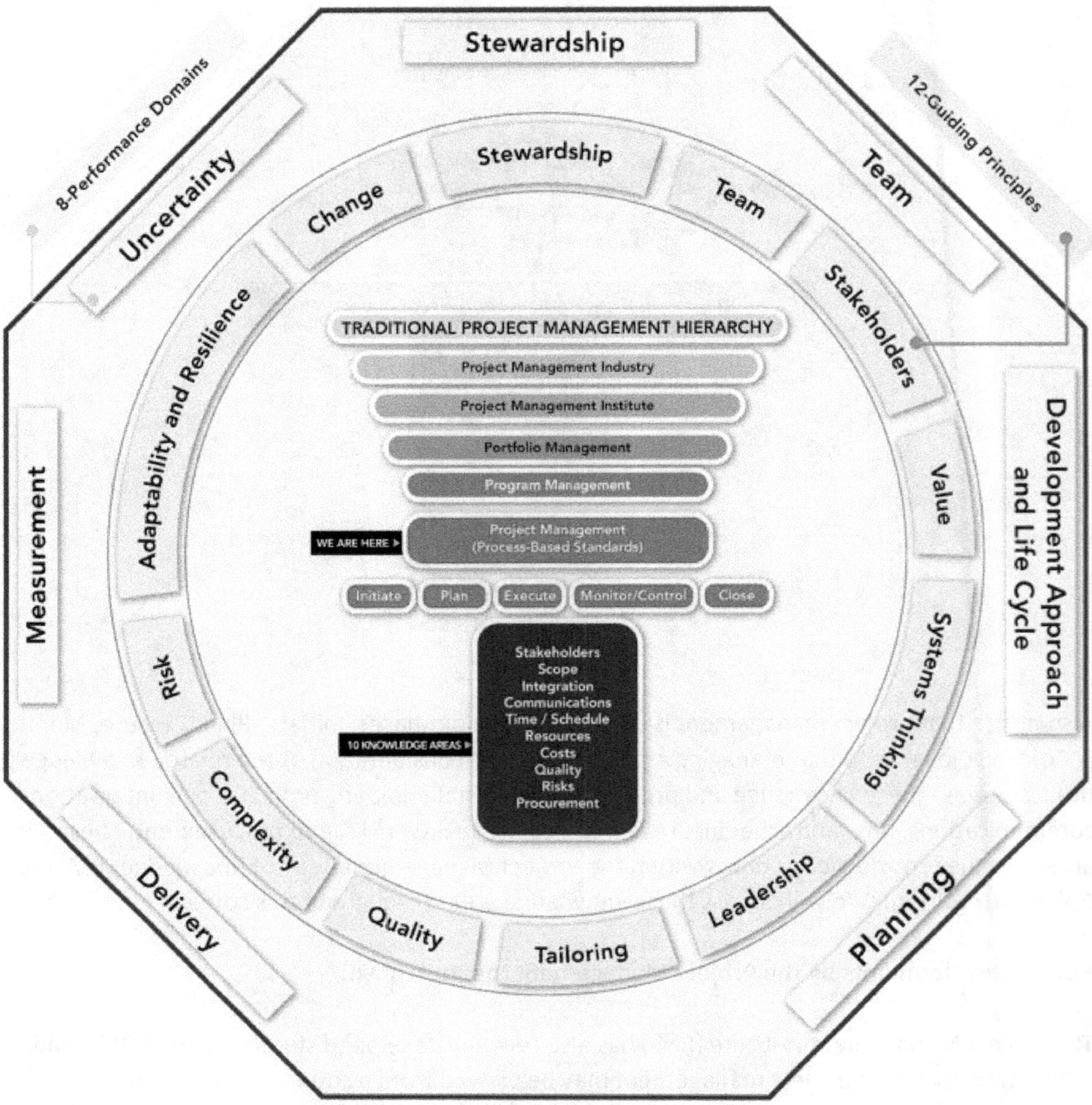

What to do:	Other actions:
☐ Review the Project Management Institute (PMI)® website (www.pmi.org).	☐ ________________
☐ Identify some of the projects you have been involved in or observed.	☐ ________________
☐ Brainstorm issues, problems, and training needs related to those projects.	☐ ________________
	☐ ________________

Remember

- ✓ A project is a temporary endeavor undertaken to create a product, service, or result.
- ✓ Project management differs from conventional management in that projects have a finite life cycle with a specific beginning and end.
- ✓ The process of managing a project is divided into phases: Initiate, Plan, Do, Close.

Enhance Your Learning

Review activities associated with the Project Management Institute (PMI)® at:

Introduction to Project Management Institute (PMI)®

PMI Certifications

PMI - PMBOK® Guide & Standards

Reinforce Your Learning

Typical Project Management Problems. Consider typical problems, issues, challenges, and needs related to project management that you have experienced or observed. Take a few minutes to list them below.

Contemporary Project Management #1.2

Become Familiar with the Project Management Institute (PMI)®

Project Management Institute (PMI)®

The project management industry is a global collection of project management professionals and organizations and the activities associated with their work. The **Project Management Institute (PMI)®**, founded in 1969 and with over 500,000 members worldwide, is a premier organization for professional project managers. PMI sets standards and delivers seminars and educational programs, and it offers professional certifications.

***A Guide to the Project Management Body of Knowledge* (PMBOK)®**

The PMI® organization works with volunteers to create industry standards, such as *A Guide to the Project Management Body of Knowledge (PMBOK)®*. This guide is in its 7th edition, released in the fall of 2021. However, most of the content in the 6th edition is still relevant and referred to in the 7th edition.

Past editions of the *PMBOK® Guide* (including the 6th edition) have represented a process-based standard for project management. In these guides, inputs, tools, and techniques, and outputs related to the project management processes are discussed, and where the outputs of one process can be the inputs to another process.

Due to the continually evolving landscape of project management, the Project Management Institute (PMI)® introduced the 7th edition which incorporated, in addition to the traditional process-based standards, an orientation towards a principles-based standard with a focus on outcomes instead of deliverables and incorporated a broader systems-view of project management.

PMI Certifications

PMI offers specialized certification in several areas of project management to support individuals in developing skills in related areas. Like a degree or other focused training, a PMI certification shows that the holder has the skills necessary to do a professional job. The following are some of the most common certifications PMI offers:

- Portfolio Project Manager (PfMP)®
- Program Management Professional (PgMP)®
- Project Management Professional (PMP)®
- Certified Associate Project Manager (CAPM)®
- Business Analyst (PMI-PBA)®
- Agile Certified Practitioner (PMI-ACP)®
- Risk Management Professional (PMI-RMP)®
- Scheduling Professional (PMI-SP)®

The **Portfolio Management Professional (PfMP)**® certification is for professionals who manage groups of projects as they relate to an organization's strategic goals. A PfMP® can manage one or more portfolios.

Requirements for the PfMP® include a panel review; passing a 4-hour, 170-question multiple-choice exam; and **either** a bachelor's degree (or global equivalent) with a minimum four years (6,000 hours) of portfolio management experience and eight years (96 months) of professional business experience within the last 15 years, **or** a high school diploma (or global equivalent) with a minimum of seven years (10,500 hours) of portfolio management experience, and eight years (96 months) of professional business experience, within the last 15 years.

The **Program Management Professional (PgMP)**® certification shows the holder is a senior-level practitioner who has the skills to manage related projects and to coordinate them where possible. Program-level management facilitates a synergism among projects that would otherwise not be possible.

Requirements for the PgMP® certification include a bachelor's degree (or global equivalent) with a minimum of four years (6,000 hours) of project management experience **and** seven years (10,500 hours) program management experience within the last 15 years; a panel review; and passing a 4-hour, 170-question, multiple-choice exam.

The Project Management Professional (PMP)® certification shows the holder has the skills and knowledge to manage projects from beginning to end. It requires passing an exam that tests a candidate's working knowledge of professional practices. The certification usually helps the holder earn more money.

Requirements for the PMP® certification include 35 hours of professional project management education; passing a 4-hour, 200-question, multiple-choice exam; and either work experience to include a bachelor's degree (or global equivalent) and a minimum of three years (4,500 hours) leading projects, or a high school diploma (or global equivalent) and a minimum of five years (7,500 hours) leading projects.

The Certified Associate Project Manager (CAPM)® is for people who are involved or associated with projects but not currently working as a project manager. Examples are a project team member, a project stakeholder, or a subject matter expert (SME) who works with project teams. This certification shows the holder is aware of project management terminology and methodologies.

Requirements for the CAPM® includes a high school diploma, associate degree (or global equivalent); passing a 3-hour, 150-question, multiple-choice exam; and **either** at least 23 hours of project management education without professional work experience; **or** at least 1,500 hours of experience on a project team without project management education.

The **PMI Professional in Business Analysis (PMI-PBA)®** certification addresses the need for professionals who have the skills to identify and document project requirements. Project requirements are the foundation of the project scope, which is used to develop the project plan. The ability to capture requirements accurately and efficiently is a significant benefit to a project team. Areas of focus include outputs and projected business results.

Requirements for the PMI-PBA® include 35 hours of business analysis education, passing a 4-hour, 200-question multiple-choice exam; and **either** a bachelor's degree (or global equivalent) with minimum three years (4,500 hours) of business analysis experience and 2,000 hours working on project teams within the last eight years; **or** a high school diploma (or global equivalent) with a minimum five years (7,500 hours) of business analysis experience and 2,000 hours working on project teams within the last eight years.

The **PMI-Agile Certified Practitioner (PMI-ACP)®** certification shows the holder has knowledge of the principles and techniques associated with the Agile project management method. The education covers various approaches to Agile, including Lean, Scrum, Kanban, extreme programming (XP), Crystal, and test-driven development (TDD).

Requirements for the PMI-ACP® include a minimum of 21 hours of professional education in Agile practices; 2,000 hours working on project teams in the last five years; **or** 1,500 hours working on Agile project teams or with Agile methods in the previous three years.

The **PMI Risk Management Professional (PMI-RMP)®** certification focuses on the skills needed to identify and mitigate negative risks while leveraging positive risks. As project environments become increasingly complex, risk management expertise has become more necessary to project success, and it is a specialized skill. Organizations that practice professional risk management are more successful at completing projects within their constraints.

Requirements for the PMI-RMP® certification include **either** a bachelor's degree (or global equivalent) with a minimum of 30 hours in project risk management; **or** a bachelor's degree (or global equivalent) with a minimum 3,000 hours of work experience in project risk management within the last five years.

The PMI Scheduling Professional (PMI-SP)® certification shows the holder has the skills to manage all aspects of project scheduling. Task duration estimates, task relationships, interactions with shared resources, and global concerns are all considerations in developing and maintaining a schedule.

Requirements for the PMI-SP® certification include passing a 3.5 hour, 170-question multiple-choice exam; either a bachelor's degree (or global equivalent) with a minimum 30 hours in project scheduling or a high school diploma (or global equivalent) with a minimum 40 hours in project scheduling; and either a bachelor's degree (or global equivalent) with minimum 3,500 hours of work experience in project scheduling within the last five years or a high school diploma (or global equivalent) with a minimum 5,000 hours of work experience in project scheduling within the previous five years.

PMP® Examination

The PMP® examination is the most popular by number of people certified and validates a candidate's knowledge in the domains of people, process, and business environment. According to PMI, the following is an approximate breakdown of the percentage of questions in each of the three groupings:

Domain	Percentage of Items on Exam
People	42%
Process	50%
Business Environment	8%
TOTAL	100%

While various industries may use different terminology, the test focuses on the universal knowledge areas necessary to manage projects successfully in any organization. For example, consulting engineers may use the term "subcontracting" as opposed to "procurement planning." But across industries, good management practices are the same and the differences are semantic.

After Certification

After attaining the Project Management Professional Certification, a PMP® must maintain status by satisfying the requirements of the Professional Development Program. Certified project managers must complete no less than 60 Professional Development Units (PDUs) within a three-year cycle and agree to continue to adhere to the PMI Code of Professional Conduct.

You can find more information about the Professional Development Program and Professional Development Units on the PMI website.

What to do:

- ☐ Discuss project management certifications with your manager for applicability.
- ☐ Find a certified associate and ask them about the advantages of being certified.
- ☐ Pursue one of the certifications, if appropriate.

Other actions:

- ☐ ______________________
- ☐ ______________________
- ☐ ______________________
- ☐ ______________________

Remember

- ✓ PMI offers many educational opportunities and several types of professional certifications. Some of these may be of benefit to you in your job.
- ✓ Explore the PMI website and talk to people in your organization who are familiar with PMI.
- ✓ Consider some projects in your organization and the kinds of skills needed to manage those projects.

Enhance Your Learning

The following link provides more information about the Project Management Professional (PMP)®:

https://www.pmi.org/certifications/project-management-pmp

Reinforce Your Learning

Benefits of Project Management Certifications. Consider projects you have been associated with or observed. Describe how the skills gained in PMI certifications may have helped prevent problems or challenges on these projects?

Contemporary Project Management #1.3

Know the Benefits of Using a Project Management Methodology

The payoff of following a project management methodology is great. Organizations benefit from the efficient and effective creation of a new product, service, or result. Stakeholders benefit from achieving their business and operational goals. Project team members benefit through the satisfaction of a job well done and by gaining experience that will help them on the next project and in building their careers.

On the flip side, those who do not use a project management methodology are more likely to fail in delivering a product, service, or result. According to *Entrepreneur Magazine*, here are **five mistakes** that **cause projects to fail** (2015):

1. Eyes bigger than your budget.
2. Working with the wrong talent.
3. Continuing to pursue bad ideas.
4. Death by committee.
5. Keeping the final approver in the dark.

You can avoid or mitigate these problems by adhering to an established project management methodology. However, there are a number of popular project management methodologies employed in business and industry today. Most follow a system of practices, techniques, procedures, and rules used by those who work in various disciplines. Some of the popular project management methodologies include:

- **Agile** fosters iterative development and team collaboration.
- **Scrum** is a flexible iterative cross-functional "sprint event" approach.
- **Kanban** focuses on visibility of work-in-progress.
- **Lean** focuses on streamlining and getting rid of waste.
- **Waterfall** is a linear sequential approach where progress flows downward.
- **Six Sigma** aims to improve quality by reducing errors.
- **Prince2** defines inputs and outputs for every stage of a project.
- **PMI/PMBOK®** involves using five process groups (initiating, planning, executing, monitoring and controlling, and closing).

The Methodology Does Not Ensure Success

No matter the methodology, size, or complexity, how forgiving or tight the schedule, the skills of the team members, or the commitment of the stakeholders, all projects have one thing in common. This essential factor is that without careful planning, a project will be plagued by one or more of the following problems:

- Delays and missed deadlines
- Wasted time and resources
- Cost overruns
- Inadequate quality
- Conflicts

Poor project planning can sink a project. However, you can avoid or mitigate these problems by adhering to an established project management methodology. Here are a few examples of the benefits of using a project management methodology:

- Project goals will be clearer.
- Duplication of effort which could result in cost overruns will be lessened.
- Potential issues can be identified before causing unnecessary problems.
- Potential negative events can be anticipated, and the risk planned for.
- Schedules and deadlines can be maintained.
- Potential litigation can be avoided from negative events or issues.
- Morale can be maintained by building teams with the right people.
- Workloads can be balanced.

Almost any major project will cost more, take more time, and have more problems than it would have if effectively planned. This means following a proven project management methodology. Planning is at the heart of all successful projects.

What to do:	**Other actions:**
☐ Consider the benefits of following a step-by-step project management methodology.	☐ ______________
☐ Identify the sources/kinds of information you need access to.	☐ ______________
☐ Reflect on the risks in your project and what their impact could be.	☐ ______________
	☐ ______________

Remember

✓ Following a project management methodology improves communication on a project. Everyone has the same expectations and shares the same information.

✓ We have all seen what happens when projects go wrong. Problems are less likely to occur when everyone follows the same project management process, and when problems arise, they are easier to understood and correct.

✓ Communication is a critical part of any project.

Enhance Your Learning

Watch the following 3-minute video that overviews the importance of project planning by Andy Kaufman. Leading cause of project failure: Poor project planning.

Kaufman, A. (2010). *Leading cause of project failure: Poor project planning.*		Available at: http://www.youtube.com/watch?v=MwQ_5JfDIJs

Reinforce Your Learning

Identify Causes of Project Failure. In your experience, what are the most common causes of project failure?

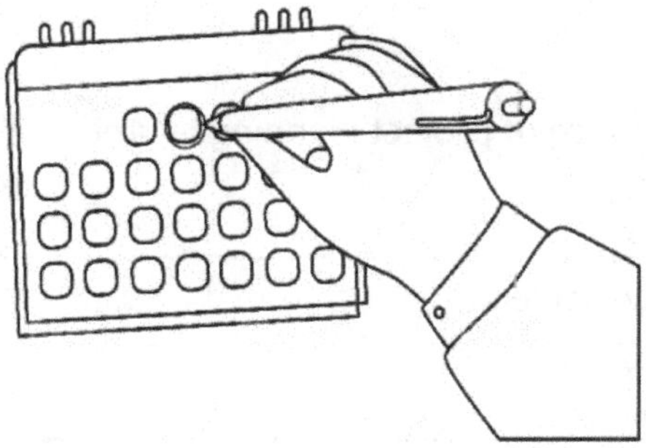

Contemporary Project Management #1.4

Be Aware of How Projects Begin, the People Involved, and Quality

Organizations move forward based on the success of the projects they choose to undertake and how successfully they complete them. Projects start as ideas championed by various people in an organization, and these people typically expect high-quality results when ideas become projects.

In this concept, we will discuss how ideas become projects, the people involved in projects, and how project managers are responsible for delivering quality.

Project Ideas

Ideas for projects come from many sources, and projects can be internal to an organization or external. For example, you work in your organization's information technology department and your department provides project services to develop and install business software for the organization. Or your organization might contract a project with another organization to build commercial office space.

Anyone can suggest an idea for a project. For example, an idea might come from a senior manager who wants to support a new business objective, an operations manager who needs more space or to improve a business process, or from a technical person who has an idea for a better way of doing something.

Organizations gather project ideas and then evaluate them to identify the projects that will bring the greatest benefits. Most organizations do not have the resources to develop all projects at once, so they plan and prioritize projects to achieve the maximum benefits to the business.

How Organizations Review Project Ideas

Organizations review ideas for projects in many ways, but most include an assessment of the benefits the project could bring to the organization. Some organizations do formal studies. For example, they may do a feasibility or return-on-investment (ROI) study. The purpose is to better describe the project, and so to better calculate the resources, costs, and time needed to complete the project. Or, some organizations may do studies during the first phase of the project, which is the planning phase, to clearly define the project's requirements and costs.

Once an idea for a project is approved, larger organizations may go a step further and formally compare one project to another to determine which project has priority. For example, an organization may decide that Project A will result in added revenues immediately, while Project B will not produce new revenue until the following year. The organization may then choose to do Project A first, and once it is underway or complete, they will do Project B.

People Involved in a Project

Each project has a sponsor who formally presents an idea for a project and approves the initial budget for the project. Sponsors are typically senior managers, and they may not have much day-to-day involvement in the project after it begins.

Approved projects have someone named as the primary client. The client is one of the stakeholders but is responsible for the business oversight of the project as it progresses. For example, the client may be an operational manager whose business area will have the most to gain when the project is complete. Or the client could be a manager who works for an outside organization the project has contracted with to provide a product, service, or result. The client has the final say in any necessary business decisions as the project progresses. For example, the client must give the final approval for any changes to the project once it is underway because changes usually affect project costs.

People who helped champion the project idea may become project stakeholders. According to PMI, a stakeholder is anyone affected by a project. For example, besides the project's client, stakeholders could include the sponsor, other executives, senior managers, operational managers, end-users, and even members of the public.

Like the client, stakeholders have an interest in the project's product, service, or result, and they must be kept informed of status and issues as the project progresses. Stakeholders often assist the client in making project business decisions, for example, analyzing proposed changes, helping solve problems with vendors, or determining a business solution for a project risk.

The project manager is responsible for project integration and putting all the pieces of a project into a cohesive whole. This includes planning, managing (monitoring and controlling), and communicating all day-to-day project activities throughout the complete life cycle of the project. The project manager is a professional who serves as the focal point of the project, the person who needs to have a handle on everything happening, and who communicates project information with the client and other stakeholders, the team, and any external people involved in the project.

The project team is responsible for doing the work of the project. The team may or may not formally report to the project manager, though team members are accountable for the tasks assigned by the project manager.

Sometimes the project manager picks team members, but sometimes team members are assigned to the project by the organization. They may work on the project full-time or part-time, depending on the project's needs for the skills they provide.

Quality

Quality is a goal in all organizations. One definition of *quality* is that it measures the degree to which a process, product, service, or result conforms to established requirements.

Quality in a project is often assessed by the stakeholders based on their expectations about the product, service, or result the project will provide. For example, an operations manager in a warehouse may want a new software system to better support the warehouse's business processes.

Stakeholder expectations are not always clear to the project manager and team at the start of the project. But the project may not be judged to be successful if it does not satisfy the expectations.

To deliver a successful project, the project manager must work to understand all stakeholder expectations, ensure the project meets those expectations or if not, that the stakeholders are aware of the reasons why not.

The project manager is responsible for defining quality standards during the planning phase, and for monitoring and controlling quality over the life of the project.

Projects are judged to be successful in terms of quality when the following criteria are satisfied:

- Business objectives are met
- Scope requirements are met
- Completed within budget
- Schedule is met
- Client is satisfied

What to do:	Other actions:
☐ Identify the criteria your organization uses to judge project success.	☐ ____________________
☐ Define what quality means for projects you (or your organization) have completed.	☐ ____________________
☐ Complete an Idea Appraisal Worksheet for a project idea that you or someone in your organization is considering.	☐ ____________________
	☐ ____________________

Remember

- ✓ Almost any project will cost more, take more time, and create more problems than it would have if effective project management had been used.
- ✓ At the beginning of a project, reviewing historical records and seeking advice from those who managed similar projects in the past is a good idea.
- ✓ The potential payoff of following a step-by-step planning process is high.
- ✓ Evaluate the merits of various project ideas before investing effort and resources. Consider ranking ideas and potential projects to identify those with the greatest benefit to the organization.

Enhance Your Learning

Review the article by ProjectManagement.com to learn more about the reasons why projects fail.

Project Management.com. (2021). *Top 10 Reasons Why Projects Fail.*	Available at: https://project-management.com/top-10-reasons-why-projects-fail/

Reinforce Your Learning

Project Idea Appraisal Worksheet. Projects start with ideas. A manager may wonder if relocating a warehouse would save money; a product manager may want to attract new business by adding features to a software product, or you may need to plan a business trip.

Identify a project that you or someone in your organization is considering. With that project in mind, complete the **Idea Appraisal Worksheet** that follows.

For more sophisticated project ideas, you may need build a business case to warrant further examination. If so, there are a number of free Project Management templates available. Here is a link to a free **Business Case Template** available from Project Management Docs:

> https://psu.instructure.com/courses/2185007/pages/project-management-docs-templates-%7C-focus-on-application?module_item_id=35373675

Project Idea Appraisal Worksheet

Use this worksheet to vet or evaluate an idea for a project by focusing on the benefits the completed project would give the stakeholders and the organization.

In a few words, what is the project idea?

What do you hope to accomplish with a project based on this idea?

What is the project's tangible deliverable? This is a product, service, or result.

What is your rough estimate for how much the project might cost?

How long might the project take?

What kind of resources—people skills and material resources—would you need?

If needed, who might be willing to back the project financially?

Who is the client (the primary stakeholder) and who are the other stakeholders?

So, do you want to pursue this project? Why, or why not?

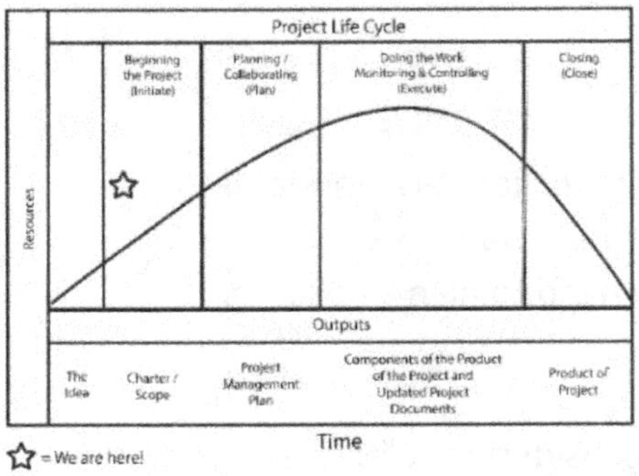

Contemporary Project Management #1.5

Explore the Project Life Cycle

The Project Life Cycle illustration below shows the phases, the graduated use of resources, and the outputs in the life cycle of a typical project.

The phases in the project life cycle include beginning, planning, doing the work while monitoring and controlling, and closing.

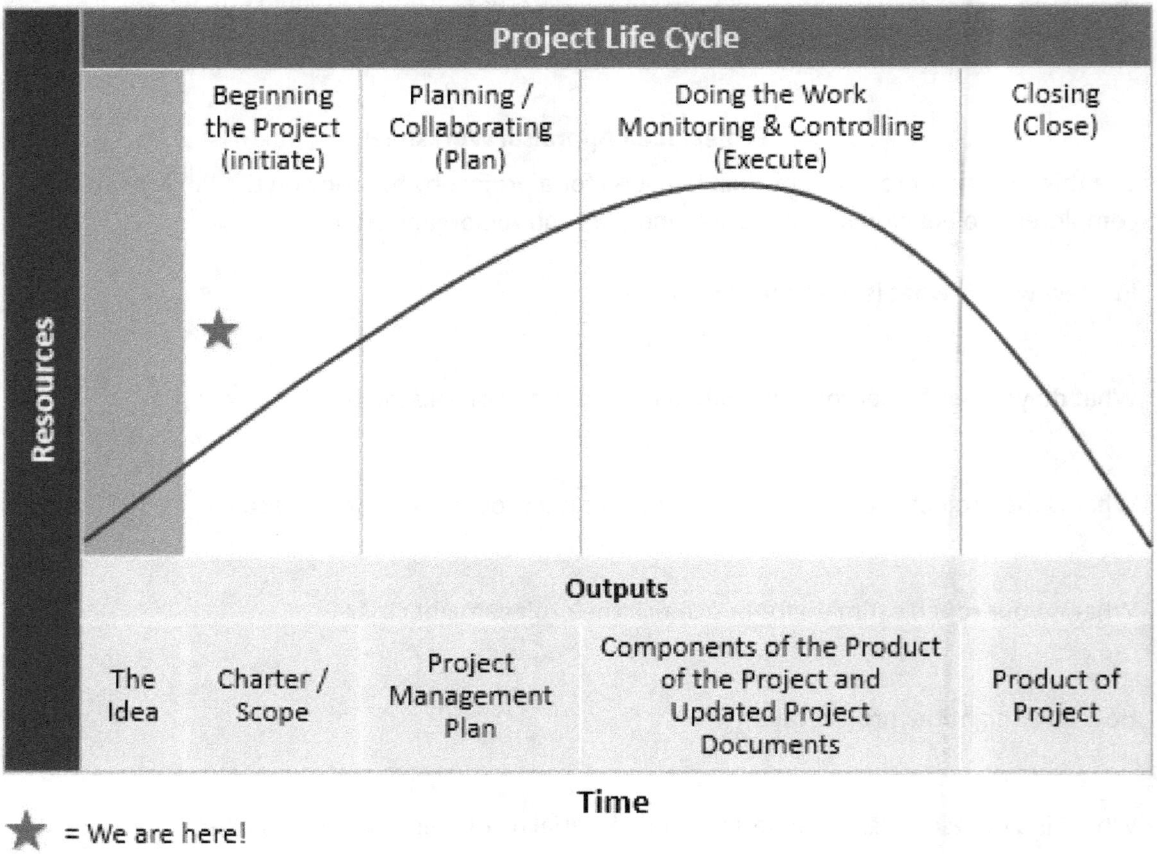

The curved line indicates the number of resources typically expended in each phase. It shows that the beginning of a project uses limited resources to establish the requirements and that more resources are needed as the project progresses. The execution phase requires the most resources. As the project approaches the end, resource needs decrease because many activities are complete. The exact shape of the curve can vary, depending on the type of project.

Each phase has processes associated with it and essential project documents.

A **project manager** is essential for all projects. This individual must have many skills. They include the ability to **develop project plans, allocate resources, resolve conflicts**, identify potential conflicts of interests, promote **safety**, monitor progress against the project plan, report progress, and assure **quality** within traditional project constraints, or what we call the **triangle of balance**.

As illustrated below, the ***triangle of balance*** includes the project scope, time/schedule, and cost/budget..

Project Management Triangle

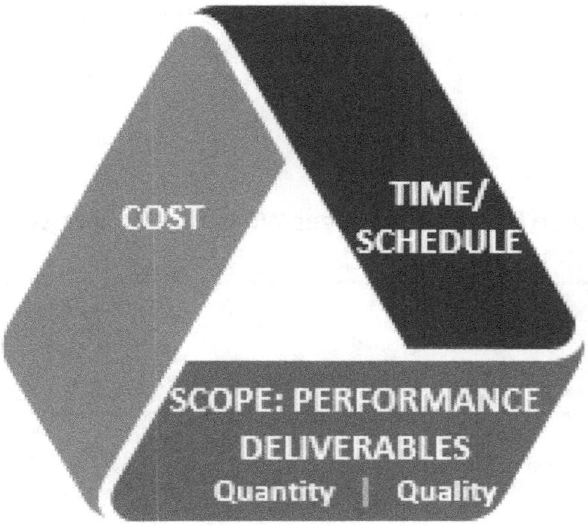

A project manager may have technical knowledge about the project area, and this knowledge may be useful when communicating with various team members. However, a project manager must focus on managing all aspects of the project and not on doing technical work, which is the responsibility of team members.

Managing a project includes monitoring and controlling project tasks and the schedule, identifying problems and leading the team in developing solutions, motivating the team when the going gets tough, and communicating with project stakeholders.

What to do:	Other actions:
☐ For a project you know about, determine the current project phase.	☐ ____________________
☐ Recognize that project managers must relinquish technical involvement in a project and focus on their project management responsibilities.	☐ ____________________
☐ Be mindful of traditional project constraints, or what is called the triangle of balance: project scope, time/schedule, and cost/budget.	☐ ____________________
	☐ ____________________
	☐ ____________________

Remember

- ✓ The project life cycle is a powerful framework for understanding, organizing, and developing a project. Phases provide an orderly way of envisioning the activities, deliverables, and resource requirements of a project as it progresses in time.
- ✓ When you work on a project, you will typically find that the phases are recursive. In other words, they go back and forth. For example, the team may have completed the planning phase and be working on executing the project, but then an issue may arise that requires the team to revisit the plan and modify it.
- ✓ The *triangle of balance* includes the project scope, time/schedule, and cost/budget.

Enhance Your Learning

Watch the following 4-minute video by Jennifer Bridges Whitt. The Project Management Life Cycle.

Whitt, J. (2014). *The project Management Lifecycle..*	Available at: https://www.youtube.com/watch?v=POuGKD3xLqs&t=7s

Reinforce Your Learning

Project Life Cycle Phases. Consider a project that you worked on or observed. Based on your experience, describe the purpose and deliverables of each phase.

Phase	Description and Purpose	Deliverables
Begin the project: **Initiate**		
Plan the project: **Plan**		
Do the project work: **Execute (Monitor and Control)**		
Close the project: **Close**		

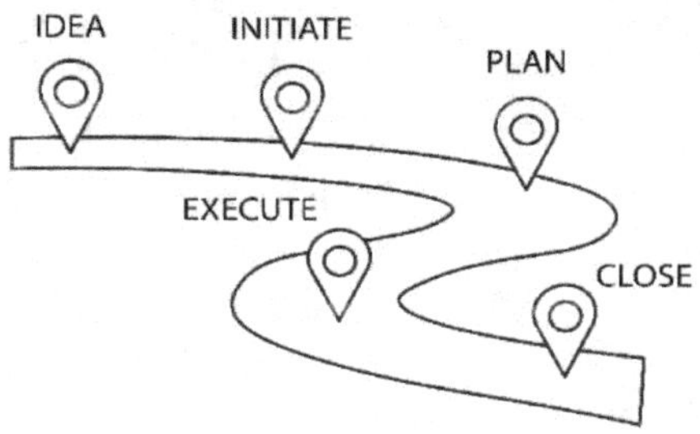

Contemporary Project Management #1.6

Delve Deeper into Project Phases

As previously discussed, after the idea phase, project activities are typically divided into several major phases: initiating, planning, executing, and closing. These phases represent the project life cycle from beginning to end. After an organization vets a project idea and decides to pursue the project, the project Initiate phase begins.

The following is a high-level description of each project phase and associated deliverables. The chart begins with a project idea and then describes the four phases: Initiate, Plan, Execute, and Close, along with the deliverables for each phase.

Be aware that project management is not a linear process. Many project activities will proceed in parallel, and sometimes with overlap and repetition.

Phase	Description/Purpose	Deliverables/Tools
Idea	• Preliminary and general description of a project idea and review of potential benefits.	• Idea appraisal • Decision: yes or no

Phase	Description/Purpose	Deliverables/Tools
Initiate: Define project requirements and scope	• Identify and document business needs and requirements • Define project scope, goals, objectives • Identify project client, sponsor, and other stakeholders • Develop initial estimates of effort, cost, schedule, and risk • Identify deliverables and acceptance criteria	• Project Description • Project Charter • Feasibility Analysis • Project Scope Statement • Stakeholder Analysis Matrix

Phase	Description/Purpose	Deliverables/Tools
Plan: Plan project from beginning to end	• Define tasks, dependencies, and duration • Refine resource requirements • Develop work plan, milestones, and schedule • Define the project control process, standards, and procedures • Develop a risk management plan • Identify communication needs • Refine and manage expectations	• Schedule Mgmt. Plan • Work Breakdown Structure • Status Report Format • Time/Schedule-Tracking Format • Risk Analysis and Management Plan • Stakeholder Comm. Matrix and Plan • Project Budget • Benefit-Cost Analysis

Phase	Description/Purpose	Deliverables/Tools
Execute: Do the work while monitoring and controlling the project	• Track actual progress to planned progress • Assess and report status • Coordinate activities and resources • Implement corrective actions and update estimates and project schedule • Manage and support individual and team activities • Manage project priorities • Manage and resolve problems and conflicts • Manage changes to project scope • Assess the quality of deliverables • Communicate work assignments, progress, and issues • Evaluate the performance of individuals and teams	• Updated Project Management Plan Activities • Updated Risk Management Plan • Status Reports • Feedback Forms • Change Order Log • Issues Log • Interim Performance Reviews

Phase	Description/Purpose	Deliverables/Tools
Close: Close and report success level to sponsor and stakeholders	• Deliver project results • Formally end the project • Evaluate the performance of the project teams, team members, and the project manager • Evaluate the project, draw conclusions, and document changes and lessons learned • Identify areas of potential improvement for future projects	• Completed Performance Appraisals • Project Assessment Report • Updated Project Documents and Records • Project Deliverables and End Results

For smaller projects, a project manager may decide to combine the initiation and planning phases into a single planning stage. Larger, more complex projects typically keep the definition and planning phases separate, to provide better control of project activities.

Most projects, especially in the definition and planning phases, involve the discovery of new information. If you ignore this information and do not incorporate it into the project life cycle, you will compromise project quality, time, and resources.

Just as important, you will be unable to take advantage of the information. Most projects involve gradually assembling project plans, team members, and organizational resources. When you begin working on a project, you must develop an initial plan. However, this plan is based on available knowledge and sometimes unrealistic expectations. You must expect that some adjustments and modifications to the plan will be needed as you and the team learn more about the project.

What to do:	Other actions:
☐ Vet preliminary project ideas with stakeholders.	☐ ______________
☐ Understand that the authorization to pursue a project is provided in a project charter.	☐ ______________
☐ Recognize that many project activities proceed in parallel and have some overlap and repetition.	☐ ______________

Remember

- ✓ Most projects, especially in the definition and planning phases, involve the discovery of new information.
- ✓ Each phase of a project serves a unique purpose. For example, you define the project work in the planning phase.
- ✓ You will produce specific documents in each phase. For example, a Schedule Management Plan is one of the documents you will develop in the planning phase.

Enhance Your Learning

Watch the following 4-minute video to learn more about the phases of a project by KnowledgeHut .

Knowledgehut. (2019). *4 Stages of Project Life Cycle | Phases of Project Management Life Cycle.*	Available at:. https://www.youtube.com/watch?v=N3N9-RLSbvo

Reinforce Your Learning

Project Phases. Review the phases of the project life cycle and the associated document outputs of each phase. Then list two or three elements you are unsure about or find most challenging and describe why they are challenging.

____

____

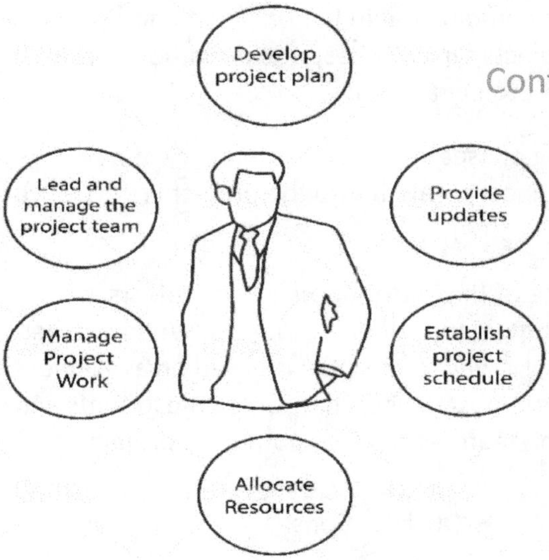

Contemporary Project Management #1.7

Understand Project Manager Responsibilities

Project managers are responsible for everything that happens on a project from the time they are assigned until they close the project. On large projects, this can be a long time—months, or even a year or more. Project managers must fulfill these responsibilities while also being available to handle unexpected and urgent situations that threaten project success.

The following is a list of typical responsibilities associated with managing a project. It is not meant to be inclusive, but to show that project managers have a variety of duties.

Project managers are responsible for all these activities throughout the life of a project, though some tend to happen during a particular phase, while others require constant attention. For example, contract negotiation is often early in a project, while communication is a daily job.

Check those areas with which you have experience.

- ☐ Kickoff a project
- ☐ Communicate status and problems and prepare formal project reports
- ☐ Represent/coordinate the project to others in the organization
- ☐ Market the project internally and externally
- ☐ Develop subcontract and joint venture agreement if needed
- ☐ Negotiation and execute contracts
- ☐ Develop a project plan and assign work
- ☐ Allocate resources
- ☐ Manage the project work
- ☐ Manage quality control and quality assurance
- ☐ Resolve conflicts
- ☐ Ensure that billing and collections are done
- ☐ Ensure that project-related personnel policies, particularly those dealing with conduct, conflicts of interest, business confidentiality, and safety are in place and used

Time Management

One of the challenges for project managers is coordinating all the things they must do so that nothing is forgotten, and no deadlines are missed. A time management system can be invaluable by providing a project manager with a way to schedule their work.

Some activities are short-term, whereas others take longer to complete. Daily work, such as making calls and answering questions, requires setting aside time each day. Other activities, for example, conducting meetings and drafting reports, can be scheduled at regular intervals such as weekly or semi-weekly.

Some activities require work over an extended period. For example, on large and complex projects, a project manager may need to set aside time over a month or more to prepare proposals, develop policies and procedures, or do a quality assessment.

What to do:	Other actions:
☐ List your typical daily tasks.	☐ ____________
☐ List the weekly and long-term duties you perform.	☐ ____________
☐ Investigate time management systems that may help you plan and schedule your work.	☐ ____________

Remember

- ✓ Every project manager has tasks and duties they must fulfill.
- ✓ These responsibilities occur throughout the life of a project at varying degrees.
- ✓ Project managers must consider short-term and long-term actions.

Enhance Your Learning

Watch the 7-minute video by Jennifer Bridges Whitt which focuses on daily and weekly job duties for project managers: Top 10 Project Management Responsibilities.

Whitt, J. (2018). *Top 10 Project Management Responsibilities*.	Available at: https://www.youtube.com/watch?v=vMWJyELLSQ4

Reinforce Your Learning

Project Manager Responsibilities. The following are tips for daily project management success developed by Jennifer Bridges, PMP. Based on your experience, can you add to the list?

- Make meetings productive and fun to keep morale and engagement high.
- Make calls personal and intentional, to build relationships and not waste people's time.
- Leave meetings with closure and action items so team members can hit the ground running.
- Resolve issues because a lingering question will not go away by itself, and as the project progresses, it will likely worsen.
- Clarify ambiguity to ensure everyone knows what needs doing, and current project status.
- Ask questions, because by staying curious and engaging your project team, you will learn and lead a better project.

Contemporary Project Management #1.8

Consider the Attributes of Successful Project Managers

Effective project managers must do many things well, which means they must have the necessary skills. But they must also have attributes that support these skills.

Attributes are qualities and characteristics that describe a person or thing. They can be vague and difficult to define. For example, caring and being positive are hard to quantify. Although most people know when someone cares or has a can-do attitude, it can be hard to describe.

Project managers must have the same attributes as any successful leader. For example, project managers must:

- Respect the people they work with and treat them accordingly.
- Understand people and how they function in teams.
- Appreciate the skills of the team members and trust them to do their jobs.
- Care what happens on the project and be willing to address problems that threaten it.
- Work diligently to ensure the project is completed according to plan.
- Care about the organization's goals.

Over time, successful project managers develop such attributes—respect, understanding, appreciation, trust, caring, courage to confront problems, and a strong work ethic—by building on their strengths and acknowledging and working to overcome their weaknesses.

The following are some of the behaviors expected from project managers. As you review the list, consider what qualities or characteristics a project manager might need to be effective and check those items for which you would like to build your competence.

Successful project managers know how to:

- ☐ Sacrifice technical involvement for management responsibilities
- ☐ Take responsibility for what team members do, both good and bad
- ☐ Sacrifice ego and promote team involvement and recognition
- ☐ Trust the skills of the team members
- ☐ Coordinate and work comfortably with various technical disciplines
- ☐ Share information with team members and stakeholders
- ☐ Keep stakeholders informed
- ☐ Delegate responsibility and authority while keeping accountability
- ☐ Work with and through others, rather than as an individual contributor
- ☐ Safeguard against allowing personal feelings or biases to interfere with managing the project
- ☐ Help the team, even when doing so conflicts with the project manager's own work and obligations

What to do:	**Other actions:**
☐ Identify the qualities and characteristics you typically display in a project.	☐ ____________________
☐ Identify your attributes that are strengths and build on them.	☐ ____________________
☐ Determine what attributes you could improve and find resources to help you grow.	☐ ____________________
	☐ ____________________

Remember

- ✓ A project manager must do many things well.
- ✓ Attributes are the qualities and characteristics someone or something has, for example, caring or having a can-do attitude.
- ✓ Project managers must have realistic views of their strengths and weaknesses.

Enhance Your Learning

Review the following article by Nutcache (2022), to learn ten attributes of an effective project manager.

Nutcache. (2022). *10 Attributes of an Effective Project Manager.*	Available at: https://www.nutcache.com/blog/10-attributes-effective-project-manager/

Reinforce Your Learning

Project Manager Attributes. Based on your experience, which of the following behaviors are your strengths, and which are your weaknesses? Write an "S" in the space provided to indicate a strength and a "W" to indicate a weakness. Then jot down several actions you might take to overcome the weaknesses.

_____ 1. Effective communication skills

_____ 2. Leadership and delegation skills

_____ 3. Good problem solving and decision-making skills

_____ 4. Proven technical expertise/skills

_____ 5. Good team builder

_____ 6. Confident and cool under pressure

_____ 7. Good influence and negotiation skills

Contemporary Project Management #1.9

Realize the Diverse Roles Project Managers Fulfill

Project managers must fulfill several functional roles as the project progresses and moves through its life cycle.

Below are descriptions of the critical roles project managers satisfy. Various roles will be more prominent at one time or another depending on what part of the project life cycle a team is working on, the problems or challenges the project faces, and what may be required to bring the project to successful completion.

For example, a project manager may fulfill a visionary role when the project is first being considered, and more of a problem analyzer role when the project is being worked on and problems arise.

Communicator
A communicator is skilled in sending and receiving messages and information in a variety of situations and media, such as presentation, interpersonal relationships, and writing. In this role, the project manager must communicate stakeholder needs to the team and communicate team progress and issues to stakeholders.

Visionary
A visionary can foresee products, services, and results not yet apparent to others. The visionary is not psychic, but they listen and gather information and work to assess how things are going. They pay attention and communicate what they see. This role requires a "big-picture person," one who sees opportunities and possibilities and who can help others see them as well.

Supporter
A supporter takes ownership of project work and deliverables and encourages others to do likewise. A supporter promotes the project internally and externally.

Motivator
A motivator helps create a positive climate and has a positive impact on work performance.

Planner
A planner helps define and structure the steps and tasks required to complete a project successfully.

Organizer and Resource Coordinator
This role plans for resources, ensuring that the right resources get to the right places at the right time and in the right quantities.

Team Leader
A team leader is responsible for the day-to-day activities of the project team and has a crucial role in influencing team members, stakeholders, and project outcomes in a positive way.

Problem Analyzer and Decision Maker
A problem analyzer identifies problems and determines their causes, brainstorms, advances possible solutions for team discussion, and recommends fixes in accordance with project goals.

Facilitator
A facilitator helps implement the project vision, is service-minded, and can focus on the big picture while respecting the contribution of each team member and supporting the team in achieving their goals.

What to do:	Other actions:
☐ Identify the roles you typically fulfill in projects.	☐ ____________
☐ Identify the roles that are your strengths and build on them.	☐ ____________
☐ Determine what roles you could improve and find resources that could help you develop your skills.	☐ ____________

Remember

- ✓ Project managers play many roles throughout the life cycle of a project.
- ✓ Peers can serve as free career development resources.
- ✓ You are responsible for your career and professional development.

Enhance Your Learning

Watch the following 17-minute video by Jason Dodd of PM Perspective (2014), to learn more about the role of the project manager by PM Perspective.

Dodd, J. (2014). *The Role of The Project Manager.*	Available at: https://www.youtube.com/watch?v=dUhJuB69ZBo

Reinforce Your Learning

Project Manager Roles. Based on your experience, check the items you feel are essential. Can you add anything?

A communicator:

- ❑ Conducts project reviews and walk-throughs.
- ❑ Does not interpret "no news" as "good news."
- ❑ Presents and reports periodically according to a schedule and as requested.
- ❑ Ensures that all reports and presentations are clear and well organized.
- ❑ Uses specific and clear terminology when communicating and no buzz words.
- ❑ Meets with others throughout the project to ensure a full understanding of stakeholder needs and concerns.
- ❑ Follows up with stakeholders after the project to ensure their satisfaction.
- ❑ Identifies all groups that may be affected by the project and seeks their input.
- ❑ Holds regularly scheduled and productive meetings with the project team and ensures representation from all areas affected by the project. Discusses status, resolves issues, and shares information.
- ❑ Informs management and stakeholders promptly when problems arise.
- ❑ Speaks on behalf of management and stakeholders in project meetings.
- ❑ Other:

A visionary:

- ❑ Possesses a business orientation.
- ❑ Considers project goals in a cross-functional manner.
- ❑ Thinks not only in concrete terms but in a conceptual context.
- ❑ Reviews the overall picture.
- ❑ Stays ahead of the project team.
- ❑ Develops a preliminary study with the project team to identify business opportunities, project requirements, project scope, and project benefits.
- ❑ Works closely with stakeholders to define project objectives and key results/goals.
- ❑ Acts today to prevent tomorrow's problems and addresses future requirements.
- ❑ Communicates factors necessary for success.
- ❑ Views the project not just in the present but also in the context of how business and technology will perform in the future.
- ❑ Anticipates and plans for the impact of the project on related systems and areas.
- ❑ Considers both short- and long-term implications of the project.
- ❑ Ensures a common understanding of the project objectives by all stakeholders and team members.
- ❑ Ensures a common understanding of the project scope by all stakeholders and team members.
- ❑ Other:

A supporter:

- ❑ Uses relationships with people from different areas in the organization to resolve issues and help guide team members.
- ❑ Assumes ownership of the project.
- ❑ Publicizes and markets the project internally and externally.

- ❑ Builds business ownership by seeking extensive business participation during the initiating and planning phases of a project.
- ❑ Encourages stakeholders and team members to assume ownership.
- ❑ Empowers others to achieve project objectives.
- ❑ Represents the stakeholder and team perspectives in the company.
- ❑ Other:

A motivator:

- ❑ Manages performance issues.
- ❑ Keeps an open-door policy and wants to hear ideas and problems.
- ❑ Positively recognizes individual and team accomplishments and results.
- ❑ Exerts whatever effort is needed to achieve project objectives.
- ❑ Manages expectations by ensuring that what the team promises gets delivered.
- ❑ Provides rewards aimed at the needs and interests of team members.
- ❑ Identifies with project quality and success.
- ❑ Rewards and recognizes people as milestones are achieved.
- ❑ Presents a positive image to set the tone for the project team.
- ❑ Other:

A planner:

- ❑ Develops and proposes strategies.
- ❑ Develops the project management plan and determines assignments.
- ❑ Establishes lines of responsibility and accountability with stakeholders.
- ❑ Understands business needs/requirements, time, and cost pressures.
- ❑ Balances ideal technical solutions and project scope against business deadlines and priorities.
- ❑ Sets and supports high standards of quality for all team members.
- ❑ Breaks a large or complex project into meaningful sub-projects.
- ❑ Develops a project plan that includes resource needs, budget, and time. Involves project team members in the detailed planning of the project.
- ❑ Maintains a detailed baseline plan, showing what to do and by what time.
- ❑ Establishes priorities.
- ❑ Sets objectives in cooperation with the project team.
- ❑ Assesses risks and develops an effective risk management plan.
- ❑ Other:

A team leader:

- ❑ Works to ensure stakeholder satisfaction.
- ❑ Models appropriate behavior for team members.
- ❑ Facilitates project work and project meetings.
- ❑ Leads the project team and stakeholders to effectively accomplish project objectives and adjusts as situations change.
- ❑ Networks with others, both inside and outside the organization.
- ❑ Builds a capable team.
- ❑ Provides direction for the project by influence and not force.
- ❑ Makes tough decisions after gathering all the necessary input and facts.
- ❑ Adapts to situations as they change.
- ❑ Keeps learning and growing and is self-confident.
- ❑ Shows flexibility and is willing to change when appropriate.
- ❑ Brings out the best in all team members.
- ❑ Defines clear roles and performance expectations for team members.

- ❑ Is objective when dealing with conflict.
- ❑ Focuses team members on project objectives and deliverables.
- ❑ Identifies schedule and budget/cost issues of requested scope changes and communicates them to stakeholders.
- ❑ Accepts responsibility for resolving project issues promptly.
- ❑ Works to influence stakeholders and team members in a positive way.
- ❑ Maintains accountability throughout the project and holds others accountable for meeting their goals.
- ❑ Other:

A problem analyzer and decision-maker:

- ❑ Determines what data is essential and analyzes it efficiently.
- ❑ Sets priorities to ensure the project is completed within budget, on time, and within quality standards.
- ❑ Focuses on problems and solutions, not on who is at fault.
- ❑ Cultivates the proper atmosphere for practical problem-solving.
- ❑ Emphasizes quality in all decisions.
- ❑ Makes tough decisions without procrastinating.
- ❑ Identifies problems early enough so they can be dealt with before they develop into time-consuming issues.
- ❑ Works with financial staff to manage the project's costs and budget.
- ❑ Develops innovative and creative approaches to problems.
- ❑ Takes calculated and prudent risks.
- ❑ Takes corrective action to keep the project on schedule and within budget.
- ❑ Confronts problems with others in a timely and direct manner.
- ❑ Other:

A facilitator:

- ❑ Delivers services and products to stakeholders successfully.
- ❑ Shows skill and is comfortable in dealing with ambiguity.
- ❑ Acts to overcome obstacles and limitations.
- ❑ Develops an atmosphere where all team members are comfortable and involved in the project openly and honestly.
- ❑ Considers the implications of finances on the project and recognizes the significance of return on investment.
- ❑ Shows awareness of the time required by team members to do jobs.
- ❑ Provides team members with assignments and training to support their growth.
- ❑ Gives direct, specific, constructive, and timely feedback to team members.
- ❑ Other:

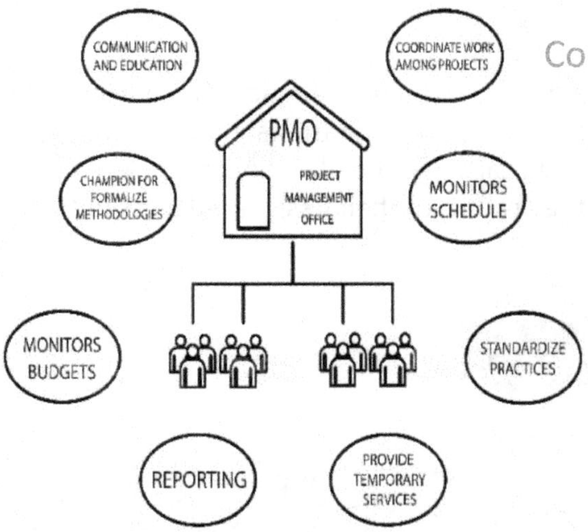

Contemporary Project Management #1.10

Examine the Merits of a Project Management Office

When you manage a project, you will find it helpful to understand how your organization is structured. Organizations may be designed or structured differently to manage projects.

Traditional Functional Organizational Structure

In the traditional organization, every group of related activities is assigned to a single functional manager and department. Each department head reports to a higher-level manager. Each department employs people who have similar skills and resources.

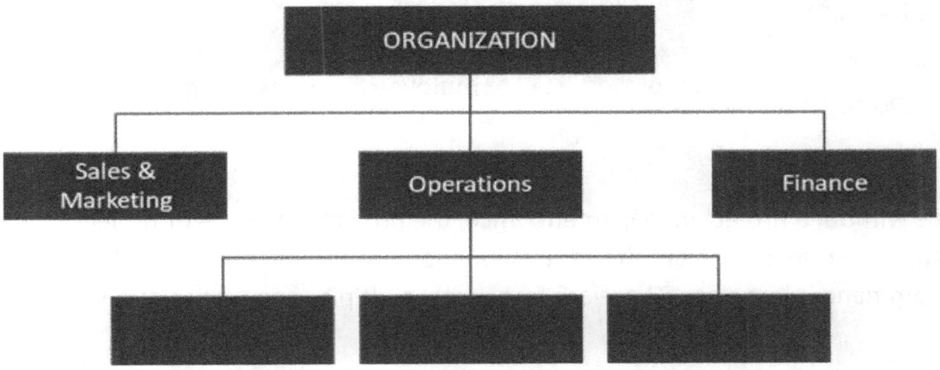

A **major drawback** for functional organizations in dealing with projects is cross-departmental communication and coordination presents challenges. Therefore, many projects are limited to the boundaries of the functional department.

A typical line and staff organization builds on the idea of the functional organization. The line consists of those employees who produce the organization's products and services. The staff consists of those who support the line employees such as accounting, maintenance, human resources, and the like.

In the line and staff organization, the chain of command is clear. While the chain of command can provide order and control, it may work against the needs of those who manage cross-functional projects. Project managers in line and staff organizations may experience challenges because they need to use a variety of types of power and influence to get their work done effectively.

Matrix Organizational Structure

The matrix organization is less traditional and is set up to do projects. For example, in an engineering firm, the designated project managers—or project coordinators—call upon the talents of employees from functional departments—or disciplines such as mechanical or electrical engineering—for a designated time period. After the project is completed, they return to their designated units until they are needed for another project.

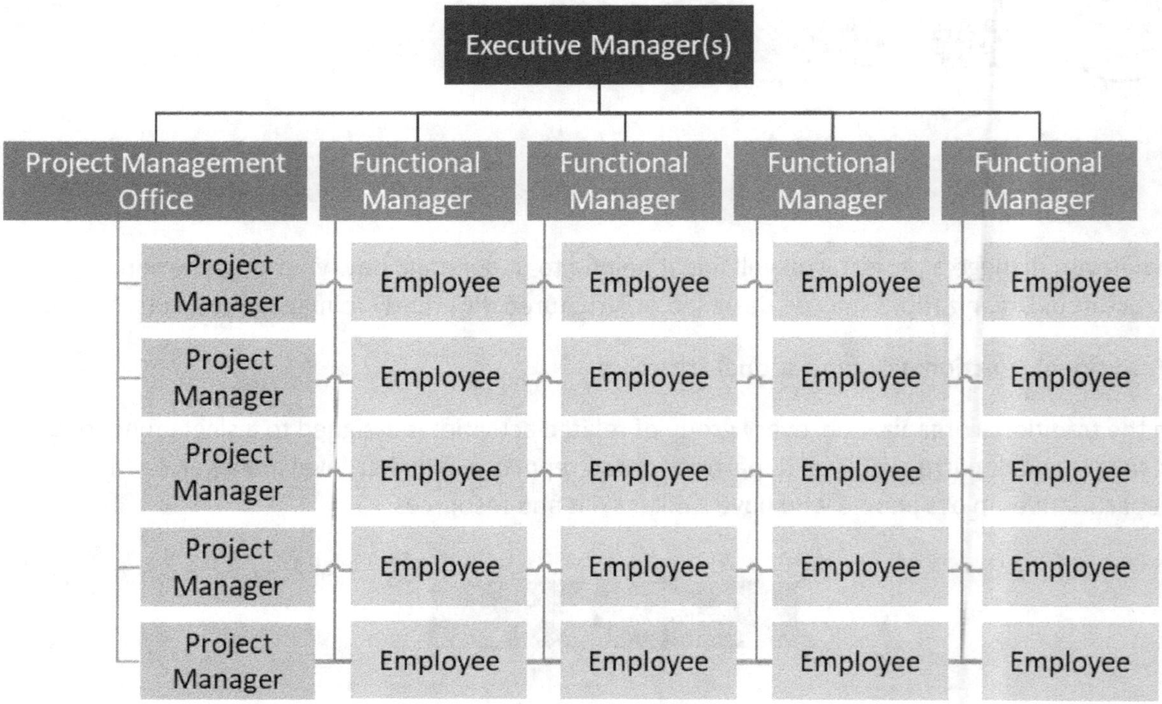

In matrix firms without a project management office, the project managers or project coordinators are often selected from the various functional areas and serve a dual role where the chain of command is less clear. There are frequently multiple reporting relationships.

Projectized Organizational Structure

Organizations can enhance their project management process by adopting organization-wide methods. The advantages of doing this are many. It comes down to the basic idea of not expending resources to reinvent the wheel. For example, when people move from one project to another, they would not need training on project procedures, and from one project to the next, people in the organization would be familiar with how projects are reported.

The projectized organization is often used on larger and longer-term projects where the project manager has a great deal of authority over the projects and over the project team members. Many organizations have segmented their project management activities under a discrete **Project Management Office** (PMO) with its own project executive.

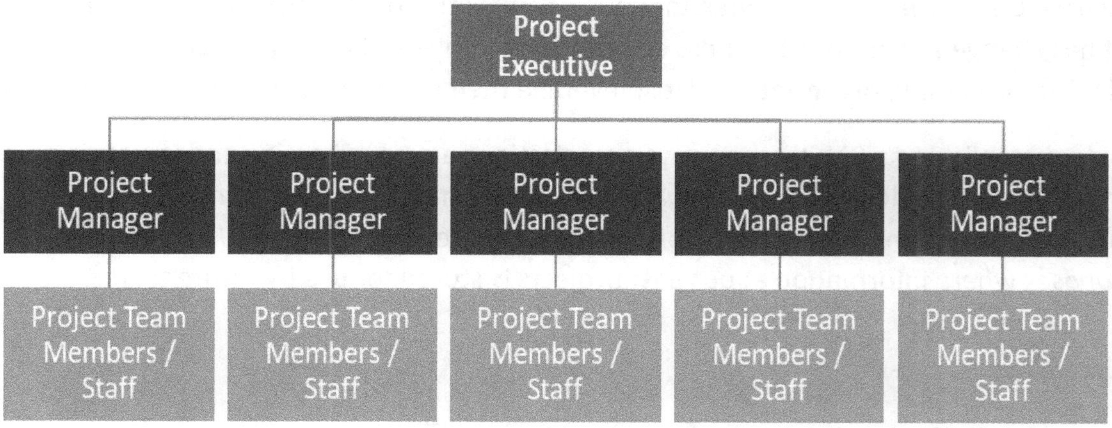

In a projectized organization all the team members work for the project manager who has control over the resources. Typically, the project manager has close ties to the client and can accommodate changes quickly. Team members are usually assigned full-time to the project and often co-located. A centralized **project management office** (PMO) can facilitate organization-wide project management in a variety of ways. For example, the PMO in an organization can:

- Coordinate work among projects
- Standardize project management practices and methods
- Conduct project management training
- Provide temporary project management services
- Champion the value of a formalized project management methodology throughout the organization

A PMO promotes consistent project management practices and methods and the use of standard project management tools; assures standardization in project execution and delivery; provides a resource for project management expertise; and coordinates multinational project management knowledge (Kerzner, 2017).

There is no single way of doing every project. By definition, each project is unique. However, a standard methodology for how to conduct projects can be established. For example, the recognized phases of the project life cycle—initiate, plan, execute, perform work while monitoring and controlling it, and close—are common to all projects. This general framework gives everyone involved in a project a roadmap to follow, even though the specific activities within each phase may not all be the same from one project to another.

Another area of commonality among projects is the **standardization** of processes for project execution and delivery. For example, a PMO can help establish practices that might include a consistent set of criteria for project approval, determination of what project management software to use, a consistent set of success criteria, and how and when a project can be closed.

Another way a PMO can serve an organization is by providing subject matter expertise. Having a PMO staffed with people who have project management experience in a variety of environments provides a significant resource for an organization. A pool of project management professionals can be instrumental in enabling an organization to consistently deliver successful projects.

Coordinating project knowledge can be cost-effective in other ways. For example, there once may have been little need to understand the complexities of conducting a global project, but that has changed. In today's business environment, more and more projects span the globe, and to be successful, organizations must leverage their comprehensive project knowledge.

Besides establishing a PMO, other ways organizations are coordinating and leveraging project best practices include the use of *lessons learned*, which is a way to document how things could have been done better after a project is complete, and *enterprise knowledge databases where* information about past projects is stored for use by future teams.

What to do:	**Other actions:**
☐ Visit a PMO or read an article about one to learn how a real-world PMO operates.	☐ ______________________
☐ Search online for two examples of a PMO. Does anything you learn apply to your environment?	☐ ______________________
☐ Consider the services a PMO could provide. How could your organization benefit from having one?	☐ ______________________

Remember

- ✓ The concept of a PMO is a centralized collection of project management expertise.
- ✓ A PMO promotes the maturation of project management best practices and methods within an organization.
- ✓ Having a PMO can improve the likelihood of projects being completed successfully.

Enhance Your Learning

View the 4-minute video which describes different organizational structures employed for managing projects.

Kumar, K. (2016). *Organization structure Influence project management / Functional Mati Projectized organization.*	Available at: https://www.youtube.com/watch?v=pbtqqZm8Ak0

Reinforce Your Learning

Project Management Office (PMO). Consider the following case study of a project within an organization that did not have a PMO.

Pipeline Education, Inc. has been growing over the last two years. They have added new employees and are struggling to keep up with requests for new online courses. Pipeline promoted one of their technical staff, Diane, who knew a lot about online education, to be a project manager. However, she had never taught a course and was not familiar with how to plan and run a project. The other project managers did not have time to train Diane or assist her. She did the best she could with her limited exposure to project management methods.

In a couple of months, Diane's project fell behind schedule. Then the organization discovered that the work Diane's team completed so far had quality issues. The software tools were not the best for online learning, the courses did not meet ADA requirements, the team's analytical tools to measure learning outcomes had deficiencies, and the methods to plan and schedule work did not meet the organization's standards.

Below, identify and describe at least two or three ways a PMO could have helped address the problems in Diane's project.

Summary

A Final Word: Contemporary Project Management

You have completed this chapter on Contemporary Project Management. You were introduced to the Project Management Institute (PMI)®, their certifications, and A Guide to the Project Management Body of Knowledge (*PMBOK*)®.

You were introduced to the characteristics of a project, the life cycle of a project, and the project management process. You also had a brief look at responsibilities, attributes, and diverse roles of a project manager as well as the potential merits of a project management office.

Information you reviewed in the chapter include:

- Distinguishing project management from conventional management
- Describing the phases of a project from beginning to end
- Understanding the benefits of using a project management methodology
- Recognizing the attributes successful project managers bring to their projects
- Being aware of the critical roles project managers fulfill

In the next Project Mastery chapter, *we will focus on* on Defining Project Scope.

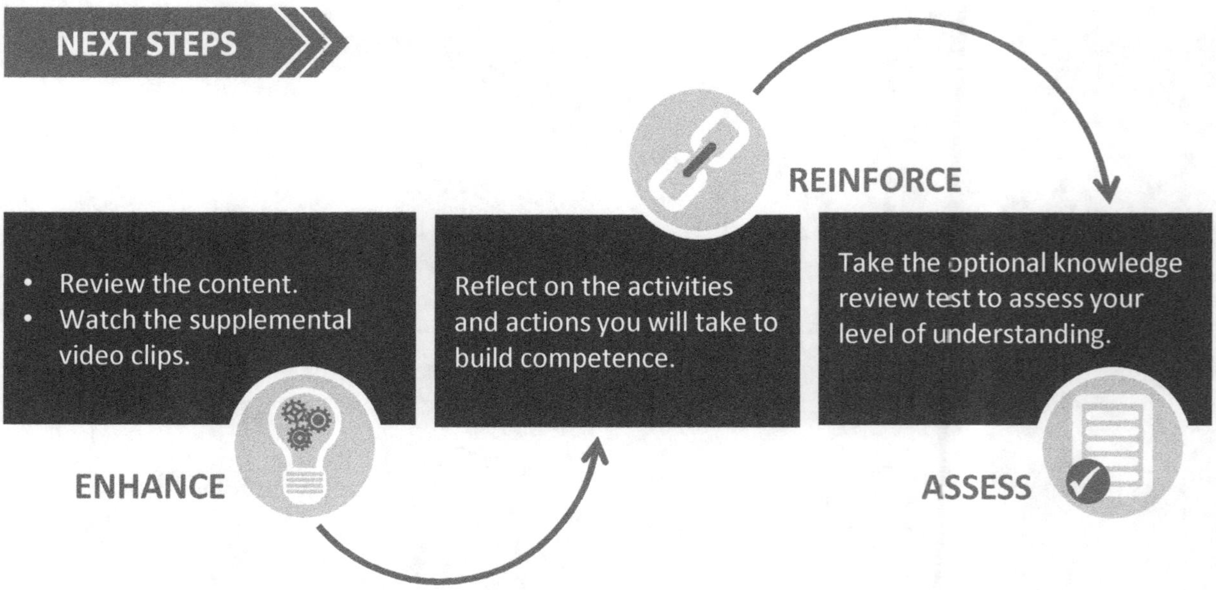

Contemporary Project Management—Recap Checklist This course describes project management and its benefits, reviews the project management life cycle, describes the functions a project manager performs, and examines the challenges that affect most projects at one time or another with a focus on the human side.
1. Define Project and Project Management □ Review the Project Management Institute (PMI) website (www.pmi.org). □ Identify some projects you have been involved in or observed. □ Brainstorm issues, problems, and training needs related to those projects.
2. Become Familiar with the Project Management Institute (PMI) □ Discuss project management certifications with your manager for applicability. □ Find a certified associate and ask them about the advantages of being certified. Pursue one of the certifications, if appropriate.
3. Know the Benefits of Using a Project Management Methodology □ Consider the benefits of following a step-by-step planning process. □ Identify the sources/kinds of information you need access to. □ Reflect on the risks in your project and what their impact could be.
4. Be Aware of How Projects Begin, the People Involved, and Quality □ Identify the criteria your organization uses to judge project success. □ Define what quality means for projects you (or your organization) have completed. □ Complete an Idea Appraisal Worksheet for a project that you or someone in your organization is considering.
5. Explore the Project Life Cycle □ For a project you know about, determine the current project phase. □ Recognize that project managers must relinquish technical involvement in a project and focus on their project management responsibilities. □ Be mindful of traditional project constraints or what is called the *triangle of balance*: scope, time/schedule, and cost/budget.
6. Delve Deeper into Project Phases □ Vet preliminary project ideas with stakeholders. □ Understand that the authorization to pursue a project is provided in a project charter. □ Recognize that many project activities proceed in parallel and have some overlap and repetition.
7. Understand Project Manager Responsibilities □ List your typical daily tasks. □ List the weekly and long-term duties you perform. □ Investigate time management systems that may help you plan and schedule your work.
8. Consider the Attributes of Successful Project Managers □ Identify the qualities and characteristics you typically display in a project. □ Identify your attributes that are strengths and build on them. □ Determine what attributes you could improve and find resources that could help you grow.
9. Realize the Diverse Roles Project Managers Fulfill □ Identify the roles you typically fulfill in projects. □ Identify the roles that are your strengths and build on them. □ Determine what roles you could improve and find resources that could help you develop your skills.
10. Examine the Merits of a Project Management Office □ Visit a PMO or read an article about one to learn how a real-world PMO operates. □ Search online for two examples of a PMO. Does anything you learn apply to your environment? □ Consider the services a PMO could provide. How could your organization benefit from having one?

Knowledge Review Test

Contemporary Project Management - Questions

1. Which is not a characteristic of a project?
   A. Temporary.
   B. Definite beginning and end.
   C. Interrelated activities.
   D. Repeats itself every month.

2. Wes is a new project manager who has never managed a project before and has been selected to plan a new project. It would be best in this situation to rely on _____ during the planning to improve his chance of success?
   A. his intuition
   B. his training
   C. historical records
   D. responsibility charts

3. The Project Management Institute was founded in what year?
   A. 1941
   B. 1939
   C. 1969
   D. 1991
   E. 2001

4. What certification designation offered by the Project Management Institute (PMI) shows the holder has the skills and knowledge to manage projects from beginning to end?
   A. Certified Associate Project Manager (CAPM®)
   B. Program Management Professional (PgMP®)
   C. Agile Certified Practitioner (PMI-ACP®)
   D. Project Management Professional (PMP)®

5. For a project to be successful, which of the following criteria must be met?
   A. Scope requirements.
   B. Client satisfaction.
   C. Project schedule.
   D. Completed on budget.
   E. All the answers are correct.

6. The CEO of your organization asked you to work on a variety of strategic business activities aimed at building, sustaining, and advancing the initiatives across your organization. What does this best describe?
   A. Portfolio analysis of the organization.
   B. Project management information system.
   C. System for value delivery.
   D. Agile and Scrum analysis.

7. Organizations review ideas for projects in many ways, but most include:
   A. A description of the project.
   B. An assessment of the benefits.
   C. A comparison against other projects.
   D. A sponsor.
   E. All the answers are correct.

8. Your project should provide value to your customers. Which of the following does not describe value?
   A. The ability to use unique features or functions of a product.
   B. The anticipated use of scrap caused by selecting an easy solution instead of a better solution that would take longer.
   C. Financial metrics, such as the benefits less the cost of achieving those benefits.
   D. The worth, importance, or usefulness of something.

9. Which of the following is not a stage in the project management cycle?
   A. Defining project scope.
   B. Planning the project.
   C. Allocating project resources.
   D. Implementing, monitoring, and controlling.
   E. Closing the project.

10. The final result or consequence of a process or project that focuses on the benefits and value that the project was undertaken to deliver is known as:
   A. Outcome.
   B. Value.
   C. Deliverable.
   D. Product.

11. Project integration means:
   A. Familiarizing team members with the project.
   B. Assigning team members to teams.
   C. Putting all the pieces of a project into a cohesive whole.
   D. Putting the pieces into a program.

12. When the project manager delivers a project charter, it should be in which phase or stage of project management?

   A. Planning the project.
   B. Implementing, monitoring, controlling.
   C. Initiation or beginning of the project.
   D. Closing the project.

13. Project managers do not address which of the following project management activities?

   A. Planning the project.
   B. Executing the work.
   C. Tracking programs.
   D. Documenting performance.
   E. None of the answers are correct.

14. A project stakeholder may include:

A. End users.
B. Suppliers.
C. Citizens.
D. All the answers are correct.

15. An effective project manager possesses a wide variety of skills and attributes. Which of the following statements does not set a good example of a Project Manager Attribute?

A. Delegates responsibility, but not accountability.
B. Sticks to his own work and obligations and encourages each team member to do the same.
C. Shares information with team members.
D. Is comfortable coordinating activities in various technical disciplines.

16. Projects have the least attention in what form of organization?

A. Functional.
B. Matrix.
C. Expeditor.
D. Coordinator.

17. One of the main advantages of a matrix organization is:

A. Improved project manager control over resources.
B. More than one boss for project teams.
C. Communication is easier.
D. Reporting is easier.

18. Which of the following is not considered a key role of the project manager?

A. Motivator.
B. Team Leader.
C. Reviewer.
D. Facilitator.

19. Which of the following helps with coordinating efforts between projects, standardizing the project management methodology, conducting project management training, providing temporary project management services internally, and championing the value of formalized project management methodology throughout an organization?

A. Your project sponsors.
B. A project management office (PMO).
C. Your stakeholders.
D. The Project Management Institute (PMI.)

20. Wes has little project experience, but he has been assigned as the project coordinator of a several new projects. Because he will be working in a matrix organization to complete his projects, he can expect communication to be:

A. Simple .
B. Open and honest.
C. Complex.
D. Hard to automate.

Contemporary Project Management —Answer Sheet

1.	___ A	___ B	___ C	_X_ D	___ E
2.	___ A	___ B	_X_ C	___ D	___ E
3.	___ A	___ B	_X_ C	___ D	___ E
4.	___ A	___ B	___ C	_X_ D	___ E
5.	___ A	___ B	___ C	___ D	_X_ E
6.	___ A	___ B	_X_ C	___ D	___ E
7.	___ A	___ B	___ C	___ D	_X_ E
8.	___ A	_X_ B	___ C	___ D	___ E
9.	___ A	___ B	_X_ C	___ D	___ E
10.	_X_ A	___ B	___ C	___ D	___ E
11.	___ A	___ B	_X_ C	___ D	___ E
12.	___ A	___ B	_X_ C	___ D	___ E
13.	___ A	___ B	___ C	___ D	_X_ E
14.	___ A	___ B	___ C	_X_ D	___ E
15.	___ A	_X_ B	___ C	___ D	___ E
16.	_X_ A	___ B	___ C	___ D	___ E
17.	_X_ A	___ B	___ C	___ D	___ E
18.	___ A	___ B	_X_ C	___ D	___ E
19.	___ A	_X_ B	___ C	___ D	___ E
20.	___ A	___ B	_X_ C	___ D	___ E

Why do so many professionals say they are project managing, when what they are actually doing is firefighting?

—Colin Bentley

Chapter 2

Defining Project Scope

Overview

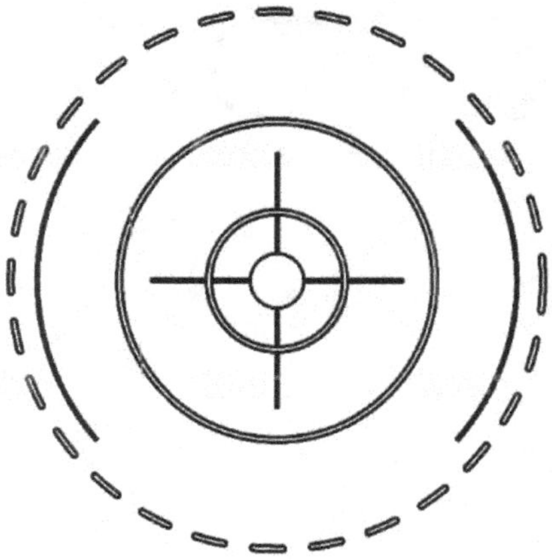

"Do not repeat the tactics which have gained you one victory, but let your methods be regulated by the infinite variety of circumstances."

—Sun Tzu c. 490 BC

When it comes to completing projects, what you do not know can hurt you and your organization. Doing the work might be 90 percent of the job, but project planning will impact the entire project and help ensure success. A project well done is always a project well planned.

The topics in this chapter introduce you to many of the issues involved with project planning and discuss how to avoid project pitfalls.

In this chapter, you will learn how to:

- Conduct a stakeholder analysis
- Define the project scope
- Recognize common causes of project failure
- Evaluate the critical factors associated with starting projects and your readiness
- Develop a Work Breakdown Structure (WBS)

The competencies associated with this chapter include:

Planning and Evaluation **Resource Management** **Communication**

Take Your Temperature for Defining Project Scope

The following assessment will help you identify what you know about defining the scope of a project and the areas where you can improve. With you and your organization in mind, read each statement below, and consider how much you agree with it. Use the numbers from 1 to 10, where **1** means you **strongly disagree** and **10** means you **strongly agree**.

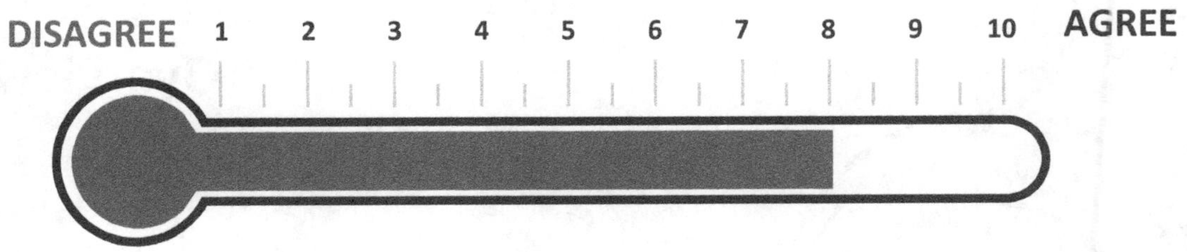

_____ 1. I know how to identify project stakeholders.

_____ 2. I can determine stakeholder requirements.

_____ 3. I am good at identifying project deliverables.

_____ 4. I can prepare a project charter.

_____ 5. I understand some of the common reasons for project failure.

_____ 6.I know the critical factors to consider before starting a project.

_____ 7. I am confident in my ability to define the scope of a project.

_____ 8. I am familiar with PMI's Scope Management Model.

_____ 9. I understand what is meant by managing a project's critical first 10 percent.

_____ 10. I am good at developing a work breakdown structure (WBS).

_____ **Total**

Take a moment to reflect on the following before you move forward in the chapter.
In your work, what problems have you experienced or observed related to defining a project scope?

Take a few minutes to also reflect on your self-assessment. List two or three areas you want to improve.

As you progress through this chapter consider what actions you can take to demonstrate competence in the following three competency areas:

1. **Planning and Evaluation**. Establishes policies, guidelines, plans, and priorities; plans and coordinates with others; aligns required resources; monitors progress and evaluates outcomes; improves organizational efficiency and effectiveness.

2. **Resource Management**. Demonstrates awareness of technical resources; knows how to apply resources to achieve desired outcomes.

3. **Communication**. Makes clear and effective presentations to individuals and groups; listens to others; communicates effectively in writing; can critically review and comprehend information written by others.

Defining Project Scope #2.1

Identify Project Stakeholders

After a project idea is accepted, the next steps are to identify the stakeholders, determine how to communicate with them, and then to define and formalize the project's scope. These efforts contribute to the overall goal of developing the project plan.

The Project Life Cycle illustration below shows these jobs are part of the Initiate phase of the project life cycle.

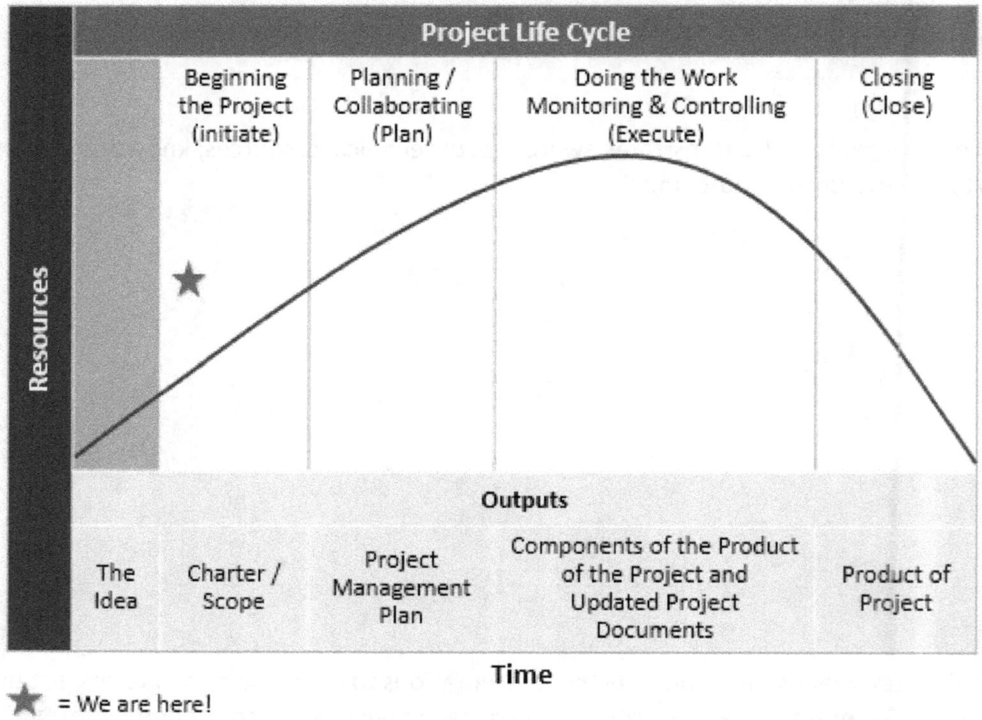

Also, be aware that changes in the project will occur throughout the project's life cycle. When they do, you must document them and revise the project scope and plan accordingly, and then communicate the changes to everyone involved.

Project Knowledge Areas and the Project Plan

The **Project Management Knowledge Areas** illustration below shows how the Project Management Plan enables integration among knowledge areas. The management plan is the mechanism that pulls all the knowledge area plans together.

Development of the Project Management Plan is accomplished by using what is called *progressive elaboration* or *rolling wave planning*. It means the Project Management Plan will require numerous iterations to stay current. As the project progresses, you must update the changes to the various knowledge area plans.

Project Management Knowledge Areas

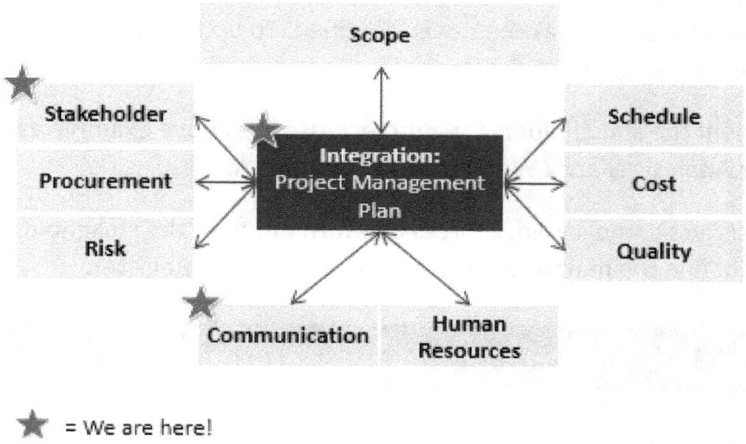

Identify Stakeholders Roles in the Project

A **stakeholder** can be anyone interested in a project's outcome. According to PMI, a *stakeholder* is "an individual, group, or organization that may affect, be affected by, or perceive itself to be affected by a decision, activity, or outcome of a project, program, or portfolio." (*PMBOK®Guide)-Sixth edition*. p. 723.

For example, in addition to the sponsor and client, stakeholders can be customers, members of the project team, project managers, executives, operational managers, end-users, and members of the public.

You must know who the project stakeholders are, what their roles will be, and determine how to communicate with them appropriately. Your goal is to communicate with all project stakeholders as early as possible, so their perspectives are included in the planning process.
PMBOK is a registered mark of the Project Management Institute, Inc.

Proper identification of stakeholders is a significant activity within the planning process. A Stakeholder Analysis Matrix is one way to identify the various stakeholders and their potential impact on the project.

Notice on the following **Stakeholder Analysis Matrix** example that the project manager has identified the impact each stakeholder may have on the project.

Stakeholder Analysis Matrix Example

Stakeholder Name	Stake/Interest	Potential Impact	Stakeholder Management Approach
Stakeholder 1	Financial, ROI	Can approve or disapprove project funding	Meet early in the project regarding approval criteria. Likelihood of approval given specific metrics.
Stakeholder 2	Operations, processes may be impacted	Can provide supportive or negative voice	Include from the beginning to ensure this perspective is represented, and projected outcomes are positive.
Stakeholder *n*...			

PMBOK is a registered mark of the Project Management Institute, Inc.

Developing a Stakeholder Analysis Matrix and associated planning is sometimes an evolving process. For example, after a few iterations of the Project Management Plan, you might need to revise parts of it because a new stakeholder is identified. To update the plan, you would address any new tasks, risks, or resources.

To formally document the stakeholders for a project, use the above example as a guide. You will find a **Stakeholder Analysis Matrix Worksheet** in Appendix A.

To keep the analysis less complicated, sources of information for the stakeholder analysis are typically not identified in the matrix, but rather in a **Stakeholder Register.**

Example Stakeholder Register

Stakeholder Name	Power / Interest Grid Position (Influence)	Org.	Org. Position	Location	Project Role	Contact Info.
Rebecca Smith	1	XYZ Rules	President	Washington, DC	Sponsor	rsmith@xyzrules.com
George Santos	4	Supplies Inc	Sales Rep.	Harrisburg, PA	N/A	sales@sales.com
Stakeholder n...						

Proper identification of stakeholders is a significant activity within the planning process. You want to be sure to include all potential stakeholders as early in the process as possible, so their perspectives are included in the planning process. It is also helpful to understand how much **power** and **interest** key stakeholder have related to a particular project as well as their desired level of engagement.

The intent of the Power/Interest, Power/Influence, or Influence/Impact **grids** are conceptually the same. You want to determine what kind of stakeholder(s) you are dealing with and how much attention you will pay to them. This analysis will contribute to the communication plan of how to communicate with each type of stakeholder, considering the content and frequency of the communications. If they are directly impacted by the product of the project, they should be in the regular communication circle of the project. If there is little change to their daily lives because of the outcome of the project, you would not want to bother them with frequent and detailed communications about the project. The following graphic is an illustration of a **Power/Interest Grid**.

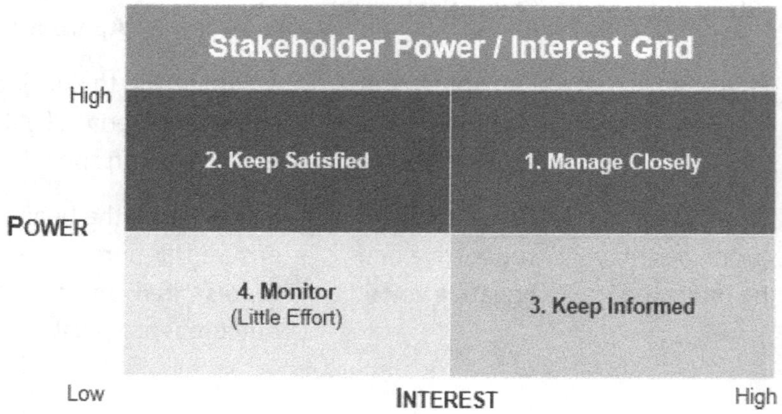

Another perspective of a stakeholder is their **level of engagement** in the project. It is important that key stakeholders are as **active** about the project as you had anticipated. There could be reasons for their lack of engagement that you need to know about as early as possible. First, determine what level of engagement you anticipate the stakeholder will have with the comings and goings of the project. As the project progresses, use the grid to assess the actual level of engagement as compared to the anticipated level of engagement. This periodic review can tell you that an important stakeholder has **drifted away** from the project, perhaps for negative reasons you should know about. The **Stakeholder Engagement Grid** provides a simple tool to keep this in check.

Example Stakeholder Engagement Grid					
Stakeholder	Unaware	Against	Neutral	For	Promote
Jim	C			P	
Jack		C	P		
Susan				P, C	
Ralph					P, C

P = Planned level of engagement (Plan Desired Actions)
C = Current level of engagement (Monitor)

Managing Communication

Communication can be one of the biggest challenges in any project. Managing communication from start to finish requires planning and discipline. The planning includes identifying the stakeholders, the types of information to be communicated, how the information is to be conveyed, and the frequency.

Communication is typically less complicated on small projects and so may not require as much planning and structure. But for larger projects, you need to plan who needs what information and how will you, as the project manager, deliver it to them.

Initial questions to consider might include: Who needs to know the project status? How will project status be communicated, for example, reports, meetings, or both? Who needs to know about project concerns or problems? Who can approve changes to the project? Does the project have a member directory? Is there a communication matrix for the project? If a stakeholder communication matrix is not available, it may be beneficial to develop one.

You must follow through with the communication plan throughout the life of a project.

What to do:	**Other actions:**
☐ Identify the stakeholders associated with your project.	☐ ______________________
☐ Create a Stakeholder Analysis Matrix.	☐ ______________________
☐ Assess the level of interest and engagement each stakeholder will have in the project and their potential impact on the project.	☐ ______________________

Remember

- ✓ Stakeholders may be anyone interested in a project's outcome.
- ✓ Identifying and classifying stakeholders is a significant process to ensure that all stakeholders receive appropriate project communication.
- ✓ Stakeholders may change during the life of a project. When changes occur, you must update the stakeholder documentation.
- ✓ Communication is a critical factor in determining the probability of project success.

Enhance Your Learning

Watch the following 12-minute video by Arham Faraaz to learn more about identifying project stakeholders.

Faraaz, A. (2019). *Identify stakeholders / Project Stakeholder Management.*	Available at: https://www.youtube.com/watch?v=0xa2Q2XU2nM

Reinforce Your Learning

Stakeholder Analysis Matrix. Consider how the stakeholders in a project you worked on were engaged in the project and how their participation affected the success of the project. Then consider an upcoming project or one you are familiar with and complete the stakeholder matrix below. A Stakeholder Worksheet is included as Appendix A.

Stakeholder Analysis Matrix			
Stakeholder Name	**Stake/Interest**	**Potential Impact**	**Stakeholder Management Approach**
Stakeholder 1			
Stakeholder 2			
Stakeholder 3			
Stakeholder 4			

Defining Project Scope #2.2

Determine Stakeholder Requirements

Defining and meeting the needs of the client and other stakeholders can be challenging. Various stakeholders may have requirements and priorities that differ one from another, and your job as a project manager is to help reconcile such differences.

Also understand that stakeholders can be *internal*, part of your organization, and *external*, outside your organization. You must meet the needs of both internal and external stakeholders. If you do not meet the needs of external stakeholders, the client may not return for repeat business, and your organization's reputation could suffer.

In a competitive bid situation, a potential client may issue a Request for Proposal (RFP). Anyone who has been involved in responding to competitive bids knows that RFPs can range from a few general ideas to specific and lengthy requirements.

After issuing an RFP, a potential client may hold a meeting for all bidders and other stakeholders, so people can ask questions. This is typically seen as the "fairest way" to deal with questions because everyone present hears the same information.

Whether an RFP is issued or not, a project manager must collect as much information as possible about stakeholder requirements. This strategy will help produce the details needed to define the project scope and the initial plan.

How to Determine Stakeholder Requirements

The first step in determining stakeholder requirements is to communicate and engage with them as early as possible. Do this as soon as you start work on the project. Start by identifying the key stakeholders, in other words, the stakeholders with the most power and influence, and document their requirements and what outcomes are most important to them.

Collecting stakeholder requirements is an essential part of completing a project charter and defining the project scope. The requirements need to be identified, recorded, and analyzed. Otherwise, you risk being unsuccessful because you will not understand what is to be accomplished, or you will misunderstand and attempt to provide something that is not wanted.

You can communicate with stakeholders and collect their requirements in a variety of ways. Here are some typical examples:

- Interview
- Focus groups
- Surveys and questionnaires
- Direct observation
- Pilot efforts and prototyping
- Historical information, data, and documents

You use the information gathered from these activities to identify and document stakeholder business requirements, constraints, assumptions, and expectations.

How Stakeholders May Judge a Project

The following is a summary of typical ways stakeholders judge how well a project meets their needs.

Quality
+ Technical quality
+ Timeliness
+ Effectiveness
+ Dependability/Reliability
+ Cooperation
+ Communication
+ Performance

Accuracy
+ Process
+ Information
+ Linking

Focus on Stakeholder Concerns
+ Their problems
+ Their convenience
+ Their perceptions

Promptness
+ Answering telephones
+ Information relay
+ Solutions
+ Keeping promises
+ Answers to questions
+ Progress reports

Attention to Convenience
+ Assuming responsibility for the outcome
+ Sensible, logical solutions
+ Reducing irritants
+ Efficient processes/solutions

What to do:	Other actions:
☐ Identify stakeholders and their requirements early in the project.	☐ ____________________
☐ Ask people who will do the work for their input.	☐ ____________________
☐ Provide opportunities for stakeholders to be engaged in the project.	☐ ____________________

Remember

✓ Consider the diverse needs of your stakeholders and be receptive to their feedback.
✓ Understand the concept of internal and external stakeholders.
✓ Stakeholder engagement includes attracting and involving individuals, groups, and organizations that may be affected by a project or may affect the project.
✓ Identify the key stakeholders and what outcomes are most important to them. Provide opportunities for them to be involved in defining the project scope.

Enhance Your Learning

Review the following website by Harry Hall, The Project Risk Coach, to learn ways to engage stakeholders.

Hall, H. (2022). *10 Ways to Engage Project Stakeholders.*	Available at: https://projectriskcoach.com/engage-project-stakeholders/

Reinforce Your Learning

Stakeholder Engagement. Consider a project you were involved in or observed and recall how you worked with the stakeholders. Then answer the following questions:

Did you identify key stakeholders early in the project?

Did you provide opportunities for stakeholders to talk with each other and with you?

Did you listen to stakeholders and seek to understand their requirements?

Did you ask knowledgeable people for their input?

Defining Project Scope #2.3

Specify Project Deliverables

Many inputs and outputs are associated with the completion of a project. Understandably, the most essential deliverable that stakeholders expect is the project's product, service, or result satisfactorily completed on time and within budget.

But to successfully plan, control, and deliver the project's product, service, or result, you must formalize project information, decisions, and changes by creating and communicating many documents. You must do this throughout the project's life cycle. You typically provide project documents to stakeholders and team members.

Early in a project, one of your first jobs as a project manager is to identify the types of documents the stakeholders and team need, how you will provide the information, for example, in paper or online reports, and the schedule, for example, weekly.

These document **deliverables** (both intermediate and final) serve as **communication tools**.

The following is an overview of the types of project documents you will use in most projects:

- **Project plan** or **work breakdown structure (WBS)**. It defines the project deliverables and the tasks/activities needed to complete each deliverable.
- **Weekly project meeting minutes**. The meeting minutes document formal project discussions and provide a record of issues/problems, actions, changes, and plans.
- **Project status, informed update,** or **variance report**. Use status reporting to aid the team in communicating progress, obstacles, and other pertinent information as the project progresses, and for you to communicate this information to stakeholders. Milestones are established and communicated in these reports.
- **Timesheet report**. This report documents each individual's hours worked on a project, and consolidated reporting shows time worked by selected project groups or functions.
- **Risk management plan**. This plan specifies project contingencies and how the team will avoid or cope with them. This plan helps the project manager and team minimize an organization's exposure to legal, financial, and public relations issues.
- **Project budget** or **financial report**. The organization uses this information to track and control project costs.
- **Communication plan**. Developing a plan for how to communicate project information among team members and among the team and stakeholders is an effective way to set up efficient channels for the exchange of project information, help maintain team morale, and support better client/stakeholder relations.
- **Lessons Learned.** After the team completes the project's product, service, or result that meets the quality level agreed on, this report describes the experience gained, so that future projects may benefit from the knowledge. The report describes both successes and failures. Additionally, the Project Management Institute (PMI) recommends that project managers capture lessons learned throughout the project life cycle. This encourages continuous improvement and leads to a more successful project.

- **Final report**. This report wraps up the project. It provides any final documentation to the people who need it, such as the team, client/stakeholders, and any other people in the organization or the public.

A stakeholder register can be an additional elaboration of a stakeholder analysis matrix that includes attributes for each stakeholder, including what deliverables are most important to each stakeholder or stakeholder group.

What to do:	Other actions:
☐ List the documents that will be used in most projects.	☐ ______________________
☐ Establish channels of communication for the exchange of project information.	☐ ______________________
☐ Communicate progress, obstacles, and other pertinent information on an ongoing basis as appropriate.	☐ ______________________
	☐ ______________________

Remember

✓ Be proactive in providing information to stakeholders.
✓ Many inputs and outputs are associated with a project, and project managers typically provide these to team members and stakeholders.
✓ The most crucial deliverable that client/stakeholders expect at the end of the project is the product, service, or result satisfactorily completed on time and within budget.

Enhance Your Learning

Watch the following 3-minute video by Chris Burke of PM Academy that describes how to identify project deliverables.

Burke, C. (2021). *How to Create Effective Project Deliverables.*	Available at: https://www.youtube.com/watch?v=7HADacRHSFs

Reinforce Your Learning

Stakeholder/Project Deliverables and Communication Tools Matrix. Assume you must develop a stakeholder deliverables matrix like the sample below. For each stakeholder listed, place an "X" in the spaces provided to indicate the deliverables you believe the stakeholder may expect or benefit from

Stakeholders / Project Deliverables and Communication Tools	A. Project plan or WBS	B. Weekly meeting minutes	C. Project status report	D. Timesheet report	E. Risk management plan	F. Project budget report	G. Communication plan	H. Final report	I. Final project or service	J. Others:
Project Managers										
Client Project Managers										
Administration										
Discipline Managers										
Team Members										
Regulatory Agency										
Public										

Defining Project Scope #2.4

Draft Project Charter

According to *PMBOK® Guide*, A ***project charter*** is a document issued by the project initiator or sponsor that formally authorizes the existence of a project and provides the project manager with the authority to apply organizational resources to project activities. Often the ***project charter*** delineates roles and responsibilities, outlines major project deliverables, identifies the main stakeholders, and specifies the authority of the project manager.

A project charter is especially useful for larger projects. For smaller projects, a formal quote, purchase order, or Statement of Work (SOW) may suffice.

The project charter is typically a succinct formal high-level document that conveys the overarching purpose of a project and the expected product, service, or result. The document usually identifies the project manager and authorizes the expenditure of funds for the project.

The project sponsor usually authorizes or signs off on the project charter.

A project charter typically includes the following kinds of information. This listing may be used as a **template** for developing a project charter for a project of your choosing.

- Project title:
- Project manager:
- Purpose or justification:
- Objectives (address a need, improve a process, fix a problem, pursue an opportunity):
- Success criteria (improve process efficiency by X%, reduce costs annually by $ X, increase sales by X percent, reduce customer complaints by X percent):
- High-level project description:
- High-level requirements:
- Product characteristics:
- Project constraints and assumptions:
- Summary milestone schedule:
- Summary budget:
- Project approval requirements (success metrics, who decides, who signs off):
- Project manager authorizations (up to (x) $ signature; up to (y) $ change controls):
- Client or Sponsor (name, title, date):

As stated previously, the **project charter** formally recognizes the existence of the project. It gives the project manager the authority to proceed with the project, so it should be issued by a person or group with the authority to sponsor the project.

What to do:	Other actions:
☐ Determine what level of authority you have to initiate projects in your organization.	☐ ____________
☐ Discuss whether a quote, purchase order, or statement of work (SOW) is sufficient authorization to initiate projects.	☐ ____________
☐ Identify who in your organization has the authority to sponsor or authorize large projects.	☐ ____________
	☐ ____________

Remember

- ✓ A Project Charter formally recognizes the existence of the project and gives the project manager authority to begin the project.
- ✓ A project charter is issued by a person or group with the authority to sponsor the project.
- ✓ For smaller projects, a quote, purchase order, or statement of work may suffice.

Enhance Your Learning

Watch the following 10-minute video by Adriana Girdler to learn more about preparing a project charter.

Girdler, A. (2020). *Project Charter Guide: HOW TO WRITE A PROJECT MANAGEMENT CHARTER.*	Available at: https://www.youtube.com/watch?v=FIPeKqrvNCs

Reinforce Your Learning

Project Charter. Examples of projects in the workplace include installing equipment, developing or conducting training, developing a new product, planning a significant event, and relocating an office or modifying building facilities. Select one of these examples or another of you choosing. Then complete a Project Charter for the project using as a template the kinds of information listed on the previous page.

Defining Project Scope #2.5

Learn Common Causes of Project Failure

Are you aware of some of the classic project failures that have occurred? They include the Challenger Space Shuttle Disaster, New Coke, and even the Leaning Tower Pisa. Can you think of others?

Many lawsuits are filed against businesses and individuals because project issues were glossed over, or the planning was poor. Just as planning a project is critical to its success, so is revising the plan as changes occur over the life of the project.

In this concept, we will discuss reasons why projects fail. Projects are complicated endeavors, and especially large projects with extended schedules. An example is when a project team involves people who may not have worked together before, and they are doing something new.

Projects fail for many reasons, with inadequate planning and poor communication being among the most significant issues. The following are some other typical reasons:

- Business requirements not identified or communicated
- Project not adequately vetted or authorized
- Project expectations not aligned with organizational goals
- Priorities, objectives, and scope constantly changing
- Assumptions poorly defined and unsubstantiated
- Feasibility analysis not fully developed or inaccurate
- Tools and systems outdated and inaccessible
- Team collaboration inadequate; team conflicts
- Project leadership inexperienced and ineffective
- Resources untrained and misaligned

Also, recognize that people's actions cause the majority of project problems. It has been suggested that **people's** actions make up about **90 percent of the risks** in a typical project's Risk Management Plan (Bourne, 2009).

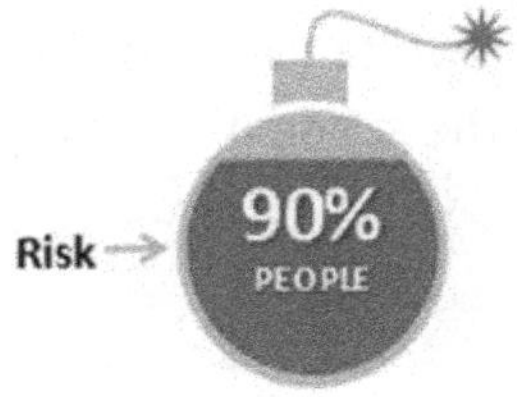

Among the biggest reasons projects fail are **changes in an organization's priorities** and **changes in project objectives** (*PMBOK® Guide*, 2013). However, when you consider the myriad of stakeholders with their individual perspectives and expectations, it is not surprising that inaccurate requirements gathering, not defining risks, and communication are not far behind.

But no matter how many requirements, or how many competing requirements among stakeholders, and no matter how many changes to the requirements occur as the project moves forward, the project manager is responsible for ensuring that all requirements are documented and approved by the stakeholders. The project manager is also responsible for ensuring that all requirements are accounted for in the original project plan and that the project scope and plan are revised when requirement changes occur.

Communication and Project Failure

The 2015 *Pulse of the Profession* from the Project Management Institute (PMI) reports that inadequate/poor **communication** is among the top five reasons why projects fail. However, one could argue that incomplete/inaccurate requirements and undefined risks are linked to inadequate/poor communication.

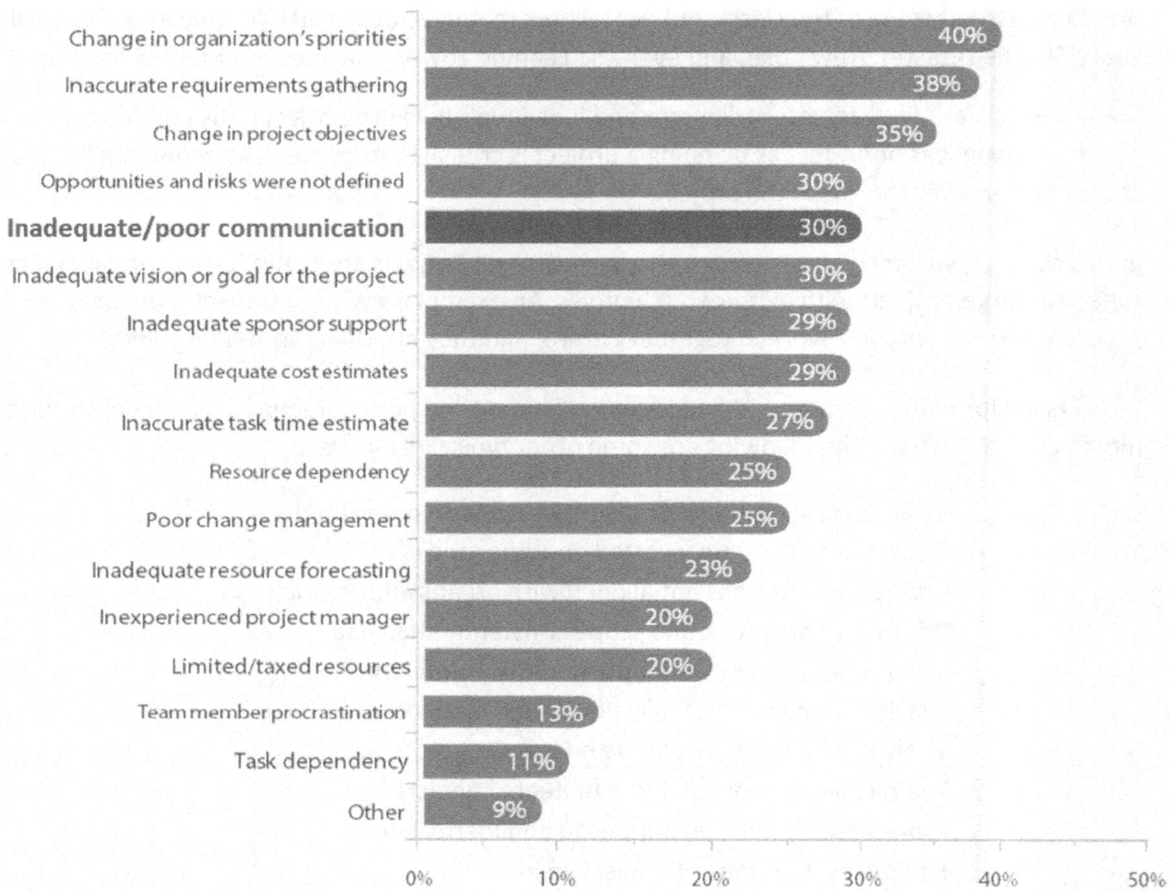

All these issues point to the importance of the initial project-planning phase and the need to accurately assess and define the scope of the project.

What to do:	Other actions:
☐ Recognize that people's actions may account for the majority of project problems.	☐ ______________
☐ Consider if you are getting or have all project requirements.	☐ ______________
☐ Consider the risks for your project and what could go wrong, and the impact on the project.	☐ ______________

Remember

- ✓ Among the biggest reasons projects fail are changes in an organization's priorities and changes in project objectives. This is followed by incomplete/inaccurate requirements gathering, not defining risks, and poor communication.
- ✓ People's actions make up about 90 percent of the risks accounted for in a typical project's Risk Management Plan.
- ✓ A project well done is always a project well begun.

Enhance Your Learning

Watch the following 9-minute video by Jennifer Whitt to learn more about project failure. :

Whitt. J. (2012). *Why Projects Fail: Top 10 Reasons for Project Failure.*	Available at: https://www.youtube.com/watch?v=CTHHiBNXJ6w

Reinforce Your Learning

Why Projects Fail. Think of a project you have been involved with or observed that failed to achieve its goals. List three reasons for the poor performance and describe what the project manager could have been done differently.

1.

2.

3.

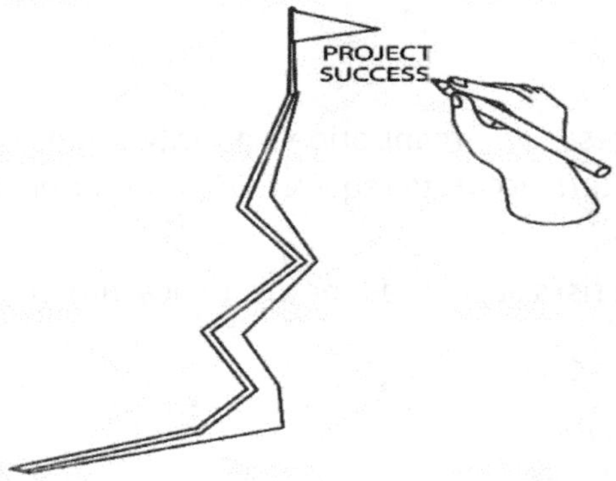

Defining Project Scope #2.6

Prepare for Project Success

In the previous concept, we discussed several reasons for project failure. In this concept, we will revisit the reasons and transform them into proactive actions project managers can take to help ensure project success.

You will benefit from these actions if you use them to prepare the groundwork for the project before you start work. These actions are the things you need to do in preparation for starting the project.

Many of these actions will also be productive at any time in the project life cycle when changes occur, and so you must revise or update project information. For example, if you learn of an additional business requirement while the project is in progress, you must go back to the stakeholders to define the requirement and get written approval from the client for the change.

Think of the information in this concept as a type of checklist. For example, the first activity discusses the necessity of gathering stakeholder requirements and why, but it does not describe how to do it. To learn more about *how* to do these activities, for example, how to gather requirements, refer to other concepts in this course.

The following is a discussion of 10 critical project activities that you as the project manager are responsible for completing.

1. Identify and document business and client/stakeholder requirements.

Be sure there is a strong business need for the project and that the stakeholders can support this need.

As a first step, gather the high-level requirements for the project, outline what you know already, and document what you think you are going to create. This will help you understand the background and context of the project.

You may need to gather requirements from a variety of sources, and it may take several meetings to achieve the final requirements. For example, for a software development project, two stakeholders, a call center manager and a marketing manager, give their high-level requirements. Then they may ask that the people gathering the requirements interview end-users in their organizations (subject matter experts) to get more details. You will need to ensure that these interviews are completed and that the requirements "hang together." If not, the people responsible for documenting the requirements must go back and review the situation with the operational managers to work out the differences and get stakeholder agreement on the final requirements.

If business needs or objectives change during the project, be sure to address this with the client/stakeholders so that you understand if and how the project plan needs to be revised.

2. Vet the project with the stakeholders and obtain proper authorization.
Work with stakeholders to outline a list of the project's expected benefits. With stakeholder buy-in and using the requirements information already developed in the first activity, project developers can then move forward with creating a high-level project description, goals, statement of work, anticipated outcomes, and justification.

You can then communicate this information to the project sponsor or group with the authority to approve the project. Be sure to get the approval in writing.

3. Align project expectations with organizational goals.

For larger projects, you will need to define the project's goals in a project charter. As we have discussed, the charter is a document that describes a project's purpose, and that gives a project manager authorization to start work. For smaller projects, you may need only a signed contract, statement of work, or purchase order.

Whether you use a project charter or one of the other methods, get the document in writing.

The project plan, as it evolves, should further explain the goals of the project and how they align, fit, or link to the organization's goals.

4. Manage project goals and objectives by clearly defining the initial scope and also define a change management process.

High-level goals and objectives are typically defined in the project charter. However, some will most likely change as the project progresses. The project manager and team refine the scope as the project evolves. The initial scope provides a benchmark as to what is and what is not included in the project from which the project manager can begin to show what the project's goal, deliverables, or finished product will look like at the end. As changes occur, they must be authorized, documented, and communicated using an established change process, so that everyone involved is kept informed.

Without such a process, scope creep may occur throughout the project, and this will lead to cost overruns, missed deadlines, and conflicts that are difficult to resolve.

A change management process lays out a framework for making changes, and the process must be communicated and understood by stakeholders because they are the ones who will approve the changes, along with changes in the budget and schedule.

5. Substantiate and verify project assumptions in the initial project scope statement.

Using the project requirements and any other information gained from the stakeholders, document project assumptions in the initial project scope statement. This information defines what the project must deliver and typically also provides information related to the cost, timing (milestones), and quality of the work products.

If no requirements were gathered, meet with key stakeholders to determine their requirements, so that assumptions, expectations, and the determinants of success can be verified, documented, and managed.

Again, scope creep is one of the most common reasons projects run over budget and deliver late. The client/stakeholders may forget the extra work and effort you have put in, insisting that you have achieved what they asked for originally.

6. Conduct a feasibility and risk analysis.

Many things can go wrong on a project. Some projects may benefit from pilot testing, but for others, that may not be practical. Taking the time to consider what your project may need is one of the best things you can do when you start a project.

Conducting an upfront and objective feasibility and risk analysis can help alleviate potential problems. Reference this information in the initial scope document and use it to highlight areas of concern and where prevention and mitigation strategies are needed. Involving key stakeholders and outlining what action you would take, should the worst happen, is of benefit to everyone involved.

7. Research and secure the proper tools and systems.

What are the right tools to use for a project? This depends on your project, your team, your client, key stakeholders, and your budget. Incorporating unnecessary tools may be costly and not give you the results you want.

Here are some commonly used tools:

- Resource planning and management: Float, Resource Guru, Hub Planner
- Project planning and timeline management: Microsoft Project, Gantt Pro, Workfront
- Collaboration with stakeholders: Google Sheets, Meeting Sphere, Confluence
- Communication with your team and key stakeholders: Zoom, Slack, Workspaces
- Managing project tasks: Trello, Jira, Monday.com

The best advice is to *keep things simple*. Consult with your team, client, and stakeholders. Avoid overcomplicating things as you can always integrate selected tools later if needed.

8. Provide adequate team building and development guidance.

If you have the luxury of selecting your team members, choose wisely. When you consider who might be available to work on a project, do not just look at availability. Think about what skills and experience levels are needed to deliver your project successfully.

Teams typically go through stages of development (form, storm, norm, and perform). Anyone who has worked in a group understands that building a high-performing team will take time and guidance.

9. Match project needs with effective project leadership.

Remember, 70 to 80 percent of employees quit their boss, not their job. A project manager must not only have some ability or technical skills in the project area, but must also be effective at managing, motivating, and leading team members. Leaders need to fit with the culture of the project team, client, and key stakeholders, and leaders should be selected that have experience completing projects of similar design, scope, and complexity without being overwhelmed.

10. Orient, align, and coach project workforce resources.

A project staffing plan defines who will be on a project team and how much of a time commitment each person will make. In developing this plan, the project manager typically negotiates with team members and their supervisors on how much time each team member can devote to the project. In other cases, a project manager may not have much say in who is assigned to a project.

Establish an onboarding and orientation plan to help new team members understand expectations and their roles and responsibilities. In some cases, individual skill assessments and specialized training may be needed. Communication is key. Ask for input, train as needed, provide guidance and coaching when needed, and tell your team when they are performing well.

What to do:	Other actions:
☐ Establish and agree on project boundaries before starting a project.	☐ ____________________
☐ Document project expectations for cost, scope, timing, and quality.	☐ ____________________
☐ Identify the actions you should take before starting a project.	☐ ____________________

Remember

- ✓ Identify and communicate business needs and requirements.
- ✓ Communicate with stakeholders and obtain proper authorization.
- ✓ Align project expectations with your organization's goals.
- ✓ Manage changes by defining the initial scope and set up a change management process.
- ✓ Substantiate and verify project assumptions in the initial project scope statement.
- ✓ Conduct a feasibility study and a risk analysis.
- ✓ Research and secure the appropriate methods, tools, and systems.
- ✓ Provide team-building and development guidance.
- ✓ Match project needs with effective project leadership.
- ✓ Orient, align, and coach project workforce resources.

Enhance Your Learning

Review the article *"Top Things You Should Consider Before Starting a Project"* by Project Vanguards.

Project Vanguards. (2021). *Top Things You Should Consider before Starting a Project.*	Available at: https://projectvanguards.com/top-things-you-should-consider-before-starting-a-project/

Reinforce Your Learning

Critical Factors to Consider Before Starting a Project. Consider a project you have been involved with or observed that failed to achieve its goals. List three things you would do differently at the beginning of the project to achieve better results.

1. ____

____

2. ____

____

3. ____

Define Project Scope

When planning a project, the problem areas are always the same: costs, time, human resource failings, and the unknown. Thus, the more you know about the project before it begins, even before you develop action plans, the better.

According to *PMBOK®Guide*, a ***project scope statement*** is the description of the project scope, major deliverables, assumptions, and constraints. It is a document that describes what is in a project and what is not. The ***project scope*** is the work performed to deliver a product, service, or result with the specifies features and functions. A ***scope of work*** is typically a narrative description of the project.

The client/stakeholders typically delegate the creation of the initial scope statement document to the project manager. The project team assists the project manager in developing this document.

The initial scope statement document is often referred to as the ***scope baseline.*** Revisions are commonly referred to as a ***progressive elaboration*** of the scope. This term highlights that the characteristics of a project are determined incrementally and continually refined as the project progresses.

Project plan development takes the outputs from the project scope—and other planning processes—and combines them into a coherent and consistent document called the Project Management Plan. We will describe it in greater detail in later courses.

Prepare to Write a Project Scope Statement

Here are some critical questions you must answer before you start writing a project scope document:

- What is in the project?
- What is not in the project?
- How should the work be done?
- How much work should be done?
- What requirements do the client/stakeholders have for the project?

<u>How the Project Scope Fits with Other Project Documents and Steps</u>

The chart that follows illustrates the typical steps you will follow in creating the scope document. Notice that the Project Scope Statement is a deliverable in the Initiate phase.

Review each step listed under Description/Purpose and consider the reasons for each one.

Phase	Description/Purpose	Deliverables/Tools
Idea	• Preliminary description and analysis to determine whether a project idea has merit	• Idea appraisal • Decision: yes or no

Phase	Description/Purpose	Deliverables/Tools
Initiate: Define Project Scope	• Identify and document business need(s) • Define project scope, goals, and objectives • Identify project client, sponsor, stakeholders, and project organization • Develop initial estimates of effort, cost, schedule, and risk • Identify deliverables and acceptance criteria	• Project Description • Project Charter • Feasibility Analysis • Project Scope Statement • Stakeholder Analysis Matrix

Expect Changes to Occur as a Project Progresses

Recall that *project management* is a professional process that helps you manage organizational resources to achieve project goals within performance quality requirements, schedule, and budget or cost constraints.

As we have discussed, at the beginning of a project, you will define the project scope. But you can expect that some parts of the project will change as work progresses. If the changes affect the project scope, you must revise or refine the scope document to reflect the changes and communicate that information to the team and stakeholders.

But suppose you cannot deliver the expected or changed requirements within the established cost and scheduling constraints. In that case, as the project manager, you must meet with the appropriate stakeholders and determine how to modify expectations for the project and revise the project planning documents accordingly. Well-designed project scope statements can help you identify issues and problems and help you communicate them to stakeholders.

Audits of project results indicate that one of the most frequent reasons for project failure is an inadequate project scope document. This is because a poorly executed scope document does not describe what the client/stakeholders expect you to deliver. The project scope is the first significant step in defining and managing the project, so a poorly defined scope threatens the success of the project from the beginning.

As a project manager, it is your job to anticipate where a project might go wrong and to take corrective action when necessary. You cannot let your organization's investment slip through the cracks due to poor planning, mismanagement, or unexpected conditions. One of the best ways to stop this from happening is to develop a comprehensive project scope, keep it updated as the project progresses, and communicate the updated scope to the stakeholders and team.

Quality

The scope defines the final deliverable needed to satisfy project requirements. It is a rule of thumb that the client will want the deliverable—the product, service, or result—to be quality work. Therefore, the scope must also define the standards that define acceptable quality.

The client/stakeholders, project manager, and team members must understand and agree to the quality standards. Completing the work in a way that does not satisfy expectations is an almost guaranteed source of conflict and often leads to project failure.

Though they may not be specific, most clients/stakeholders tend to mean that they want high-quality when they say they want quality. Most project managers define *quality* as "meeting the requirements," though some managers may not take that a step further and make a distinction between "high quality" and "low quality." Take the time to understand and set up specific quality standards in the project scope document. It will help alleviate such miscommunication and head off misunderstandings and possibly failure.

Gold Plating

Adding extra features or components to a project beyond the original scope requirements is referred to as *gold plating.* While there can be a tendency for project managers to do extra things to please clients/stakeholders, be aware that adding to the requirements may affect the schedule and add to the project's cost. An example of *gold plating* would be if the initial requirements are for a Chevy and that is what the client is paying for, and then the project manager adds extras to give the client a Cadillac.

The initial, upfront effort a project manager takes to clearly define expectations, requirements, and standards in a project scope document will reduce the potential for conflict as the various project activities progress.

Communication

Communication is a concern and a priority in all project activities. As a project manager, you must communicate clearly. You cannot expect stakeholders to understand the jargon involved in the project. Project managers must take sufficient time and develop enough detail to fully complete scope documents that are understandable to all.

What to do:	**Other actions:**
☐ Meet with the stakeholders often to ensure you understand their requirements.	☐ ______________
☐ Communicate your understanding of the deliverables to the stakeholders.	☐ ______________
☐ Resolve any disagreement among stakeholders early in the planning phase and agree on the quality criteria for project success.	☐ ______________
	☐ ______________

Remember

- ✓ Project plan development takes the outputs from the project scope—and other planning processes—and combines them into a coherent and consistent document— the Project Management Plan. We will discuss it in later courses.
- ✓ Project scope planning involves creating the documents that define the project scope, and you may need to revise or refine it as the project progresses and changes occur.
- ✓ A narrative description of the work that must be done is called the Scope of Work.
- ✓ When planning a project, the problem areas are always the same: costs, time, human resource failings, and the unknown.

Enhance Your Learning

View the following 17-minute video by People Rich about scope management.

People Rich. (2012). *Project Management 2 – Scope Management.*	Available at: https://youtu.be/xjawa7wnd-w

Reinforce Your Learning

Items to Include in a Project Scope Statement. Consider the information you need to write a project scope document. Below are two project scope statements for constructing a house. Indicate the one you think is better and why. Then make suggestions about how to improve the project scope document.

Statement A: The construction project will include the following:

- Clear the land.
- Build the house.
- Landscape the area around the house.

Each of these steps should be done in a way that meets quality requirements.

Statement B: The contractor will:

- Clear the land to ensure that no tree stumps are visible above ground level.
- Build the house according to the attached design and material specifications the contractor and customer have agreed on.
- Landscape the entire property that surrounds the house once the building is complete. The landscaping will include two inches of # 2 sifted topsoil over the whole area and planting "select turf" grass seed.

Suggestions for improvement:

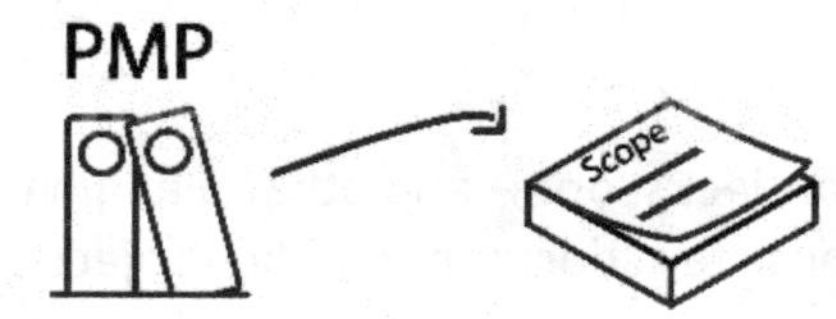

Defining Project Scope #2.8

Review PMI's Scope Management Model

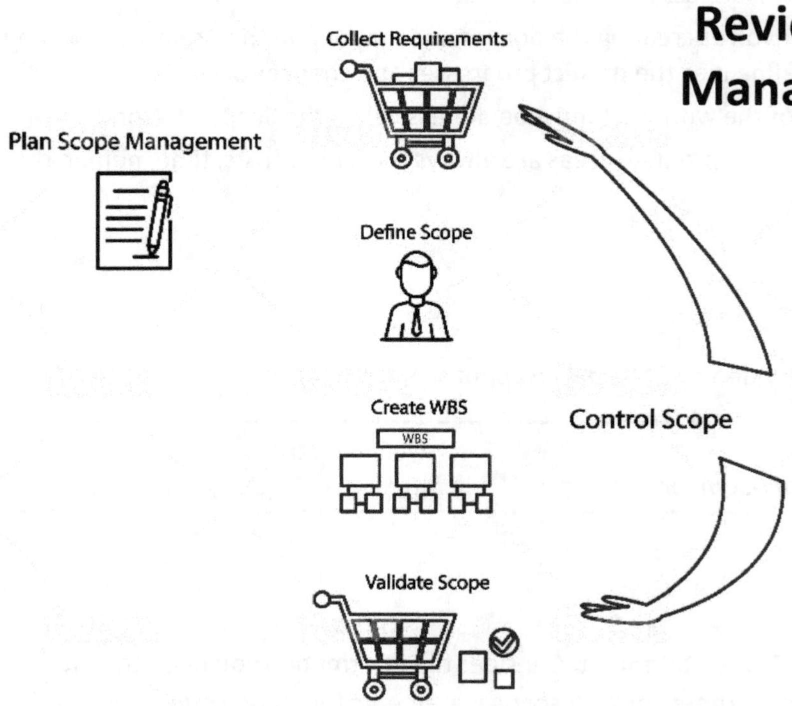

Project scope is at the heart of project management. As discussed previously, it is a description of the work that must be completed to achieve the project's goal, a new product, service, or result, and with the required features and functions, and according to the quality standards agreed to at the beginning of the project. The completion of a project is benchmarked against the project plan.

To avoid confusion, be aware that **project scope** and **product scope** do not mean the same thing. *Product scope* is a description of the features and functions of the new product, service, or result a project intends to deliver. The delivery of a product is benchmarked against the requirements for the new product, service, or result.

The **Project Management Institute (PMI)** offers a model for the scope management processes. It includes the following steps (Adapted from *PMBOK®Guide- 6th Edition*, p. 130):

1. Plan scope management
2. Collect requirements
3. Define scope: detail project deliverables as manageable components
4. Create a Work Breakdown Structure (WBS): subdivide the project deliverables into sub-deliverables and then tasks
5. Validate scope: get agreement among the stakeholders on project scope
6. Control scope: monitor and manage changes to the original project scope

The following flow chart lists the phases of the project life cycle that we have discussed, along with the corresponding steps listed above for PMI's scope management model.

Idea: Preliminary Project Description, Appraisal, and Decision to Pursue

⇩

Initiate: Begin the Project (1. Plan Scope, 2. Collect Requirements, 3. Define Scope; plan and develop project charter and initial or baseline scope statement document)

⇩

Plan: Plan and Collaborate with Key Stakeholders (4. Create a WBS; identify and detail project tasks or product components aligned with the project scope, develop a project management work plan)

⇩

Execute: Do the Work, Monitor, and Control, (5. Validate Scope, 6. Control Scope; verify the scope and update documents, manage execution efforts, manage scope change process)

⇩

Close: Close Project (Communicate completion of final project product or project, get agreement among key stakeholders)

In practice, many project activities and tasks overlap with interdependencies, and project life cycles vary, especially in Agile and Dynamic environments.

Agile project management environments, for example, software development projects, are characterized by frequent reassessment and adaptation of plans. In situations with emerging requirements, such as software development, less time may be spent on defining the initial baseline scope and more time in establishing an iterative process for scope refinement.

Related to Agile project management is Scrum, which focuses on short communication meetings, flexible reassessment of plans, and iterative phases of work.

What to do:	Other actions:
☐ Review PMI's Scope Management Model and associated concepts.	☐ ________________
☐ Recognize that many project activities and tasks overlap with interdependencies and that project life cycles can vary.	☐ ________________
☐ Investigate how Agile and SCRUM techniques may help you manage a project.	☐ ________________

Remember

✓ The scope document is the foundation of all other components of a project management plan.
✓ The initial project management scope document is referred to as the baseline project scope.
✓ Agile project management environments are characterized by frequent reassessment and adaptation of plans. Related to Agile project management is Scrum, which focuses on short communication meetings, flexible reassessment of plans, and iterative phases of work.

Enhance Your Learning

Watch the following 3-minute video from The Crowd Training to learn more about the Project Scope Management knowledge areas.

The Crowd Training. (2016). *Drawn Out Project Scope Management..*	Available at: https://www.youtube.com/watch?v=KZn2ly_p8pY

Reinforce Your Learning

Project Scope Statement Document. The initial scope document is called the *scope baseline*. You will need to revise the scope baseline throughout the life of the project as changes occur. The revisions are referred to as *progressive elaboration*, which points to the fact that the characteristics of the project are determined incrementally and continually refined. Review the scope document template that follows and compare it to others you may be familiar with. Use this template to define the scope of an upcoming project or practice by focusing on a project that you are already familiar with.

Project Scope Statement Document

Project Title:

Introduction (overview of concept, schedule, and cost, and refined project objectives):

Description (overview of what is included and not included, and the cost-benefit analysis):

Deliverables (scope/schedule of work, deliverables, boundaries of what is and is not included, risks, and assumptions):

Resources (personnel and other resource requirements):

Costs (all estimated costs and cost-benefit details):

Schedule (all estimated tasks/activities; Milestone chart):

Acceptance Criteria (success metrics for scope, costs, and schedule):

Client / Sponsor: **Title:** **Date:**

Defining Project Scope #2.9

Understand Why Pre-planning is Critical

Project management professionals refer to the *critical first 10 percent* as the time spent planning a project from start to finish. This time is vital because everything that happens later in the project will be influenced by how well the project was planned.

The following is a summary discussion of the many responsibilities involved in planning a project. Review the summary and consider if you are prepared to meet these expectations.

Understand the demands of project planning. The work of a project manager requires you to think through every detail. You will need to ask many probing questions, starting with the who, what, where, when, and why of project planning:

- What benefits will the organization gain from the project?
- Who is the sponsor? The client? Other stakeholders?
- What product, service, or result do the client/stakeholders want?
- What are the stakeholder requirements for the project?
- What work needs to be done, in detail, from the start to the end of the project?
- Who are the people who have the skills and experience necessary to complete the project and will they be available in the time frames needed?
- What is the budget?
- Is there a deadline for when the project must be done?
- When is the project expected to start, and when will it finish?
- Why do you think you are qualified to lead the project?

As a project manager, you are responsible for ensuring that the time, money, and human resources of your organization are used wisely. Poor planning at the beginning of a project, regardless of its size, may lead to project failure and loss of time, money, and reputation.

Be aware that many planning documents are needed. As a project manager, you will develop several planning documents. You will also need to revise/update many of these documents as the project progresses and changes occur. The documents (both initial and final) typically include the following:

- Scope statement of the requirements for the project's product, service, or result, along with a description of the quality standards agreed to.
- Work breakdown structure (WBS) that defines the project deliverables and the activities/tasks needed to complete the project.
- Risk management plan that outlines all contingencies for the organization and how to avoid them or cope with them to minimize the organization's exposure to legal, financial, and/or public relations losses.
- Project budget that is an honest reflection of the demands of the project.
- Status report format and procedures the team can use to communicate progress on milestones, problems, and other pertinent information.
- Communication plan that supports team morale, client relations, and provides effective channels for the exchange of project information among all interested parties.
- Final report to the client/stakeholder.

Determine the project scope. As the project manager, you must envision the beginning and the end of a project and everything that happens in between. You must anticipate where things might go wrong and where time or money could be wasted. Thus, during the project planning stage, it is your responsibility to determine the scope and demands of the project from the creation of the contract to the collection of the last payment.

Chart the project's course: The project plan is the project's roadmap. Project managers must be efficient and effective. This is achieved by creating and executing according to a *project plan*, which is a realistic, action-oriented document with the following characteristics:

- Complete and current for the actual project, not a template plan.
- Flexible and can be adapted to extenuating and unpredictable circumstances.
- Logical and reflects the scope of the work to be done.
- Not easy to ignore and will assist everyone in understanding the project.

Break down work activities and tasks. Are you aware of the various project planning tools available? Examples of some useful tools include Work Breakdown Structures, Gantt Charts, and Network Diagrams.

Be aware that a lack of planning will lead to poor results. Without a plan, you will not know if a project is going well or if it is in trouble. Further, even in a well-planned project, things will go wrong. As a project manager, you must expect and plan for glitches along the way. The key is to plan the project carefully and realistically, which will help you avoid as many problems as possible. When glitches happen, manage them so they do not grow into issues that can damage the project.

Budget and Control project costs. Your job as a project manager is to budget thoroughly and carefully, and then to monitor the budget as work progresses. You must estimate expenses and create a spending forecast. Think of it as a cash flow road map. You will not know where you are or where you are going without it.

As a project manager, you are the first line of defense in your organization's daily battle with marketplace competitors. To succeed, you need to ensure profitability. To do that, you need to know how much money is required to meet expectations and complete the project on time.

Start with a comprehensive analysis of the project and keep asking these questions: How much will be spent; where will it go, and where will it come from?

Avoid legal issues. Careful project planning is the best defense against litigation. Project managers must keep project objectives in mind, and this includes consideration of possible legal problems. It is often possible to avoid or minimize your organization's exposure to costly lawsuits by adding the proper legal clauses in project documents during the planning stage. Here, as in many areas, knowledge is power.

Are you aware of the many legal clauses to consider as you plan a project? Do you know how to incorporate them when appropriate?

Be prepared: No project is risk free! You plan for risk by developing a realistic assessment of potential problem areas and responses.

In a risk assessment, you answer two questions: Can your project team complete the deliverables, as promised, and what might get in the way of you presenting these deliverables to the client? Factors to consider include—but are not limited to—contract problems, vendor and/or subcontractor nonperformance, and the unavailability of resources.

Evaluate critical factors before starting a project. As a project manager, you are asked to make judgment calls throughout the proposal process. Your first judgment call is a "gut" reaction to a project opportunity or a Request for Proposal (RFP). You decide if it is worth investing the effort to progress to a pre-proposal face-to-face session with the potential client and a formal Go, No-Go decision process.

What to do:	Other actions:
☐ Be aware of what can go wrong as a result of poor project planning.	☐ ______________________
☐ Ensure that the project planning process meets client/stakeholder expectations and requirements.	☐ ______________________
☐ Recognize that careful project planning is the best defense against litigation.	☐ ______________________

Remember

- ✓ When planning a project, the problem areas are always the same: costs, time, human resource failings, and the unknown. Thus, the more you know about the project before it begins, even before you develop action plans, the better.
- ✓ Every project, regardless of its size, scope, and budget, will face its share of setbacks. The key is to avoid problems that can wreck reputations, client relationships, the project, and even the organization itself.
- ✓ As the project manager, you must ensure that your project has a budget to cover all costs.
- ✓ The budget is one of the key considerations in project success.
- ✓ Often it is possible to avoid or minimize your organization's exposure to lawsuits by including proper legal clauses in documents created during the planning phase.
- ✓ The project plan is a guide. Involving others in the planning leads to more buy-in later in the project.

Enhance Your Learning

Watch the following 5-minute video by Jennifer Whitt to learn more about the project planning process. your work:

Whitt, J. (2013). *Project Planning Process: 5 Steps to Project Management Planning.*	Available at: https://www.youtube.com/watch?v=Do8iykQKMfU

Reinforce Your Learning

Critical Factors to Consider When Planning and Implementing Projects. Consider a project you are familiar with. Assume you are evaluating the critical factors discussed below to assess whether you can successfully complete the project. Review each of the statements and check the column that indicates your level of agreement.

How confident am I?	very little	some what	very much
I have planned and managed projects like this in terms of size, scope, and difficulty.			
I am comfortable that the scope of this project has been clearly defined and communicated to the team and stakeholders.			
I believe the budget established for this project is adequate and reflects reality. I can complete the project at or under its cap.			
The budget accurately depicts costs for the specific people who will work on the project.			
The proper people are on the project team. They have the right skills to meet the demands of the project.			
The team will stay together for the duration of the project.			
The project plan has clearly defined milestones, target dates, and deadlines. Ample time was built into the plan to allow for satisfactory completion of the project, meaning it will be on time and adhere to the budget.			

Work Breakdown Structure

Defining Project Scope #2.10

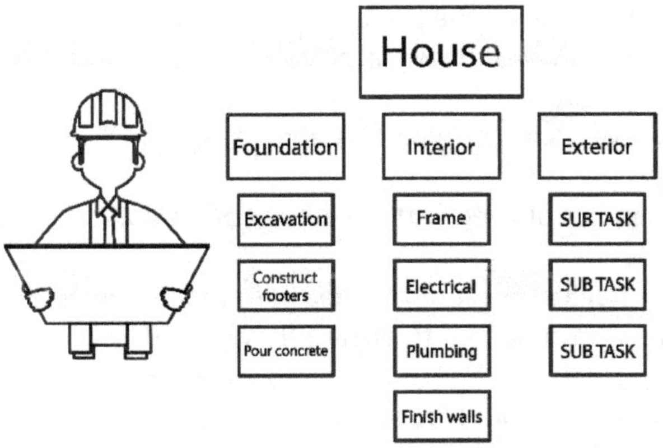

Develop Work Breakdown Structure

After the scope statement document, the next step in a project is to identify and define the work needed to achieve the project's product, service, or result.

You define the project's work by preparing a **Work Breakdown Structure** (WBS).

A WBS is a vital tool for project managers. In some cases, the WBS can be as simple as listing the tasks necessary to complete a project.

For more complex projects, a WBS is a detailed hierarchy of deliverables and sub-deliverables and the tasks needed to accomplish the deliverables.

A WBS structure provides management with a database for planning, executing, monitoring, and controlling the project work. The hierarchical structure also furnishes information appropriate to each level.

The following illustration is an example of a hierarchical **Work Breakdown Structure** (WBS) for planning a wedding:

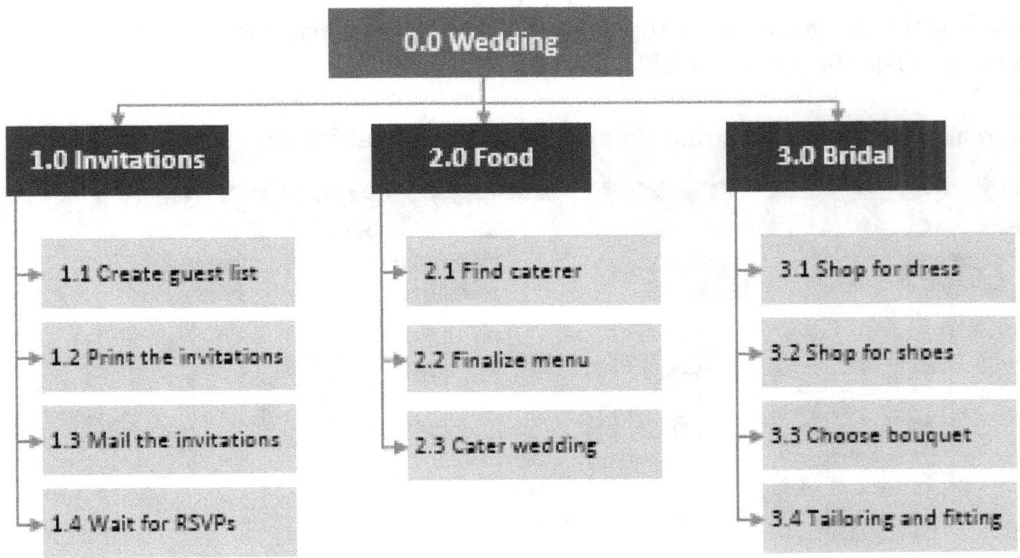

Start by Brainstorming Ideas for the WBS

Ideas for what is needed in a project's WBS are often generated in a **brainstorming** session facilitated by the project manager. It involves all the relevant people on the project team. Include those who know what needs to be accomplished and those who will do the work, and someone to take minutes.

The following are guidelines for running a productive brainstorming session:

- Ideas must be recorded publicly and as offered with no editing or paraphrasing.
- No one can comment (positive or negative) about the ideas offered. If needed, the minute taker may ask to have the idea repeated to ensure they record it correctly.
- The objective of a brainstorming session is idea quantity. After idea generation, the group evaluates, clarifies, and classifies the ideas to consolidate the list and reach some agreement on the best ideas.
- When necessary, the group can research information.

How to Develop a WBS

Work from the top down. Identify the deliverable and put the name in the box at the top. If it is a WBS for an entire project, put the project name in the box. In the example above, the deliverable is Wedding.

Then define the significant deliverables to achieve the project deliverable. In the Wedding example, these are Invitations, Food, and Bridal.

Note: If you want to make a WBS for a deliverable in the project, put the name of that deliverable in the box. You will likely also want to add its number. For more information about numbering, see the numbering section below.

Break each major deliverable into sub-deliverables. After you define the major deliverables, you will then identify the sub-deliverables necessary to accomplish each one. The total number of sub-deliverables depends on the complexity of the higher-level task.

Then break the sub-deliverables into tasks. Tasks are the lowest work level on a WBS. They represent work that cannot or does not need to be broken down further. A task must complete the sub-deliverable or the deliverable one step above in the hierarchy.

Keep repeating this process until the sub-deliverable details are small enough to manage, and one person can complete the sub-deliverable.

This hierarchical structure is illustrated in the **WBS** example that follows.

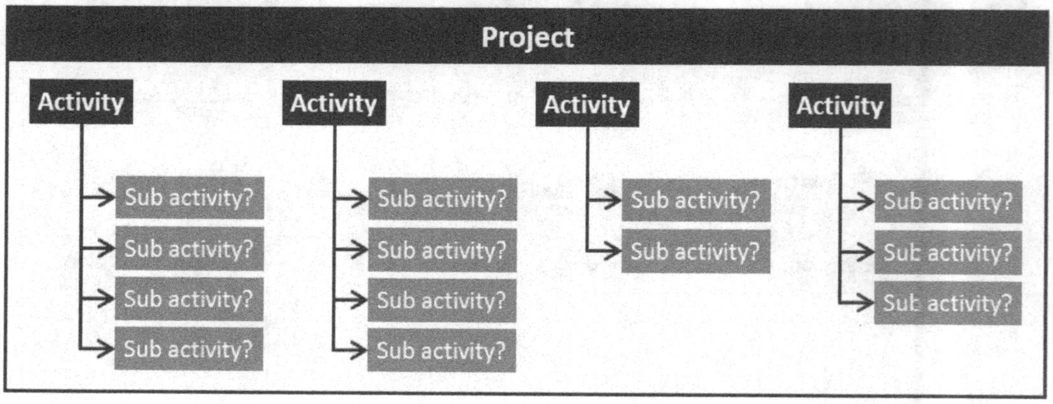

Work specifications can be written for each activity and sub-activity in the project.

You can also number each level in a WBS. See the WBS illustration below in the section on numbering.

If needed, you can develop a second WBS for some of the more complex sub-deliverables.

Work Packages

Sub-deliverables are then organized by **work packages**. These are typically the **lowest level breakdown** for a project manager. The work packages are grouped by the type of work involved, such as hardware, programming, testing, and so on.

The groupings are **cost accounts**. For example, hardware might be a cost account. This way of grouping by type of work supports a system of monitoring project progress by work and responsibility.

Here is a general **rule of thumb** for the size of a work package: **no work packet more than 80 hours of work effort or less than 8 hours of effort.**

The project manager defines the following information for each work package:

- Description of work (what)
- Time to complete the work package (how long)
- A time-phased budget to complete the work package (cost)
- Resources needed to complete a work package (how much)
- A single person responsible for the work (who)
- Monitoring points for measuring progress (control)

Number the Levels on a WBS

For larger projects with multiple levels, it is helpful to assign numbers to the sub-levels. Numbering is particularly useful when you use software tools.

WBS Level 1 — **WBS Level 1 (1.0):** Major project deliverables that are about same level of effort; equal 100%.

WBS Level 2 — **WBS Level 2 (1.1)**: Typically the primary sub-deliverables at which reporting is done.

WBS Level 3 — **WBS Level 3 (1.1.1)**: Work package level where cost and schedule can be estimated; 8hr. – 80 hr. rule.

WBS Level 4 — **WBS Level 4 (1.1.1.1)**: Activities that have to be done; where; who; when; how much $$$.

What to do:	Other actions:
☐ Identify project activities and alternatives by conducting a brainstorming session.	☐ ______
☐ Prepare a WBS that defines all the deliverables and work in the project.	☐ ______
☐ Break each deliverable into sub-deliverables. The number of sub-deliverables depends on the complexity of the higher-level task.	☐ ______

Remember

- ✓ Brainstorming is a useful tool for identifying project activities, generating alternatives, and obtaining input from stakeholders and team members who will do the work.
- ✓ Set a goal to meet or exceed stakeholder expectations and requirements.
- ✓ Prepare a WBS to include all the deliverables and work in a project. Then break each deliverable into sub-deliverables. The number of sub-deliverables depends on the complexity of the higher-level task. Repeat until the sub-deliverable details are small enough to manage, and one person can complete the sub-deliverable.

Enhance Your Learning

Watch the following 8-minute video by Dartmouth University about the development of a multi-level work breakdown structure.

Dartmouth University. (2014). *Five-Level Work Breakdown Structure.*	Available at: https://www.youtube.com/watch?v=SqcuLsyFr-o

Reinforce Your Learning

Develop a Work Breakdown Structure. Pick a work-related project such as organizing a celebration and segment it into its major activities. Then list the sub-activities under each major activity. Imagine an outline format, or you could make a chart like the wedding example at the beginning of this concept.

Project: ______________________________

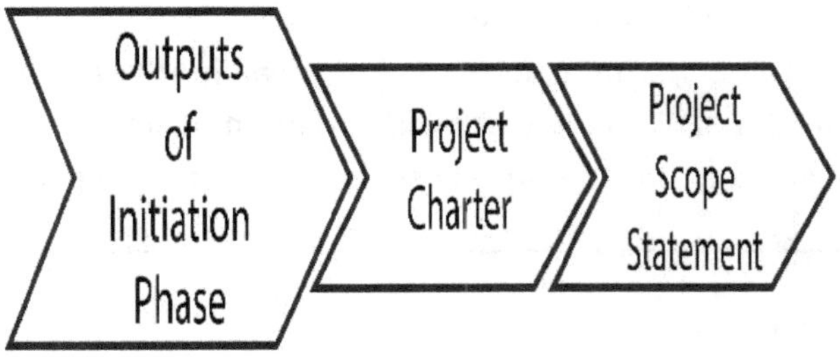

Summary

A Final Word: Defining Project Scope

You have completed a chapter on Defining Project Scope.

Information you reviewed in this chapter include:

- Conducting a stakeholder analysis
- Understanding the benefits and obstacle in defining a project scope
- Recognizing common causes of project failure
- Evaluating the critical factors associated with starting projects and your readiness
- Developing a Work Breakdown Structure (WBS)

In the next Project Mastery chapter, we will focus on Planning and Scheduling Projects.

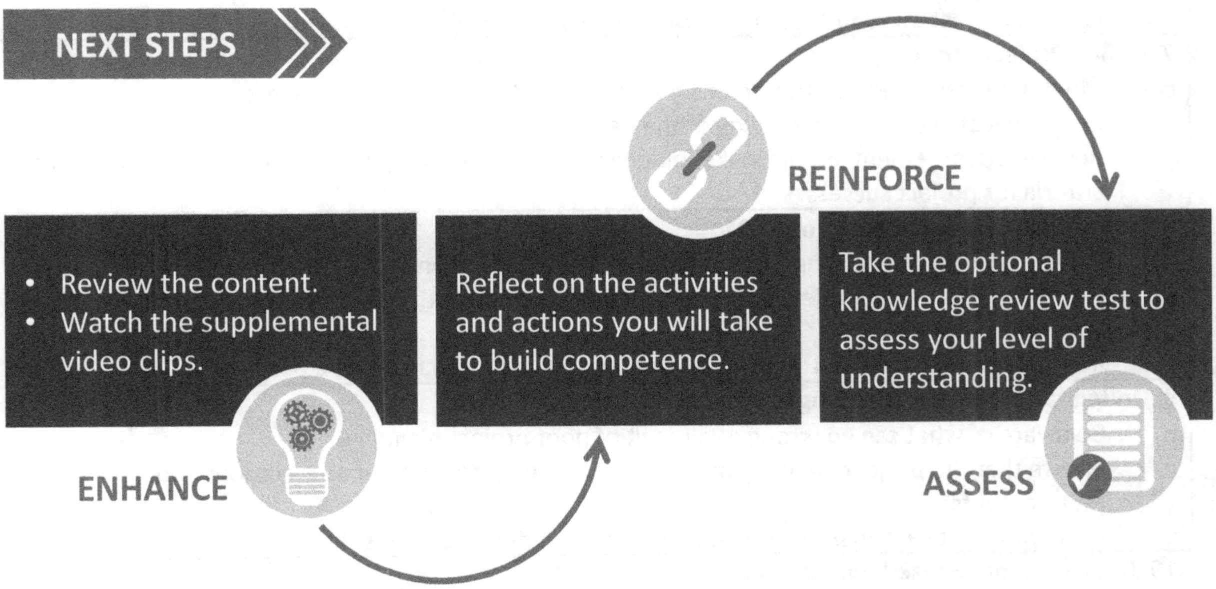

Defining Project Scope—Recap Checklist
When it comes to the successful completion of projects, what you do not know can hurt you. Doing the work might be 90 percent of the job, but the initial planning work will impact the whole project. A project well done is always a project well begun. The project scope describes what is in the project and what is not.
1. **Identify Project Stakeholders** □ Identify the stakeholders associated with your project. □ Create a Stakeholder Analysis Matrix. □ Assess the level of interest and engagement each stakeholder will have in the project and their potential impact on the project.
2. **Determine Stakeholder Requirements** □ Identify stakeholders and their requirements early in the project. □ Ask people who will do the work for their input. □ Provide opportunities for stakeholders to be engaged in the project.
3. **Specify Project Deliverables** □ List the documents that will be used in most projects. □ Establish channels of communication for the exchange of project information. □ Communicate progress, obstacles, and other pertinent information on an ongoing basis as appropriate.
4. **Draft Project Charter** □ Determine what level of authority you have to initiate projects in your organization. □ Discuss whether a quote, purchase order, or statement of work (SOW) is sufficient authorization to initiate projects. □ Identify who in your organization has the authority to sponsor or authorize larger projects.
5. **Learn Common Causes of Project Failure** □ Recognize that people's actions may account for the majority of project problems. □ Consider if you are getting or have all the project requirements. □ Consider the risks for your project and what could go wrong, and the impact on the project.
6. **Prepare for Project Success** □ Establish and agree on project boundaries before starting a project. □ Document project expectations for cost, scope, timing, and quality. □ Identify the actions you should take before starting a project.
7. **Define Project Scope** □ Meet with stakeholders often to ensure you understand their requirements. □ Communicate your understanding of the deliverables to the stakeholders. □ Resolve disagreement among stakeholders early in the planning phase and agree on the quality criteria for project success.
8. **Review PMI's Scope Management Model** □ Review PMI's Scope Management Model and associated concepts. □ Recognize that many project activities and tasks overlap with interdependencies and that project life cycles can vary. □ Investigate how Agile and SCRUM techniques may help you manage a project.
9. **Understand Why Pre-planning is Critical** □ Be aware of what can go wrong as a result of poor project planning. □ Ensure that the project planning process meets client/stakeholder expectations and requirements. □ Recognize that careful project planning is the best defense against litigation.
10. **Develop Work Breakdown Structure** □ Identify project activities and alternatives by conducting a brainstorming session. □ Prepare a WBS that defines all the deliverables and work in the project. □ Break each deliverable into sub-deliverables. The number of sub-deliverables depends on the complexity of the higher-level task.

Knowledge Review Test—Questions

Defining Project Scope

1. Which of the following is not part of the stakeholder management effort?

A. Give them extras.
B. Identify them.
C. Determine their needs and expectations.
D. Manage their expectations.

2. It is important that key stakeholders are as active about the project as you had anticipated. There could be reasons for their lack of engagement which may include:

A. Unaware of project activities occurring.
B. Against selected project actions.
C. Neutral as to what is occurring in a project.
D. Satisfied with level of activity.
E. All the answers are correct.

3. You can communicate with stakeholders and collect their requirements in a variety of ways, including:

A. Interview.
B. Focus groups.
C. Surveys and questionnaires.
D. Direct observation.
E. All the answers are correct.

4. Which statement about stakeholder expectations is not true?

A. The product or service should be usable as intended.
B. Contractual provisions, including schedule and performance standards, should be met.
C. Changes should be made immediately, and all stakeholder wants should be satisfied without bureaucratic hassle.
D. The seller should assume the responsibility of understanding stakeholder needs and wants and address them effectively.

5. Which of the following documents is NOT typically used in most projects?

A. Work breakdown structure (WBS).
B. Timesheet report.
C. Risk management plan.
D. Organization's annual report.
E. Lessons Learned.

6. Goals and objectives are correctly stated as "SMART" if they follow this format

A. Measurable and time specific.
B. Specific, measurable, and realistic.
C. Specific, measurable, achievable, realistic, and timely.
D. Timely, measurable, verifiable, and achievable.
E. None of the answers are correct.

7. A project manager is employed by an organization and is responsible for the furnishing of the completed facility. One of the first things that the project manager for this project should do is to prepare a

A. Work breakdown structure.
B. Budget baseline.
C. Project charter.
D. Project plan.

8. A project charter formally authorizes the work to be done and is especially useful for larger projects. For smaller projects what may suffice?

A. Formal quotation.
B. Purchase order.
C. Statement of Work (SOW).
D. All the answers are correct.

9. One of the biggest reasons that projects fail is:

A. Resources untrained and misaligned.
B. Changes in an organization's priorities.
C. Tools and systems outdated and inaccessible.
D. Assumptions poorly defined and unsubstantiated.
E. Team collaboration inadequate and team conflicts.

10. People's actions make up about what percent of the risks in a typical project's Risk Management Plan?

A. 20 %
B. 50 %
C. 70 %
D. 90 %
E. 100 %

11. Each project phase is marked by the completion of one or more deliverables. The deliverable for the initiation phase of the project is:

A. Project plan.
B. Lessons learned document.
C. Project charter and scope statement.
D. Resource spreadsheet.

12. Ideally, the project manager should be selected and assigned:

A. During the initiating phase or process.
B. During the project planning process.
C. At the end of the concept phase of the project life cycle.
D. Prior to the beginning and the development phase of the project life cycle.

13. The document that describes the objectives, work content, deliverables, and an end product of a project is the:

A. Project charter.
B. Product description.
C. Scope statement.
D. WBS.

14. Adding extra features or components to a project beyond the original scope requirements is referred to as:

A. Quality enhancement.
B. Exceeding customer expectations.
C. Gold plating.
D. Excellent customer service.
E. Scope enhancement.

15. The PMI Scope Management Model involves which of the following steps?

A. Initiation.
B. Planning.
C. Definition.
D. Verification.
E. Change Control.
F. All of the answers are correct.

16. Agile project management is characterized by:

A. Detailing project deliverables as manageable components.
B. Overlapping of many project activities and tasks.
C. Monitoring and managing changes to the original project scope.
D. Frequent reassessment and adaptation of plans.

17. Project management professionals refer to the critical first 10 percent as:

A. The most difficult time in a project.
B. The identification of the ten most critical elements of a project.
C. Planning the first ten percent of a project.
D. The time spent planning a project from start to finish.

18. You are a project manager for a new building project and have just received a project charter. You can create a detailed project schedule only after?

A. A budget is created.
B. A WBS is crested.
C. A project control plan is created.
D. A project plan is created.

19. What is a "Work Breakdown Structure"?

A. A list of deliverables.
B. A map of the entire project.
C. A list of cost accounts.
D. A hierarchical structure.
E. All the answers are correct.

20. A project is plagued by changes to the project charter. The person with primary responsibility to decide what changes are necessary is:

A. Project manager.
B. Team leader.
C. Management.
D. Stakeholder.
E. All the answers are correct.

Defining Project Scope — Answer Sheet

1. _X__ A ___B ___C ___D ___E
2. ___A ___B ___C ___D _X_ E
3. ___A ___B ___C ___D _X_ E
4. ___A ___B _X_ C ___D ___E
5. ___A ___B ___C _X_ D ___E
6. ___A ___B _X_ C ___D ___E
7. ___A ___B _X_ C ___D ___E
8. ___A ___B ___C _X_ D ___E
9. ___A _X_ B ___C ___D ___E
10. ___A ___B ___C _X_ D ___E
11. ___A ___B _X_ C ___D ___E
12. _X_ A ___B ___C ___D ___E
13. ___A ___B _X_ C ___D ___E
14. ___A ___B _X_ C ___D ___E
15. ___A ___B ___C ___D _X_ E
16. ___A ___B ___C _X_ D ___E
17. ___A ___B ___C _X_ D ___E
18. ___A _X_ B ___C ___D ___E
19. ___A ___B ___C ___D _X_ E
20. ___A ___B _X_ C ___D ___E

Chapter 3

Planning and Scheduling Projects

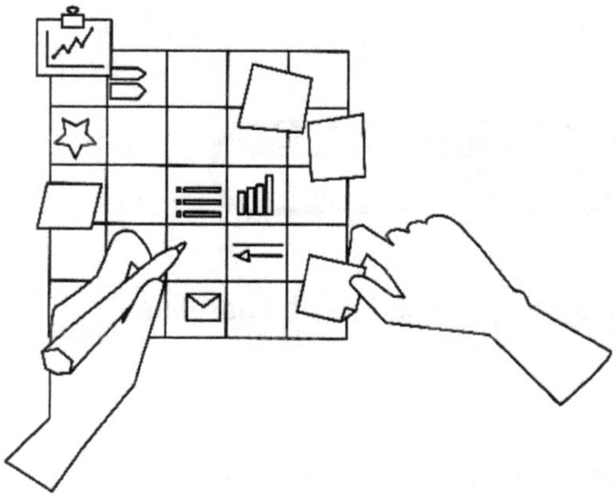

Overview

> ***"Strategy without tactics is the slowest route to victory. Tactics without strategy is the noise before defeat."***
>
> —Sun Tzu c. 490 BC

Your project plan guides you in managing project costs, resources, schedules, and risks. Thus, effective planning is key to success.

The Project Management Institute (PMI)® produces an annual global survey about portfolio, program, and project management. *PULSE OF THE PROFESSION* ®: Capturing the Value of Project Management, 2015, included the results of this question: In your estimation, what percentage of the projects completed within your organization in the past twelve months **finished within their initial scheduled times**? The answer was **50 percent.** This means that half of all projects take longer than expected. You can read more about the survey results at Pulse (PMI, 2015).

Planning requires that you identify all project activities and tasks, sequence them correctly, and estimate schedules and resources accurately. Be aware that planning is an iterative process. Throughout a project, you will revise your plan as more details become available, tasks are refined, and project changes are approved.

This chapter introduces tools and techniques that help you plan and schedule projects. It also reviews the skills needed to develop and maintain plans, and it discusses common pitfalls.
In this chapter, you will learn how to:

- Describe what is in a Schedule Management Plan
- Recognize commonly used project scheduling tools and their applications
- Develop a project schedule
- Explain what is meant by a schedule baseline
- Avoid planning and scheduling pitfalls

The competencies associated with this chapter include:

Planning and Evaluation	Resource Management	Problem Solving

PMI and *PULSE OF THE PROFESSION are* registered marks of the Project Management Institute, Inc.

Take Your Temperature for Utilizing Project Planning and Scheduling Projects

The following assessment will help you identify what you know about project planning and scheduling tools and the areas where you can improve. With you and your organization in mind, read each statement below, and consider how much you agree with it. Use the numbers from 1 to 10, where **1** means you **strongly disagree** and **10** means you **strongly agree**.

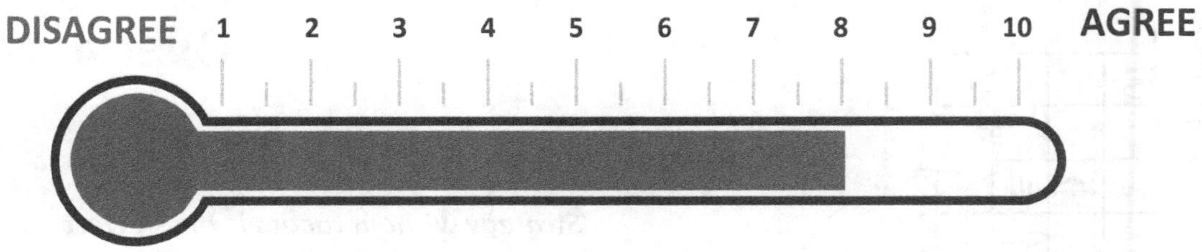

_____ 1. I can outline components for building a project schedule and baseline.

_____ 2. I know how to sequence project tasks.

_____ 3. I understand how to determine the effort needed for tasks.

_____ 4. I am confident in my ability to set project milestones.

_____ 5. I can develop Gantt charts.

_____ 6. I can schedule activities using a spreadsheet.

_____ 7. I know how to develop a project network diagram.

_____ 8. I can calculate a project's critical path.

_____ 9. I can describe the Program Evaluation and Review Technique (PERT).

_____ 10. I know how to avoid common project planning and scheduling pitfalls.

_____ **Total**

Take a moment to reflect on the following before you move forward in the chapter.
In your work, what problems have you experienced or observed related to planning and scheduling projects at work?

Take a few minutes to also reflect on your self-assessment. List two or three areas you want to improve.

As you progress through this chapter consider what actions you can take to demonstrate competence in the following three competency areas:

1. **Planning and Evaluation.** Establishes policies, guidelines, plans, and priorities; plans and coordinates with others; aligns required resources; monitors progress and evaluates outcomes; improves organizational efficiency and effectiveness.

2. **Resource Management.** Demonstrates awareness of technical resources; knows how to apply resources to achieve desired outcomes.

2. **Problem Solving.** Recognizes and defines problems; analyzes relevant information; encourages alternative solutions, develops plans to solve problems.

Planning and Scheduling Projects #3.1

Build the Project Schedule

After you initiate a project, define the scope, and start assembling the project's work activities, it is time to develop the project schedule. Scheduling, along with determining the level of effort for tasks and activities, is one component of a project management work plan.

The **Project Life Cycle** illustration below shows the phases involved in a project life cycle, the graduated use of resources, and the project outputs. This course focuses on the **planning** phase.

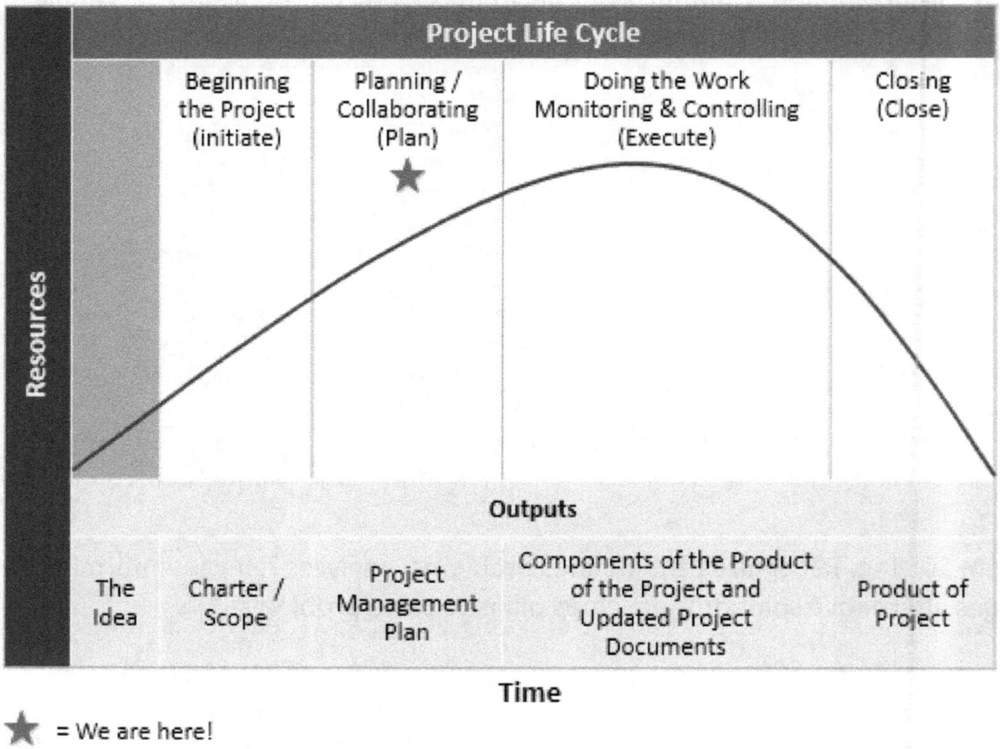

A ***Project Management Plan,*** often referred to simply as the ***Project Plan,*** addresses how tasks and resources will be planned, executed, monitored, and controlled.

The initial project management plan is referred to as the ***baseline schedule and plan***. A *baseline* is a fixed schedule that represents the standard used to measure the performance of a project.

In project management, there are **three baselines—schedule** baseline, **cost** baseline, and **scope** baseline. The combination of all three baselines is often referred to as the **performance measurement baseline**.

The Project Management Plan Integrates Knowledge Areas

The illustration below shows how the Project Management Plan enables integration among a project's knowledge areas.

This course focuses on the **schedule knowledge area**.

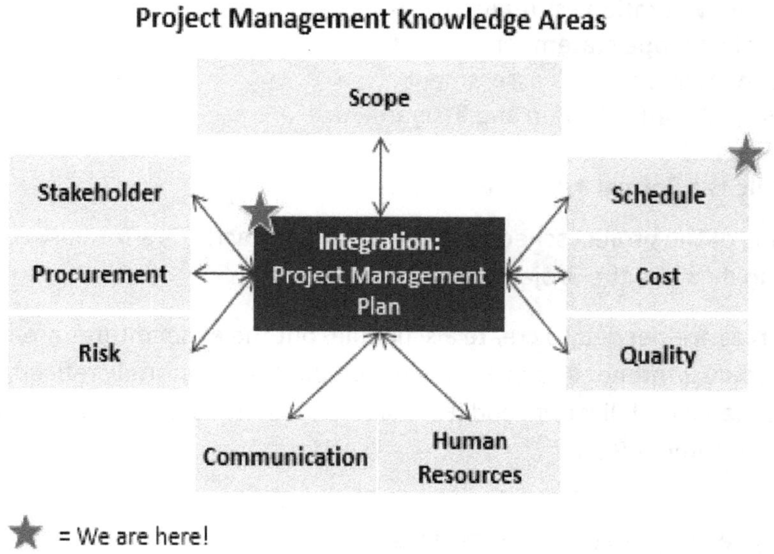

The Project Planning Phase

The following illustration shows some of the major steps and project deliverables in the planning phase.

Phase	Description/Purpose	Deliverables/Tools
Plan: Plan from beginning to end of the project	• Define tasks, activities, inter-dependencies, and activity times • Refine resource requirements • Develop work plan, milestones, and schedule • Define project control process, standards, and procedures • Develop risk management plan • Identify communication needs • Refine and manage expectations	• Schedule Management Plan • Work Breakdown Structure • Status Report Format • Time/Schedule -Tracking Format • Risk Analysis and Management Plan • Stakeholder Communication Matrix • Communication Plan • Project Budget • Benefit-Cost Analysis

Components Needed for Building a Schedule

The components or inputs for planning and developing a schedule include several kinds of information, including the following:

- Activity list and attributes
- Resource list and calendar
- Activity duration estimates
- Project scope statement
- Constraints and risk assessment
- Project staff selection and assignments

A Project Schedule is a Formal Agreement

It is useful to think of the project schedule as a kind of contract. It is a formal agreement between the client/stakeholders and the project manager and team.

Effort and resources are needed to create a schedule, but the expenditures are worthwhile because the schedule is intended to provide a document that accurately reflects what the project team is capable of delivering, and it considers the schedule, cost, and scope as specified in the Project Management Plan.

Changes Will Happen as the Project Progresses

Changes occur in almost all projects. The larger and more complex the project, the more confident you can be that changes will occur; some may be significant.

Baseline revisions show the impact of the changes on the original schedule, cost, and scope.

A periodic review of the *original scope* or *scope baseline* includes a comparison of the original scope of the project plus or minus any scope changes. In other words, a comparison allows you to see the gap between the baseline initially agreed to and any revised baseline.

Project managers must show clients/stakeholders the impact a change may have on a project *in advance* of actually making the change. That way, the client/stakeholders can weigh the disadvantages (schedule, cost, or scope) against the benefits of including the changes in the project. Using this information, they can also make an informed cost-benefit analysis of the proposed change, if needed.

How to Select Planning and Scheduling Tools

Numerous methods/approaches and tools are available to assist you in planning and scheduling the activities defined in a Work Breakdown Structure (WBS).

Some of the most useful include a Gantt chart, milestone chart, the deterministic (known relationships) approach or network, and the probabilistic approach or network.

For small projects, a Gantt/Bar Chart and a Milestone Chart are sufficient. For large projects, project managers use a combination of all four tools.

The following chart reviews the criteria to consider when you evaluate and select tools for a project and provides a high/low assessment of the methods/approaches listed above.

Criteria for Selecting Scheduling Tools (*** = High   * = Low)**

				Network Approaches	
	Gantt/Bar Chart	Milestone Chart	Small Project Tool	Deterministic/ Critical Path (CPM)	Probabilistic/ Program Evaluation Review Technique (PERT)
Ease of preparation	*****	*****	***	**	*
Ease of updating	*****	*****	**	*	*
Use with simple projects	*****	*****	*****	*	*
Use with complex projects	*****	*****	**	*****	*****
Useful as a communication tool	*****	*****	***	***	**

What to do:

- ☐ Identify project methods/approaches and tools you are familiar with and those to learn more about.
- ☐ Consider a baseline schedule as a formal agreement.
- ☐ Review the baseline schedule to validate that the project team can provide the deliverables as scheduled.

Other actions:

- ☐ ____________________
- ☐ ____________________
- ☐ ____________________

Remember

✓ Various approaches and tools are available to help you plan a project. Project managers should select the ones they are comfortable using and that best serve their project and project team needs.

✓ A published project schedule is a baseline. It is what you will be measured against, and you should consider it to be a type of contractual or formal agreement.

✓ Change controls are the control mechanism to bridge the gap between the original baseline schedule and the actual schedule.

✓ When the schedule changes, all associated project documents must be updated to ensure appropriate project communication and that the project is monitored and controlled appropriately.

✓ The initial project management plan is referred to as the baseline schedule and plan.

Enhance Your Learning

Watch the following 3-min. video to learn more about project scheduling from ProjectManager.com.

Bridges, J. (2018). *What is Project Scheduling?*	Available at: https://www.youtube.com/watch?v=WNWSQOynrl0

Reinforce Your Learning

Planning and Scheduling Challenges. List three planning and scheduling challenges and identify methods or tools that could help improve the situation.

Planning and Scheduling Challenges	Helpful Project Planning Tools
1.	
2.	
3.	

Sequence Project Tasks and Activities

People sometimes use task and activity interchangeably; however, there are differences. When managing a project, task refers to a unit of work to be done. Activity refers to an event or happening. According to the Project Management Institute's *PMBOK ® Guide,* an activity is a distinct, scheduled segment of work. Stated differently, a task is the "what," and an activity is the "how and when."

Project managers construct project schedules based on practical methods, or "rules of thumb," that are not guaranteed to be optimal or perfect but are useful.

Rules of thumb come from a variety of sources. Some examples include project manager professional judgment; the experience of subject matter experts (SMEs), and ideas from the project team and stakeholders.

Project manager judgment is based on knowledge and experience. For example, knowledge may come from studying for the certifications offered by PMI. The prior experience might include having worked on projects like the current project or referring to a previous schedule management plan. Project managers may also be fortunate enough to have some documented "lessons learned" from an earlier project.

SMEs (Subject Matter Experts) may be in the best position to identify and sequence current activities and work packages.

Project **meetings** are generally an excellent place to discuss ideas and conduct brainstorming sessions because the meetings typically include knowledgeable team members, SMEs, stakeholders, and others who may be affected by the project.

One goal of sequencing project tasks and activities is to identify the logical sequence. This includes the relationship between tasks, such as "complete A before B can start." Another goal is to identify any constraints.

The four types of dependencies are: mandatory, discretionary, external, and internal. Mandatory dependencies may occur because of physical limitations. A successor activity may not physically be able to finish until another predecessor activity is complete.

Legal constraints may also require a particular sequence of activities.

What to do:	Other actions:
☐ Review a sequence of project activities and validate they are accurate.	☐ ______________________
☐ Pay attention to how one activity relates to another (predecessor and successor).	☐ ______________________
☐ Inform stakeholders that you reviewed the sequence of activities and that you agree or want to recommend changes.	☐ ______________________

Remember

- ✓ You must plan/schedule the sequence of activities so that you have the right resources available at the appropriate times.
- ✓ Precedence diagramming is a tool project managers use to determine the type of relationship between two activities.
- ✓ The four major types of dependencies are: mandatory, discretionary, external, and internal.

Enhance Your Learning

View the following 6-minute video from Andy Kaufman about activity sequencing:. The project example in the video assumes he is going to take a family trip to Florida to watch a space launch.

Kaufman, A. (2015). *Sequence Activities Demonstration.*	Available at: https://www.youtube.com/watch?v=H-1Ab30_rrM

Reinforce Your Learning

Activities Template. Use the template below to list activities involved in making copies.

Based on your experience as a subject matter expert, list all the activities required to complete the work of making copies. For example, the first activity would be to gather the documents you want to copy, which is shown as the first activity below.

For each activity, identify the predecessor activity and the successor activity, and any constraints. After completing your activity list, review it with a colleague.

Project Name: Making Copies

Activity List

ID#	Activity Name and Description	Predecessor Activity	Successor Activity	Constraints
1	Gather documents for hard copy.			

Planning and Scheduling Projects #3.3

Determine Effort for Tasks

Effort often refers to the number of labor units or man-hours needed to complete a task or activity. However, *effort* can also refer to human resources, material, or financial project needs.

In this concept, we will focus on estimating and planning the time and resource effort required to complete project work.

Work Packages

Recall, as part of developing a WBS, you will define work packages. A rule of thumb is no work packet is more than 80 hours of work or less than 8 hours.

The following is the information you will define for each work package:

- Work to be done (what)
- Time to complete a work package (how long)
- Time-phased budget to complete a work package (cost)
- Resources needed to complete a work package (how much)
- A single person responsible for the unit of work (who)
- Monitoring points for measuring progress

Planning Techniques

For complex projects, a WBS requires thought and planning to determine the realistic effort required to complete work. Two ways of evaluating work effort are work decomposition and what is called the rolling wave or progressive elaboration method.

Decomposition means taking a deliverable, task, activity, process, or event and breaking it down into smaller, more manageable components to provide better control.

The rolling wave or ***progressive elaboration*** method means walking through the activities or work packages over many iterations. Each iteration may identify something previously missed, refine what was determined earlier, or validate that all work packages are accounted for. This is a litmus test for the "no rocks left unturned" approach.

Scheduling Tools and Techniques

The following are some of tools and techniques for estimating the time and resource effort needed on a project:

- Expert judgment
- Alternative analysis

- Published estimating data
- Bottom-up estimating
- Group decision-making
- Reserve analysis
- Deterministic scheduling approach
- Probabilistic scheduling approach

Expert judgment can come from people with experience in the kind of work to be estimated. Experienced people can help predict the time and resources needed to do a piece of work, help pinpoint the necessary skills, and identify people who may have those skills.

Alternative analysis considers ways the work could be done. For example, an activity planned initially to be done manually, such as a data collection task, might be automated. That way, a lesser-skilled resource may have the ability to work on activities originally assigned to someone highly skilled.

Other alternatives include types or sizes of machines, varying output rates, or make-rent-or-buy considerations. As an example, depending on the need for equipment and the lead time needed to purchase it, a rental approach may save time and money.

Published estimating data is one source of information for use in bottom-up estimating. It can come from a variety of sources, for example, production rates, unit costs, material costs, and equipment costs from across the globe.

No matter the source, you must consider how the data relate to your project's environment. For example, you do not want to use data that are no longer accurate/up to date or that do not closely reflect the current situation.

Bottom-up estimating leverages the experience of the project's SMEs and others knowledgeable about the activities. Working knowledge is instrumental in identifying not only the effort required to complete work, but also any dependencies between work units. Each work unit is estimated individually, and then all the individual estimates are added together to arrive at a total.

This contrasts with a *top-down approach*, where the completion date may be imposed on the project team. Bottom-up is considered the most accurate method in estimating cost and duration. However, be aware that it requires the most time.

Group decision-making techniques include the concepts of multiple perspectives, knowledge areas, people close to the processes, and buy-in of the capability to meet the estimates. It is essential that project team members and stakeholders are aware of what is being committed to from a schedule standpoint. They may have input that could change projections, and the sooner you know that the better.

Reserve analysis involves the concept of including a "buffer" reserve to account for estimate uncertainties. They can be referred to as contingency reserves or time reserves. The reserve analysis is based on the level of confidence the estimators have in their work, and it considers the similarity with prior experiences. It is important for reserves to be included in the schedule and identified as such. It is also helpful to have organizational standards as to how reserves should be handled for consistency across projects.

Deterministic scheduling approach. If you have confidence in the information and data that is being estimated and used in scheduling, a deterministic scheduling approach may be useful. In this approach, you know what outcomes are precisely determined through known relationships. In such methods, a given input will always produce the same output. Stated differently, project schedules are networks of tasks, activities, or resources connected to each other with known dependencies that precisely and accurately describe the tasks, activities, or resources.

Probabilistic or three-point scheduling approach. If you have developed a project schedule or budget, you know it is often difficult to make accurate estimates. The best strategy is to estimate the individual times and costs and then add them together to get the total time and cost. Doing this will also help you in tracking progress and monitoring costs. However, this can be difficult. Even under the best conditions, it may be impossible to get a realistic schedule.

Experience has shown that you can improve the accuracy of an estimate by using the **probabilistic** or **three-point estimating** approach. This approach has been proven in a wide range of projects. It considers the **best-case** scenario and the **worst-case** scenario and balances them with the **most likely** scenario.

Making Probabilistic or Three-Point Scheduling Estimates

The following formula is commonly used in making probabilistic estimates (**E**):

$$E = (O + 4M + P)/6$$

(O) is the best-case scenario/most optimistic estimate, added to four times the most likely estimate (**M**). Add the worst-case scenario/most pessimistic estimate (**P**). Then divide that number by 6. This approach is equally valid for both the schedule and the budget.

The following are guidelines for using the probabilistic approach:

- Consider it when you do not have better information. If better information is available from previous projects or from input others may give, use that information.
- Confirm the best case and worst-case scenarios with people who have experience in the same type of project. This will help improve the accuracy of your estimates.
- Combine these time estimates with your network diagram to form the basis of your schedule.
- When needed, use this same approach to estimate costs and develop a budget.

Complete an Activity List with Resources and Time Estimates Chart

Use the outputs from an activity duration estimation process to create an Activity List with Resources and Time to Complete chart. It lists the activities, resources (who), and the time to complete (duration) estimates, along with any constraints.

Activity List w/ Resources and Time to Complete Estimates

ID#	Activity Name	Resources (Who?)	Predecessor Activity	Successor Activity	Time to Complete	Constraints

After you identify the activities, sequence them, add resources, and provide an estimated duration, you will be ready to develop a baselines schedule for the project.

In the next concept, we will discuss how to convert an activities list into a project work schedule.

What to do:	Other actions:
☐ Identify your best sources of information for determining the effort required to complete activities and tasks.	☐ ________________
☐ Consider using this same approach to estimate costs and develop a budget.	☐ ________________
☐ Use the probabilistic or three-point estimating approach when you do not have better information.	☐ ________________

Remember

- ✓ If you have accurate and up to date information, use it instead of the probabilistic approach. The more accurate the information, the better your schedule and budget.
- ✓ Remember that estimates are only estimates. To detect potential issues as early as possible, you must monitor the project as work progresses.
- ✓ Probabilistic or three-point estimating requires that you provide estimates for the least amount of time or effort, the most likely amount of time or effort, and the most amount of time or effort.
- ✓ When your project has work similar to work done in another project, refer to the schedules developed for that project. They may be more accurate than your estimates.

Enhance Your Learning

Watch this 2-minute video explaining Level of Effort by Janis.

Janis. (2020). *What is Level of Effort?*	Available at: https://www.youtube.com/watch?v=u5JJOiSIaSI

Reinforce Your Learning

Activity List with Resources and Time Estimates. Return to your activities list for making copies. Using the Activity List with Resources and Time to Complete Estimates template below, identify the resources and estimated time needed to complete each of your copy making activities.

Activity List with Resources and Time to Complete Estimates

ID #	Activity Name	Resources (Who)	Predecessor Activity	Successor Activity	Time to Complete (Duration)	Constraints

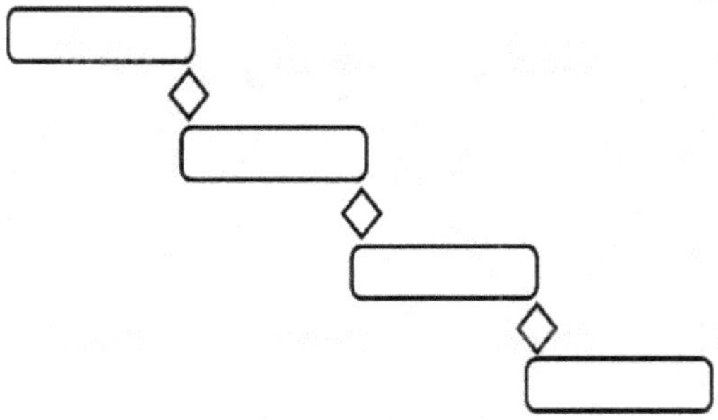

Planning and Scheduling Projects #3.4

Set Project Milestones

For your project to be successful, you will need to establish challenging yet realistic and achievable milestones. The milestones will help you visualize your project's key points and important junctions.

You can think of **milestones** in several ways and use them according to your project's needs.

A *project milestone* can be:

- A task of zero duration that shows an essential achievement in a project.
- A significant point in the project that defines where you are at any given point, and that is verifiable by others.
- Any point in the project that requires approval **or** a decision before proceeding to the next point.
- Points that represent a clear sequence of events that incrementally build until your project is complete.

How to Develop Milestones

Project managers develop milestones in various ways. Some project managers start by planning forward from the present into the future; some start with the end in mind and work backwards, and some plan in a random way, typically using sticky notes.

No matter the method you choose to plan your project milestones, here are the typical steps you will follow:

1. Define/name your milestones
2. Create links
3. Assign dates
4. Attach tasks
5. Allocate tasks

For larger and more complex projects with multiple components, you may need to determine strategies to link components and develop a series of milestone charts.

Project Milestone Charts

A project milestone chart lists project activities and deadline dates. You can include other information as needed for your project. For example, you might add the team members involved in each activity and completion dates as they occur.

Milestone charts are easy to construct and provide informational highlights for the project. They focus on responsibilities and project status.

The following is a sample milestone chart that shows significant points in a project:

Project Milestone Chart

Task/Activity	By	Deadline	Completed
1. Study	HAJ	9/30	9/15
2. Design	HAJ	10/15	10/15
3. Purchase	DMM	10/31	10/31
4. Contract	DWT	11/15	
5. Test	DMM	11/30	
6. Operate	JRZ	12/15	
7. Accept	WED	12/31	

What to do:

☐ Review a milestone chart that lists significant events or decision points with deadline dates.
☐ Select major events that incrementally build until your project is complete.
☐ Specify what and who needs to approve/decide before proceeding to the next point.

Other actions:

☐ ______________________________
☐ ______________________________
☐ ______________________________

Remember

✓ A project milestone chart consists of a list of activities with deadline dates, and you can include additional information as needed.
✓ Milestone charts provide informational highlights.
✓ The focus of a milestone chart is on responsibilities and project status.

Enhance Your Learning

Watch the following 4-minute video by Mike Clayton to learn more about milestone planning.

Clayton, M. (2017). What is Milestone Planning?	Available at: https://www.youtube.com/watch?v=aJcd1rQ0Z9w

Reinforce Your Learning

Milestones. Milestones are critical points in a project life cycle. How would you logically re-order the following list of milestones?

- Use it
- Deliver it
- Test it
- Do it
- Prepare it

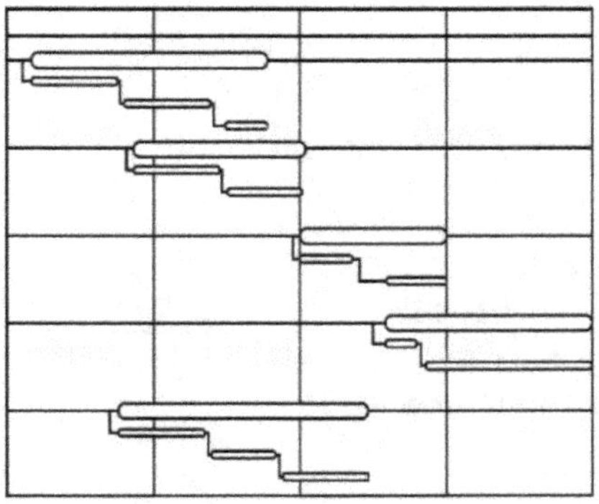

Planning and Scheduling Projects #3.5

Develop Gantt Charts

Gantt or bar charts are popular because they present an easy-to-understand, clear picture of a project's progress and status. The format is typically a two-dimensional representation of the project schedule, with activities down the rows and time across the columns. This tool gets its name from Henry Gantt, who introduced this way of charting in the early 1900s.

Project managers use Gantt charts throughout a project's life cycle for planning, resource scheduling, and status reporting. Because Gantt charts provide a clear overview of the project schedule, they are often posted on the walls of project offices.

The major **weakness** of Gantt or bar charts is the **absence of dependency** relationships among project activities. If, for example**, float** or **slack** is used on an early activity in the network chain, it cannot be used on a later activity in the same chain, and this dependency is not shown on a bar chart.

Although some computer software will develop bar charts with dependency lines, the lines soon become difficult to read and thus defeat the simplicity of the bar chart.

Note that the bar chart is derived from the project network and **not** the other way around.

You can expand a basic Gantt chart to include other information such as who is responsible and the proposed budget for each task or activity.

Gantt charts can also show actual expenses and times as well as other accumulated information. These charts can be as simple or complex as needed.

Gantt Chart Example

The following **Schedule Management Work Plan** example illustrates how a Gantt or bar chart depicts the time duration for each activity listed. You list activities in the rows and time across the top. For simplicity, the time in this example is represented in months.

Schedule Management Plan

ID#	Activity	1	2	3	4	5	6	7	8	9	10	11	12
1	Gather documents for hard copy	X											

Computer Generated Gantt Chart Example

For complex projects, project management software is useful for making Gantt charts. For example, we made the following Gantt chart for the Making Copies project using Microsoft Project software:

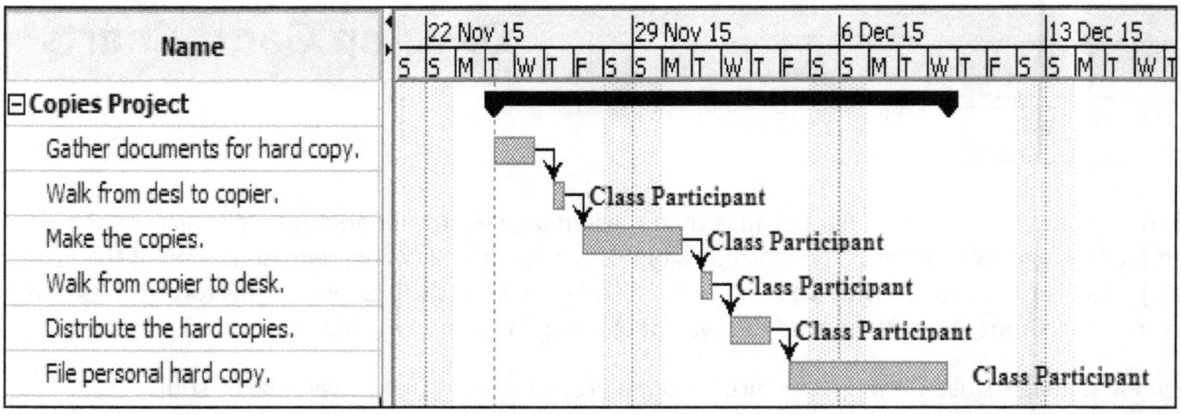

After you finish your Gantt chart, you will have a useful tool to help you manage your project. You can use it to track progress, see when the next activity should start, determine how much money has been spent, and help identify problems such as schedule and cost overruns.

What to do:

- ☐ Select a simple work-related project and create a WBS outlining the major activities or tasks.
- ☐ Develop a Gantt or bar chart by listing the major tasks along with start dates and the length of time for each task.
- ☐ Try producing a Gantt chart using one of the many project scheduling software tools.

Other actions:

- ☐ ______________________
- ☐ ______________________
- ☐ ______________________

Remember

- ✓ The Gantt chart is a flexible tool that you can use to assign responsibility and track costs, completed (and partially completed) work, overruns, and other project information.
- ✓ Defining activities is a process you will work on throughout the life of the project in iterations that become more and more detailed.

Enhance Your Learning

Watch the following 10-minute video on how to make a Gantt Chart by Mike Clayton

Clayton, M. (2022). *How to Create a Gantt Chart in 9 Easy Steps.*	Available at: https://www.youtube.com/watch?v=FX70rWEE8eY

Reinforce Your Learning

Schedule Management Plan. Assume you are planning a retirement celebration or another project you are familiar with. Develop a Schedule Management Plan using the Gantt chart template that follows. List the project's major activities or tasks, along with start dates and the length of time for each task in months. You will find a Schedule Management Plan worksheet in Appendix Part B to copy and use on other projects.

Schedule Management Plan

Project Name: ______________________

ID#	Activity		1	2	3	4	5	6	7	8	9	10	11	12

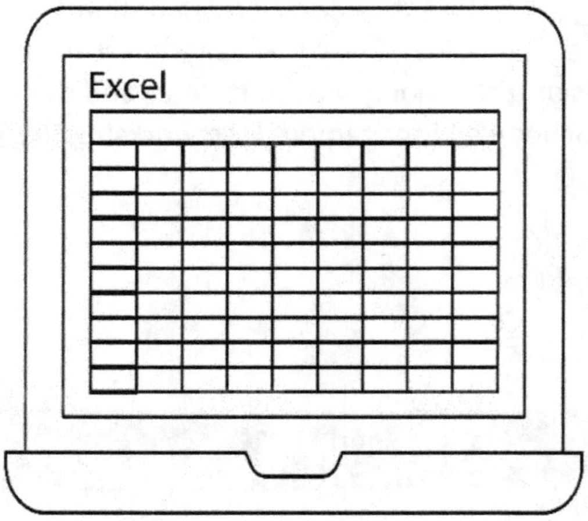

Use a Spreadsheet for Scheduling Small Projects

For small projects that you manage carefully and that have participants who feel part of a team, you can use a spreadsheet to manage the project. This is because you will not need as much detail as you would in a larger project.

As an example, you might manage a small project using an Excel spreadsheet. In this concept, we will review several sample spreadsheets. You can also find many spreadsheet examples and templates online.

Even in a small project, your emphasis is still on the project deliverables, but you can manage them with a simplified WBS matrix. A WBS work package becomes a task activity assigned to one person or one organizational unit.

For this method, you list the task activities on the left side of the spreadsheet, followed by the people/function or work units involved in the project. Then the estimate level of effort time, which is typically tracked monthly in hours, is designated horizontally across the spreadsheet, followed by the same people/function or work units.

Specific individuals/functions or work units track their Level of Effort hours and enter their hours into the spreadsheet monthly.

The following is an example of an interesting "Small Projects Tool" spreadsheet where monthly tracked "Level of Effort" hours are entered into the spreadsheet, first "in aggregate" for the month by the people/function or work units under "Level of Effort per Month" and subsequently segmented separately under the "Level of Effort by Task." The spreadsheet may be programmed to automatically calculate and display the "Level of Effort" total hours on a monthly basis by both the Individual/function or work unit, and by the specific Tasks for the small project.

As an example, the following "Small Projects Tool" spreadsheet was used for a *Schedule and Level of Effort Analysis Chart for a Hypothetical Building Renovation Project*: ***Wes's Shop***.

ABC Contractors

Small Projects Tool

Schedule and Level of Effort Analysis Chart

Project: ***Wes's Shop***

TASK ACTIVITIES	MONTH 1	2	3	4	5	6	7	8	9	10	11	12	LEVEL OF EFFORT BY TASK SERIES (IN HOURS): A. Project Manager	B. Department Head	C.Work Scheduler	D. HR Specialist	E. Project Designer	F.Technician/CAD	G.Purchasing	H. Material Expediter	I. Fabricators	J. Accounting	K. Technical Editor	L. Clerk Typist	M. Shipping Manager	N. Marketing	O.	TOTAL HOURS
TASK SERIES																												
01.0 - Project Management													50	2												2		
02.0 - Quote													4	4	4		2					2	2	2		2		
03.0 - Pre-project													4	4	2	2												
04.0 - Contract negotiations													4	4												4		
05.0 - Kickoff													4	4	4	2	4	2	2	2	2					4		
06.0 - Planning/Scheduling													4	2	16		4	4	16	8								
07.0 - Design													4	4			16	16			4							
08.0 - Execution													4	2	4		8	4	4	8	16				4			
09.0 - Shipping/Delivery													2		4				4	8	16	4			8			
10.0 - Closeout													4	2	2	2	2	2				16	2	2	4	4		
11.0 - Warranty Phase													2										2			2		
12.0 - Lessons Learned													2	2	2		2				2					2		
LEVEL OF EFFORT PER MONTH (IN HOURS)																												
A. Project Manager	8	8	8	8	8	8	8	8	8	8	8		88															
B. Department Head		8	4	4	4	2	4		2		2			30														
C. Work Scheduler		6	8	8		2	2	4	4	2	2				38													
D. HR Specialist		2	2							2						6												
E. Project Designer		2	8	4	4	4	4	4	2	4	2						38											
F. Technician/CAD		2	2	2	4	4	4	4	2	4								28										
G. Purchasing			2	4	4	4	4	2	2	4									26									
H. Material Expediter		2	4	4		4	4	4	4											26								
I. Fabricators		2		2	2	8	8	8	8		2										40							
J. Accounting		2						4	16													22						
K. Technical Editor		2								2	2												6					
L. Clerk Typist		2								2														4				
M. Shipping Manager							2	2	8	4															16			
N. Marketing	2	4	6							6	2															20		
O.																											0	
TOTAL HOURS	10	42	44	36	26	36	40	40	56	38	20	0																388

What to do:

- ☐ List the tasks for a small project you are familiar with or have observed.
- ☐ Identify and list individuals/functions or work units responsible for preforming the work.
- ☐ Develop a spreadsheet for your small project.

Other actions:

- ☐ ______________________
- ☐ ______________________
- ☐ ______________________

Remember

- ✓ For small projects, consider using a spreadsheet to identify tasks, assign responsibility, and track costs.
- ✓ Numerous spreadsheet templates are available online.

Enhance Your Learning

View the following 3-minute video to learn more about five spreadsheet templates.

Bridges, J. (2017). *Top 5 Excel Project Templates - Project Management Training..*	Available at: https://www.youtube.com/watch?v=BAY78mJr9cQ

Also evaluate the Small Projects Tool build in Excel that displays a Schedule and Level of Effort Analysis Chart by using the empty Excel template.

Donahue, W. (2020). *Small Projects Tool - EMPTY SPREADSHEET.*	Available at: https://app.box.com/s/gq1t096g0cs51bbi4dwzi681xutzrh72

Reinforce Your Learning

Use a Spreadsheet or Small Projects Tool.

Part 1. For the spreadsheet shown for Wes's Shop, information in the shaded area is missing. Can you fill in the missing information so that all of the rows & columns add up? (Be careful that there is no unique answer.)

Part 2. On your own, use the "Small Projects Tool" spreadsheet worksheet that follows for a simple construction or fabrication project (or project of your choice). Customize the tasks, identify the specific individuals (functions) or organizational units, estimate each one's monthly time in hours by task activity and practice completing the Excel spreadsheet.

Part 1. Wes's Shop

Small Projects Tool
Schedule and Level of Effort Analysis Chart

Project:

TASK ACTIVITIES	MONTH 1	2	3	4	5	6	7	8	9	10	11	12	LEVEL OF EFFORT BY TASK SERIES (IN HOURS): A. Project Manager	B. Department Head	C. Work Scheduler	D. HR Specialist	E. Project Designer	F. Technician/CAD	G. Purchasing	H. Material Expediter	I. Fabricators	J. Accounting	K. Technical Editor	L. Clerk Typist	M. Shipping Manager	N. Sales/Marketing	O.	TOTAL HOURS
TASK SERIES																												
01.0 - Project Management													50	2												2		
02.0 - Quote													4	4	4		2					2	2	2		2		
03.0 - Pre-project													4	4	2	2												
04.0 - Contract negotiations													4	4												4		
05.0 - Kickoff													4	4	4	2	4	2	2	2	2					4		
06.0 - Planning/Scheduling													4	2	16													
07.0 - Design													4	4														
08.0 - Execution													4	2	4													
09.0 - Shipping/Delivery													2		4													
10.0 - Closeout													4	2	2													
11.0 - Warranty Phase													2															
12.0 - Lessons Learned													2	2	2													
LEVEL OF EFFORT PER MONTH (IN HOURS)																												
A. Project Manager	8	8	8	8	8	8	8	8	8	8	8		88															
B. Department Head		8	4	4	4	2	4		2		2			30														
C. Work Scheduler		6	8	8		2	2	4	4	2	2				38													
D. HR Specialist		2	2							2						6												
E. Project Designer		2	8	4	4	4	4	4	2	4	2						38											
F. Technician/CAD		2	2	2	4	4	4	4	2	4								28										
G. Purchasing			2	4	4	4	4	2	2	4									26									
H. Material Expediter		2	4	4		4	4	4	4											26								
I. Fabricators		2		2	2	8	8	8	8		2										40							
J. Accounting		2						4	16													22						
K. Technical Editor		2								2	2												6					
L. Clerk Typist		2								2														4				
M. Shipping Manager							2	2	8	4															16			
N. Sales/Marketing	2	4	6							6	2															20		
O.																											0	
TOTAL HOURS	10	42	44	36	26	36	40	40	56	38	20	0																388

The information in the shaded area above is missing. Can you fill in the missing information so that all the rows & columns add up? (Be aware that there is no unique answer.)

Part 2. Your Own Project https://app.box.com/s/gq1t096g0cs51bbi4dwzi681xutzrh72

Small Projects Tool

Schedule and Level of Effort Analysis Chart

Project:

		MONTH												LEVEL OF EFFORT BY TASK SERIES (IN HOURS)															
	TASK ACTIVITIES	1	2	3	4	5	6	7	8	9	10	11	12	A. Project Manager	B.	C.	D.	E.	F.	G.	H.	I.	J.	K.	L.	M.	N.	O.	TOTAL HOURS
TASK SERIES	0100 - Quote																												
	0200 - Pre-project																												
	0300 - Contract negotiations																												
	0400 - Kickoff																												
	0500 - Planning/Scheduling																												
	0600 - Design																												
	0700 - Execution																												
	0800 - Shipping/Delivery																												
	0900 - Closeout																												
	1000 - Warranty Phase																												
	1100 - Lessons Learned																												
LEVEL OF EFFORT PER MONTH (IN HOURS)	A. Project Manager													0															
	B.														0														
	C.															0													
	D.																0												
	E.																	0											
	F.																		0										
	G.																			0									
	H.																				0								
	I.																					0							
	J.																						0						
	K.																							0					
	L.																								0				
	M.																									0			
	N.																										0		
	O.																											0	
	TOTAL HOURS																												0

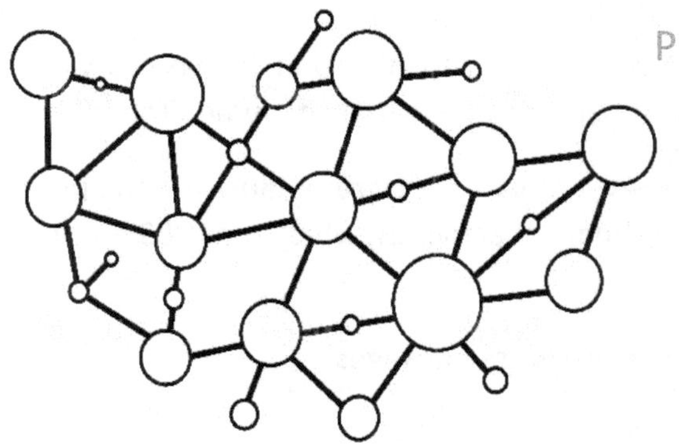

Planning and Scheduling Projects #3.7

Develop a Project Network Diagram

A ***project network*** is a tool for planning, scheduling, and monitoring project progress. You develop the network using the information collected for the WBS.

The network diagram is like a flowchart of the project work. It depicts the project activities that must be completed, the logical sequences, the interdependencies of the activities, and, in most cases, the time for the activities to start and finish along with the longest path through the network—the critical path.

The project network is the framework for the project information system that project managers use to make decisions about time, cost, and performance.

The project network can take time to develop, and this may become costly. Some people ask, are networks worth the struggle and cost? The answer is definitely yes. The exception would be when a project is considered trivial or is short in duration.

People can easily understand a network diagram because it displays the flow and sequence of work through the project.

Once you develop a network diagram for your project, you can easily modify it when unexpected events occur. For example, if materials for an activity are delayed, you can quickly assess the impact and revise the project in only a few minutes with the use of a computer. Then you can rapidly communicate the revisions to the client/stakeholders and project team.

The project network provides added information and insights. It is the basis for scheduling labor and equipment. It enhances communication that melds all managers and groups together in meeting the time, cost, and performance objectives of the project. It provides an estimate of project duration rather than picking a project completion date or using someone's preferred completion date. The network gives the times when activities can start and finish and when they can be delayed. It provides the basis for budgeting the cash flow of the project. It identifies critical activities, which are activities that should not be delayed if the project is to be completed as planned. It highlights what activities to consider if you need to compress the project to meet a deadline.

There are other reasons that project networks can be worth their weight in gold, but in essence, having a project network minimizes surprises by getting the plan out early, and it supports corrective feedback.

A commonly heard statement from practitioners is that the project network represents three-quarters of the planning process. Perhaps this is an exaggeration, but it signals the perceived importance of the network to project managers in the field.

How to Build a Network

You build a network using nodes (boxes) and arrows. Nodes depict activities, and arrows show dependency and project flow. The activity represents one or more tasks that consume time.

The network process is similar to the WBS process. However, network diagrams illustrate the project schedule by identifying dependencies, sequencing, and timing of activities, which the WBS is not designed to do.

The primary inputs for developing a project network plan are **work packages**.

Two Approaches to Project Networks: AON and AOA

There are **two approaches** to developing project networks: **activity-on-node (AON)** and **activity-on-arrow (AOA)**.

Both methods use two building blocks—the arrow and the node. Their names derive from the fact that the first uses a node to depict an activity, while the second uses an arrow to depict an activity.

From the first use of these two approaches in the late 1950s, practitioners have offered many enhancements. However, these basic approaches have withstood the test of time and still prevail, with only minor variations in form.

In practice, the AON method is most often used in project planning. But there are good reasons for project managers to be proficient in both methods. For example, departments and organizations typically have their preferred approach and are loyal to software already purchased and in use. As a project manager, you should be able to use AON or AOA.

Constructing a Project Network Using the AON or Precedence Diagramming Method (PDM)

The AON method is also referred to as the **precedence diagramming method (PDM)**.

For the AON or PDM method, a project schedule network diagram uses boxes or circles, referred to as nodes, to represent activities, and it connects them with arrows that show the dependencies.

This method addresses the types of relationships between activities. These include **Finish-to-start** (FS), **Finish-to-finish** (FF), **Start-to-start** (SS), and **Start-to-finish** (SF). This identifies when one activity can start or finish in relation to another. This type of network diagramming does not employ the use of dummy activities.

In the **FS relationship** illustrated below, Activity 1 (predecessor) must finish before Activity 2 (successor) can start. In the **SF relationship** illustrated below, Activity 7 cannot start until Activity 8 is completed.

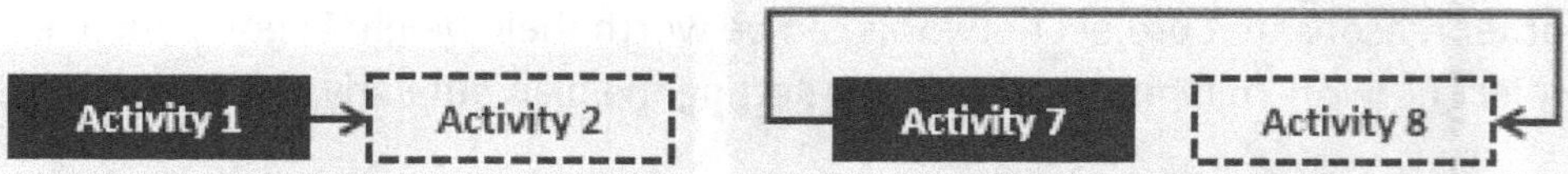

Finish-to-Start (FS) Example

Start-to-Finish (SF) Example

In the **SS relationship** illustrated below, Activity 5 must start before Activity 6 can start. As an example, you cannot drive your car until after you start it. In the **FF relationship** illustrated below, Activity 4 cannot finish before Activity 3 is finished. For example, you typically do not turn off your car until after you put it in Park or Neutral.

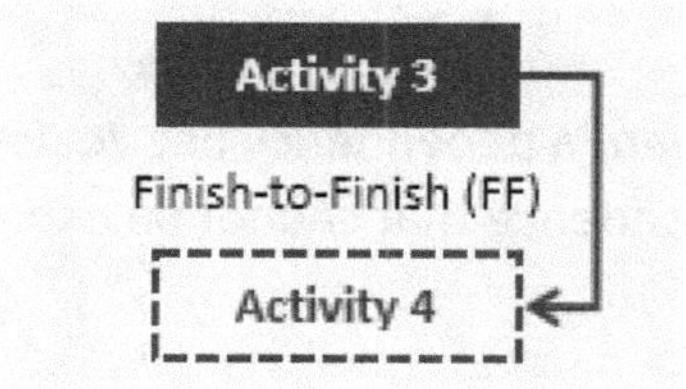

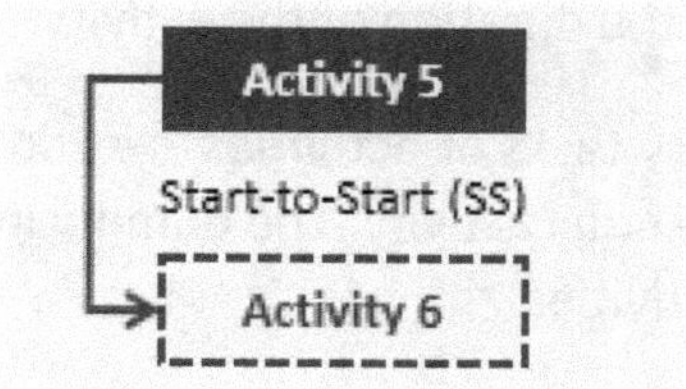

Another consideration of timing has to do with **lead** and **lag times**. **Lag time** is illustrated in the example below. It shows Activity 2 can start *after* Activity 1 has started.

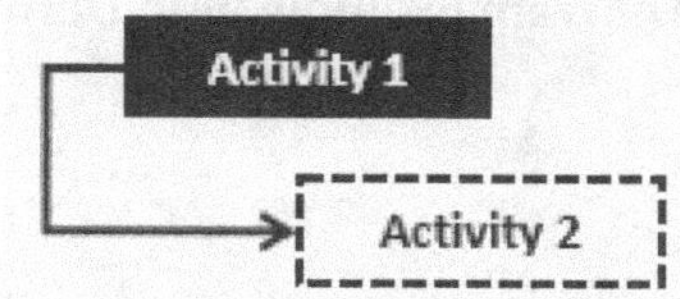

Lag Time Example

Lead time is when Activity 2 can start *before* Activity 1 has finished. As an example: Laying the foundation of a house could be Activity 1 and framing the walls could be Activity 2, whereby you could start framing the walls independently while the foundation is being built and put the framed walls in place once the foundation is completed.

In the example that follows, laying the foundation of a house could be Activity 1 and water-sealing the foundation could be Activity 2. The water sealing could start with the first part of the foundation; while the remainder of the foundation is laid, the completed portion could already be receiving water sealing.

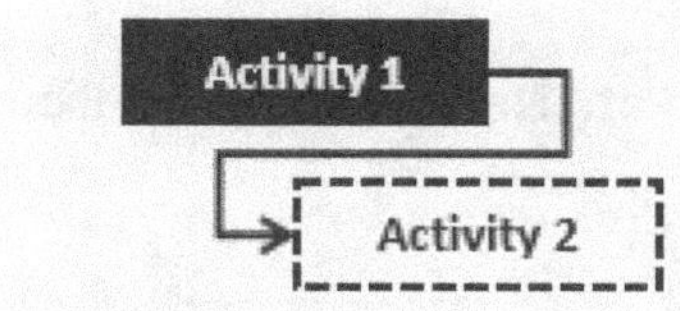

Lead Time Example

Remember, a predecessor task or activity must be completed before the next activity in the project can be done. Return to the Activity Template and review the sequence of activities. Consider the tools we have discussed: precedence diagramming, dependency determination, and lead/lag time.

Also, consider there are several types of dependencies: mandatory, discretionary, external, and internal. We need to first determine if any activity must be done before any of the others. After you list and sequence the activities, you can show the flow of activities.

Constructing a Project Network Using the AOA or Arrow Diagramming Method (ADM)

The **AOA** method is also referred to as the **arrow diagramming method** (ADM).

In AOA or ADM networks, activities are represented by arrows. Precedence relationships between activities are represented by numbered circles connected by one or more arrows. The length of the arrow represents the duration of the activity.

Sometimes **dummy tasks or activities** are added to represent a dependency between tasks. The dummy is not an actual activity. The dummy indicates precedence that cannot be expressed using only the actual activities.

You indicate the dummy activity by using a dashed line.

A dummy task often has a completion time of 0.

The following network diagram illustrates the use of a dummy activity. Notice the dashed line.

AOA Network Diagram Example with Dummy Activity

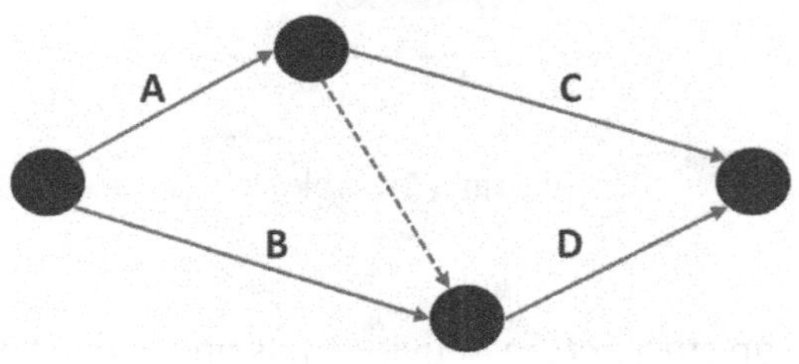

PERT and CPM

The activity diagramming method is often used for **Program Evaluation and Review Technique (PERT)** and **Critical Path Method (CPM)** analysis. However, the use of AOA or ADM as a common project management practice has declined with the adoption of computer-based scheduling tools.

The network diagram below illustrates the concept of the **critical path**. A project's critical path is the path of activities that take the longest time to complete. In this example, the critical path is the route in the middle with a total completion time of 12 weeks.

Critical Path Network Diagram Example

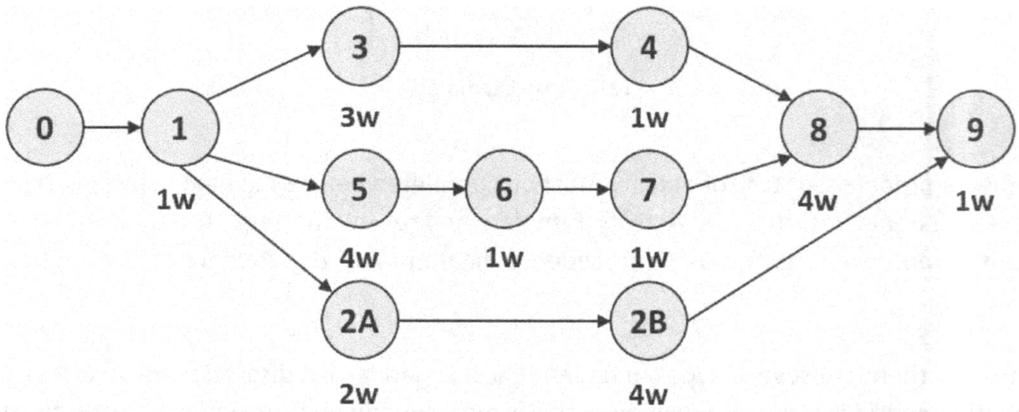

Summary of Rules for Developing a Network Diagram

The following **eight rules** apply when developing a project network:

1. Networks flow typically from left to right.
2. An activity cannot begin until all preceding connected activities have been completed.
3. Arrows on networks indicate precedence and flow. Arrows can cross over each other.
4. Each activity should have a unique identification number.
5. An activity identification number must be a larger (higher) number than any activity number that precedes it.
6. Looping is *not* allowed. In other words, recycling through a set of activities cannot occur.
7. Conditional statements are *not* allowed. In other words, do not use, "If successful, do something; if not, do nothing."
8. Experience suggests that when there are multiple starts, a common start node can be used to indicate a clear project beginning on the network. Similarly, a single project end node can be used to indicate a clear ending.

Computer Software

Be aware that all the concepts presented to this point are best applied with the use of computer software. This is also true for the statistical techniques available to a project manager in assessing project risk, such as **PERT** (**Program Evaluation and Review Technique**) or **probabilistic** network scheduling.

What to do:	Other actions:
☐ Review the sequence of activities for a project you are involved with or have observed to validate them.	☐ ____________________
☐ Pay attention to the relationships (predecessor and successor) between activities.	☐ ____________________
☐ Create a simple network diagram for a work-related project or event.	☐ ____________________

Remember

✓ A network diagram is a visual representation of a project schedule that uses nodes and arrows to show the relationship between activities.

✓ A network diagram shows the sequence of activities and considers the relationships between activities.

✓ Precedence diagramming is one tool you use to determine the type of relationship between two activities. There are four major types of dependencies: mandatory, discretionary, external, and internal.

✓ Developing a schedule is critical for project success. It serves as a tool for time management, and you also will use the schedule to track and measure project progress.

✓ You can create a network diagram for the overall project. If needed, you can also create a network diagram for an activity area within a project.

✓ When time is critical to your project, use a network diagram to identify the critical path. Then, working with your team, brainstorm ways to shorten the critical path.

Enhance Your Learning

Watch the following 7-minute video from Creative Education to view a simple example for preparing a network diagram.

Creative Education. (2020). *How to draw a Project Network Diagram*.	Available at: https://www.youtube.com/watch?v=pWkhR67bkXw

Reinforce Your Learning

Network Diagrams—Making Copies. Review the activities for the Making Copies project previously discussed. Use the activities list you made or the list below and sketch a network diagram for the project.

Activity #	Activity Description
1	Gather documents to copy.
2	Walk from desk to copier.
3	Make the copies.
4	Walk from copier to desk.
5	Distribute the copies.
6	File personal copy.

Sketch a network diagram for the Making Copies project:

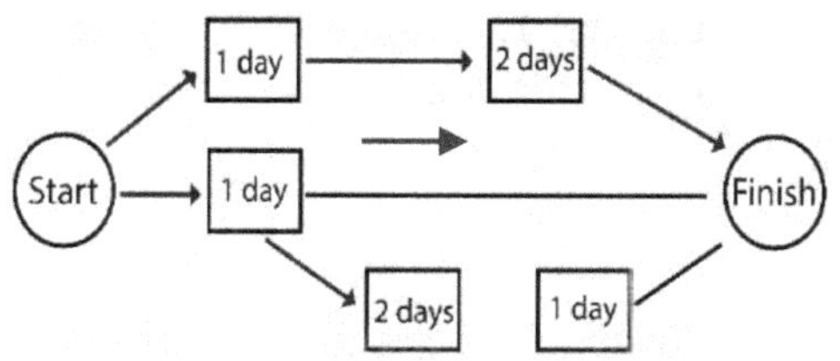

Critical Path = 4 Days

Planning and Scheduling Projects #3.8

Calculate the Critical Path

Earlier in this course, we touched on how you can use a network diagram to determine a project's **critical path**. Recall that a network diagram is a visual representation of the work to be done in a project, and the critical path shows you the time (the longest path of work) needed to complete the project.

In this concept, we will go into more detail about how project managers use the critical path in project management. The critical path is of crucial concern because to manage a project effectively you must know the sequence of work activities that are the longest path (have the longest duration). It shows the time needed to complete the project.

Assume you have a project with many interconnected activities, and you want to complete the project in the least time possible because the organization has a deadline.

However, as you learned, subsequent activities cannot be completed until the work in the predecessor activities is finished. Unless you have a plan that clearly shows the relationships between the activities, you run the risk of false starts and unnecessary delays. You might have people starting before the necessary previous work is finished or sitting idle and waiting to start because they do not know the predecessor work is finished.

But even with a good plan firmly in place, you still will not know how long the project will take until you determine **the length of the critical path**. The *critical path* is the longest path through the network, and this includes both dependent and independent tasks.

The following is an example of a network diagram showing that the longest path of work is activities 0, 1, 4, 5, 6, and 7. It is the project's critical path.

Critical Path Example

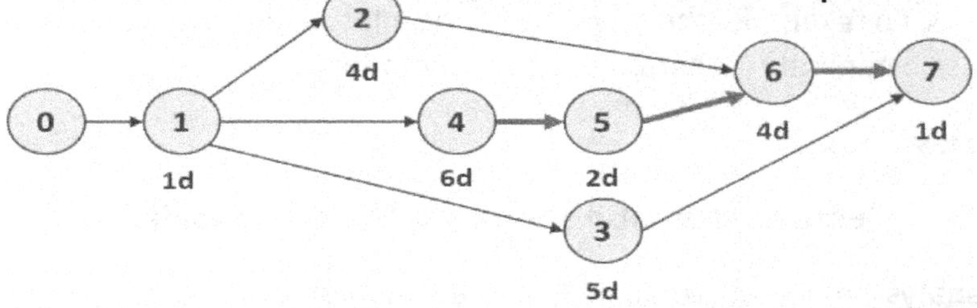

Start by noting your time estimates for each task in your network diagram. For example, you can see in the diagram above that after Activity 1 is finished, it will take four days to finish Activity 2 and six days to finish Activity 4.

Once you have your time estimates noted on your diagram, you determine the critical path by adding all the times on each of the paths. The path with the most time is the project's critical path. In this example, it is the path for Activities 0, 1, 4, 5, 6, and 7.

The Critical Path and Project Deadlines

Still using the network diagram illustrated above, the only way to shorten the length of the project is to shorten the time in the critical path (Activities 0, 1, 4, 5, 6, and 7).

Suppose the organization changes the deadline for the project by reducing the time for the project by two days. You might somehow be able to shorten the path along Activities 1, 3, and 7 by five days, but that would not shorten the project length at all. The only way to shorten the project length is by reducing its critical path.

Compressing a Schedule

Sometimes the estimated time to complete a project is longer than the time requirements imposed by the client/stakeholders. If that case, you may need to pursue other courses of action to compress the schedule.

The following is a discussion of several ways to compress a schedule. You might want to do this, for example, if you are asked to complete the project by a specific date. However, be aware that all these methods involve risk. You must ensure that the risk will not endanger the project to such an extent that problems incurred from compressing the schedule outweigh the benefits of completing the project by a specific date.

Slack or Float. *Slack* or *float* is the amount of time that some tasks in a network may be delayed without causing a delay to subsequent tasks. Identifying such tasks and removing the slack time may be an easy way to compress some aspects of a project. Related slack or float, the ***Forward pass*** is a technique to move forward through network diagram to determining project duration and finding the critical path or Free Float of the project. ***Backward pass*** represents moving backward to calculate late start or to find if there is any slack in the activity.

Crashing. This method of schedule compression uses extra resources to complete critical activities ahead of schedule. Only activities on the critical path are eligible for crashing because they affect the time needed for the project. Because crashing uses more resources, it is often more expensive and may not be compatible with the project budget.

Fast Tracking. This method of schedule compression involves rearranging the schedule so that activities that would typically be sequential are performed at the same time.

Assigning Limited Overtime. This method allows team members to work a limited amount of overtime so that critical activities can be completed earlier. The overtime must be limited to prevent team members from burn out. Too much overtime can also increase project costs.

Using Shortcuts. This method allows team members to take shortcuts rather than following exact project specifications. For example, the team might use a less expensive, generic resource instead of a name brand, or use a computer program to design a model.

Modeling Techniques

There are two modeling techniques: **what-if scenario analysis** and **simulation**.

What-if scenario analysis modeling could include the variables most likely to change and consideration of what those changes could mean to the project if they were to materialize. An example question might be what would the schedule look like if we added a person with a particular skill set? To run a model, you would plug that resource into the WBS, assign appropriate activities, and then review the overall impact it has on the project schedule.

Simulation modeling requires computer software to simulate project time, cost, or resource availability using **Monte Carlo techniques** that rely on computational algorithms and random sampling to obtain numerical results such as the probability or critically index for an activity in the schedule.

A simulation generates the **probability** of an activity or path **becoming critical**. The software uses a simple triangular distribution to represent the range and average of each activity duration.

Simulating each activity duration distribution for a project gives one set of activity values used to compute the critical path. This process is repeated hundreds of times to determine the criticality of any activity or path.

You can simulate cost by using the high and low estimates of costs for each activity and each simulation trial. By using a resource-constrained program with the simulation program, potential resource conflicts are identified and assessed.

What to do:	Other actions:
☐ Use the critical path to focus on activities that directly impact the project schedule.	☐ ________________
☐ Estimates are only estimates. Monitor project progress to detect potential issues as early as possible.	☐ ________________
☐ If time is an important aspect of the project, look for ways to shorten the critical path.	☐ ________________

Remember

- ✓ The critical path is the time needed to complete a project. In other words, you cannot complete the project sooner. The critical path is significant in every project.
- ✓ Consider your confidence level for project estimates.
- ✓ Sometimes the estimated time to complete a project is longer than the time the client/stakeholders require. If that happens, you may need to compress the schedule.

Enhance Your Learning

Watch the following 12-minute video by Andy Kaufman: How to Calculate Critical Path.

Kaufman, A. (2017). *How to Calculate Critical Path: Project Management Professional (PMP)® Exam Prep.*	Available at: https://www.youtube.com/watch?v=0Lo4zsB-bjE

Watch the following 7-minute video by Scott Christianson for a simplified explanation of project management networks and the forward and backward pass.

Christianson, S. (2019). *Project Management Networks Part 2: Forward and Backward Pass.*	Available at: https://www.youtube.com/watch?v=Momnar3Mr9Q

Reinforce Your Learning

Calculate Critical Path. Using the network diagram example below, calculate the number of days needed to complete the project.

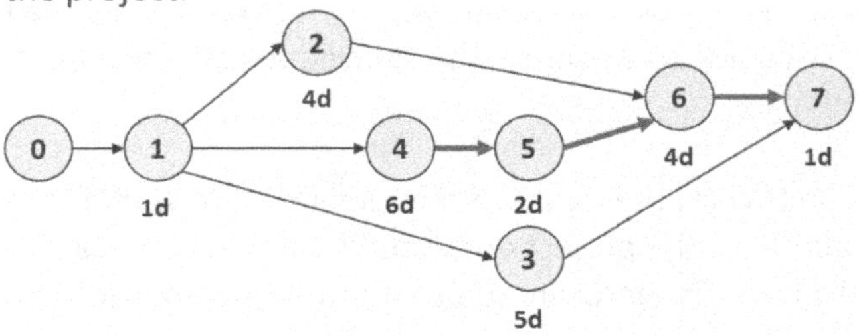

Task	Optimistic (O)	Most Likely (M)	Pessimistic (P)
Task A	2 Wks	4 Wks	5 Wks
Task B	1 Wks	2 Wks	3 Wks
Task C	2 Wks	3 Wks	4 Wks
Task B	3 Wks	5 Wks	8 Wks
Completion	8 Wks	14 Wks	20 Wks

Review the Program Evaluation Review Technique (PERT)

If you have ever developed either a project schedule or a project budget, you know it's almost impossible to come up with accurate overall estimates for either one. A simpler and better way is to estimate the individual times and costs and then add these to get your total time and cost. Doing this also helps you develop better tools for tracking your progress and monitoring your costs.

However, even for simple projects this too can be difficult. Under the best of conditions, it's often impossible to come up with a realistic schedule. Experience has shown that you can improve the accuracy of your estimate by using an approach that has proven itself in a wide range of projects. This approach takes into consideration the "best case scenario" and the "worst case scenario" and balances these two estimates with a healthy dose of the "most likely" scenario. This is a probabilistic method similar to PERT. The formula for this is shown here:

$$E = (O + 4M + P)/6$$

In this estimate approach (**E**), take your "best case scenario" or most optimistic estimate (**O**) and add it to four times the most likely estimate (**M**). Add in your "worst case scenario" or most pessimistic estimate (**P**). Divide that number by 6 to come up with the estimate. This approach is equally valid for both the schedule and the budget. Obviously for complex projects a more sophisticated approach is needed.

In 1958 the Special Office of the Navy and the Booze, Allen, and Hamilton consulting firm developed Program Evaluation Review Technique (PERT) to schedule the more than 3,300 contractors of the Polaris submarine project and to cover the uncertainty of activity time estimates.

PERT is almost identical to the critical path method (CPM) except PERT assumes each activity duration has a range that follows a statistical distribution. PERT uses three-time estimates for each activity. In other words, each activity duration can range from an optimistic time to a pessimistic time, and a weighted average can be computed for each activity.

Because project activities usually represent work and work tends to stay behind once it gets behind, the PERT developers chose an approximation of the *beta distribution* to represent activity durations. This distribution is flexible and can accommodate empirical data that do not follow a normal distribution. The activity durations can be skewed toward the high or low end of the data range.

Activity durations skewed toward the right are representative of work that tends to stay late once it is behind. The distribution for the project duration, the sum of the weighted averages of the activities on the critical path(s), is represented by a normal (symmetrical) distribution.

How to Compute the Probability of Meeting Various Project Durations

Knowing the weighted average and variances for each activity allows the project planner to compute the probability of meeting various project durations.

We describe the computation steps in the example below. The jargon may seem difficult for those not familiar with statistics, but the process is relatively easy to follow once you have worked through a few examples.

The weighted average activity time is computed using the following formula:

$$te = (a+4m+b)/6$$

where:

te = weighted average activity time
a = optimistic activity time (1 chance in 100 of completing the activity earlier under normal conditions)
b = pessimistic activity time (1 chance in 100 of completing the activity later under normal conditions)
m = most likely activity time

When three-time estimates have been specified, this equation is used to compute the weighted average duration for each activity. The average (deterministic) value is placed on the project network as in the CPM method, and the early, late, slack, and project completion times are computed as they are in the CPM method.

The variability in the activity time estimates is approximated by the equations below. The first represents the standard deviation for the activity, and the second represents the standard deviation for the project. Note the standard deviation of the activity is squared in this equation; this is also called variance. This sum includes only activities on the critical path(s) or path being reviewed.

Standard deviation for activity: $\sigma te = (b - a)/6$

Standard deviation for project: $\sigma TE = \sqrt{\sum(\sigma te)2}$

Finally, the average project duration (TE) is the sum of all the average activity times along the critical path (sum of te), and it follows a normal distribution. Knowing the average project duration and the variances of activities allows the probability of completing the project (or segment of the project) by a specific time to be computed using standard statistical tables.

The equation below is used to compute the Z value found in statistical tables (Z = number of standard deviations from the mean), which, in turn, tells the probability of completing the project in the time specified.

$$Z = (TS - TE)/ \sqrt{\sum(\sigma te)\ 2}$$

where:

TE = critical path duration
TS = scheduled project duration
Z = probability (of meeting scheduled duration) found in standard statistical tables

You can make this same type of calculation for any path or segment of a path in the network. When such probabilities are available to management, trade-off decisions can be made to accept or reduce the risk associated with a particular project duration.

At least two choices are available. First, management can spend money upfront to change conditions that will reduce the duration of one or more activities on the critical path.

A more prudent, second alternative would be to allocate money to a contingency fund and wait to see how the project progresses as it is implemented.

Program Evaluation Review Technique (PERT) Example

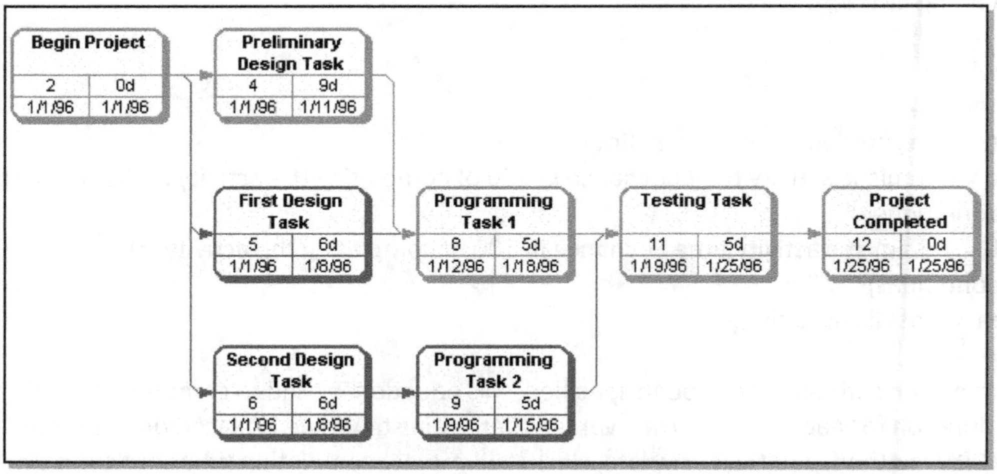

PERT Simulations

This analysis technique requires computer software to simulate project time, cost, and resource availability using the Monte Carlo technique. For example, using the same time estimates developed for PERT, simulation generates the probability of any activity or path becoming critical.

The software uses a simple triangular distribution to represent the range and average of each activity duration. Simulating each activity duration distribution for a project gives one set of activity values used to compute the critical path. This process is repeated hundreds of times to determine the criticality of any activity or path.

Cost can be simulated similarly by using the high and low estimates of costs for each activity and each simulation trial. By using a resource constrained program with the duration PERT simulation program, potential resource conflicts are identified and assessed.

Risk retention or transfer decisions are made using information gained from time, cost, and resource simulation.

PERT and PERT simulation are useful in extremely important projects that have a great deal of inherent uncertainty and reasonably accurate time estimates for activities. Numerous software packages are available. Here are a few examples:

- Clarizen
- Microsoft Project
- Primavera SureTrak
- Project Libre
- ZOHO Projects

What to do:	Other actions:
☐ Identify a project where you are uncertain of activity time estimates.	☐ ____________________
☐ Review the PERT technique and consider if it may be appropriate to use.	☐ ____________________
☐ Ask those experienced with using project scheduling software to recommend a product.	☐ ____________________

Remember

- ✓ PERT is almost identical to the critical path method (CPM) technique, except it assumes each activity duration has a range that follows a statistical distribution.
- ✓ PERT uses three-time estimates for each activity. Basically, this means each activity duration can range from an optimistic time to a pessimistic time, and a weighted average can be computed for each activity.

Enhance Your Learning

Watch the following 5-min. video by Mike Clayton to learn more about how to create a PERT Chart

Clayton, M. (2017). *What is a PERT Chart? Project Management in Under 5.*	Available at: https://www.youtube.com/watch?v=i160aaBX7mE

Reinforce Your Learning

Retirement Celebration Planner. Try your hand at estimating the time it will take to plan a formal event like a retirement celebration. Use the probabilistic formula to calculate (in months) the result. Then compare your estimate with others.

$$E = (O + 4M + P)/6$$

Optimistic: ____________________

Pessimistic: ____________________

Most Likely: ____________________

Estimate = ____________________

Planning and Scheduling Projects #3.10

Be Aware of Common Scheduling Pitfalls and Tips for Success

As we have discussed, one of the most challenging functions of a project manager is planning a complete and realistic project schedule. The plan lists the work to be done, by whom, and when it starts and finishes.

Without adequate planning and scheduling work, a project manager cannot manage a project because they will not know what work should be in progress or completed at any given time, and when foundational work is not done, later work cannot follow. Without an adequate schedule, the project manager may not even know this is happening and so cannot take corrective action. Without an adequate plan and schedule, the project manager cannot communicate with stakeholders and team members.

In this concept, we discuss common pitfalls to avoid, followed by 15 tips for planning and scheduling a successful project.

Common Pitfalls. Below are 15 mistakes project managers sometimes make. Any of them can cause a project to flounder and possibly fail.

1. Not adequately defining the project's purpose.
2. Not getting requirements in writing.
3. Making up or getting estimates from unreliable or unqualified sources.
4. Neglecting to reserve necessary resources.
5. Assuming needed staff members will be available and trained.
6. Failing to unite the team.
7. Not scheduling time for client and stakeholder reviews and changes.
8. Being too rigid and refusing to adapt when problems arise.
9. Underestimating project risk and assuming you can control all activities in the schedule.
10. Forgetting the roles client and stakeholder play in keeping to the schedule and the need to obtain approvals and sign-offs.
11. Assuming all preliminary groundwork needed for the next task is done.
12. Not scheduling adequate time for outside agency reviews and changes.
13. Not scheduling time for communication and coordination tasks such as phone calls, meetings, and client relationship building during and after the project.
14. Not providing adequate contingency plans in case deadlines are not met.
15. Assuming that because you are a project manager, others will respect your authority, accept your schedule, and participate accordingly.

Tips for Success

The following are 15 tips that will help you plan and schedule a project and bring it to a successful conclusion.

1. Perform due diligence to define requirements: work, people, time, and costs.
2. Ensure the right resources and subject matter experts provide estimates.
3. Create or find templates for project documents and use them.
4. Create a milestone for each project deliverable.
5. Prepare in calendar time, not in the number of workdays.
6. Build time into the schedule for slippage and revisions.
7. Involve your team in defining tasks required to produce each deliverable.
8. Create task criteria to accurately measure and track progress.
9. Break down the time needed to complete tasks and activities into 40 hours or less.
10. Assign every task to a specific team member.
11. Consider whether your team members will be assigned projects in addition to yours.
12. Ensure that the project schedule is updated as soon as possible.
13. Establish and communicate task and milestone naming conventions to avoid confusion.
14. Automate the reporting process.
15. Ask questions, and then listen.

What to do:	Other actions:
☐ Do not underestimate the importance of a complete and realistic schedule.	☐ ________________
☐ Review the planning and scheduling pitfalls. Are you making any of these mistakes?	☐ ________________
☐ Determine what tips for success you can apply in your work.	☐ ________________

Remember

- ✓ One of the most challenging functions of a project manager is the creation of a good project schedule.
- ✓ Project managers sometimes underestimate the importance of planning and scheduling project work.
- ✓ Project managers must plan and schedule as accurately as possible and learn from past lessons and utilize the wisdom of others.

Enhance Your Learning

Watch the following 12-minute video by Tony Zink to obtain project management scheduling tips.

Zink, T. (2017). *Project Management Tips: Working With Tasks and Milestones in Your Project Schedule.*	Available at: https://www.youtube.com/watch?v=tP4wIlWjsqU

Reinforce Your Learning

Scheduling Tips. From the list of tips, select the three you consider most important.

1.

2.

3.

Summary

A Final Word: *Project Planning and Scheduling*

Consider what actions you can take to better understand your influence on a project schedule. All roles within a project have the potential to impact the schedule, both positively and negatively. Consider how your role may impact the project schedule and time management.

You have completed a chapter on Project Planning and Scheduling.

Information you reviewed in the chapter include:

- Describing what is included in a Schedule Management Plan
- Discussing common project scheduling tools and their applications
- Developing a project schedule
- Explaining what is meant by a schedule baseline
- Identifying common planning and scheduling pitfalls

In the next *Project Mastery* chapter, we will focus on Monitoring and Controlling Projects.

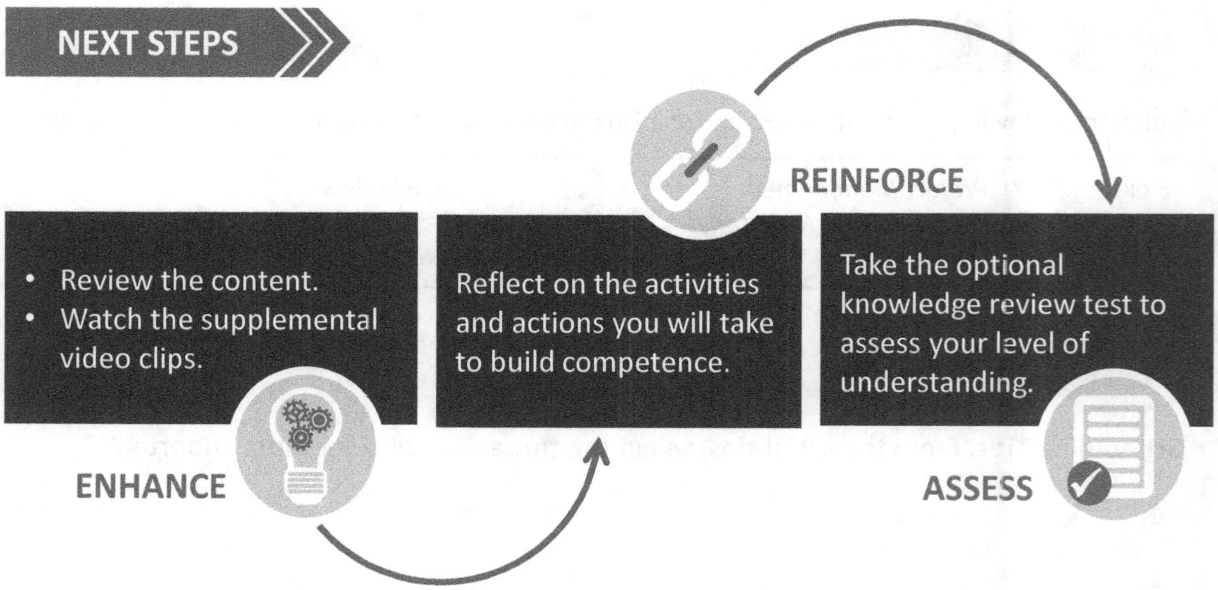

Planning and Scheduling Projects—Recap Checklist
Half of all projects take longer than expected, so proper planning is critical. Good planning requires that you identify all activities and tasks, sequence them correctly, and accurately estimate schedules and resources.
1. **Build the Project Schedule** □ Identify project methods/approaches and tools you are familiar with and those to learn more about. □ Consider a baseline schedule as a formal agreement. □ Review the baseline schedule to validate that the project team can provide the deliverables as scheduled.
2. **Sequence Project Tasks and Activities** □ Review a sequence of project activities and validate they are accurate. □ Pay attention to how one activity relates to another (predecessor and successor). □ Inform stakeholders you reviewed the sequence of activities, and you agree or want to recommend changes.
3. **Determine Effort for Tasks** □ Identify your best sources of information for determining the effort required for completing activities and tasks. □ Consider using this same approach to estimate costs and develop a budget. □ Use the probabilistic or three-point estimating approach when you do not have better information.
4. Set Project Milestones □ Review a milestone chart that lists significant events or decision points with deadline dates. □ Select major events that incrementally build until your project is complete. □ Specify what and who needs to approve/decide before proceeding to the next point.
5. Develop Gantt Charts □ Select a simple work-related project and create a WBS outlining the major activities or tasks. □ Develop a Gantt or bar chart by listing the major tasks along with start dates and the length of time for each task. □ Try producing a Gantt chart using one of the many project scheduling software tools.
6. Use a Spreadsheet for Scheduling Small Projects □ List the tasks for a small project you are familiar with or have observed. □ Identify and list the individuals/functions or work units responsible for preforming the work. □ Develop a spreadsheet for your small project.
7. **Develop a Project Network Diagram** □ Review the sequence of activities for a project you are involved with or have observed to validate them. □ Pay particular attention to the relationships (predecessor and successor) between activities. □ Create a simple network diagram for a work-related project or event.
8. **Calculate the Critical Path** □ Use the critical path to focus on activities that directly impact the project schedule. □ Estimates are only estimates. Monitor project progress to detect potential issues as soon as possible. □ If time is an important aspect of the project, look for ways to shorten the critical path.
9. **Review the Program Evaluation Review Technique (PERT)** □ Identify a project where you are uncertain of activity time estimates. □ Review the PERT technique and consider if it may be appropriate to use. □ Ask those experienced with using project scheduling software to recommend a product.
10. **Be Aware of Common Scheduling Pitfalls and Tips for Success** □ Do not underestimate the importance of a complete and realistic schedule. □ Review the planning and scheduling pitfalls. Are you making any of these mistakes? □ Determine what tips for success you can apply in your work.

Knowledge Review Test—Questions

Planning and Scheduling Projects

1. According to a PMI study, what percentage of the projects typically completed within an organization in the past 12 months finished within their initial scheduled times?
   A. 25%
   B. 35%
   C. 50 %
   D. 65 %
   E. 75 %

2. In project management, there are typically three baselines. Which of the following is NOT one of the baselines?
   A. Schedule baseline.
   B. Cost baseline.
   C. Scope baseline.
   D. Quality baseline..

3. In project management "SME" stands for:
   A. Simultaneous Method Estimate.
   B. Subject Matter Expert.
   C. Similar Metal Extract.
   D. Simple Math Equation.

4. Project managers construct project schedules based on practical methods or "rules of thumb" which come from a variety of sources. Which of the following is NOT one of the typical sources:
   A. Project manager professional judgment.
   B. The experience of subject matter experts.
   C. Ideas from the project team and stakeholders.
   D. Friends and relatives.

5. Effort often refers to the number of labor units or man-hours needed to complete a task or activity. However, effort can also refer to other items. Which of the following is NOT one of the typical items?
   A. Work breakdown structure.
   B. Human resources.
   C. Material.
   D. Financial project needs.

6. Experience has shown that you can improve the accuracy of an estimate by using the probabilistic or three-point estimating approach. This approach has been proven in a wide range of projects. Which of the following is NOT one of the scenarios?
   A. Best-case scenario.
   B. Ideal-case scenario.
   C. Worst-case scenario.
   D. Most likely scenario.

7. The components or inputs for planning and developing a schedule typically include several kinds of information. Which of the following is NOT one of the kinds of information?
   A. Activity list and attributes.
   B. Resource list and calendar.
   C. Activity duration estimates.
   D. Actuary tables.
   E. Project scope statement.

8. A project milestone can be:
   A. A task of zero duration that shows an essential achievement in a project.
   B. Any point in the project that requires approval or a decision before proceeding.
   C. Points that represent a clear sequence of events that incrementally build until your project is complete.
   D. All the answers are correct.

9. In project management, decomposition means:
   A. The process by which dead organic substances are broken down into simpler organic or inorganic matter.
   B. Taking a deliverable, task, activity, process, or event and breaking it down into smaller, more manageable components to provide better control.
   C. The process of simplifying a single chemical entity into two or more fragments.
   D. A reaction in which a compound splits up into two or more simpler substances.

10. Which of the following COULD NOT be a component of a Gantt chart?
   A. Horizontal bars.
   B. A numbered list of tasks.
   C. A timeline.
   D. Expenses.
   E. Plotted points.

11. Which of the following tools is the best to use to determine the longest time the project will take?
   A. Bar chart.
   B. Work breakdown structure.
   C. Network diagram.
   D. Project scope document.

12. Which of the following items can be tracked using a spreadsheet approach?
   A. Project tasks.
   B. Project budgets.
   C. Project timesheets.
   D. Project issues.
   E. All the answers are correct.

13. The relationship between standard deviation and risk can be best described by which of the following?
   A. Standard deviation tells you if the estimate is correct.
   B. There is no relationship.
   C. The standard deviation tells you the worst-case scenario..
   D. The standard deviation tells you the level of uncertainty.

14. An "activity" in a project network is something that requires:
    A. Planning.
    B. Time.
    C. A deliverable.
    D. A precedent.
    E. An outcome.
15. Which term refers to the least time needed to complete a project?
    A. The Network Diagram.
    B. The Work Breakdown Structure.
    C. The Critical Path.
    D. The Project Scope.
    E. None of the answers refer to the least time needed to complete the project.
16. In project management, the float of an activity is determined by:
    A. Determining the wait time between activities.
    B. Conducting a Monte Carlo analysis.
    C. Determining lead and lag.
    D. Determining the length of time an activity can be delayed without delaying the critical path of a project.
17. PERT stands for:
    A. Project Estimation Review Tactics.
    B. Program Evaluation and Retrieval Techniques.
    C. Possible Estimation of Regression Tabulation.
    D. Program Evaluation Review Technique.
    E. None of the answers are correct.
18. Dummy tasks are sometimes added to represent a dependency on what type of network diagram:
    A. PDM.
    B. ADM or AOA.
    C. IBM.
    D. LAN.
19. Which item that follows is NOT a good thing to do?
    A. Do your due diligence to define requirements: work, people, time, and costs.
    B. Ensure the right resources and subject matter experts provide estimates.
    C. Assume needed staff members will be available and trained.
    D. Create or find templates for project documents and use them.
    E. Create a milestone for each project deliverable.
20. Which item that follows is a good thing to do?
    A. Do your due diligence to define requirements: work, people, time, and costs.
    B. Create task criteria to accurately measure and track progress.
    C. Break down the time needed to complete tasks and activities into 40 hours or less
    D. Assign tasks to specific team members.
    E. All the answers are correct.

Planning and Scheduling Projects — Answer Sheet

1.	___ A	___ B	_X_ C	___ D	___ E
2.	___ A	___ B	___ C	_X_ D	___ E
3.	___ A	_X_ B	___ C	___ D	___ E
4.	___ A	___ B	___ C	_X_ D	___ E
5.	_X_ A	___ B	___ C	___ D	___ E
6.	___ A	_X_ B	___ C	___ D	___ E
7.	___ A	___ B	___ C	_X_ D	___ E
8.	___ A	___ B	___ C	_X_ D	___ E
9.	___ A	_X_ B	___ C	___ D	___ E
10.	___ A	___ B	___ C	___ D	_X_ E
11.	___ A	___ B	_X_ C	___ D	___ E
12.	___ A	___ B	___ C	___ D	_X_ E
13.	___ A	___ B	___ C	_X_ D	___ E
14.	___ A	_X_ B	___ C	___ D	___ E
15.	___ A	___ B	_X_ C	___ D	___ E
16.	___ A	___ B	___ C	_X_ D	___ E
17.	___ A	___ B	___ C	_X_ D	___ E
18.	___ A	_X_ B	___ C	___ D	___ E
19.	___ A	___ B	_X_ C	___ D	___ E
20.	___ A	___ B	___ C	___ D	_X_ E

Once the project management plan is baselined, it may only be changed when a change request is generated and approved through the Perform Integrated Change Control process.

—PMBOK® Guide, 5th Edition

Chapter 4

Monitoring and Controlling Projects

Overview

"All courses of action are risky, so prudence is not in avoiding danger (it's impossible) but calculating risk and acting decisively. Make mistakes of ambition and not mistakes of sloth. Develop the strength to do bold things, not the strength to suffer."

—Niccolo Machiavelli, The Prince

Scope changes and scope creep can undoubtedly impact a project. According to the Pulse of the Profession®: Capturing the Value of Project Management: 2015, published by the Project Management Institute (PMI), scope changes occurred on 44% of the projects. How do changes impact the project? How many aspects of the project are impacted (cost, schedule, quality, etc.)? The myriad of dynamics associated with the execution phase of a project is why monitoring and controlling is an absolute fundamental.

However, just as big of a problem can occur when project managers spend time planning the project but fall far short in monitoring and controlling the project's execution. In other words, planning is vital, but success also depends on monitoring and controlling the project work as it progresses.

Project managers must monitor and control several factors throughout the project life cycle. Developing project control baselines and gaining knowledge of tools and techniques to manage key aspects of the project are critical to its success.

In this chapter, you will learn how to:

- Manage scope variation from all project perspectives
- Identify and develop the key components of a project management plan
- Recognize the aspects of a project that most need monitoring and controlling
- Monitor and control a project
- Appreciate the benefits of early detection and correction of project metrics (scope, schedule, and cost) problems

The competencies covered in this chapter include:

Resource Management	**Management Controls**	***Oral and Written Communication***

Take Your Temperature for Monitoring and Controlling Projects

The following assessment will help you identify what you know about monitoring and controlling projects and the areas where you can improve. With you and your organization in mind, read each statement below and think about how much you agree with it. Use the numbers from 1 to 10, where **1** means you **strongly disagree** and **10** means you **strongly agree**.

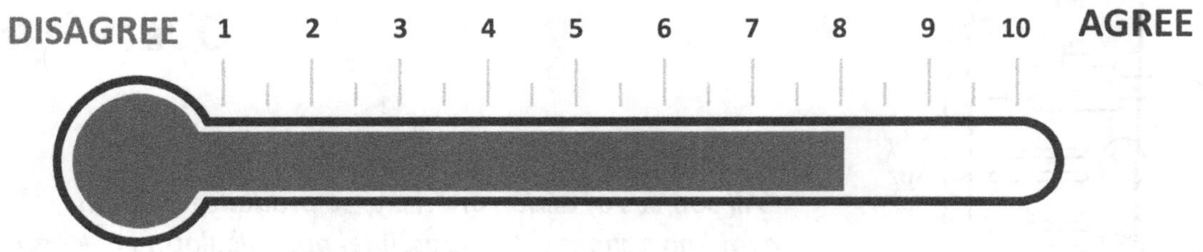

_____ 1. I can differentiate monitoring from controlling.

_____ 2. I know how to manage the critical first 10 percent of a project.

_____ 3. I can explain the components of a project plan.

_____ 4. I can monitor and control scope creep.

_____ 5. I have the skills to monitor and control project resources.

_____ 6. I can identify the components of risk planning and management.

_____ 7. I am confident in the use of quality monitoring and control tools.

_____ 8. I can deal with quality issues and early project termination.

_____ 9. I am good at monitoring and controlling the communication process.

_____ 10. I know how to monitor and control the change management process.

_____ **Total**

Take a moment to reflect on the following before you move forward in the chapter.
In your work, what problems have you experienced or observed related to monitoring and controlling projects at work?

Take a few minutes to also reflect on your self-assessment. List two or three areas you want to improve.

As you progress through this chapter consider what actions you can take to demonstrate competence in the following three competency areas:

1. **Resource Management.** Demonstrates awareness of technical resources; knows how to apply resources to achieve desired outcomes.

2. **Management Controls.** Ensures the integrity of the organization's processes; promotes ethical and effective practices.

3. **Oral and Written Communication.** Makes clear and effective presentations to individuals and groups; listens to others; communicates effectively in writing; can critically review and comprehend information written by others.

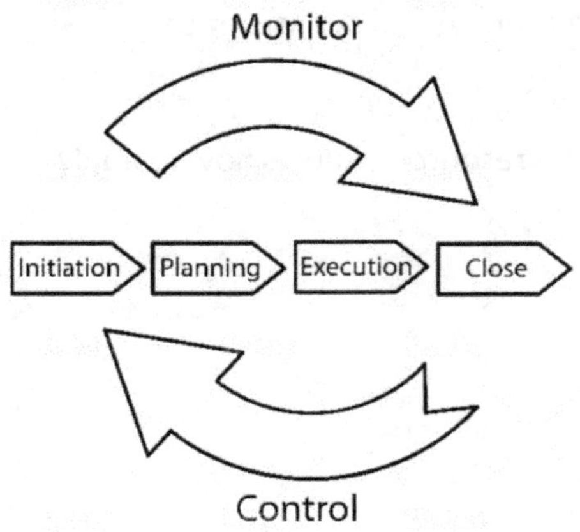

Differentiate Between Monitoring and Controlling a Project

A project will go through several phases in its life cycle: initiating, planning, executing, and closing.

This course focuses on the **execution** phase which is **doing the work**. The Project Life Cycle illustration below shows where we are in the life cycle. Execution includes both monitoring and controlling projects. **Monitoring** entails watching over the project, collecting, reporting, and disseminating project performance data and information; whereas **controlling** entails exercising influence over the project, taking action to analyze, evaluate alternatives, and recommend corrective action.

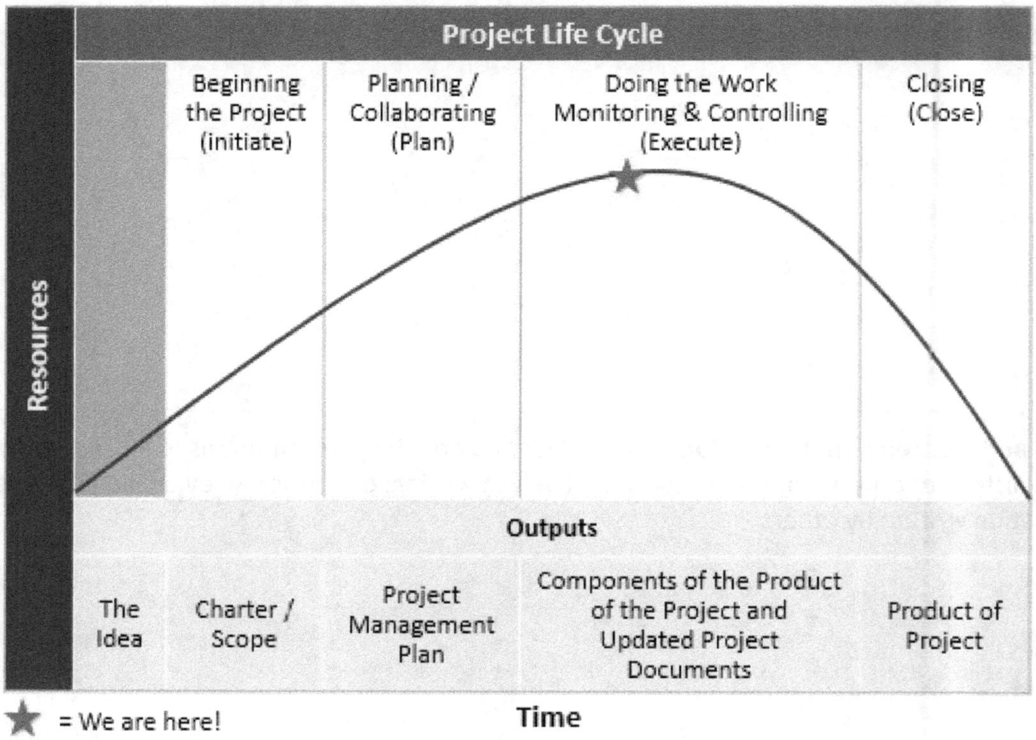

Execution is the phase where project work is done and where planning meets the realities associated with doing the work. It is also where most project funding will be spent.

In the execution phase, you follow your plan and do the work to build the project deliverables. But there is another critical job project managers must perform during the execution phase. In addition to doing the work, project managers must also monitor and control the project and make adjustments as surprises and changes occur

According to the *PMBOK® Guide*, *monitoring* involves collecting project performance data, producing performance measures, and reporting and communicating performance information. *Controlling* involves comparing actual performance with planned performance, analyzing variances, assessing trends to effect process improvements, evaluating possible alternatives, and recommending appropriate corrective action as needed.

Project Management Knowledge Areas

The execution phase involves all knowledge areas, as illustrated below. We will discuss specific tools and techniques used to determine how well a project is progressing. The intent of these tools is to facilitate early detection of anything that may have a detrimental impact on the progress. The earlier the issues are detected, the sooner the project manager and team will be in a position to address them. Some impacts could be significant and require effort to rectify.

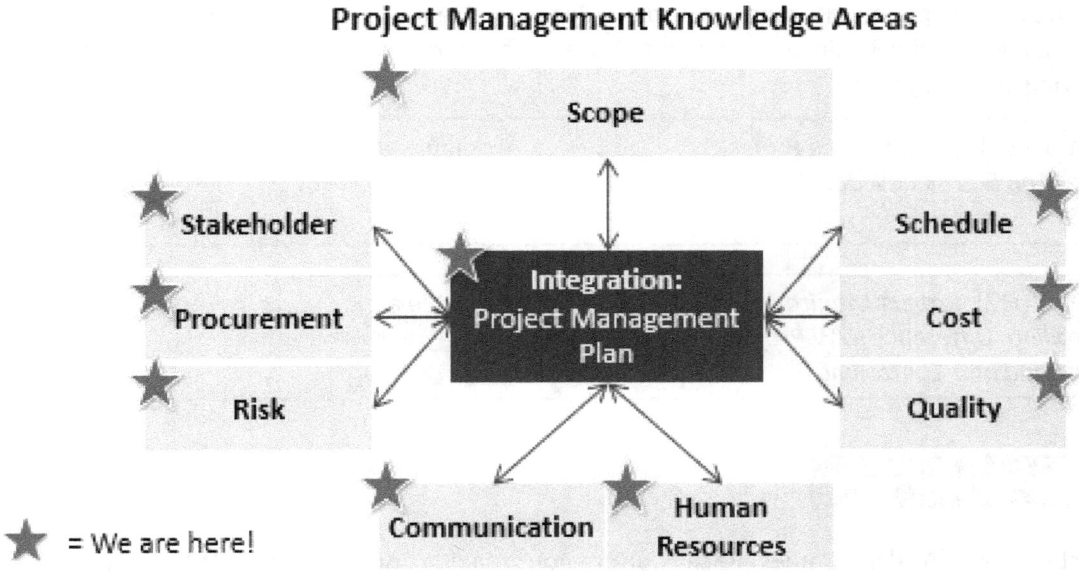

Change Management

Besides the knowledge areas in the above illustration, we will also discuss **change management**. During the execution phase, you will perform activities associated with all knowledge areas, and as the project progresses, you can expect there will be **variation** from the original **schedule baseline**, **cost baseline**, and **scope baseline**. You must ensure that all such variations are discovered and adequately accounted for.

What to do:	Other actions:
☐ Review the project charter and scope statement for updates and changes.	☐ ________________
☐ Be mindful that there will be changes to the project's deliverables.	☐ ________________
☐ Check the project planning for monitoring and controlling guidelines.	☐ ________________

Remember

- ✓ During the execution phase, there will be changes to project deliverables. You must account for these changes in all parts of the project and in the project's documentation.
- ✓ Project documents are tools to help the project be successful.
- ✓ The project manager and team must keep project documents current, and they must accurately reflect what is actually happening with the project.
- ✓ Monitoring and controlling involve all aspects of a project.
- ✓ Every project will experience changes. Monitoring and controlling can help keep the project on track.

Enhance Your Learning

Watch the following 7-minute video by ProjectManager to learn more about controlling project scope. Then watch the 3-min. video by IPM Academy focused on the difference between project monitoring and controlling.

Whitt, J. (2013). *Controlling Project Scope: Top 5 Project Scope Control Tips.*	Available at: https://www.youtube.com/watch?v=9A80QdJPR9U
Akbar. (2022). *Project Monitoring vs Controlling, Difference B/W Project Monitoring and Controlling.*	Available at: https://www.youtube.com/watch?v=XTo99IsVS14&t=1s

Reinforce Your Learning

Project Changes: Making Copies. Recall our previous Making Copies exercise. What kind of surprises or changes could affect your project plan? For example, what if the copy machine is broken? What areas of the project management plan will be impacted?

Project Changes: New Home. Assume you are building a new home. After construction has started, you decide to add a two-car garage with a stone face front. You also want to advance your hobby of making pottery by having a small kiln in the back right corner. What areas of the project management plan will be impacted by these changes?

Monitoring and Controlling Projects #4.2

Start the Execution Phase by Reviewing the Critical First 10 Percent

In the Defining Project Scope course, we discussed the **critical first 10 percent of a project**. It is the *time spent planning a project from beginning to end.* In the execution phase, we must not only do the work planned during the first 10 percent of the project, but also ensure that the team understands the plan and regularly revisit the original baseline plan for changes.

The project plan describes what is to be done and who is expected to do it. This contributes toward the development and verification of the project's cost, schedule, and scope. Any variation from the original baseline scope is typically considered a **change request** and must be documented accordingly.

A project manager wears many hats. In addition to changes in the technical aspects of a project that need monitoring and controlling, there will also be changes in the people aspects that require monitoring and controlling. Human performance must be managed, conflicts resolved, and team members supported.

All project elements from beginning to end must be implemented, monitored, and controlled. This includes keeping all project components synchronized as changes to the scope, schedule, budget, resources, and deliverables occur.

The following list describes some of the critical jobs that must be done during a project's **Execution Phase** and associated deliverables:

Phase	Description/Purpose	Deliverables/Tools
Do Implement, Monitor, and Control Project	• Compare actual progress to planned progress • Assess and report status • Coordinate activities and resources • Implement corrective actions and update estimates and project schedule • Manage and support individual and team activities • Manage project priorities • Manage and resolve problems and conflicts • Manage changes to project scope • Assess quality of deliverables • Communicate work assignments, progress, and issues • Evaluate performance of individuals and teams	• Updated Project Management Plan Activities • Updated Risk Management Plan • Status Reports • Feedback Forms • Change Order Log • Issues Log • Interim Performance Reviews

The Critical First 10 Percent of a Project

It may seem counterintuitive, but project managers must begin the execution phase by focusing on the first 10 percent of the project, which is the time when the following planning work occurs:

- Developing the project management plan, which is also referred to as the project execution plan or simply the project plan
- Establishing design criteria
- Setting the kick-off meeting
- Defining the initial monitoring and control parameters

The design criteria and the kick-off meeting are integral parts of the Project Management Plan, but worthy of separate consideration. With comprehensive and detailed initial planning, followed by communication of the project plan and initial monitoring to ensure everything is understood and is being implemented, project success should follow, assuming the proper monitoring and control is maintained.

However, project managers sometimes neglect to monitor and control the first 10 percent of a project when they begin work on the execution phase. They mistakenly assume that everyone understands the project plan and that everyone will execute the project according to the plan. But the opposite is often true. Some team members, for example, may go off on a tangent, get ahead of themselves and the schedule, or spend too much time and detail on conceptual items. Project managers must monitor and control all project work, and this includes ensuring that the team understands the plan.

Tips to Execute a Successful Project

The following are some tips for successfully executing a project:

- Plan, and always think ahead
- Standardize
- Communicate
- Get it right the first time
- Monitor and control
- Establish a change control process

Plan, and also think ahead. Planning is critical, but at the same time, you must always think ahead and look for surprises. Did the department just get another big project that may affect your project's resources? What can be done to improve efficiency? Is someone, perhaps a team member of a stakeholder, not buying into the project? How will an upcoming holiday or the illness of a team member impact the work?

Standardize. Defined repeatable processes are one of the best ways to bring order to chaos. A simple example is using a template to create a spreadsheet as opposed to starting from scratch and having to research ideas and procedures. The project manager has more to do than to reinvent the wheel. Standardization is an important technique you can use to reduce errors and increase efficiency. Standardization also frees the project manager to think ahead.

Get it right the first time. How often does your project need rework? If you are like many project managers, you may think it is too often. Look for the reasons this is happening and ask hard questions. For example, could the need to reword parts of the project relate to the lack of clear instructions about what needs to be done?

Monitor and control. Does having to rework part of a project relate to a lack of early and ongoing monitoring before the damage is done? This often happens when a project manager thinks they are "too busy" to manage. Remember, monitoring is management.

Change control. Does rework relate to the project manager's propensity to have it done their way even if it was not incorrect in the first place? The project manager must avoid indulging in personal whims. Nothing frustrates and demoralizes team members more than to have what they have done ripped apart and done over, and in some cases, done over again and again, just to satisfy the project manager's personal preference. Think about a recent report draft, layout, design, or drawing. How many times has it been reworked? Why was it reworked? This point is worthy of being emphasized again and again because it is so vital to project success.

What to do:

- ☐ For a project you worked on, did the team follow the project plan at the start of the execution phase?
- ☐ Revisit the tips to execute a successful project. Do you agree that monitoring is management?
- ☐ Do you often ask people to re-do work? If so, what could you do to help avoid that problem?

Other actions:

- ☐ ______________________________
- ☐ ______________________________
- ☐ ______________________________

Remember

- ✓ Project integration is not easy and can be time-consuming. Make it worthwhile.
- ✓ Project integration is an iterative process. If the scope changes, other aspects of the project will also change and must be accounted for.
- ✓ The impact of project changes can ripple throughout various parts of the project.

Enhance Your Learning

Watch the following 6-minute video by Dr. Mike Clayton to learn more about project integration.

Clayton, M. (2019). *Project Management in Under 5: What is Project Integration?*

Available at: hhttps://www.youtube.com/watch?v=cHUP61KU3xE

Reinforce Your Learning

New Home: Project Management Plan Updates. Recall the New Home project from earlier. Consider how each of the following project management plan areas will be impacted by your change of adding a garage after the project had already started.

Scope:

Human Resources:

Schedule:

Stakeholder Communication:

Cost and Procurement:

Quality and Risk:

Review the Components of a Project Plan

According to the *PMBOK® Guide, a* ***project management plan*** *or* ***project plan*** is a document that describes how a project will be executed, monitored, controlled, and closed.

In this concept, we will review some foundational topics having to do with the project plan: knowledge areas, changes in a project, and team performance. In later concepts, we will cover these topics in more detail as they relate to monitoring and controlling a project.

Project Management Knowledge Areas

A project management plan typically integrates all the project's component plans. PMI refers to the components as **project management knowledge areas**, as illustrated below.

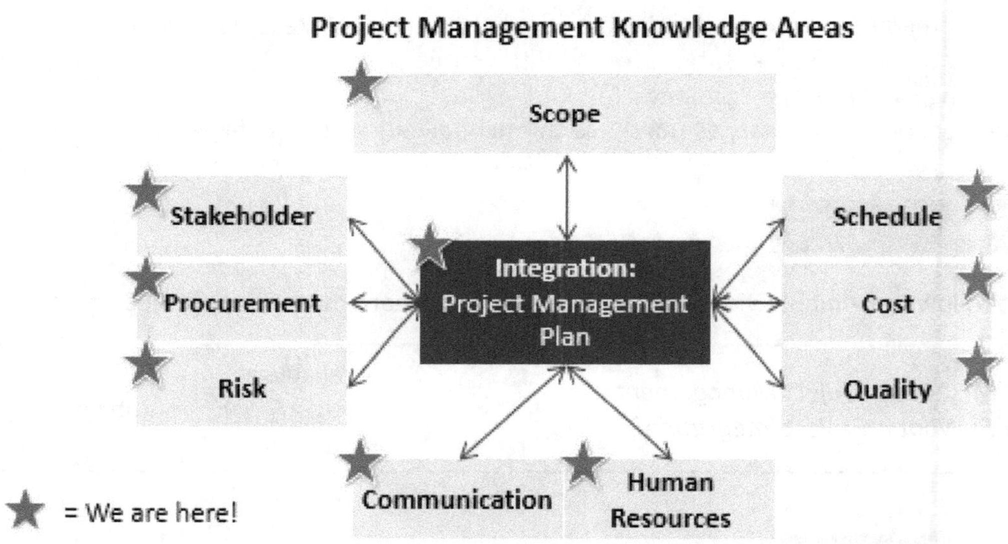

The project management plan is a structure for integrating the components and a guide for executing the project's activities. The purpose of the plan is to describe how a successful project outcome will be achieved, who will be involved in the project, and how the project will be measured and communicated. As a project manager, you must plan what you will do, how you will achieve it, and the people who will be involved.

Project management knowledge areas are described more fully in the *PMBOK® Guide.* For smaller projects, detailed plans for each knowledge area or component may or may not be needed.

Project Changes

Once a project plan is approved, the project can go forward. But as the project progresses, changes and issues will occur. Project managers must track and document them.

The following **change request log** and **issues log** examples illustrate the types of information to track and document. You will also find worksheets for these logs in Appendices A and B.

Change Request Log Example

Request Date	Change Request Reference Number	Requestor Name	Requestor Contact Info.	Request Description	Why / Benefit	Est Cost Impact	Status (APP / REJ)	Status Date	User Accept Name	User Accept Date

Issues Log Example

Issue Ref. #	Issue Description	Priority (H, M, L)	Reported by	Assigned to	Status (open / closed)	Status Date	Comments

Changes to the Original Project Plan Baseline

As issues arise and changes are proposed, the project manager must prepare a revised scope and project management plan, compare them to the original scope and plan, and communicate the differences to the client/stakeholders. This is key in controlling project change.

The purpose is to highlight the differences between the original baseline and what would become the revised baseline if the changes/issues are approved. In other words, baseline revisions show the impact of change on the original schedule, cost, and scope. Seeing the difference, the client/stakeholders will understand the potential effects the changes/issues would have on the project. They can weigh the costs of the proposed changes (in the schedule, cost, or scope) against the benefits, and make an informed decision to approve the changes or not. If needed, they can do a cost-benefit analysis to help them reach a decision.

Project Team Performance

Human performance is another crucial aspect of the project management plan. Project managers sometimes have limited opportunities in terms of deciding who will participate on a project team. Further, the project manager may not be familiar with the team's skills and any previous performance issues.

Performance reviews from previous projects can help determine if any team member issues need to be monitored and controlled in the current project. If any team members faced challenges in the past, the project manager must work to understand the problem, monitor current performance, and proactively take corrective action as needed.

The corrective action could be as simple as having a one-on-one meeting with a team member considered at risk. The project manager may learn, for example, that a person is experiencing difficulty because they are relatively new to their position or the organization, or they may be dealing with technology new to them, or they may have a conflict with someone on the team.

Once the project manager understands the source of the problem, the manager must develop remedies, such as training and coaching. Following that, the project manager must monitor performance to ensure tasks are completed as scheduled.

What to do:	Other actions:
☐ Document and communicate updates to your plan: why, who, what, when, and how.	☐ ____________
☐ Review your baseline project management plan and set up triggers to help monitor your project to ensure it is going as planned.	☐ ____________
☐ Control your project by proactively taking corrective measures to get back on track.	☐ ____________

Remember

- ✓ Recognize that a project rarely, if ever, goes precisely according to the original plan.
- ✓ The published project schedule is the baseline. It is what you will be measured against, and it should be considered contractual.
- ✓ Change controls are the control mechanism to bridge the gap between the original baseline schedule and the actual schedule.
- ✓ When the schedule changes, all associated project documents must be updated.
- ✓ The project manager must respond quickly to issues and risks as they arise in the project and report status promptly.
- ✓ You can set up automated reminders or triggers to alert stakeholders of project changes.
- ✓ Performance shortfalls or problems in the team must be addressed as soon as possible after they occur.

Enhance Your Learning

Watch the following 6-minute video by ProjectManager to learn more about how to successfully execute a project plan.

Bridges, J. (2019). *How to Successfully Execute a Plan.*	Available at: https://www.youtube.com/watch?v=802yQd8TNf8&t=1s

Reinforce Your Learning

Baseline Comparisons: Making Copies. Below is the original budget (schedule and costs baseline) for our previously discussed Making Copies project.

Budget (baseline):

Activity	Estimated Duration (Sec.)	Comment	Resource
Gather hard documents to copy	35		You
Walk from desk to copier	105		You
Make 10 copies	71		You
Walk from copier to desk	90		You
Distribute the hard copies	475		You
File personal hard copy	53		You
Total Projected Schedule	829		
Minutes	13.82		
BUD Summary:			
Labor:	5.54		
Material:	0.10		
Equipment:	0.05		
Total Projected Cost:	$ 5.69		

The illustration below shows detail for the labor, material, and equipment costs.

Supporting Calculations:

40	Hours per weekend worked	**Labor**	
52	Weeks per year paid	13.82	Minutes of Labor
2,080	Hours per year paid	0.40	Per minute labor rate
50,000	Annual Salary	5.54	Labor costs
24.04	Rate per hour		
0.40	Rate per minute	**Material**	
		0.01	Cost per piece of paper
		10	Number of copies
		0.10	Cost of material
		Equipment	
		0.005	Copy / click cost
		10	Number of copies
		0.05	Cost of equipment

Project Changes. Some budget variations occurred during the execution of the copying activities.

First, on the way to the copier, you were stopped in the hall about an unrelated issue, which added to the walk time to the copier. Second, while you were at the copier, it ran out of paper, and then the new paper got jammed.

Actual:			
Activity	**Estimated Duration (Sec.)**	**Comment**	**Resource**
Gather hard documents to copy	35		Me
Walk from desk to copier	105		Me
Caught in the hallway	300	(5 minutes)	Us
Make 10 copies	71		Me
Out of paper	60		Me
Paper jam (5 pages lost)	120	(2 minutes)	Me
Copier guru helped	120	(2 minutes)	Joe
Walk from copier to desk	90		Me
Distribute the hard copies	475		Me
File personal hard copy	53		Me
Total Projected Schedule	1,429		
Minutes	23.82		
ACT Summary:			
Labor:	9.54		
Material:	0.15		
Equipment:	0.08		
Total Projected Cost:	$ 9.77		

Actual activities

So, the result was that you spent additional time because of loading paper and clearing the paper jam.

To clear the paper, you received help from a resource not included in the baseline. Time for this person (we assume the same salary as yours) was added.

Also, neither the paper used nor the additional cycles (clicks) on the copier were in the baseline, so we added them to the actual schedule and costs.

The illustration that follows shows the detail for how the help received and additional clicks were calculated.

Actual: Supporting Calculations:			
40	Hours per weekend worked	**Labor**	
52	Weeks per year paid	23.82	Minutes of Labor
2,080	Hours per year paid	0.40	Per minute labor rate
50,000	Annual Salary	9.54	Labor costs
24.04	Rate per hour		
0.40	Rate per minute	**Material**	
		0.01	Cost per piece of paper
		15	Number of copies
		0.10	Cost of material
		Equipment	
		0.005	Copy / click cost
		15	Number of copies
		0.075	Cost of equipment

Actual Supporting Calculations

The data for the calculations have been provided, so the impact of the variations on the performance results is visible. In the world of project management, this would be considered a failed project. The project failed on two of the three constraints: The schedule was elongated, and the costs were over budget.

As you can see, the impacts are significant, considering the minimal timeframe in the example. More complex issues have the same types of considerations, and the significance of the increased time/schedule and costs have a higher impact proportionally.

Discussion

This example is for a small project. There was no time to monitor and control the Making Copies project, but the concepts of having a baseline and actual data collection along the way still apply. With a project schedule in days instead of in seconds, you can see how the monitoring and control would be accomplished if similar unanticipated activities were to occur. Corrective actions could then be activated.

Activities and attributes (exhaustive list, resources, durations, relationships, and costs) **are the core foundation of the schedule.** Anything that impacts any one of these could influence the delivery of the project's product, service, or result.

Humans are eternal optimists; we generally assume a best-case scenario. For life in general, this may work well; for projects, not so much.

When estimating, you must consider any risks that could challenge the duration you are projecting. You must also be comfortable with what you are being asked to schedule.

Anything new represents a risk to a project. A new technology, a new project team, a new external entity—anything unfamiliar adds risk. Be sure to consider all the "new" in your confidence factor and provide adequate reserves (buffers) for each.

The world of projects is subject to the domino effect. When one project shares a resource with a second project, and the first project starts to go off schedule, it typically will have a detrimental impact on the second project. Now the resource is under pressure to perform more (spend more time/effort) on the first project without slowing down the second project. This situation often impacts performance on both projects.

Activity:
Reflect on this Making Copies project example, then consider a project that you are familiar with that experienced variations from the plan. Briefly describe below what was done or should have been done and list your conclusions below:

Monitoring and Controlling Projects #4.4

Monitor and Control Scope Creep

Recall, the **project scope** describes what is included in the project and what is not included. The **scope identifies the deliverable(s)**, which could be a product, service, or result. It will not exist until the project is completed.

The scope drives the activities list, risks, resources, costs, and schedule. Because of the potential impact of a scope change to the project plan, as the project progresses, the project manager must monitor, control, and verify all activities for variation.

The **scope** also dictates **what must be done** to produce the project's product, service, or result. As scope variations occur, they are bound to impact the schedule, costs, risks, human resources, and potentially all aspects of the project baseline (original schedule and resources).

It is imperative that the client/stakeholders clearly understand the scope of the project because they are the people who must review and approve changes to the scope. As we have discussed, once any changes are reviewed and authorized, the project can continue, and the project manager will then monitor and control project work as per a revised plan.

Uncontrolled Project Changes

PMI's The *Pulse of the Profession®: Capturing the Value of Project Management: 2015* included survey results for this question: In your estimation, what **percentage of the projects completed** within your organization in the past 12 months **experienced scope creep** or **uncontrolled changes** to the project's scope? The answer was **44 percent**! You can read more about the *Pulse of the Profession* at www.pmi.org/pulse

Uncontrolled changes that impact the scope can occur at any point after a project begins. Uncontrolled changes are often referred to as **scope creep**, **scope leap**, or **gold plating**. All these terms refer to scope changes after the original scope baseline has been established. They often occur when the scope of a project is not adequately defined, documented, and controlled.

Gold plating means intentionally adding extra features or functions to the project that were not included in the original scope statement. Gold plating is common and typically done by team members to show their abilities, or by the project manager to please the client/stakeholders.

How to Manage Changes in Project Baselines

The **scope baseline** is the **original scope** that subsequent project plans use as input. As part of the development of a project communication management plan, the scope baseline is used to determine who needs to be included in project communications. **If the scope changes** via a scope change control, the **communication plan** should be updated accordingly. Additionally, **corrective action** should be implemented to bring future project performance into line with the project plan.

For reference purposes, we will identify the original scope baseline as scope baseline 1. Then as changes occur, we will refer to them as scope baseline 2, scope baseline 3, and so on.

What determines whether a new scope baseline must be established depends on the organization and the project. Guidelines can detail how scope baseline levels progress from one to another. For example, a guideline could require a new baseline if the scope is changed by x additional items added, the costs associated increase by x %, or the schedule is padded by x days. This can be taken a step further by requiring an additional authorization level when any such threshold is reached.

Review the illustration below. It begins by listing the **project's original scope**, which is **scope baseline** 1. Scope changes cause a 5 percent increase in project cost. Now, per the thresholds, we must have a VP-level signature to authorize the scope changes and subsequent increase in cost. With that signature, we are now at **scope baseline 2**. The threshold and additional signature requirement call attention to what is deemed critical. In this example, the cost reaching a threshold was the trigger for an additional review of the project.

Scope Baseline Thresholds			
Scope Baseline Reference Number	Cost Increase %	Schedule Increase %	Required Approval Signatures
1	Original	Original	Sponsor
2	≤5	≤2	VP
3	>5 and ≤10	>2 and ≤5	Senior VP
4	>10	>5	President

Again, the particular thresholds and threshold levels will vary from organization to organization, and possibly from one project to another within an organization. However, the concept of scope monitoring and controlling is consistent from the standpoint of establishing thresholds and having mechanisms in place to enforce them. You should set thresholds at levels where timely corrective action can be taken.

Scope Management Plan and the Scope Verification Process

For large complex projects, the **project management plan** should include a **scope management plan**. It outlines how the original scope was determined, how it can be changed, and how scope monitoring and controlling activities will be conducted.

The **scope verification process** provides a mechanism for establishing and documenting the level and extent of completion should unexpected events occur. One example would be the termination of a project due to loss of funding.

Scope verification also involves reviewing work products and results to ensure that all were completed satisfactorily and formally accepted. A common way to validate scope changes is to do an **inspection**.

Requirements Traceability Matrix

For some projects, the scope may change because the project **requirements** change or evolve. This is often the case with software projects or the innovative design of new products and services. Some circumstances may merit the **creation of prototypes** to establish requirements.

In these situations, it is beneficial to develop a **requirements traceability matrix** for use in the monitoring and control efforts. It will help you ensure that all requirements have been satisfied.

The matrix also helps to provide verification of the project deliverables. The matrix documents signoff of the requirements associated with the deliverables (if they are not the deliverables themselves).

The following is an example of a requirements traceability matrix. This is just one format for this type of matrix. The exact information to include depends on the needs of the project and the organization.

The format may also vary according to the complexity of the project.

Requirements Traceability Matrix Example

Requirements Traceability Matrix						
Project Name:				Company:		
Project Manager:				Department		
Project Sponsor:						
Requirement Identification Number	Requirement Description	Requirement Document Reference	Test Case Reference	User Signoff	Signoff Date	Comments

Monitoring and Controlling Change

Change requests and unexpected **project issues** are a natural part of the process for monitoring and controlling scope. The project management plan and the change management plan address how changes are to be managed. This includes what type of changes are acceptable and who has the authority to accept or reject proposed changes. This may include authorization levels, depending on how the cost and schedule are impacted by the change. All this activity should be captured in the change request log and the issues log.

The **change request log**, which we described earlier, should include all change requests, whether eventually approved or not. This log can be used to determine how much change request activity is currently going on to project how much more change activity will be needed. Project issues may or may not rise to the level of making formal changes. As we have discussed, the consequences of the changes (additional scope, cost, and schedule) must be clearly understood by those approving the changes.

The people who **approve changes** must understand there are tradeoffs between allowing the changes and the impact of the changes on scope, cost, and schedule. If a higher-than-anticipated number of change requests keep coming, it may be time to add reserves for the activity and re-baseline the project accordingly.

Applicable **work performance data** should be updated according to the results of inspection and validation activities. This should include both human performance and enterprise resources involved with the project.

Updated project documents include any portion of the project that is impacted by the scope monitoring and controlling activities. This could include any documents associated with integration, scope, schedule, cost, resources, quality, communications, risk, procurement, or stakeholders.

What to do:	Other actions:
☐ Review the requirements for your project and validate them to this point in the project.	☐ ______________
☐ Review the project change request log to ensure appropriate actions have been taken.	☐ ______________
☐ Review the scope baseline thresholds to ensure compliance. Create thresholds if they do not exist.	☐ ______________

Remember

✓ Requirements are one of the best resources for monitoring and controlling a project's scope.

✓ Inspection is an effective way to validate scope.

✓ Scope baseline thresholds are a way to establish warnings.

Enhance Your Learning

View the following 8-minute video by RiskDoctorVideo about scope creep: *The Wasa–a true story of scope creep.*

RiskDoctorVideo. (2015). *The Wasa - a true story of scope creep.*	Available at: https://www.youtube.com/watch?v=kmJ59yyYza4

Reinforce Your Learning

Scope Creep. Consider a project where you experienced or observed scope creep. With the project in mind, respond to the questions below.

Project name: ______________________________

Did the scope creep affect the cost of the project? ______________________________

Did the time to do any tasks change? If so, did the project schedule change?

Did the scope creep affect the quality of the project's product, service, or result?

Control Resources

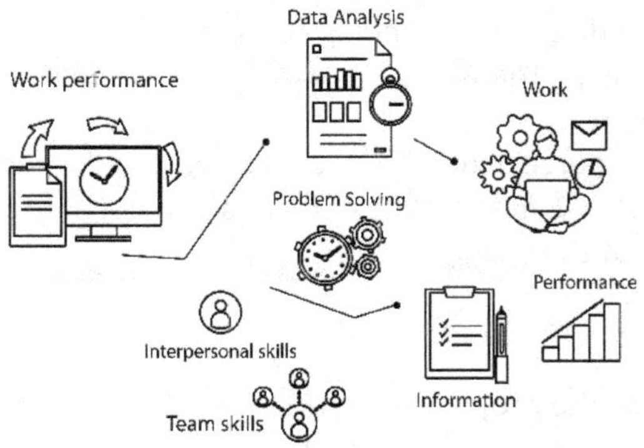

Monitor and Control Project Resources

Everything in business is governed by money. Even the most amazing project must stay within budget, be cost-effective, and justify expenditures. Knowing how to plan, track, and explain project costs are necessary skills for project managers because every project manager, at some point, will need to develop, approve, and communicate project costs and budgets.

The project sponsor/client must approve a project budget before the project starts, and then, as the project progresses, the project manager must keep the project within its budget. If a budget issue develops, the project manager must understand the cause and explain the problem.

We will go into more detail about project costs and budgeting in a later course. In this concept, we will discuss monitoring and controlling project resources because resources are significant costs in any project.

Resources must be available on time for a project to be completed successfully. Thus, planning for when resources will be needed is a crucial aspect of running a successful project.

Once the initial planning is done and the project progresses, the project manager must ensure that resources are allocated as planned.

Monitoring activities provide insight into when resource availability may be in question. Controlling activities facilitate the corrective actions needed to deal with problems.

We will examine **resources** from two perspectives: **human** (people) and **enterprise** (other than people).

Human Resource Availability

The **human resource management plan** is part of the project management plan, or at least has some elements within it. It addresses how human resource monitoring and controlling will be done.

Project resource assignments and **resource calendars** can be referenced and shared among projects.

Resource monitoring ensures that the intended allocation of resources occurs and that the project stays on schedule. For example, if a person with a specialized skill such as a nurse is allocated 50 percent to your project, but is only working 25 percent of that time, your project may fall behind schedule because the tasks will take twice as long to complete. Further, if the tasks are predecessors, the successor tasks will also push to later dates.

Besides affecting the schedule, a problem like this could affect the resource calendar, costs, and scope. The resource calendar could change not only because of changes in specific task assignments but also because of the associated predecessor or successor tasks, and that would impact future resource allocations. Costs may change because of the delay in tasks being completed. In this example, the resource allocation change would not change the scope.

Alternatives to solving the problem of a change in a resource allocation should be reviewed, and the project manager should measure the trade-offs for each. Whatever the decisions, the updated project documents need to capture the ramifications of the change.

Enterprise Resource Availability

Enterprise resources must also be available on time for the project to be successful. For example, if a specialized piece of equipment is needed to run samples for a project, and the samples are the predecessor for a lengthy analysis process, then delays in the samples will delay the analysis and potentially the entire project schedule.

By monitoring and controlling such equipment resources, a project manager would likely discover the problem and have time to find a solution before the project schedule was affected.

How to Monitor and Control Project Resources

Project performance reports track how a project is performing against its intended schedule, scope, and cost baselines. For example, if a resource spends 50 percent of their time on the project as allocated, but all their tasks take longer than anticipated, it would be prudent to re-estimate the remaining tasks. This would be considered a re-baseline of the schedule because of longer times than initially planned to complete tasks.

Observation and discussion can be effective ways of determining how resources are being used. This can include interactions in project meetings, but it may also include a less formal approach of chatting with the project team and stakeholders about how things are going, both the good and the not-so-good. An informal interaction provides an opportunity for someone to discuss an issue they feel uncomfortable talking about in front of a group.

Conflict management is a necessary part of all projects. For example, problems will occur if various team members do not agree on how to accomplish a task or do not agree with the project plan. A related issue is that a team member could do something in a way that causes others to spend more time than projected, which could lead to a problem in the schedule.

As a project manager, you must identify the source of conflict and find a solution that not only keeps the project on schedule, but that is agreeable to the people involved. Brainstorming is an excellent way to moderate conflict and encourage the team to reach a consensus on the best solution to an issue.

Conflict can occur between two individuals, such as when one person does not agree with another, or among several individuals. The symptoms of conflict can take many forms; however, **poor communication** is often the root cause.

Conflict often involves content, process, or interpersonal issues. **Content conflicts** are based on the **subject matter** of team discussions. For example, people on the team may have differences about an idea discussed in a team meeting. **Process conflicts** occur when people disagree about **how** the team should accomplish a task. For example, some people may want to do a job a certain way while others do not. **Interpersonal conflicts** are based on **relationships** between people. For example, one team member may dislike another person's communication style.

Research shows that people approach conflict in various ways. Kenneth Thomas and Ralph Kilmann, well-known for their work in conflict resolution, have observed that people tend to respond to conflict with **assertiveness**, where they strive to satisfy their own concerns, or they respond with **cooperativeness**, where they endeavor to satisfy the concerns of others.

People also tend toward particular styles of assertiveness or cooperativeness. The five basic styles, according to Thomas and Kilmann, are:

- **Competing** (assertive and uncooperative). People using this style take a firm stance on getting their needs met, and often draw on power, persuasion, or rank.
- **Accommodating** (unassertive and cooperative). People using this style sacrifice their own concerns to satisfy the concerns of others.
- **Compromising** (between competing and accommodating). People using this style look for the middle ground. They address issues more directly than with *avoiding,* but not to the extent as with *collaborating*. Concessions are often called for.
- **Avoiding** (unassertive and uncooperative). People using this style defer by circumventing, postponing, or withdrawing from tense situations. Nothing is gained from this approach.
- **Collaborating** (assertive and cooperative). People using this style try to bring all parties together to satisfy all concerns. These people are leaders who look at what each person needs and has to offer and attempt to end with a win-win for all. This problem-solving technique is the best way to resolve a conflict, but it is not always easy, especially when emotions run high.

To address conflicts more consistently across projects, organizations sometimes develop conflict management policies and procedures. This helps ensure that disputes are resolved productively. Different types of conflict may require different approaches.

Interpersonal skills are essential for all team members. The better the team can communicate within itself, the more effective the entire team will be. The project manager also must communicate effectively with stakeholders.

Project team members do not always report directly to the project manager. When they do not, the project manager must rely on persuasiveness and be creative in engaging the team and in helping them make the project a priority.

Coaching and training can significantly impact the success of a project. While traditional training can increase productivity by over 20 percent, training and coaching combined can result in nearly a **90 percent increase in productivity** (as cited in TCS, 2012). When you know how to coach for performance improvement, you help support more productive employees and a more profitable organization.

What to do:	Other actions:
☐ Monitor resources to ensure they are allocated according to the project plan. If not, how could the problem be resolved?	☐ ____________________
☐ Consider how much time you spend observing and discussing issues with others.	☐ ____________________
☐ Identify an interpersonal skill that someone you work with does well and compliment the person.	☐ ____________________

Remember

- ✓ Monitoring activities provide insight into when resource availability may be in question, and controlling activities facilitate the corrective actions needed to deal with problems.
- ✓ Being assertive and cooperative is the best way to resolve conflict, but it is not always easy, especially when emotions run high.
- ✓ Training can increase productivity over 20 percent, but training and coaching combined can result in nearly a 90 percent increase in productivity.

Enhance Your Learning

View the following six-minute video by ProjectManager about conflict resolution.

Whitt, J. (2013). *Conflict Resolution Training: How To Manage Team Conflict In Under 6 Minutes!*	Available at: https://www.youtube.com/watch?v=PHJ8eybXJdw

Reinforce Your Learning

The Project Team. A project team member who reports to you describes a project meeting where conflict arose because some people were unable to work the planned amount of time due to unexpected vacation scheduling. This led to an adjustment in the project's original schedule baseline. The meeting minutes noted "resource unavailability" as the cause.

What are your initial thoughts about this sequence of events?

What tools did we discuss that you might use in this situation?

What would you do?

Coaching Worksheet: Coaching Session. Assume that you plan on conducting a coaching session with the people whose unexpected vacation required adjusting the project's original schedule baseline. To prepare for the coaching session role play, complete the Coaching Worksheet below.

Coaching Worksheet

Coaching Situation: ____

Communication and Problem-Solving Steps:

1. Describe the problem in a non-confrontational way.

2. Ask for the team member's assistance in solving the problem.

3. Brainstorm causes of the problem or contributing factors.

4. Identify possible solutions.

5. Agree upon a specific action plan.

6. Set a follow-up meeting.

Identify Components of Risk Planning and Management

Risk is a significant element of any project and **can be found during all phases** of the project management life cycle. In the initial planning phase of a project, the original scope, cost, and schedule baseline plans describe the project risks and also the contingency plans to mitigate the impact on the project if any of the risks were to happen.

In the project's execution phase, the project manager must **monitor and control** the identified risks, while also being aware that new issues and risks will arise as changes to the project occur. **Unidentified risks** that arise often result in **workarounds**. Changes may impact the scope, cost, and schedule baselines, and project managers must not only evaluate how the changes affect the project, but also consider the changes from a risk perspective.

Risk Management Tools and Techniques

The **risk register**, **risk mitigation** and **contingency plan**, and the **risk breakdown structure** are three foundational references that project managers use to determine the current risk in a project. Risk assessment and risk audit are somewhat related, and we will address them together later in this concept. Project managers use these references to evaluate the original risks, as well as any new risks identified when project changes are introduced.

You can review a **risk register**, illustrated below, to determine if any expected risks have occurred. It highlights the most important risks according to ratings defined by the organization or project. In our example, "resources not available" is identified as a critical risk. Pay the most attention to critical risks. As part of monitoring and controlling risks, you can also assess the planning effort by reviewing the status during the execution phase.

Risk Register Example

Ref. #	Description	1. Probability	2. Impact	3. Probability Score	4. Impact Score	5. Severity Score	6. Severity Description
1	Resources not available.	H	H	5	5	25	Critical
2						0	
n						0	

Working with a risk mitigation and contingency plan, illustrated below, we can determine if the risk has occurred or not. Then we will need additional information, such as why the mitigation plan did not prevent the risk from occurring. Was it because the plan was ineffective, or did something about the risk change to make the mitigation plan less effective?

In this example, the project manager would start by determining whether the contracted resources were put on retainer as the mitigation plan requires. If so, then why has resource availability negatively impacted the schedule? Could it be that the retained resources did not have the necessary skills, or was the problem something else?

A secondary question would be whether or not the contingency plan worked. If the shortage of resources had to do with skills outside of the retained resources' capabilities, then there is no way to engage the contracted resources as stated in the contingency plan. The mitigation plan was flawed because it did not identify the right resources to be retained.

Risk Mitigation and Contingency Plan Example

Ref. #	Risk Description	Severity	Status	Mitigation Plan	Trigger	Contingency Plan
1	Resources not available.	Critical	Red	Have contracted resources retained.	Calendar review, resource unavailable.	Engage contracted resources.

This may also be a good lesson learned from the standpoint of breaking the types of resources into more finite groups, which in this example would have made the mitigation and contingency plans more useful. It may also have helped the project manager to better define the critical skills needed for the project.

The other references for **risk assessment** and **auditing** are the **change request log** and **issues log**. You can use these logs to discover whether any risks introduced because of an issue or change request have been added to the risk register and analyzed for how critical the risk is to the project. If the risk is determined to be critical, then appropriate mitigation and contingency plans need to be put into place.

For an audit, the same type of review, as described above, could be conducted at a later date to determine whether the risk did occur and whether the mitigation and contingency plans worked as expected.

Variance and trend analysis is a technique project managers use to help them track the number of risks that have occurred and their impact on the project. Risks can be recorded through the change request process by noting that a change request is being generated because of a risk not identified in the planning phase.

For example, the number of change requests caused by risks may be exorbitant if the number is more than a predetermined percentage of the total number of change requests. If the risk assessment had been done correctly, you would expect a small number of risk-related change requests within all change requests.

One reason for monitoring and controlling this information is to determine, as early as possible, whether the risk assessment is broken or flawed. If so, you may decide to do a full-blown reassessment of the risks associated with the activities not yet completed. You want to avoid continuous surprises from risk-related change requests throughout the rest of the project. Change requests are disruptive to the flow of the project; too many can generate uneasiness among the project team because the scope keeps changing.

Technical performance measurement is another technique for managing risk. It can take many forms. As an example, assume that one of a project's deliverables is a reduction in the average wait time of system users by 20 percent or more. In this case, the team might measure computer server processing rates, and there are a few other things we would need to do.

One is to capture user wait time before the project and use it as a benchmark. Then we must capture user wait time after the product is ready for testing. This is the only way to ensure that the basis of success is quantitatively derived using "before" and "after" data. Do not make this decision by asking qualitative questions like, "Do the users *think* the system is running faster or slower than before?"

Reserve analysis is a risk management technique that helps project managers assess whether the monetary reserves set aside for contingency purposes will suffice, given the recent history of the project. Consider an extension of the discussion above about risks causing more change requests than anticipated. One metric impacted by change requests is cost.

As more and more change requests are approved, the consequences of the changes may consume the reserves. Assume the cost reserve created was 15 percent of the total cost of the project. As each change request is added, its projected cost must be deducted from the reserve. It could have been predetermined that once the cost reserve reached a certain threshold, it would require additional levels of authorizing entities to approve.

What to do:	Other actions:
☐ Review the issues and change request log examples and consider how risks are tracked.	☐ ________
☐ Review the risks in your project and consider what tools you could use to improve risk management.	☐ ________
☐ Review your project's monetary reserves and consider recent project performance. Is the reserve likely to be adequate?	☐ ________

Remember

✓ The risk register, mitigation plan, and contingency plan are useful resources for risks monitoring and controlling activities.
✓ The issues and change request logs are useful resources for risk assessment and audit.
✓ It is helpful to have quantitatively supported project results whenever possible.

Enhance Your Learning

View the following 3-minute video by Simplilearn about identifying, monitoring and controlling risk.

Simplilearn. (2012). *Monitor and Control Risk / Project Risk Management.*	Available at: https://www.youtube.com/watch?v=TdWd_4glh_0

Reinforce Your Learning

Your New Home: Revisited. Recall the new home exercise from earlier where you are building a new home, and after the work had already started, you decided to add a garage to the project. In the previous exercise, you were asked to identify the part(s) of your project management plan that would be impacted by adding the garage.

Using your response(s) from that previous exercise, what **risk management tools** and **techniques** could you or the contractor use to effectively monitor and control risk in your project? Explain how you would use these resources.

Monitoring and Controlling Projects #4.7

Review Quality Monitoring and Control Tools

You either get better or worse. These words are as true in business as they are in sports. In today's business environments, work systems and processes have received a great deal of attention because they are central to judging and managing quality.

Quality is a variable and can only result when **variation** is **understood** and **controlled** as part of project management. **Quality assurance** relates to how a process is performed or how a product is made, **quality control** is more the inspection aspect of quality management.

Quality **cannot be inspected** into products and services. The "find it and fix it" approach to quality is ineffective.

How you monitor and control quality is primarily determined by the kind of processes in a project environment. In environments that *do not include* repetitive processes, quality monitoring focuses on **performance measurements** or the **metrics of the project,** including scope, schedule, and cost. In environments *that include repetitive processes*, additional quality monitoring and control are needed.

In this concept, we will discuss enterprise environmental factors and organizational processes to better understand what elements of quality must be included in a quality management plan. We will also review some tools and techniques project managers typically use to monitor and control quality. These are metrics, quality checklists, work performance data, change request logs, and issues logs.

Process Management

A major management strategy for quality is problem **prevention** through the application of statistical quality control and process management before and during an operation.

Process management is nothing more than good business management, and good business management involves problem management. Problem management entails:

- Organized identification of problems
- Formal improvement of quality and processes by solving problems
- Planned avoidance of problems

Baseline Performance and Entitlement Performance

Each function in an organization performs at some measurable level that is often referred to as baseline performance. Examples are unit cost, cycle time, and time to complete new designs.

Organizations make investments in equipment, software, designs, processes, and people. These investments entitle an organization to an "as designed" level of performance, most often substantially higher than baseline. In this sense, entitlement and optimization are synonymous.

The following are early pioneers in the quality/continuous improvement journey:

- Shewart: Statistical Process Control charts (SPC)
- Deming: Total Quality Management (TQM); Plan-Do-Check-Act
- Juran: fitness of use; 85 percent of quality problems are under management control
- Crosby: zero defects; quality is "free"
- Ishikawa: cause and effect diagram (fishbone); quality circles
- Taguchi: design of experiments
- Galvin, Gumpta, Harry, Schroeder, Zinkgraf (Motorola): Six-Sigma
- Ohno (Toyota): Lean; Total Production Systems (TPS)

Quality metrics vary from project to project. We will walk through a set of metrics associated with the Making Copies example you are familiar with. These metrics are related to the improvement of time to produce copies and the reduction of their cost. This is relevant in most environments because valuable resources can be wasted on something simple such as making copies or recovering documents sent to the printer.

Quality checklists are customized to a situation. The intent is to remove as much variation from an activity or process as possible.

For example, below is a sample checklist for making copies. The purpose is to ensure that everything at the beginning of the process is correct. This should eliminate some of the things that previously introduced variation to the process.

Making Copies Checklist Example

Copier Checklist:

___ Copier is clear and available.

  ___Previous job documents not present

___ Control panel is correct.

  ___Set to appropriate number of copies.

  ___Set to the correct paper size.

  ___Set to correct simplex/duplex setting.

  ___Green light appears for making copies.

Work performance data considers how frequently problems occur. In our making copies example, we collected data using a check sheet, as illustrated on the page that follows. It shows the most frequent copier problems, and how much work time is lost because of those problems. Examining this data may also reveal additional problems not initially identified in the planning process.

A **change request log**, as we discussed earlier in the course, is maintained to ensure that any ideas for improvement (change requests) are logged for follow-through and control. Part of the follow-through is in the form of the request actually being fulfilled. The other part is validating that what was done to satisfy the request actually satisfied the request.

Deliverables include anything to be delivered within the scope, or the change requests, of the project. Not only should a deliverable be put in place, but it should be done correctly. After the deliverable is in place, you should verify that the anticipated benefits are being realized. For example, as we progress toward optimization of the copier process, all improvements that are discovered and implemented will be verified to ensure the gains are realized and the anticipated benefits validated.

Tools and Techniques for Managing Quality

Tools and **techniques** for monitoring and controlling **quality** include:

- Five basic quality tools
- Statistical sampling
- Inspection
- Approved change request review

The following is a discussion of **five commonly used quality tools**: check sheet, pareto diagram, cause-and-effect diagram, affinity diagram, and process flow map.

Check sheets help you determine what types of problems, or defects, are occurring and how frequently. A check sheet that addresses the defect or problem of spending too many resources making hard-copy documents is illustrated below.

In conversation with stakeholders, the check sheet was put at the copier so users could record appropriate information while going about their daily work. This version of a check sheet includes the amount of time lost for each occurrence to determine how much of the organization's resources is affected by this process. The actual format of the check sheet is determined by the situation.

The check sheet example below is the result of a project deciding to record issues, or defects, types, minutes lost, and the number of occurrences.

Making Copies Check Sheet Example for Recording Issues (Defects)

Copier Checksheet							
Defect Types	Time Lost (minutes)	Mon	Tue	Wed	Thu	Fri	Total
Printer is down.	100	1	2		2		5
Printer jam, fixed.	20	2		2		1	5
Printer jam, someone else fixed.	30		1	1	1		3
No paper, had to load.	85	2	4	4	5	2	17
No paper, none available	50			1		1	2
Ink, had to load.	25	1					1
Ink, none available.	60			1			1
Total	370	6	7	9	8	4	34

Please record time lost for each occurrence.

Using this information, we can produce different versions of a **Pareto Chart**. The first version we will look at is the number of occurrences with each defect type. This enables the team to focus on defects that occur most frequently. Addressing the most frequent occurrences will typically give you the biggest bang for your buck.

The illustration below is a Pareto Chart for the most frequently occurring defect.

As you can see, Load Paper is the most reoccurring defect by far. It is understandable why this activity would occur regularly.

There could be some time spent to find a more efficient method of loading the paper, especially if a new method would reduce the chances of a paper jam.

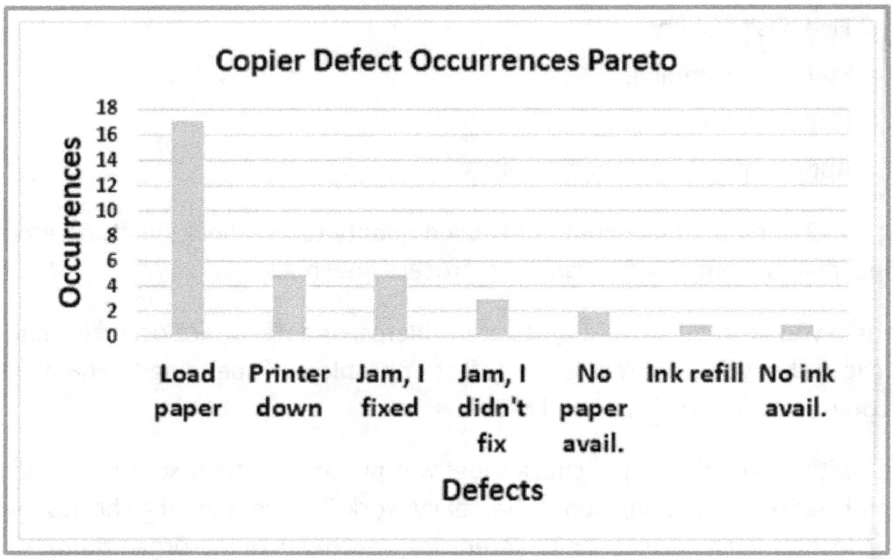

If we were to look at the data differently because **the problem is the amount of time and money put into producing hard copy**, we can take the same data and **create a pareto chart for the time lost in dealing with the defect.** We now have a new number one defect, which is Printer down.

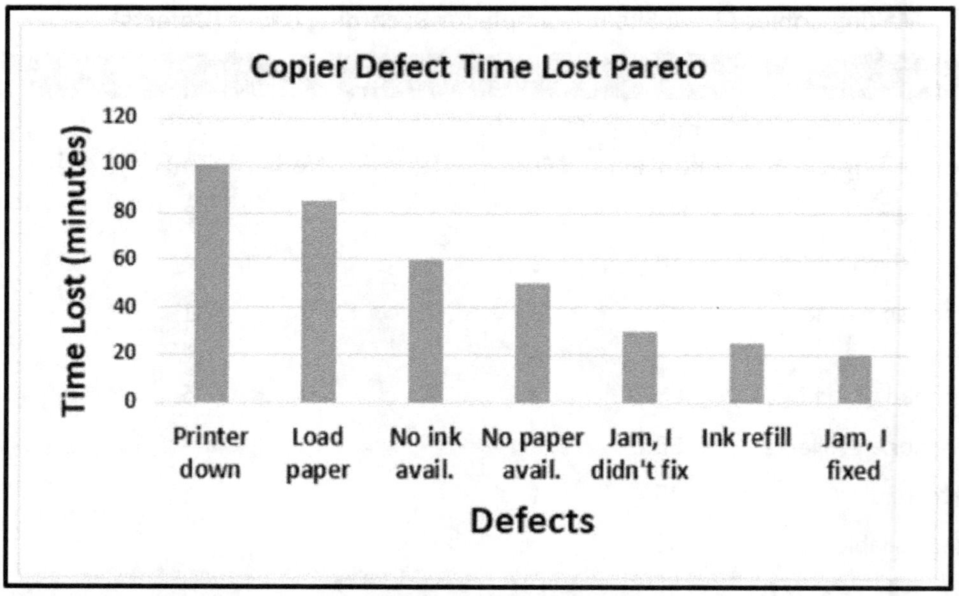

In our copier scenario, the project team had already created a **cause-and-effect diagram (**also known as an **Ishikawa** or **Fishbone** diagram**),** shown on the next page, during the planning phase of the project. We can use that diagram to validate that all root causes have been identified and decide how we might best take advantage of that prior work.

Copies Cause-and-Effect Diagram Example

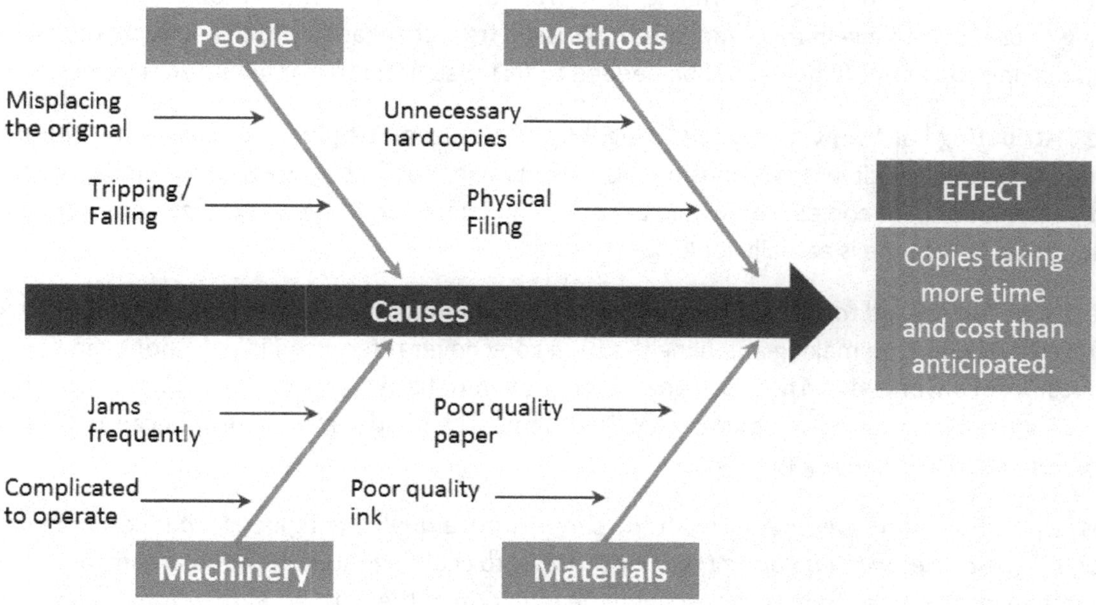

Understanding what connects activities or the assumptions behind some activities can be helpful. These are methods (procedures people follow), as opposed to machinery concerns.

In our example, we will consider copying activities that result in unnecessary hard copies and physical filing root causes. We will use an **affinity diagram** to think through some considerations associated with these activities. The affinity diagram is a method which can help you gather information and data and organize them into groups or themes based on their relationships.

We will start by identifying the related activities of making a hard copy, filing a hard copy, and distributing a hard copy, as illustrated in the affinity diagram that follows.

Affinity Diagram Example

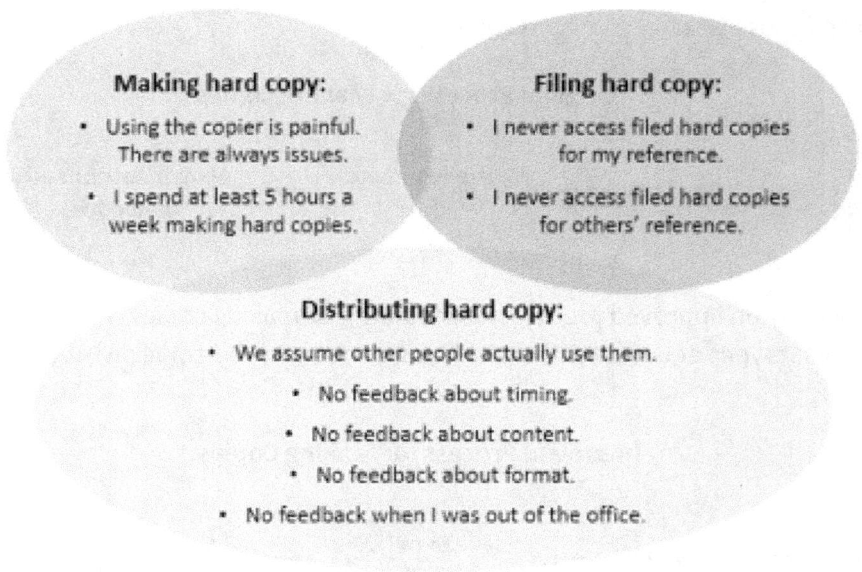

The **making hard copies** discussion brought out a couple of thoughts. First, using a copier is typically not a user-friendly experience. It often leads to frustration because of the problems encountered in trying to use it. We have initially addressed this issue from the standpoint of analyzing the problems with making copies and what would be needed to fix those problems or minimize their occurrence.

The **distributing hard copy** discussion brought up a few ideas. The person doing the distributing was just assuming that recipients were using them—but never got feedback about the timing, content, or the format of hard copies. You would expect compliments or, at the very least, complaints, unless, maybe, no one is actually looking at them . . . .

The discussion related to **filing hard copy** led to the idea that there was no apparent need for future reference. The person making and filing the hard copy never referenced it later, not even by request from anyone else. There are environments where hard copy must be retained for some time. This project was not one of those environments. The filing was done only because it was a common practice in the organization.

This line of discussion may seem almost too simple from a quality perspective. It is common for one problem *(resources wasted during the copy process)* to challenge something upstream, or downstream, that may dramatically change the direction of the path forward. It is also common for an outside pair of eyes with no history of how things have always been done to ask questions that may stir insights from those with a working knowledge of the process. This can open doors for changes that improve the process.

The checklist discussed earlier addressed the machinery portion of the root causes. The methods merit more discussion because they are a procedural issue. If the team is making unnecessary hard copies that represent a significant amount of volume for the copier, we want to address this issue first because resolving it could also resolve the issue of hard copies taking unnecessary resources. This could also spill over into the physical filing issue because if hard copies are of no value, then there is no value in filing them.

Process flow maps are an interesting way to represent current activities and also what the flow could be in the future.

Below is the original process *(current state)* for the making copies process. It considers the person doing the job and those who receive hard copies.

Original Process for Making Copies

Below is the agreed-on improved process (future state) for making copies. The discussion included guidelines on what types of documents should be scanned and the email distribution.

Improved Process for Making Copies

We have reviewed some tools used for quality monitoring and control. Our example describes how working through a process using these tools can change the way things are done. The results of using one tool may completely negate the results of another tool. In our example, the Copier Defect Check Sheet and resulting pareto charts were negated by the determination that the volume of hard copy being produced was unnecessary. That decision alone will significantly diminish the volume of hard copy and the resources needed to handle them. Since resource consumption was the "effect," the results of our exercise are successful.

Statistical sampling and inspection are tools that can be helpful for processes that create a large number of outputs, such as an assembly line. Several methods of sampling the outputs every so often and inspecting them for defects are discussed in common Six-Sigma and other statistics textbooks.

What to do:	Other actions:
☐ Review a cause-and-effect diagram for your project and validate it.	☐ ________________
☐ Review the quality management plan for specifics of monitoring and controlling quality.	☐ ________________
☐ Review a process flow for your project for potential improvements.	☐ ________________

Remember

✓ Quality considers aspects of the project and the organization's processes.
✓ Planning documents are instrumental for use in monitoring and controlling quality.
✓ Quality monitoring and controlling tools need to be customized to the situation.

Enhance Your Learning

Watch the following 3-minute video by the Arizona Western Region Public Health Training Center to learn more about affinity diagrams.

Andrade, R. (2013). *Affinity Diagrams.*	Available at: https://www.youtube.com/watch?v=jvTSsJrDZec

Reinforce Your Learning

Affinity Diagram. Pick a complex issue such as the "**Construction of a House**" or another project of your choice. Start by identifying two to four processes or ideas, as indicated in the affinity diagram below. Then ask the team to brainstorm ideas associated with those processes or ideas. The rationale of the ideas is not an issue. It is enough to just gather all the ideas. Once everyone has time to contribute, organize the ideas into groups. The groups usually become apparent, but you may need to make a few judgment calls.

This exercise can be done using the template below. Or you can use Post-it-notes to capture the team's ideas and then move them into the appropriate affinity groups on the diagram that follows.

Affinity Diagram Exercise

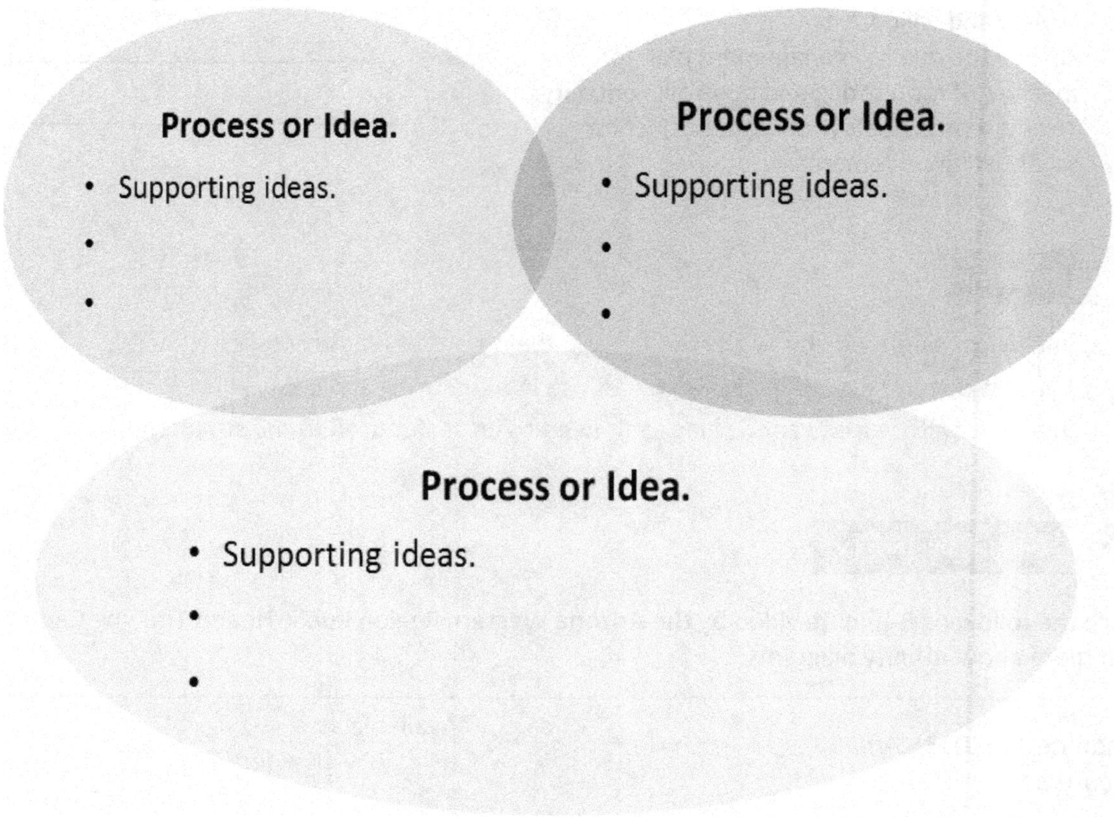

Once all the ideas are on the diagram, review the quality management plan and the project management plan to determine whether everything in the affinity diagram has been accounted for/addressed. If not, recommend appropriate adjustments.

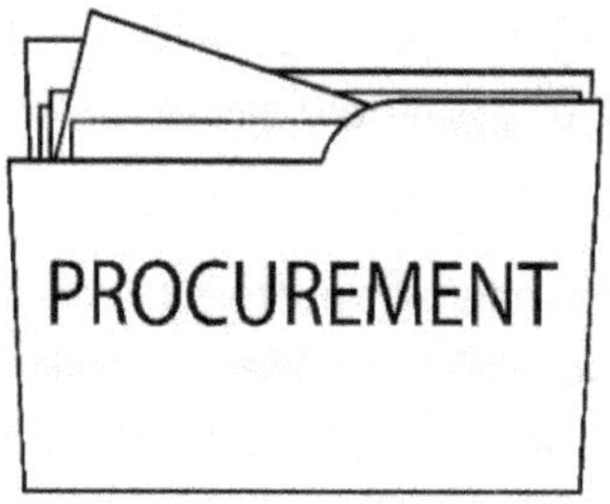

Monitoring and Controlling Projects #4.8

Monitor and Control the Procurement Process

Procurement refers to the process of obtaining goods and services, and so it is part of project management. Project managers usually work with others in an organization, such as the procurement, contracting, or legal departments, to obtain the goods and services needed for a project. At a minimum, project managers must know how to negotiate and manage contracts.

Project managers are also responsible for putting in place monitoring and controlling mechanisms to ensure reliable and effective procurement practices during the life of the project, and to see that goods and services are obtained as specified in the project plan.

In this concept, we discuss ways to keep procurement performance in line with projections.

Tools and Techniques for Procurement Monitoring and Control

The following are some of the commonly used tools and techniques for procurement monitoring and control:

- Contract change control system
- Procurement performance reviews
- Inspections and audits
- Performance reporting
- Payment systems
- Claims administration
- Record management system

A contract and change control system is often needed on large complex projects and is typically included in the change control management plan. Depending on the size of the project, this could be an additional section, or particular mention, for each contract, and it could be general guidelines for contract changes.

Each contract can have specific guidelines that justify a separate paragraph or special mention. For example, if the procurement has to do with the delivery of materials to be used in the production of goods, the guidelines may be different from a procurement contract for technical services.

The specific references or general guidelines can also define what thresholds require what types of reviews and signatures.

Procurement performance reviews should be included in the quality management plan. This would include performance requirements with specific quantification of acceptable and unacceptable performance levels. In the case of materials being procured to produce a finished product, you would need to establish procurement standards that the supplier will be measured against.

As an example, assume a project to build a new outdoor obstacle course. The materials needed are wood for the scaling walls, and other large objects to crawl over and under. The grade and texture of the wood were specified, and the project schedule agreed on.

One month into the project, the supplier advised the project manager of production problems because of a problem getting the raw materials. This looked as though it would cause the schedule to slip.

Stipulations could have been built into the change control/performance parameters that required the supplier to either find another raw material source or another supplier to provide the materials within the published project schedule. However, if there were no such provisions, then the project manager may have no option other than waiting for the material and adjusting the schedule accordingly.

Inspections and audits are also part of the quality management plan and procurement management plan. To extend the previous obstacle course example, it turns out the original supplier did return to production and shipped the materials late. Upon receipt of the materials, the project team discovered they were not finished to the specifications and rejected them because course participants may have been injured from the rough surface of the materials. Thus, the schedule suffered an additional setback because of this issue.

Depending on the agreement associated with inspections and audits, it may be determined that the product is replaced or that the supplier comes to the site and corrects the defect there. These options would be agreed upon as part of the inspection and audit supporting details.

Performance reporting can be managed in several ways. One approach is a procurement scorecard, as illustrated below. The scorecard would include the essential data associated with supplier performance. The type of performance shown in the example would be cause for alarm and corrective action.

Corrective action could be in the form of reimbursement of some cost, going to a different supplier with some funding from the original supplier, or some type of monetary penalty. All potential options should have been discussed before the procurement contract was signed.

Procurement Scorecard (reporting) Example

Procurement Scorecard – Supplier Y					
Date	Event	Cost	Quality	Timing	Comment
2/3/yyyy	Missed delivery date.	1	N/A	2	Delay caused by raw material supplier.
3/6/yyyy	Materials not to specification	1	2	2	Inferior raw materials caused inferior finished materials.
n...					

Legend: 1=Acceptable, 2=Unacceptable, N/A=Not applicable

The idea behind implementing a procurement system is that once implemented, it can streamline the procurement monitoring and control efforts. Such a system would include a **payment system, claims administration**, and a **record management** system.

A procurement system should provide a ready reference to overall procurement status at any point in time and a color-coding system is often helpful.

If business rules are established that require a specific corrective action for a rejection, these can be built into the system to ensure that appropriate steps are followed. This information would also contribute to the development of the procurement scorecard example above, but because all information is now in one system, the scorecard is an electronic by-product of the information being accurate and current.

This same information can then be rolled up to the procurement contract number level to show the current status of each. This enables monitoring and controlling at the procurement contract number level, as illustrated below.

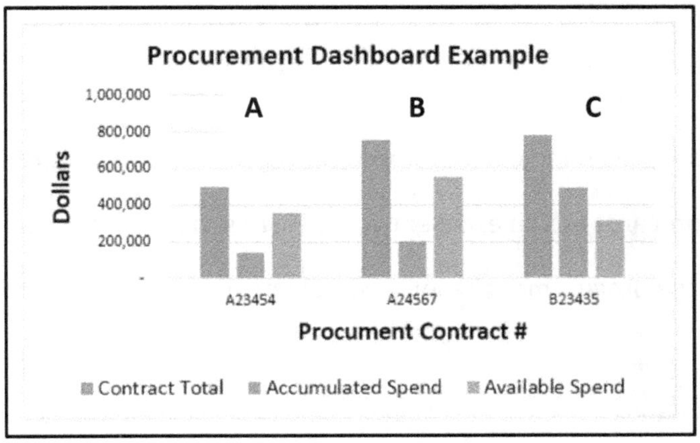

Project managers use the **dashboard to gain insights** related to procurement performance and to help determine if current procurement costs make sense compared to projections.

For example, if a particular contract was expected to have the most spending in the latter part of the project, and the project is only in its early phases now, but the contract spend is higher than anticipated, the project manager would want to find out why. If it is because planned purchases were made earlier than expected because some predecessor tasks were completed sooner than anticipated, that would be acceptable.

If, on the other hand, the amount of the earlier spend was higher than what was anticipated, there should be a review with the supplier to determine how this will be handled. If the contracted rate was based on the current market price with the anticipation that it would not increase at the rate that it has, then the project manager may need to use reserves that should have been set aside for the risk associated with the price being variable.

If the price was fixed in the contract, but the supplier is now charging more because of variables not included in the contract, that also needs to be addressed. In other words, let the supplier know you plan on paying only the contracted price.

You can also add business rules for when a contract spend is reaching a certain percentage of the total. For example, the project manager may want to know if the current spend total has reached a higher percentage than the calendar date progression of the project. This could apply to a service contract where you anticipated the same amount of service hours equally for each month. At the end of the first month, the billing might be, for example, 20 percent of the total contract when you were anticipating 16.6 percent.

If the overspend was because of additional work completed, and if the next month's billing should be less and offset the difference, then all is well. If the first month of work is taking more effort than anticipated, and it looks like this may continue, additional actions might be needed. These could include using procurement reserves, renegotiating the rate with the service provider, or possibly re-baselining the procurement plan.

What to do:	Other actions:
☐ Review the contract change control system that applies to your project.	☐ ____________
☐ Verify the procurement activity that is associated with your responsibilities within the project. Are they on track with the original baseline?	☐ ____________
☐ Check to see how suppliers are doing on your project's procurement scorecard.	☐ ____________

Remember

✓ Procurement systems can enable monitoring and control of the project's procurement process.

✓ Procurement scorecards are an excellent way to rate supplier performance.

✓ Variations from the procurement baseline should be dealt with promptly.

Enhance Your Learning

Watch the following 4-minute and 5-minute videos from ProjectManager to learn tips for project tracking and monitoring. Consider how you can apply this information in your work:

Deen, D. (2011). *Top 5 Project Tracking & Monitoring Tips for Project Managers.*	Available at: https://www.youtube.com/watch?v=CTg_9K5K6hE
Bridges, J. (2019). *How to Monitor Daily Progress as a Project Manager.*	Available at: https://www.youtube.com/watch?v=E2Qahnv1j1g

Reinforce Your Learning

Procurement Dashboard. The following is procurement information for a six-month project.

- Contract # A23454 is for equipment purchases and has a majority of the spending in the first part of the project.
- Contract #A24567 has a majority of the spending halfway through the project because of supporting activities for new equipment purchased with contract #A23454.
- Contract #B23435 is set to spend equally across all six months of the project.

Following is the procurement dashboard report created at the end of the project's third month. Assume you are the project manager and consider the questions that follow.

Procurement Dashboard Exercise – 3 Months			
Contract #	Contract Total	Accumulated Spend	Available Spend
A23454	900,000	154,000.23	745,999.77
A24567	100,000	10,000.00	90,000.00
B23435	600,000	425,345.23	174,654.77

1. What observations can you make based on the information provided?

2. What types of things or events may have contributed to what you observed?

3. Can you identify a procurement management tool or system that could be used to help with early detection of the problems?

Monitor and Control Project Communication

All such situations require an analysis of the effect on the project's scope, cost, and schedule. Several people could be involved in the analysis, and after decisions are made, you must communicate the results.

During the planning of the project, the project manager must define the **reporting structure** and communication chain, both for people internal to the organization and any who may be external.

The outputs for communication monitoring and control typically include things such as:

- Project communications in all forms
- Work performance information
- Change requests
- Updated project documents

Project Manager Responsibilities

The project manager is typically the focal point for communication with the client/stakeholders. To avoid communication problems, the project manager should also be the only person who communicates with consultants, agencies, or governing bodies. However, there can be exceptions to the rule. If your project has exceptions, document them at the beginning of the project.

Communication has a significant influence on whether projects succeed or fail. It is no exaggeration to say that effective and timely communication throughout the life of a project is crucial to its ultimate success.

The execution phase of the project is often a dynamic period in the life of a project. For example, you may discover differences associated with the original activity projections. These could be activities that were overlooked or, for example, risks with underrated impact, not to mention change requests that affect the scope.

Besides establishing clear lines of communication and ensuring that everyone is informed about what they need to know, the project manager must also consider time management for stakeholders and the project team. In this day of electronic communication, it is easy to send messages, memos, and reports to many people at one time. But too much communication going to people who do not need the information can be a significant source of wasted time.

Sometimes the opposite happens, and a project manager might not distribute communications to people who need them. For example, based on a change in a project, the project manager may forget to add a new team member to a communication list. The result would be that someone does not know what is going on, what has been added or modified, or what may need their attention. This will lead to a frustrated team and project delays.

As a project manager, you are responsible for ensuring that the communication plan is current and that everyone has the information they need. A good habit is to routinely check with stakeholders and team members to verify they have what they need to do their jobs.

In addition to electronic and paper communication, periodic **project review meetings** are intended to communicate information, both from the project manager to the team and from the team to the project manager. Information should be communicated within the context of the project scope, schedule, and budget. The stakeholders may also attend these meetings, or you might set up separate meetings with them, depending on what works best for your project.

Controlling Project Communications

Controlling a project's communication from start to finish requires planning and discipline. Such preparation includes identifying who needs what information and when. This includes the types of information to be communicated, how it will be delivered, and the frequency.

Obviously, the complexities of communication are affected by the size of the project. For larger projects, initial questions to ask might include: How will the ongoing status of the project be communicated? Who will receive project status reports? Who needs to know about project concerns and problems? Is a project team directory available? Is a communication matrix available?

Following through with planned communication occurs throughout the project life cycle. If a stakeholder communication matrix is not available, it would be beneficial to develop one.

The following is an example communication matrix. Stakeholders are listed in the lefthand column, the types of project information needed display across the top, and the cells indicate the communication medium and frequency.

Stakeholder Communication Matrix Example

Project Information / Stakeholders	Contracts	Project Plan	Informal Updates	Design Submittal	Kick-off Monitoring	Progress Reports	Scope Changes	Meeting Minutes	Project Specific Info.	Invoices	Etc.
Directors											
Project Managers											
Client Project Managers											
Administration											
Discipline Managers											
QA Managers											
Regulatory Agency											
Etc.											

Communication Medium

1. Written Reports
2. Telephone
3. Web Page
4. Meeting
5. Email
6. Project Related Information
7. Other

Frequency:

D: Daily

W: Weekly

M: Monthly

S: Schedule

O: Occurrence

Baseline Changes

Changes to the project scope, schedule, and budget must be communicated with a comparison to the original baseline and the current baseline. This includes the most current information related to change requests, resources, costs, and risks.

We have discussed monitoring and controlling tools and techniques for schedule, scope, and cost. All of them take into account the most current baseline information. But this is based on the concept that established guidelines state when an additional baseline may be required. Include the guidelines in the project, scope, schedule, cost, and change management plans.

Timely and effective communication associated with a baseline change is a big deal. A baseline change should be a flag for the client/stakeholders and team that at least one metric has changed enough to require a new baseline.

Communicating the Differences Between Baselines

You can use a project information management system to describe the differences between baselines. Below is an example of a project's original baseline information. You can see the anticipated length, or duration, of each activity. This information is saved as baseline 1

The red box with a 1 inside highlights an activity that will be changed because of a recent discovery. The duration, or how long it should take to complete, will be extended. The red box with a 2 highlights a successor activity to the activity with the 1 in the red box.

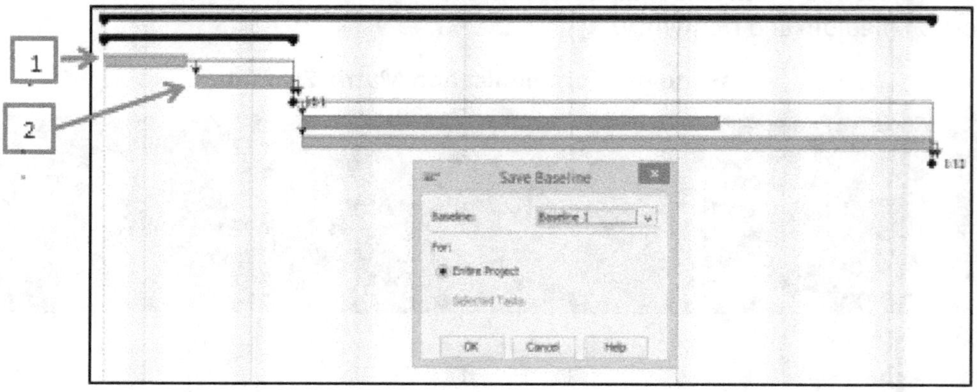

Baseline 1 Example

All purple lines show the original schedule for each activity. This provides a powerful visual that illustrates just how much the schedule was impacted. The people receiving the schedule update can see the impact and ask as many questions as needed for them to feel comfortable, they understand the current schedule.

For example, a question might be: Why was the activity elongated? As the project manager, your response might be something recently discovered was not anticipated in the original projections.

The next question might be: If that happened with this activity, how many other activities may have a similar surprise associated with them? Your response might be that the activity was more complicated than any others and that you reviewed the problem with the project team, and they are confident all the other estimates are sound.

This type of communication can be challenging, but it is necessary to keep the client/stakeholders in sync with the project, so they can make informed decisions.

When delays, or other surprises, are reported, it can degrade confidence in the project and sometimes in the project manager. An experienced project manager must do due diligence to ensure that any additional surprises are kept at a minimum. One way is to work through the root causes of each significant surprise and verify the same type of surprise is not likely to occur in another part of the project.

Clients/stakeholders consider how many surprises have been reported throughout the life of the project. They rely on their confidence in the project manager and their intuition as to whether the level of surprise is acceptable moving forward. This likely will include some discussion about this topic with the project manager, so everyone has the same expectations moving forward.

The above example discussed a schedule change, but the same concept applies to any changes introduced during the project.

Project Status Reports

One tool that you must use consistently is a **project status report**. The content and format of the report should be worked out with the project client/stakeholders in the project's planning phase, along with an agreement as to how much supporting information should be included.

Below is a sample of a typical report. We will use it as our example in the discussion that follows.

Activity List with Status Example

Name	Duration	Start	Finish	Status
⊟**Project 1**	45 days	**11/19/...**	**1/20/...**	
⊟**Task 1**	15 days	**11/19/...**	**12/9/...**	
Task 1A	10 days	11/19/1...	12/2/...	Red
Task 1B	5 days	12/3/15...	12/9/...	Green
Task 1 completed	0 days	12/9/15...	12/9/...	Green
Task 2	20 days	12/10/1...	1/6/1...	Green
Task 3	30 days	12/10/1...	1/20/...	Green
Project Complete	0 days	1/20/16...	1/20/...	Green

Schedule, scope, and cost are the three big hitters most likely to impact the state of a project. The status report should include them, as well as a snapshot of the risks. A risk assessment should be conducted periodically, but if a serious risk arises, add it to the status report even if its status is, "We must spend more time understanding the risk before we can report on it."

Your goal in reporting status is to report what the client/stakeholders need to know as simply and informatively as possible. One way of doing that is to use red, yellow, and green status indicators, as shown above. Using the same activity list from earlier in this concept where one activity was elongated, we have marked the status of the activity as red because we missed its original end date.

The other activities that depended on this activity to be finished were pushed out to later dates. The team assured us there will not be any other surprises with the remaining items, so we marked their status as green. This visual will also generate questions associated with the red item, so the client/stakeholders can determine if they understand the reason for the change.

What to do:	Other actions:
☐ Review your most current project status report. Is what you need to know included?	☐ ______________________
☐ Identify the types of communication you are receiving and determine their effectiveness. Can you suggest improvements?	☐ ______________________
☐ Consider the client/stakeholders on your project. Do you think they know the project's current status?	☐ ______________________

Remember

- ✓ Things change. Effective communication of those changes is critical.
- ✓ If you think you are not getting the information you need, let your project manager know.
- ✓ The number of project baselines can indicate the significance of changes to a prɔject.

Enhance Your Learning

Watch the following 15-minute video to learn more about communicating with stakeholders.

Akinwale, P. (1016). *Manage Communications/Stakeholder vs ControlCommunications/Stakeholder.*	Available at: https://www.youtube.com/watch?v=ZKYmRKadujg

Reinforce Your Learning

Communication Plan. Review the situation below and note your thoughts.

Jane, one of the project stakeholders, just learned about several changes to the scope of a project. The changes have caused the project to lag behind the original schedule and have increased project costs.

What forms of communications, or tools and techniques, might a project manager have used to mitigate or try to prevent this situation from occurring?

____

____

What should the project manager do to ensure that everyone is informed about project changes?

____

____

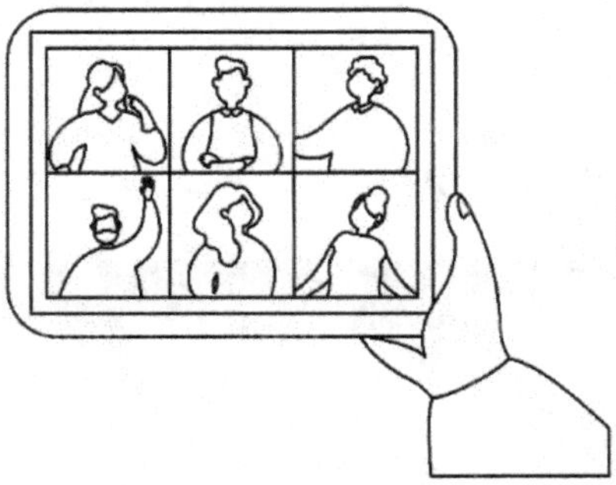

Implement Change Management Monitoring and Controlling

We will consider change management from two perspectives. The first is change management within a project; the second is change management within an organization. A project will not only experience changes as it progresses, but it will inevitably cause change within the organization. In some cases, the change can be dramatic and unsettling.

The Planning Elements Involved with Change Management

The planning elements associated with monitoring and controlling **change management** include the following:

- Project management plan
- Change management plan
- Work performance reports
- Change request log
- Issues log
- Enterprise environmental and process factors

Tools and Techniques for Managing Change

Tools and techniques for change management monitoring and control typically include:

- Expert judgment
- Change control policies, processes, and procedures, and documentation

Expert judgment draws on the experience of those most familiar with what changes. Experts help identify all factors to consider in assessing the impact of a change and how to monitor and control the change. We will walk through an example of using expert judgment later in this concept.

As we have discussed, change management within the project is primarily concerned with how things are handled when variations from the baseline occur. We mentioned using a change management plan, change control policies and procedures, change request forms, and change request logs. Another tool often integrated into larger projects is a control board.

A **change control board** helps protect a project from changes that are not advantageous. Project managers use it to help ensure that changes are incorporated into the project correctly, to ensure effective monitoring and controlling, and to oversee accountability for the results of the changes.

Change Request Example

Below is an example of how you might log a change request. While working through requirements, the team noticed there was a weakness in processing customer requests, and Jane wrote a change request aimed at correcting that problem. This improvement was not part of the original scope and will involve additional work.

Jane is the manager of this area, and she provided supporting information about why this change should be included in the scope. The change control board approved the scope change, and the project management plan has been updated accordingly.

Abbreviated Change Request Log Example

Project Change Request Log					
Request Date	Change Request Reference Number	Requestor Name	Requestor Contract Info.	Request Description	Why / Benefit
2/1/yyyy	0010	Jane Smith	Cell: 111-111-1111 Email: jsmith@org.com	Establish future state process flow for customer request processing.	Response time is too long.

The example illustrates how you might manage change. It is a format of a change request log that includes less information than the change request log we reviewed earlier in the course, and that is a template in Appendix A. You could use this format, or use the more extended version of a change request log from Appendix A.

This abbreviated example has a brief description of the change and why the change is needed, which could also be described by naming the benefit.

This is also an example of how project changes can change an organization, in this case, it will affect people's lives by changing the way they do things. This change to the project will impact three customer service representatives.

Change within an Organization

That brings us to the perspective of change management within an organization. Understanding how people do a job now and how that might change often includes mapping out the process of how things are done now (current state) and how things could be done in the future (future state).

The future state could disrupt the way things are currently done in an organization. This can be unsettling to people, particularly those who are most affected by the change. **Emotional reactions** to change are common, and **communication** is vital. As you read the list below, consider how employees might feel as they respond to the following initiatives, and the role of appropriate communication.

- New technologies
- Mergers and/or acquisitions
- Downsizing and layoffs
- Transfers or job relocations
- Leadership changes
- Reorganizations
- New policies or work rules
- New competition
- Unexpected financial or legal liabilities

The following diagram depicts some of the emotional reactions that may be experienced.

Change Process

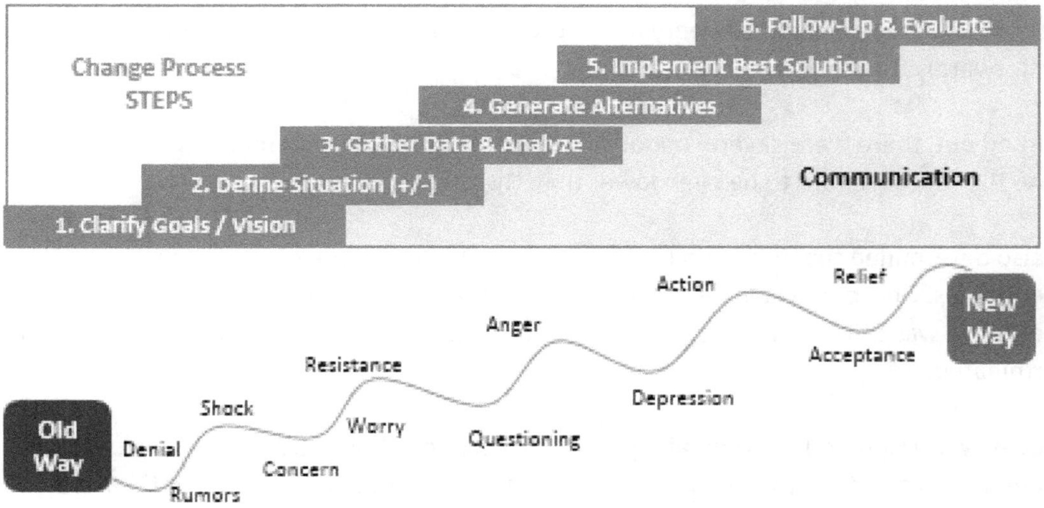

Communication about an upcoming change should be continuous. Discussions should be open, focusing on potential issues and concerns, as well as what support leaders intend to provide to facilitate change implementation. When the door is kept open for associate feedback and discussion, ambiguity is reduced while acceptance increases.

As an example, have you ever had to completely change the way you are doing something, something you may have been doing for a long time? For example, every time there is a new operating system for your computer or phone, things are not what they used to be. Something you could do easily yesterday may seem difficult today. This can be frustrating until you become familiar with the new way.

Some people may resist change, and that resistance, as well as its potential stress, must be accounted for in the change management plan. Monitoring and controlling this aspect of change management draws on soft skills.

So, back to our change request and its potential impact on the three service representatives. We need to walk through the current processes to understand what they do and how they do it. This means that you will ask a lot of questions like, "Why do you do it that way?" and "Have you considered doing it this way?" Nearly all the time, there are good reasons why they do what they do—or there used to be such reasons.

Sometimes something negative happens with a process. Corrective action is put into place so that negative thing does not happen again. Later, maybe much later, something changes upstream or downstream from that process that makes the corrective action of no value. Now, we have an activity that takes time and resources but adds no value. But this is the way that it has always been done, so we keep doing it—perhaps because we are not aware of the other change.

It is essential for those gathering the requirements, or possibly doing process flow mapping, to respect the people doing the job. They may have been doing the job the same way for years and learning why is like digging deep in someone's backyard. Walls can develop between subject matter experts and the project team when respect is not part of the process.

Not only may the original review be potentially unsettling, but as activities are transitioned to being done electronically, many manual processes are eliminated. It could be that a person who has been doing process steps 1 to 15 for many years now sees that a majority of those steps will go away. This can cause them to begin to question the need for them within the organization. This can be demoralizing and may make them leery of the project. This issue needs to be addressed with them straightforwardly.

As it turned out, there were several opportunities to improve the customer response process and decrease the response rate to be even lower than the original objective.

It was also determined that it was no longer necessary to have three customer service representatives. Only one would be needed to handle the revised workload. In this case, one representative was offered another position that had recently been vacated. The remaining position was terminated.

The process was improved by eliminating wasteful activities. However, it was unfortunate that one representative had to transition to another job, and even more unfortunate that the third representative lost their job.

Tips for Managing Change Successfully

Project managers must not only manage changes to the project, but they must manage "people change" constructively. Here are a few tips that will help make things go smoother.

Respect the people whose lives are being rearranged by the change. One way to do this is to remember how difficult it may have been for you to cope with a change in your life.

Support the people who are challenged by the change during the execution phase. This is when reality sets in as the changes are tested. For the first time, people can see the new way of doing things. The current state is transitioning to the future and the people affected are learning new procedures

Identify project champions among the stakeholders. Reach out to those most likely to benefit from the change and look for individuals are known and respected in the organization Ask these people to champion the new way of doing things by sharing their positive experiences, helpful tips for learning the new way of doing things, and the benefits they expect to gain from the change.

Maintain a positive "can-do" attitude. This can make any transition better just by itself. Find ways to show people how their lives will improve while also listening to their perspectives and working with them to solve what to them may seem like obstacles.

Change Management Documentation

The outputs for change management monitoring and controlling typically include:

- Approved change requests
- Change request log
- Updated project documents

Points of Control

Internal and external changes to a project will occur during the project's execution phase. Experts in project management say there are many points of control in managing a project.

The following 12 possible points of control are often used:

1. Milestone hit rate
2. Deliverable hit rate
3. Variance from the critical path
4. Actual costs versus estimated costs
5. Issues management
6. Number of risk items
7. Team status
8. Sponsor/management/client time commitment
9. Resource shortfall
10. Critical skills shortfall
11. Personnel turnover rate
12. Training completed versus training planned

These are not the only points of control available to you as a project manager. Nor are you expected to use all of them on every project. But project managers must rely on metrics or quantifiable items that can be tracked. This will help you identify project changes and departures from expectations, and this is how control is established and maintained.

What to do:

- ☐ Identify the information needed on your project's change request log.
- ☐ List several ways a "can do" attitude will help you manage organizational change.
- ☐ Review the 12 points of control and identify the best ones for your project.

Other actions:

- ☐ ______________________
- ☐ ______________________
- ☐ ______________________

Remember

✓ The change request log provides a dashboard look at the state of a project's change requests, depending on how you set up the format.

✓ Resistance to change is natural, but to keep the project on track, you must manage the resistance.

✓ A change control board is intended to protect a project from inappropriate change requests. It will help you evaluate trade-offs in scope, schedule, and cost.

Enhance Your Learning

View the following 5-minute video by strategy+business about change management and learn why only 54% of major change initiatives are successful.

Aguirre, D. (2014). *How to Lead Change Management.*	Available at: https://www.youtube.com/watch?v=PQ0doKfhecQ

Reinforce Your Learning

Implementing Changes

1. What is the most important thing you can do to minimize the impact of internal change? External change?

2. Supporting Organizational Change. Read the following scenario, then review the communication questions and follow the instructions for developing your action plan.

Scenario: Major Building Renovation. Assume you have been asked to spearhead a project to relocate (for two years) your business unit's facilities, including operations and administrative space. Responsibilities include integration and coordination of all department activities, as well as selected auxiliary departments housed locally. All implementation efforts must promote a collaborative environment and positive impact.

Communication Questions	Planned Actions
To whom should it be communicated?	
______________________	______________________
______________________	______________________
______________________	______________________
How should it be communicated?	
______________________	______________________
______________________	______________________
______________________	______________________
What communication methods should be used?	
______________________	______________________
______________________	______________________
______________________	______________________

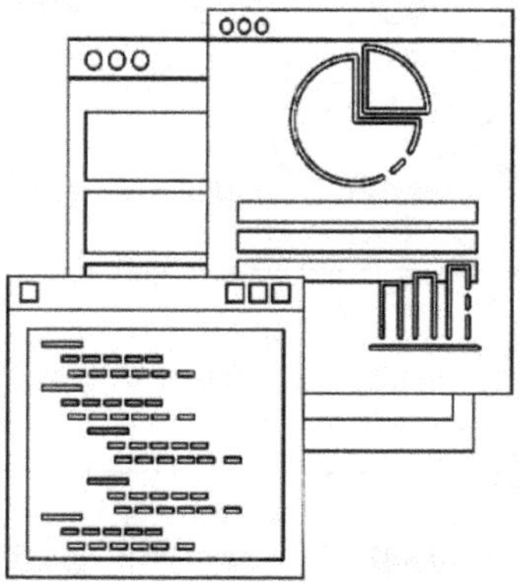

Summary

A Final Word: Monitoring and Controlling Projects

This chapter focused on monitoring and controlling all aspects of a project. We emphasized the types of things that can impact a project during the execution phase, but you can apply the concepts discussed to all phases of the project life cycle.

Information you reviewed in this chapter include:

- Appreciating how scope variation must be managed from all project perspectives
- Recognizing the key components of a Project Management Plan
- Describing all aspects of a project that need monitoring and controlling
- Understanding specific ways monitoring and controlling can be accomplished
- Expressing the advantages of early detection and correction for project metric (scope, schedule, and cost) variations

In the next two ***Project Mastery*** *parts or sections:* ***People*** and ***Resources***, we focus on the human and business sides for project management.

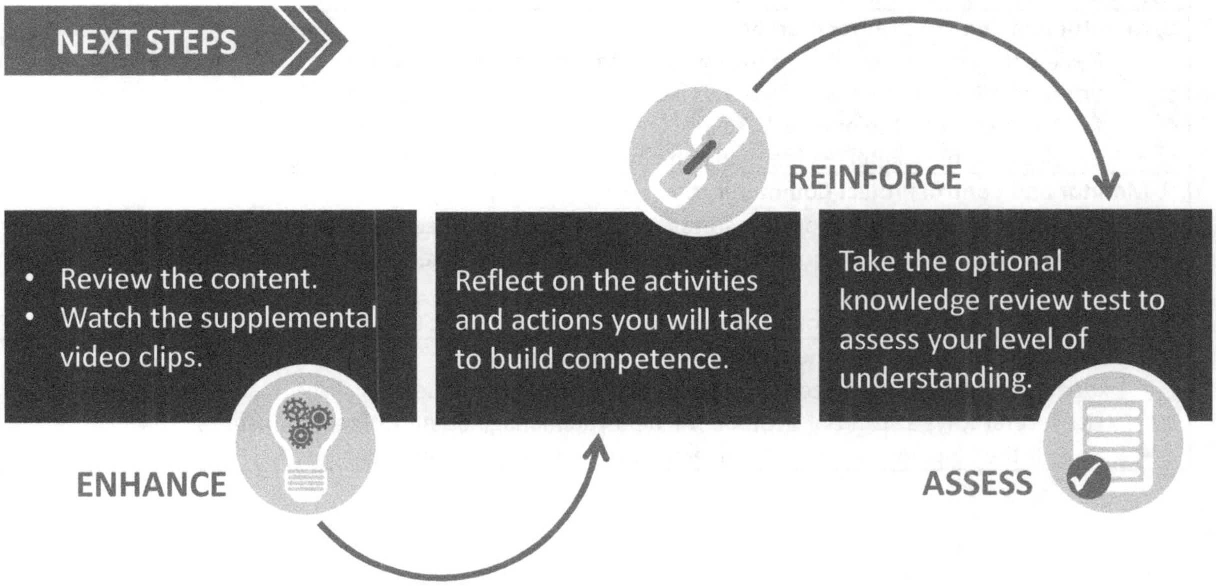

Monitoring and Controlling Projects—Recap Checklist During the execution phase of a project, the project manager must monitor and control the project to keep it on track and resolve potential problems before they impact project success.
1. **Differentiate Between Monitoring and Controlling a Project** □ Review the project charter and scope statement for updates and changes. □ Be mindful that there will be changes to the project's deliverables. □ Check the project planning for monitoring and controlling guidelines.
2. **Start the Execution Phase by Reviewing the Critical First 10 Percent** □ For a project you worked on, did the team follow the project plan at the start of the execution phase? □ Revisit the tips to execute a successful project. Do you agree that monitoring is management? □ Do you often ask people to re-do work? If so, what could you do to help avoid that problem?
3. **Review the Components of a Project Plan** □ Document and communicate updates to your plan: why, who, what, when, and how. □ Review your baseline project management plan and set up triggers to help monitor your project to ensure it is going as planned. □ Control your project by proactively taking corrective measures to get back on track.
4. Monitor and Control Scope Creep □ Review the requirements for your project and validate them to this point in the project. □ Review the project change request log to ensure appropriate actions have been taken. □ Review the scope baseline thresholds to ensure compliance. Create thresholds if they do not exist.
5. Monitor and Control Project Resources □ Monitor resources to ensure they are allocated according to the project plan. If not, how could the problem be resolved? □ Consider how much time you spend observing and discussing issues with others. □ Identify an interpersonal skill that someone you work with does well and compliment the person.
6. **Identify Components of Risk Planning and Management** □ Review the issues and change request log examples and consider how risks are tracked. □ Review the risks in your project and consider what tools you could use to improve risk management. □ Review your project's monetary reserves and consider recent project performance. Is the reserve likely to be adequate?
7. **Review Quality Monitoring and Control Tools** □ Review a cause-and-effect diagram for your project and validate it. □ Review the quality management plan for specifics of monitoring and controlling quality. □ Review a process flow for your project for potential improvements.
8. **Monitor and Control the Procurement Process** □ Review the contract change control system that applies to your project. □ Verify the procurement activity that is associated with your responsibilities within the project. Are they on track with the original baseline? □ Check to see how suppliers are doing on your project's procurement scorecard.
9. **Monitor and Control Project Communication** □ Review your most current project status report. Is what you need to know included? □ Identify the types of communication you are receiving and determine their effectiveness. Can you suggest improvements? □ Consider the client/stakeholders on your project. Do you think they know the project's current status?
10. **Implement Change Management Monitoring and Controlling** □ Identify the information needed on your project's change request log. □ List several ways a "can-do" attitude will help you manage organizational change. □ Review the 12 points of control and identify the best ones for your project.

Knowledge Review Test—Questions

Monitoring and Controlling

1. Controlling occurs:
   A. On a continual basis throughout the project.
   B. At the end of each phase of the project life cycle to assess its success in meeting project objectives.
   C. At the end of each planning process.
   D. On a continual basis throughout the executing process and as needed in the other processes.

2. Monitoring entails:
   A. Doing the work.
   B. Exercising influence over the project, taking action to analyze, evaluate alternatives, and recommend corrective action.
   C. Watching over the project, collecting, reporting, and disseminating project performance data and information.
   D. All the answers are correct.

3. During a team meeting the team determines that the client would benefit from several activities of extra work, and they decided to add them to the project. This is called:
   A. Gold plating
   B. Extra scope
   C. An authorized change request
   D. Superior customer service

4. Any variation from the original baseline scope is typically considered a ____ and must be documented accordingly:
   A. Bad thing.
   B. Variable ingredient.
   C. Change request.
   D. Gold plating element.

5. Which of the following are tips for successfully executing a project:
   A. Plan, and always think ahead.
   B. Standardize.
   C. Get it right the first time.
   D. Monitor and control.
   E. All the answers are correct.

6. The components of a project plan are important in change control because they:
   A. Provide the baseline against which changes are managed.
   B. Are expected to change throughout the project.
   C. Alert the project team to issues that may cause problems in the future.
   D. Provide information on project performance.

7. Uncontrolled changes are often referred to as:
    A. Scope creep.
    B. Scope leap.
    C. Gold plating.
    D. All the answers are correct.

8. A scope change is one that:
    A. Results in a modification of all the project baselines.
    B. Ends with a lessons learned.
    C. Changes the project's agreed-upon scope as defined by its WBS.
    D. Requires adjustments in cost, time, quality, and other objectives.

9. Which of the following is an example of quality assurance?
    A. Inspection
    B. Team training
    C. Pareto charts
    D. Fishbone diagrams

10. Which of the following is not part of a well-designed performance management system?
    A. Clarifying expectations.
    B. Helping team members succeed.
    C. Rewarding performance.
    D. Following up on poor performance.
    E. None of the answers are correct

11. Tools and techniques useful in risk identification include which of the following:
    A. Checklists.
    B. Databases.
    C. Flowcharts.
    D. Interviews.
    E. All the answers are correct.

12. Unplanned responses to negative risk events are often called:
    A. Workarounds.
    B. Risk avoidance.
    C. Workouts
    D. Risk mitigation.

13. Quality assurance is considered synonymous with prevention and quality control is often considered synonymous with inspection. Common tools and techniques for managing quality may include which of the following:
    A. Observation and audits.
    B. Cause and effect diagrams.
    C. Root cause analysis.
    D. Statistical sampling and inspection.
    E. All the answers are correct.

14. Project quality management was once thought to include only inspection or quality control. In recent years, the concept of project quality management has broadened. Which statement is not representative of the new definition of quality management?

A. Quality is designed into the product or service, not inspected into it.
B. Quality is the concern of the quality assurance staff.
C. Customers require a documented and, in some cases, registered quality assurance system.
D. National and international standards and guidelines for quality assurance systems are available.

15. The idea behind implementing a procurement system is that once implemented, it can streamline the procurement monitoring and control efforts. Such a system would include all the following except:

A. Payment system.
B. Expert witness training.
C. Claims administration.,
D. Record management system.

16. A change control system should include all the following except:

A. Paperwork, tracking systems, and approval levels necessary for authorizing changes.
B. Procedures for conducting a mid-project control system review.
C. Procedures for automatic approval of defined categories of change.
D. A description of the powers and responsibilities of the change control board.

17. The outputs for communication monitoring and control typically include all the following except:

A. Project communications in all forms.
B. Project charter.
C. Work performance information.
D. Change requests.
E. Updated project documents.

18. An Ishikawa diagram helps:

A. Explore past outcomes
B. Stimulate thinking, organize thoughts and generate discussion
C. Show team responsibilities
D. Show functional responsibilities

19. The best description of resilience is:

A. The ability to block out change.
B. The ability to deal with many changes simultaneously.
C. The ability to absorb high levels of disruptive change while displaying minimal dysfunctional behavior.
D. The ability to deal with employees resisting change and effectively handle ongoing activities of the project.

20. Which of the following conflict resolution techniques will generate the most long-lasting solution?

A. Forcing
B. Compromising
C. Smoothing
D. Problem solving

Monitoring and Controlling Projects — Answer Sheet

1.	_X_ A	___ B	___ C	___ D	___ E
2.	___ A	___ B	_X_ C	___ D	___ E
3.	_X_ A	___ B	___ C	___ D	___ E
4.	___ A	___ B	_X_ C	___ D	___ E
5.	___ A	___ B	___ C	___ D	_X_ E
6.	_X_ A	___ B	___ C	___ D	___ E
7.	___ A	___ B	___ C	_X_ D	___ E
8.	___ A	___ B	_X_ C	___ D	___ E
9.	___ A	_X_ B	___ C	___ D	___ E
10.	___ A	___ B	___ C	___ D	_X_ E
11.	___ A	___ B	___ C	___ D	_X_ E
12.	_X_ A	___ B	___ C	___ D	___ E
13.	___ A	___ B	___ C	___ D	_X_ E
14.	___ A	_X_ B	___ C	___ D	___ E
15.	___ A	_X_ B	___ C	___ D	___ E
16.	___ A	___ B	_X_ C	___ D	___ E
17.	___ A	_X_ B	___ C	___ D	___ E
18.	___ A	_X_ B	___ C	___ D	___ E
19.	___ A	___ B	_X_ C	___ D	___ E
20.	___ A	___ B	___ C	_X_ D	___ E

Part II: People

Part II: **People** focuses on the leadership and human side of project management.

PEOPLE

5. Enhancing Project Communication
6. Developing Interpersonal Project Relationships
7. Building High-Performance Project Teams
8. Leading and Managing Projects

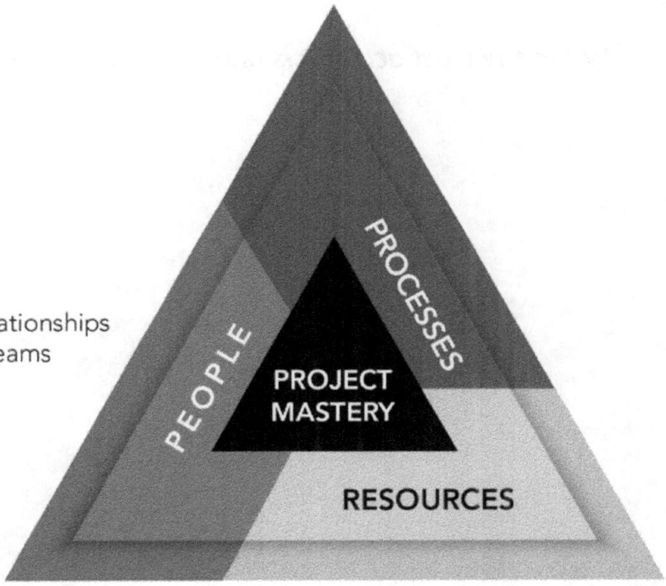

When the trust account is high, communication is easy, instant, and effective.

—Stephen R. Covey

Chapter 5

Enhancing Project Communication

Overview

> ***"An eye can threaten like a loaded and leveled gun, or can insult like hissing or kicking; or, in its altered mood, by beams of kindness make the heart dance with joy."***
>
> —Ralph Waldo Emerson

Successful project management depends on effective communication. Project managers must communicate with sponsors, clients, and other stakeholders as well as with their team, peers, and management. Without the necessary communication skills, project managers run the risk of damaging relationships and adversely affecting project outcomes.

Estimates are that leaders spend nearly 70 percent of their time communicating in one way or another. Further, each organizational level through which information filters has a drop in the percentage of information understood. Thus, what eventually moves through five organizational levels might be only 20 percent of the original message. The need for improvement at each level is apparent!

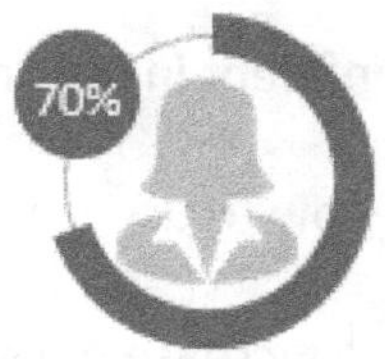

Time spent communicating

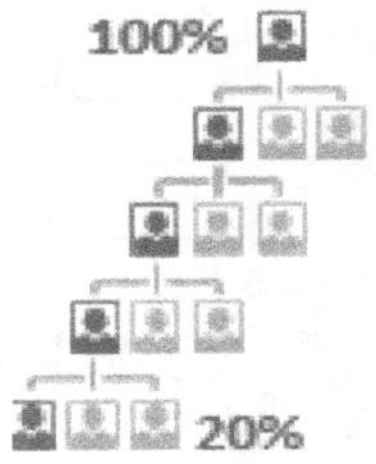

Lost meaning

This chapter examines how your interpersonal communication skills will impact your success as a project manager. You will learn about the various ways people communicate, that perception matters, and how to handle communication challenges. You will improve your communication style and therefore become a better communicator.

After completing this chapter, you will know how to:

- Identify the elements of the communication cycle
- Recognize the causes of breakdown in interpersonal communication
- Understand how nonverbal communication transmits a message
- Use active listening to enhance understanding
- Recognize the types of information a project manager must communicate and to whom

The competencies associated with this chapter include:

Oral and Written Communication	Interpersonal Relationship Building	Interpersonal Skills

Take Your Temperature for Enhancing Project Communication

The following assessment will help you identify what you know about communication and the areas where you can improve. With you and your organization in mind, read each statement below, and consider how much you agree. Use the numbers from 1 to 10, where **1** means **strongly disagree,** and **10** means **strongly agree**.

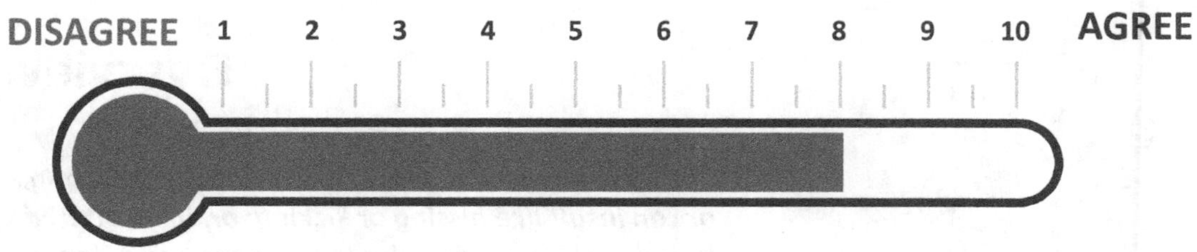

_____ 1. I can explain communication as an interpersonal process.

_____ 2. I know how to differentiate perception, communication, and understanding.

_____ 3. I understand why communication breaks down.

_____ 4. I am aware of the power of nonverbal communication.

_____ 5. I want to improve my listening skills.

_____ 6. I am confident in my ability to listen actively.

_____ 7. I know how to use common communication modes.

_____ 8. I want to increase my effectiveness in communicating.

_____ 9. I am good at email and other virtual communication methods.

_____ 10. I am skilled in communicating project and job information.

_____ **Total**

Take a moment to reflect on the following before you move forward in the chapter.
In your work, what problems have you experienced or observed related to project communication at work?

Take a few minutes to also reflect on your self-assessment. List two or three areas you want to improve.

As you progress through this chapter consider what actions you can take to demonstrate competence in the following three competency areas.

1. **Oral and Written Communication.** Makes clear and effective presentations to individuals and groups; listens to others; communicates effectively in writing; can critically review and comprehend information written by others.

2. **Interpersonal Relationship Building**. Considers and responds appropriately to the needs, feelings, and capabilities of others; seeks feedback and accurately assesses impact on others; provides helpful feedback; builds trust with others.

3. **Interpersonal Skills**. Appropriately sociable, interacts effectively with others.

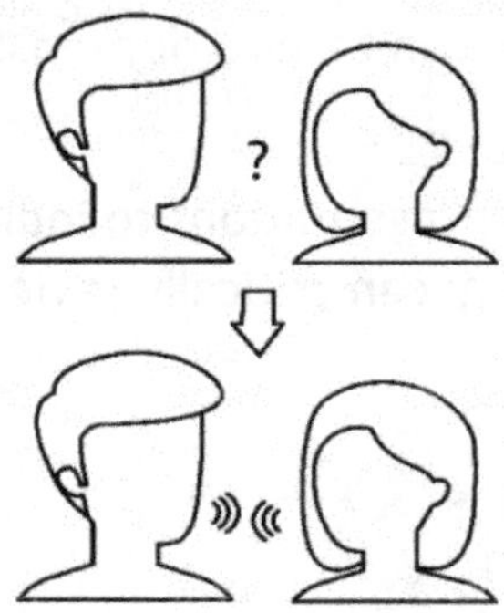

Understand Communication as an Interpersonal Process

A large part of each workday involves interactions with others. *Interpersonal communication* in a project is the exchange of information—facts, feelings, ideas—between two or more people with the goals of being understood, getting something done, and understanding others.

A significant part of interpersonal communication is what you say. However, it is also about how you say it; your tone, the pace of speaking, body language, and facial expressions. All these affect how information is perceived. Messages delivered over the phone lack visual elements, while messages sent in print or text rely almost exclusively on words.

In this concept, we discuss interpersonal communication as a process with both the sender and receiver having responsibility for communicating effectively.

What are the Elements Involved in Communication?

The following are areas to consider as you look for ways to hone your communication skills:

- **Communicators**: people who exchange information, typically a **sender** and a **receiver**.
- **Message**: ideas to be conveyed, typically concise, non-biased, and effectively delivered; expressed verbally, nonverbally, or both.
- **Feedback**: response sent by the receiver to the sender, which indicates whether the message was understood as intended; feedback can be verbal or nonverbal, or both.
- **Noise or barriers**: conditions that distort a message or prevent it from being heard, such as physical conditions (noise), language differences, technological barriers, social differences, disinterest, and inconsistencies between words and behavior.
- **Environment or context:** physical setting and time, emotional or psychological climate, and the relationship between communicators.
- **Method or means:** how a message is delivered, such as face-to-face, over the phone, by memo or email, video, or teleconferencing.

The Interpersonal Communication Process

The following graphic is a model of the interpersonal communication process.

The **sender** has a reason, or **intent,** to communicate, develops the **content** of the message, selects the method or means of communication, and adds any material needed to support the message. The sender then transmits the message.

The **receiver** listens and **decodes** the message and **interprets** it based on a variety of factors such as personal values, knowledge, experience, and perception. The receiver usually provides **feedback** to the sender, which indicates the message has been received together with any questions or comments the receiver may have.

Interpersonal Communication Process

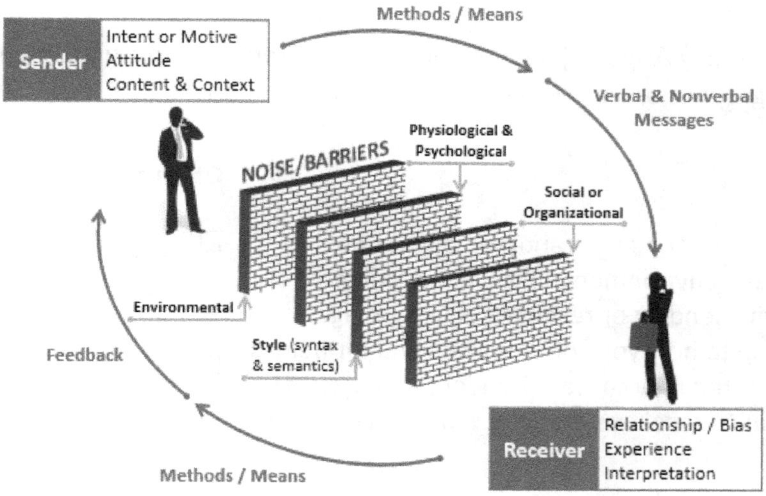

How Communication Goes Wrong

Even well-intended messages can result in misunderstandings. For a variety of reasons, problems can occur at both the sending and the receiving ends of a communication.

Information loss. Sometimes a message is vague or distorted, either because the sender does not provide enough information or because the receiver does not hear the complete message.

Assumption of understanding. Senders of messages often assume that receivers understand what was said. Likewise, receivers often think they know what was said. Neither party checks with the other to validate what was said or heard.

Excessive information. Information is not concise, and vital messages are buried in the details of the message. The sender does not realize the receiver is overloaded, and the receiver does not understand the core of what the sender wants to convey.

Speed of delivery. Speakers typically talk at a rate of 120 to 200 words per minute. Listeners can comprehend from 600 to 1,000 or more words per minute. The disparity creates opportunities for the listener to engage in mind-wandering, day-dreaming, or self-listening. Alternately, if a complex message is given rapidly, the receiver may not fully track all the details. If given too slowly, the receiver may consider the sender to be ill-prepared or not too intelligent. Or, the receiver may perceive the sender as speaking slowly because the sender does not think the receiver is "sharp" enough to take in more rapid speech.

Failure to ask for clarification. Receivers do not ask clarifying questions because they think they understand, or they feel such questions are unwelcome. Likewise, senders do not ask if receivers if they have questions or want to discuss the issue further.

Wrong communication channels. People seek information but do not ask the right questions or do not ask the right people.

Semantic confusion. Language is complex, and words often have multiple meanings. A word used by the sender to mean one thing can mean something different to the receiver.

Syntax errors. Syntax refers to the structure or sequence of language. When a sender places words in an unusual sequence, the receiver may be confused. A classic example is the Pennsylvania Dutch saying: "Throw the cow over the fence some hay." Also, the frequent use of fillers, for example, "you know" or "like," can distract the listener from the message.

Values conflict. People misunderstand each other because of personality or philosophical differences.

Stereotyping. People may respond judgmentally, draw inferences, or misunderstand because of differences in backgrounds or experience.

What to do:	Other actions:
☐ As you plan to interact with another person, be mindful of any environmental factors that might interfere with sending or receiving your message.	☐ ________________
☐ Pay attention to how you communicate and for beliefs or behaviors that may cause misunderstandings.	☐ ________________
☐ Make it clear to receivers that you want to hear their questions.	☐ ________________

Remember

✓ Communication is a two-way street. For it to be successful, information must be clearly sent and accurately received.
✓ Messages are often distorted or misunderstand because of noise or barriers that exist in the environment.
✓ Face-to-face communication involves not only words but also the tone of voice, volume, facial expressions, and body language.

Enhance Your Learning

Review the following web pages by Dynamic Flight that describes effective communication as a process.

Dynamic Flight. (2017). *Effective communication.*	Available at: http://www.dynamicflight.com/avcfibook/communication/

Reinforce Your Learning

Communication Misunderstandings. Consider a recent interaction you had with someone on your team that did not go well. Briefly describe the situation:

Review the possible causes of communication problems as discussed in this concept. Identify two reasons you think may have contributed to the unsuccessful interaction. For each reason, note what you might do to reduce the impact in future interactions.

Possible cause: ______________________

How can I reduce the impact?

Possible cause: ______________________

How can I reduce the impact?

Enhancing Project Communication #5.2

Appreciate How Perception Affects Communication

Perception is the process of acquiring, analyzing, organizing, and giving meaning to sensory information. Sensory information is what you see, hear, taste, touch, and smell. Through perception, you translate what you sense into concepts, ideas, and beliefs.

Perception is selective in both acquisition and interpretation. We do not take in all the sensory information available, and we reach conclusions based in part on information stored in memory. For example, your perceptions about your team members are influenced by what you have experienced in working with them, what you know of their backgrounds, and how other people regard them.

Perceptions are unique to each person. They are influenced by experience, place in the world, and vantage point. For example, if a project manager says they earn $80,000 a year, your perception could be "that is a lot of money." Another person's perception could be something like "it must be hard to support a family in this city on that income."

In this concept, we discuss how perception problems affect communication and how the four-element approach will help you become a better communicator.

How Does Perception Affect Communication?

Our perceptions have a significant influence on shaping our interpersonal communication, even when they are inaccurate and especially when they are unquestioned. Our perceptions affect how we speak and how we listen, and often without our awareness.

Self-Perception. Self-perception is how you perceive yourself as compared to others. It includes your self-esteem and how you see yourself in the world.

It may seem counterintuitive, but how you see yourself influences how you communicate with others, and it affects their perceptions of you. For example, an inflated self-concept may result in a person being regarded as arrogant and can lead to alienation from others. Conversely, a healthy self-concept can enhance one's ability to communicate authentically.

Self-Fulfilling Prophecies. A self-fulfilling prophecy might influence what you sense. For example, if you expect a person to behave in specific ways, you may see only the behavior that supports your expectations. This is true even when the person acts in ways that should cause you to question the accuracy of your expectations. People tend to screen available sensory information so that any behavior that does not fit expectations is screened out, and thus not taken in and processed.

Your expectations influence how you see people, how you respond to people, and in return, how people perceive and respond to you.

The Halo Effect and the Horn Effect. People tend to generalize. If you know, for example, that John has one or two admirable traits, you may assume John is good at other things as well. The tendency to generalize positive characteristics is referred to as the halo effect. Similarly, if someone has one or two traits we do not like, or we are biased against, we may assume the person has no good traits. This is referred to as the **horn** effect.

Unconscious and Conscious Bias. *Bias* is an inclination **in favor of or against** a thing, person, or group as compared with another. Having a bias involves prejudging someone or something, often with a refusal to consider the merits of other views. Bias seems to be a natural part of human thinking, and it is learned within cultural contexts.

Are you aware of your biases and how they affect your communication?

Bias can be **conscious**. In this case, we are aware that we are not giving a person or an idea a fair chance. Or, we can have an **unconscious** bias that affects our thinking and behavior without our awareness. Conscious or unconscious biases against people are not limited to ethnicity, race, gender, or age.

Most of us find it easier to reject incoming information than to question or change our beliefs. We do not hear information that conflicts with what we "know," or we dismiss the information as incorrect or irrelevant. This can interfere with openness to new ideas and, in some cases, lead us to be disrespectful when someone holds a contrary view.

In face-to-face interactions, unconscious biases are most often conveyed through facial expression, tone of voice, and body language. Without full awareness, you signal your regard for the other person through these nonverbal and paralanguage elements. The other person picks up your cues and forms assumptions about your attitude and you. It does not matter that neither party to the conversation may be aware, at a conscious level, that bias has affected what is understood and how the communicators perceive each other.

If you are aware of having a positive or negative bias toward another person, consider this as you interact with the person. It may help you minimize communication problems caused by bias.

If you are unaware of bias, it is difficult to minimize its effects, but habits of challenging yourself to listen fully to everyone and questioning your assumptions can help.

If you have a history of flawed or failed communications with another person, either one or both of you may be responding to unconscious bias. Taking a bit more care in planning before communicating with this person may lead to more satisfying interactions.

Environment and Context. Finally, the environment and context can influence perception. For example, if you work in a clothing store and hear someone shout from the ladies' dressing room, "I've got a run," your perception of the statement will most likely be different from the same exclamation shouted at a baseball game.

Use the Four-Element Approach to Improve Communication

The following are four elements to consider when you interact with others. These elements matter in any interpersonal communication, but they are particularly important when you think bias may be a barrier to understanding.

As you plan your message, consider yourself, the message, the other person, and the environment in which the communication will occur.

1. **Yourself**
   Consider your frame of mind. Are you looking forward to the discussion, or does it stress you? Are you familiar with the content or new to it? Can you identify anything in your communication style that could inhibit the other person from understanding your message? As objectively as possible, review the points you want to make and identify alternative views. Prepare to listen to the other person non-judgmentally.
2. **The Message**
   Will you communicate a simple message, or is your message complicated? As the complexity increases, so should your planning and possibly the tools you use to support the message. Complex information often needs to be expressed by more than one method of communication. Can you identify any information you want to share that is likely to meet with disagreement? Consider alternate ways of expressing your ideas. Look for words that support your message without intimidating or disrespecting the other person. Practice your message in front of a mirror and check your body language to ensure it reinforces your words.
3. **The Listener**
   Can you identify anything in the listener's typical communication style that could interfere with the intended communication? Consider the following kinds of questions. Has it been difficult in the past for you to get the listener's full attention? How did the listener express disagreement? Does the listener already have an opinion about the information? If so, is it positive or negative? If negative, how will that affect your message? Also, think about whether the timing is right for the listener. Pick a time when the person is not stressed, fatigued, or focused on other concerns.
4. **The Environment**
   Is the environment communication friendly? Or are conditions in play that may interfere with your attempts to communicate? Consider environmental noise, potential distractions, set up of space, and physiological comfort.

Plan to be a Better Communicator

If you experience frequent misunderstandings or communication breakdowns, you may benefit from better planning before an interpersonal interaction. Prepare for the communication by listing the four elements (yourself, message, listener, and environment), and for each element, identify the conditions that may affect your communication.

Plan also to use any positives—such as your knowledge of the topic, good rapport with the other person, or a supportive environment—to help minimize any negatives.

What to do:	Other actions:
☐ Consider your perceptions and reflect on how others perceive you.	☐ ______________________
☐ Challenge your assumptions and listen attentively.	☐ ______________________
☐ Use the four-element model to improve your planning for interpersonal communication.	☐ ______________________

Remember

- ✓ Many factors, such as environment, personal experience, and expectations affect perception.
- ✓ Awareness of how others perceive you is an essential first step in becoming a more effective communicator.
- ✓ We form beliefs about other people based mainly on our interactions with them. Each interaction, especially early in the relationship, adds information to a growing set of assumptions and expectations. At some point, the bits and pieces of information become fixed. We think we know what the person is like, what they will do in a given situation, and how the person views or regards us. Once formed, these perceptions can prove highly resistant to change.
- ✓ Do not give up because of problems in your first few attempts to listen more attentively and improve your communication planning. As with everything else, these skills are developed over time and with practice.

Enhance Your Learning

Watch the following 4-minute video by Jared James to learn more about perception and how you can take control.

James, J. (2011). *Jared James explains how Perception is Reality.*	Available at: https://www.youtube.com/watch?v=maUICwTzK5k

Reinforce Your Learning

The Role of Perception. Review the classic illustration below and answer the questions that follow. Show the illustration to others and ask them what they see.

What does this illustration of selective perception have to do with interpersonal communication?

What happens when people do not have the same perceptions about the context or the intent of a communication?

Enhancing Project Communication #5.3

Recognize Barriers and Why Communication Breaks Down

Successful interpersonal communication consists of a message worth hearing, a sender who expresses the message clearly, and a receiver who listens well and provides feedback to ensure the message was received accurately. This may seem simple, but many possible barriers can complicate the process. In fact, the belief that interpersonal communication is simple and straightforward is a barrier to success.

In this concept, we discuss communication barriers and how to keep them from damaging your interpersonal communication.

What are communication barriers?

The following is a review of common barriers to communication and what you can do to minimize their effects.

Physical or environmental noise. Things such as temperature, background noise, crowded space, poor lighting, the layout of space, and the presence of other people can be distracting. They may create discomfort for speakers and listeners.

Ensure you choose a place to communicate that is appropriate and conducive to the type of conversation you plan. Check for and resolve environmental issues in advance of all face-to-face communication.

Physiological conditions. If someone is fatigued, if they have just completed a demanding work shift, or have low energy due to medication, they may not be ready for an interaction.

Avoid having critical conversations with someone who isn't feeling their best. Plan for a time when the person can participate with no distractions.

Psychological or emotional conditions. These are barriers caused by expectations, attitudes, feelings about the other person, and preoccupations. For example, if you wanted to talk to Mark about a task, but you expect he would not be receptive, your expectation could negatively affect the communication. Or, if you were preoccupied with concerns about a family member, you would not be in the best frame of mind to listen to a coworker. Also, you may have difficulty communicating effectively with a person if you have an antagonistic relationship.

Assess your emotional readiness for a conversation before the interaction. If you sense a negative attitude or expectation on your part or the part of the other person, determine what you can do to rectify the problem and do that before you hold the conversation.

For example, if you are annoyed with a team member who did not complete a task on schedule, wait to talk to the person until you are in a more productive frame of mind. Or, if you think the other person may be annoyed with you, clear the air with the person before you discuss your intended topic.

Social noise. Some barriers result from cultural, gender, or organizational differences between communicators. If, for example, you speak with someone from a culture you are not familiar with, you may become self-conscious and pay more attention to your thoughts than to what the other person is saying. Many people also become self-aware when talking with someone a couple of levels higher in the organization. We become so concerned with how we are coming across that we lose track of the message.

In any situation with the potential for social noise, rely more fully on your active listening skills.

Organizational overload. Organizations can over-communicate with message after message and meeting after meeting. Or, information might be shared with people who do not need it and not with people who do need it. Information overload is a barrier to understanding because people become desensitized and cannot discern what information is essential to absorb.

Always check to ensure your messages go to the right people.

Technology. While technology such as cell phones, video conferencing, emails, and PowerPoint presentations can enhance communication, its failure or misuse can be a barrier.

Plan for appropriate back-up when you use technology to communicate. Also, be aware that ringing phones or beeping computers contribute to noise. Keep devices off or on silent when having discussions with others.

If you must receive a call or message during a discussion, tell the people you are speaking with beforehand, and apologize for the interruption.

Communication method. Choosing the wrong communication medium can be a barrier. For example, posting a memo to remind people about an event when you are targeting only a single person would contribute to information overload, and it would be inefficient. A one-on-one discussion would be more productive.

Communication style and language. When the people communicating do not share a common language—whether native language or specialized jargon—and you do not consider this, understanding can be compromised.

Regardless of language differences, it also confuses and distracts listeners when a speaker uses incorrect words or bad grammar or presents information in a flow or sequence that is hard to follow. Inserting placeholders or fillers such as "umm," or overusing one or two words, can also draw attention away from a core message.

To overcome these issues, prepare your points in advance. Know what you plan to say and how you plan to say it.

If you plan to give a speech, practice it in front of someone who can give you honest feedback. Also, practice the speech on your own, and work to remove "umm" and other fillers or placeholders.

Unchecked body language, lack of eye contact, and distracting gestures. Distracting body language can interfere with communication. For example, if you do not make eye contact with the other person, they may perceive you as distant or distrustful, and not interested in the conversation.

Make eye contact with the other person, just as you would in a conversation with a friend. Avoid movements such as fidgeting and shifting your weight from one foot to the other. Remain alert.

What to do:	**Other actions:**
☐ Hold conversations in environments free from noise.	☐ ______________
☐ Consider the psychological conditions and social noise that might be present in your interaction with others and act to alleviate such barriers.	☐ ______________
☐ Assess your communication style and consider your word choices. Could you use better words?	☐ ______________

Remember

✓ Be aware of potential barriers in all situations to ensure that your message will be received as intended.
✓ Mitigate any communication barriers that are within your control by preparing in advance.
✓ Watch for signs of over-communication or inefficient use of media.

Enhance Your Learning

Watch the following 2-minute video from Expertvillage to learn more about Non-verbal communication tools.

ExpertVillage Leaf Group. (2008). *Non-Verbal Communication Tools: Non-Verbal Communication: Paralanguage.*	Available at: https://www.youtube.com/watch?v=P5j0DFkbnE0

Reinforce Your Learning

Reducing Barriers. Consider a recent conversation with a team member and what barriers may have been present. In the space below, briefly describe who was involved, the topic of discussion, and whether you think the conversation went well.

____

____

____

Check any barriers listed below that could have been a factor in this situation. Make brief notes to describe the barrier and how you think it affected communication.

☐ Physical/environmental noise

☐ Physiological conditions

☐ Psychological and emotional conditions

☐ Social noise

☐ Communication style and language

☐ Other

Enhancing Project Communication #5.4

Be Aware of the Power of Nonverbal Communication

Typically, when we think about communication, we think about the words people say. That is, we think of **verbal language**. But two additional languages play a role in the meaning perceived in a message: **nonverbal language** and **paralanguage.**

Nonverbal language refers to what we see as we communicate with others. The following are examples:

- Facial expression, including eye contact and eye movement
- Body posture, movement, and gestures
- Use of distance, territory, and personal space, including touching behavior
- Clothing, accessories, and other aspects of physical appearance

Paralanguage refers to messages conveyed by vocal characteristics other than words. The human voice is a versatile instrument capable of expressing many emotions regardless of the actual words spoken. Key vocal characteristics include:

- Rate or speed of speaking
- Tone of voice
- Vocal rhythm, including inflection
- Diction or clarity of sound
- Volume
- Use of pauses, hesitations, and sighs

In this concept, we discuss nonverbal and paralanguage communication and what you need to know to use these "languages" to your benefit.

How Do We Communicate?

One study of the three languages found that attitudes and feelings were mainly conveyed through nonverbal elements, especially facial expression (Mehrabian, 1971, 1972).

Specifically, the study found the following percentages of meaning:

- 7 percent through the spoken words or **verbal**
- 38 percent through the way that the words were said **or paralanguage**
- 55 percent through **nonverbal language**, primarily facial expressions

In any interpersonal communication, it is virtually impossible to separate verbal and nonverbal messages. While people often intentionally choose their words, especially in sensitive communications, this is not the case with the other elements. We are often unaware of the messages we send nonverbally and through paralanguage.

Nonverbal Language

Body language. The way you hold your body will have a significant impact on how your message is received. As an example of how true this is, consider professional poker players. They work hard to avoid "tells," which is body language that might reveal a good or bad hand.

Whether you speak to a crowd, small group, or one person, the way you stand, the way you hold your hands, and how you move will all influence the way your message is perceived. For example, fidgeting could be seen as insecurity or lack of confidence and tapping a foot as impatience or anxiety.

Learn to read the body language of others. Recognize when they agree with you, disagree with you, have a question, or appear uncertain. This will help you anticipate and correct misunderstandings.

Facial expressions. Your face is the primary conveyer of feelings and attitudes to those around you. While you may not be able to fully control your facial expressions, being mindful of how you appear will help prevent misunderstanding. Similarly, paying attention to other people's facial expressions can help you better understand their feelings and attitudes. Being able to read facial expressions is especially crucial during sensitive or challenging communications.

Eye contact. How you use eye contact can convey such things as your honesty, confidence, and sincerity or the lack thereof. Communicators who use direct eye contact with occasional glances away are generally seen as being honest and straightforward. Speakers who use too little eye contact are often perceived as being less than honest or as lacking confidence in what they are saying. Staring, which involves extended direct eye contact, is usually seen as intimidating.

Proximity. Distant and non-attentive or up close and in-your-face are two extremes in a communication situation. Most cultures have a "comfort zone" that is often implicitly understood and accepted. Violating this comfort zone by being too close—or too distant— can add to the noise in communication.

Diversity. Be aware that nonverbal behavior may differ significantly from culture to culture. What is acceptable nonverbal behavior in one culture may be unacceptable in another. How much space a person is comfortable having around them or how people use hand gestures are typical differences. Observe the nonverbal behaviors of people who come from cultures different from your own. Discover the differences and try to respond accordingly.

Mixed Messages

A mixed message results when the words spoken do not match the tone of voice or facial expression. A mixed message sometimes "leaks" information about one's true feelings or attitudes. The speaker chooses words they consider acceptable or "safe" to say, but their true feelings are reflected nonverbally, often without awareness of it happening. When we sense a mixed message, we tend to attach less meaning to the words we hear than to other aspects of the message.

It is always possible that a speaker's gestures, facial expression, or tone of voice do not reveal their true feelings but are merely habits. Whether the nonverbal elements show true feelings or not, a mixed message creates confusion in the listener's mind.

When you receive a mixed message, get clarification. Use clarifying questions to probe for a fuller understanding of the speaker's message. For example, "I get the impression you're not too comfortable with this approach. Am I right?" Or, "You said you had time to go into this now, but I sense that it's not a good time for you. Would you prefer to set up another meeting?"

Think about the message you are sending and consider the body language, facial expression, and eye contact you typically use. Check first to see that your habits do not conflict with your message. For example, if you tend to keep your hands in your pockets as you speak, you will want to avoid this habit when sharing your enthusiasm about a topic.

Conversely, consider how you can use nonverbal signals to support your message. For example, when you are committed to an opinion, look the other person in the eye and stand firmly. This will show you believe what you are saying, and it will communicate a stronger message than your words alone.

Paralanguage

Do not forget about paralanguage: tone of voice, volume, speed, and inflection. All these are part of communication and can support, contradict, or weaken an intended message.

What to do:	**Other actions:**
☐ Pay attention to your communication habits, especially your nonverbal behavior and paralanguage.	☐ ______________________
☐ Pay attention to the nonverbal behavior and paralanguage of the people around you.	☐ ______________________
☐ Ask clarifying questions when you receive a mixed message.	☐ ______________________

Remember

- ✓ What you say—the words you choose—is only part of your message. How you say the words—paralanguage—also matters.
- ✓ Your body language may have the most significant impact when communicating, whether you are speaking to a crowd, group, or individual.
- ✓ For effective communication, your verbal and nonverbal messages must agree. Any incongruence will confuse the other person.
- ✓ Many nonverbal signals are based on culture. An example is how close to each other people should be when communicating. In some cultures, it is about a handshake apart. In others, a greater or lesser distance is preferred.

Enhance Your Learning

Watch the following HBR 6-minute video that overviews the importance of non-verbal communication and provides some tips.

Harvard Business Review. (2014). *Take Control of Your Nonverbal Communication.*	Available at: https://hbr.org/video/3541641444001/take-control-of-your-nonverbal-communication

Reinforce Your Learning

Paralanguage. Increase your awareness of voice inflection by considering the sentence below. Practice by saying each sentence out loud and emphasize the word in bold. Notice how the meaning changes when the only difference is the word emphasized.

- "**I** didn't tell her to do that."
- "I **didn't** tell her to do that."
- "I didn't **tell** her to do that."
- "I didn't tell **her** to do that."
- "I didn't tell her to do **that**."

Now, try some mixed messages:

- Say, "I'd love to do that," but say it with **no enthusiasm**.
- Say, "What a great idea," but say it with a **belittling tone**.
- "Of course, I don't mind," but indicate that, as a matter of fact, you **do mind** very much.

Communication Factors. Give examples of how the following could affect communication:

Eye contact:

Voice and tone, inflection, rate of speaking:

"Fillers" and non-words:

Posture and movement:

Listener involvement:

Appearance and dress:

Physical setting:

Other:

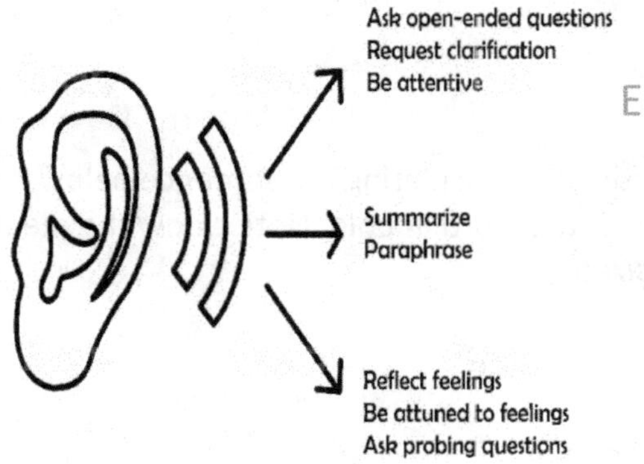

Enhancing Project Communication #5.5

Realize Listening is a Skill

You want to communicate your message, but you must also hear and understand what other people tell you. This requires the ability to listen, and the awareness that listening is more than hearing.

Even though we all spend considerable time in the role of listener, that does not necessarily mean we do it well. One reason is the difference between speaking and listening rates. We think and process ideas up to eight times faster than we speak. The average speaking rate ranges from 120 to 200 words per minute, and the listening comprehension rate ranges from 600 to 1,000 (or more) words per minute. As a result, we sometimes take mental vacations while pretending to listen. Usually, we do not miss too much, but how can we be sure?

Other reasons we may not listen well are that we do not understand the benefits of being a good listener, and we have not learned how to listen. Listening may seem like an ordinary activity, but to do it well takes skill. The first steps in developing the skill are understanding the benefits and acknowledging that listening takes knowledge and practice.

In this concept, we discuss what listening is, why it matters, and general approaches that will help you listen more effectively. In the next concept, we take the discussion a step further and describe active and reflective listening. They are skills that, with practice, will help make you a better listener.

The Benefits of Being a Good Listener

Listening is an essential skill for developing good relationships at work. It expands thinking, allows learning and growth, helps you form more accurate perceptions about others, supports better understanding, and increases retention of information.

Listening well will serve you in ferreting out workplace problems and in ensuring that you understand the opinions and concerns of your stakeholders and team. On the flip side, if you are not a good listener, you will struggle to manage your projects.

Effective listening will help you do the following things:

- Remove communication barriers and bias
- Bridge gaps in understanding before they become crises
- Keep communication channels open
- Prevent you from being cut off from the information you need to achieve results
- Develop insights that are impossible for people who "don't have time to listen"
- Enhance the support you receive from people above, around, and below you in the organization
- Cement relationships with stakeholders, your team, and your boss
- Reduce friction and help resolve conflict productively
- Assert your confidence, authority, and leadership better than words
- Keep people from going around you to someone who will listen

Paying Attention to Nonverbal Communication is Part of Effective Listening

A sometimes-neglected aspect of listening is the ability to sense and understand what is being communicated nonverbally. Listening involves *seeing* as well as hearing and can extend to the senses of touch and smell. It also includes feelings and thoughts based on *intuition* rather than on information taken in through sensory receptors.

Further, your nonverbal communication when you are a listener is just as powerful as when you are a speaker. Rather than risk sending a mixed message, be sure not only to listen but also to look as if you are.

The following are ways to communicate that you are listening nonverbally:

- Sit in a way that suggests attentiveness, for instance, face the speaker and slightly lean toward the person
- Maintain eye contact as if you were speaking with a friend
- Use supportive facial expressions and nod to show you are paying attention
- Occasionally use "uh-huh" or a similar expression to encourage the speaker to continue
- Avoid fidgeting and look alert, but comfortable

Be an Encouraging Listener

Give your undivided attention. Approach every communication with the knowledge that while communication is problematic, many misunderstandings can be overcome if you give the speaker your attention. We must pay careful attention to what a speaker is trying to convey, and that is difficult if we are thinking about or doing other things.

Listen fully. Avoid the natural temptation to plan your response while the other person talks. Otherwise, you will miss much of what the other person says. Listen fully, then consider your response.

Avoid interruptions. Interruptions break the speaker's train of thought and may serve to sidetrack the real message they want to convey. Reserve your questions and reactions until you have given the speaker a fair opportunity to express their thoughts.

Avoid assumptions. Keep an open mind and avoid making assumptions about the speaker's message. Listen without jumping to conclusions. When the speaker finishes, you can ask for more information if you need it.

Support the speaker. Picture what the speaker is trying to tell you. Try to understand the other person's point of view or perceptions.

Listen to understand **what** the person is trying to communicate as well as the purpose. Try also to understand **why** the person is communicating with you at this time.

Pay attention to the speaker's body language, gestures, and facial expressions. This is how you "listen" for feelings and attitudes. By understanding the speaker's attitudes and feelings, you will better understand their frame of reference or point of view.

Look for agreement. When people hear something, they do not immediately accept; the tendency is to stop listening and engage in a mental search for an argument that will refute what was said. This approach is not productive. First, it will annoy the speaker, and second, you may miss parts of the speaker's message. Do not get caught up in any side issues and lose track of what is significant in the speaker's message.

If you listen instead for points of agreement, many apparent disagreements may disappear. Any that remain can be resolved after the speaker has finished.

Ask questions. Asking questions is the best way to clarify things a speaker says that you do not fully understand. Wait until the speaker pauses before you ask and keep your questions on the topic. If the answer leads the conversation off-topic, take responsibility for returning to the topic.

Summarize or **restate**. An effective way of ensuring you have heard the speaker's intended message is to summarize or restate what you heard. This can be a valuable confirmation for you as well as the speaker. When the speaker is done, summarize your understanding of the message by repeating key points. Or restate the key points and ask if you have heard correctly.

If the speaker is asking you to take action, describe what you think the speaker wants you to do, and then ask if that is correct.

Take Notes. If the information you receive is lengthy or complicated, jot down the main points, so you will have a record to refer to later.

Not All Listening Situations are the Same

Do not assume that the context in which you listen does not matter. Different listening skills may be needed in different situations. For example, receiving instruction is not the same as hearing a person's reason for an action, and these are both different from listening in a heated negotiation. Recognizing the best listening behavior for a given situation and being able to use that behavior are essential to effective listening.

What to do:	Other actions:
☐ Examine and improve your nonverbal listening behavior.	☐ ________________
☐ Practice listening for feelings, attitudes, and meaning.	☐ ________________
☐ Commit to paying full attention to the speaker. Avoid interrupting or focusing on what you want to say.	☐ ________________

Remember

- ✓ Listening is not as easy as many people think, partly because our rate of speaking is five to eight times slower than our rate of listening. The difference makes it easy for our minds to stray from the speaker's message.
- ✓ You can hear words, but effective listening requires that you honor the speaker by paying full attention and listening beyond the words.
- ✓ Ask questions to clarify points, and when the message is complex, take notes.

Enhance Your Learning

Watch the following 7-minute TEDTalk to learn more about listening.

Treasure, J. (2011). *5 ways to listen better.*	Available at: https://www.youtube.com/watch?v=cSohjlYQI2A

Reinforce Your Learning

Listening Skills Self-Assessment. Complete the self-assessment that follows. It will help you become more aware of how you listen. Reviewing the results will help you know what actions to take to improve your listening skills.

Answer each question in terms of how you listen at work. Be honest, so you will clearly understand where you want to improve.

Assessment adapted from classic work developed by Shmukler, A. & Turner, D.W., Penn State University, Management Development Programs and Services and used via license.

Using the following key, circle the numbers that correspond to your choices:		
1 = Never	3 = Occasionally	5 = Most of the Time
2 = Rarely	4 = Frequently	6 = Always

1.	When possible, I control any environmental factors that might interfere with my listening (e.g., heat, lighting, noise, seating arrangements).	1	2	3	4	5	6
2.	I have reliable ways of dealing with interruptions or other listening distractions when they occur.	1	2	3	4	5	6
3.	I am comfortable asking the other person to repeat or clarify what I may have missed or did not understand.	1	2	3	4	5	6
4.	I restate or summarize information to ensure that I have heard it correctly.	1	2	3	4	5	6
5.	I am able to focus on what a person is saying whether I believe it is relevant to the point we are discussing or not.	1	2	3	4	5	6
6.	Even when I think I "know" what the other person is going to say, I continue to pay close attention and stop myself from "completing the thought" for them.	1	2	3	4	5	6
7.	When I am unfamiliar with a person's terminology, I ask questions about what is meant.	1	2	3	4	5	6
8.	I am able to listen for specific information that a person is sharing as well as for what is most important to the person.	1	2	3	4	5	6
9.	I am equally able to listen to what a person is saying with words and what he or she is saying nonverbally.	1	2	3	4	5	6
10.	Even in situations where there are obvious differences (social, cultural) between the other person and myself, I control my self-consciousness so it doesn't interfere with my attentiveness to the other person.	1	2	3	4	5	6
11.	I am able to listen to a person without judging what I am hearing or seeing.	1	2	3	4	5	6
12.	I postpone evaluating what the person has said until I am sure that I understand the message.	1	2	3	4	5	6

Using the following key, circle the numbers that correspond to your choices:

1 = Never	**3 = Occasionally**	**5 = Most of the Time**
2 = Rarely	**4 = Frequently**	**6 = Always**

13. I avoid interrupting people with questions or adding information before they have finished speaking.	1	2	3	4	5	6
14. I listen to understand what the person is really saying, rather than for what I think I need to hear.	1	2	3	4	5	6
15. When evaluating information, I consider whether the assumptions and conclusions I am reaching are valid.	1	2	3	4	5	6
16. I avoid evaluating and reaching conclusions before I have enough information.	1	2	3	4	5	6
17. I am able to allow a person to say negative things without becoming defensive.	1	2	3	4	5	6
18. I am able to control my listening response when a person says something with which I disagree or which I believe not to be true.	1	2	3	4	5	6
19. I am able to send nonverbal messages that indicate I am paying attention and listening to the other person.	1	2	3	4	5	6
20. When I am listening to lengthy, detailed information, I take notes to help jog my memory.	1	2	3	4	5	6
To score, add the totals for each column and record here:						

Add the column totals and note your Overall Score ________. Interpretation:

120 – 105 You have outstanding listening skills. Continue listening as you do now.

104 – 90 You have very good listening skills. Note items where scores are lower and work for specific improvements in those areas.

89 – 75 You have good listening skills. Note items where scores are lower and consider what you might do to change the specific behaviors that may make you less effective. At the same time, continue the behaviors that are contributing to your effectiveness as a listener.

74 – 60 Your listening skills are fair. You should take deliberate actions to improve. Identify the specific behaviors that interfere with your effectiveness and plan how you can replace these with effective behaviors. Don't forget to examine the items where you scored high and continue these effective listening behaviors.

59 and below Your listening skills are poor, especially if your score is below 45. Look at this assessment as a map on the road to better listening, but do not delay. Follow the guidelines given for scores between 74 and 60.

Enhancing Project Communication #5.6

Use Active and Reflective Listening

Interpersonal communication is complicated. The speaker and the listener each have a role. Speakers are responsible for organizing their thoughts and communicating them as clearly as possible. Listeners are responsible for actively listening, which is a skill. It means doing all you can to make the speaker's job easier.

Active listening techniques help you to not only understand a speaker's message but also to make appropriate responses. Sometimes, your response can be as simple as encouraging the speaker to continue. You can do this by using nonverbal nods, saying "uh-huh" or similar utterances, or with expressions such as "please go on." These acknowledge you are listening but do not reflect your opinion or conclusion.

In this concept, we describe active and reflective listening and how you can use these skills to make yourself a better listener.

Ask Questions

A crucial part of active listening is the ability to ask questions to clarify or confirm your understanding of what a speaker said. This alone will help you communicate with people and help speakers feel comfortable talking to you.

During any communication, you may hear something you do not understand or something you will use to make a decision or take action. In such circumstances, you *must* confirm your understanding of the message. This will help ensure that any decisions or actions you take are based on understanding facts and, in some cases, expectations.

Here are some of the benefits of asking questions:

- Let's the speaker know you have been listening
- Helps you understand what the other person has said
- Allows the other person to clarify what they said
- Enables you to check your assumptions
- Encourages the other person to provide feedback

Even though questions are powerful tools for understanding, people can feel uncomfortable with questioning a speaker. Sometimes, people are **reluctant** to admit they do not understand what a speaker said. As a result, they leave a discussion without a full understanding. This may be especially true when the speaker is a manager or an authority in the organization.

Another reason people do not question a speaker is they **assume** they know what the speaker is trying to say and so jump to a conclusion without listening to the entire message.

Finally, it is not unusual at work to be **rushed** because of upcoming responsibilities, and so people do not take the time to focus on the speaker and their message.

If you have not been in the habit of asking questions, you may feel uneasy about doing so. Do not let discomfort stop you. With a little practice, you will realize the benefits.

Active Listening Techniques

Here is a summary of the techniques that will help make you an active listener.

Clarify when you are not sure you understand either the content or the intent of a message. To clarify, ask questions such as, "Do you mean . . .?" or "Are you saying . . . ?"

Clarifying is essential when you are expected to use the information in the future. After asking for clarification, you can continue to listen actively.

Confirm when you must make a decision or take action based on what you are told. To confirm, you restate your understanding of the content of the message, the intent, or both. Such statements help safeguard that you understand both the information you heard and why you are being given the information.

Examples are: "My understanding is . . ." and "Just to be sure we are talking about the same thing . . ."

Summarize at the end of a message to help both the speaker and listener review what has been shared. Summarizing is especially useful when several ideas or bits of information have been exchanged. Summarizing is also helpful when transitioning to a new topic.

Examples are: "What we've agreed to thus far is . . ." or "I'd like to summarize what I see as your major equipment requirements . . . "

Summarizing or paraphrasing what the speaker says ensures that both people understand the interaction the same.

Re-open the discussion when additional information is needed. After the speaker has completed their communication and answered any clarifying questions, the listener may still want more information. If so, re-open the discussion to ensure the full story is told.

Reflective Listening or "Mirroring Back"

As you listen, try picturing what the speaker is trying to say. Hear their words and interpret them, and then mirror the ideas back in your own words to determine if you have received the intended meaning.

If you have not received the communication as intended, the speaker can repeat or clarify using different words.

It is often surprising how much of an intended message is missed. The technique of reflecting (mirroring back) can be invaluable in gaining a full understanding.

Here are some **patterns** or templates you can use to reflect your understanding to the speaker:

Repeat word for word.

Speaker's statement: "That makes sense."

Your reflection: "I agree. That makes sense."

Repeat, but change "I" to "you."

Speaker's statement: "I'd like to see this project succeed."

Your reflection: "You'd like to see this project succeed."

Repeat part of what the speaker said but summarize it.

Speaker's statement: "I would like the project staff to be less concerned and more positive and energetic about the change."

Your reflection: "You would like the project staff to be more supportive of the change."

Summarize in your own words.

Speaker's statement: "Trying to get here on time, I took a shortcut that wasn't short, missed a turn, and got stuck in a line of traffic."

Your reflection: "You had several traffic challenges."

Summarize in your own words and express feeling.

Speaker's statement: "I'm concerned about missing the meeting."

Your reflection: "You're worried that your viewpoint will not be represented."

Interpreting what is said or felt, of which the speaker may not be aware.

Speaker's statement: "I'm concerned about missing the project meeting."

Your reflection: "You're worried because you don't think you'll have another opportunity to act on this issue."

What to do:	Other actions:
☐ Commit to asking questions to enhance your understanding of a speaker's message.	☐ ____________________
☐ If you are too rushed to listen fully, apologize and set a time when you can actively listen.	☐ ____________________
☐ Select one of the active listening techniques that you do not use and start using it.	☐ ____________________

Remember

- ✓ The listener's responsibility is to listen actively—to do everything in their power to make the message-sending easier. In general, active listening techniques help you to not only understand the speaker's message but also to choose appropriate responses.
- ✓ Try picturing what the speaker is trying to say.
- ✓ Ask questions. Do not be reluctant to confirm your understanding or assume you do not have to listen to the other person to know what they are saying.
- ✓ Start using reflective listening to mirror back the feelings you are picking up from a speaker's message.

Enhance Your Learning

Watch the following 2-minute video that shows active listening in action.

Gordan Training. (2010). *Leadership Training - Active Listening.*	Available at: https://www.youtube.com/watch?v=ESujTCel6IM

Reinforce Your Learning

Reflective Listening Practice. Assume a team member has just made the following statement:

"The way this project team is working, I think we'll be weeks behind schedule."

For each of the following ways of reflecting your understanding to the speaker, write one or more responses you could make.

Repeat, but change "I" to "you:"

Repeat part of what is said, but summarize it:

Summarize in your own words what the speaker said:

Summarize in your own words and express feeling:

Interpret what is said or felt, of which the speaker may not be aware:

Enhancing Project Communication #5.7

Consider Differences as You Communicate

Today's workplace is more diverse than ever before. No matter where you work or what you do, you will interact with people who are different from you. As well as differing in gender, age, cultural background, native language, religious beliefs, sexual identity, ability, education, and experience, we also differ in appearance, attitude, expectations, and communication style. All these differences can lead to problems in interpersonal communication.

In this concept, we discuss being aware of differences and working to keep them from derailing communication.

Cultural Differences

Culture refers to the norms, traditions, beliefs, and customs of a group that influence the perceptions, behaviors, and interactions of its members. Be aware of how your own cultural beliefs are different from those of the people you interact with at work.

In 2013, more than **13 percent of the U.S. population was foreign-born,** and that number is increasing (Brookings, n.d.). Even if you work exclusively in the United States, you will meet people whose heritage and backgrounds are different from yours; customers, employers, suppliers, service providers, and so on.

When you interact with people from cultures different from yours, you may need to try harder to communicate effectively, and you may find some unexpected challenges in teamwork. Learn how to show respect within the cultural norms of others. You may also need to overcome some of your own biases or limiting mindsets.

Generational Differences

Today's workforce is multi-generational and consists of a broader range of ages than ever before. Although the percentages from each generation vary, five different generations are working today, as illustrated in the chart that follows.

Members of each generation bring different backgrounds, perceptions, assumptions, attitudes, and "life lessons" to the workplace.

Understanding the traits common to each of the workforce generations will help you communicate better.

Note that the periods in the chart are estimates, not absolutes. Also, while many people within a generational group share preferences and values, not everyone does. There are always exceptions. Do not make assumptions about any individual.

Generations in the Workforce

GI Generation or **Traditionalists**	People born before 1945. Some are still in the workforce.
Baby Boomers	People born between 1946 and 1964. Now retiring, but many are still in the workforce.
Generation X	People born from the mid-1960s through 1980. Make up a large portion of the workforce.
Generation Y or **Millennials**	People born from 1981 through 2000. Make up a large portion of the workforce.
Post 9/11 or **Generation Z**	People born in 2001 or later. These people are entering the workforce.

In the past, age often meant seniority, but that is not the case today. Depending on the job, education, technological know-how, and personality are just as likely to influence seniority. Thus, we often find younger people supervising older people, or even experienced older people retiring and taking on lower-level part-time jobs to stay active.

Keep Differences from Affecting Communication

Do not assume that people from **different cultures** will always understand the same words, concepts, and gestures in the same way. Further, what is acceptable in one culture may not be okay in another.

If you know you will be interacting with a person from a culture you are not familiar with, take some time to learn about basic social and business rules in that culture. This does not mean you should stereotype people, but rather you should be knowledgeable and not judgmental about cultural differences.

When interacting with someone who differs from you in terms of culture, national origin, or age, make sure to provide explicit and accurate information and check for understanding. Do not lose patience if it takes a while to get your point across and make it clear that you are available for further discussion should that be needed.

Err on the side of caution. For example, if you are unsure about whether shaking hands is acceptable, do not initiate a handshake.

Be respectful of gender and identity issues. Today's workplace consists of people from diverse family backgrounds, and thus you will most likely meet people with attitudes different from yours. Do not allow your personal views to influence your work relationships.

Keep business, business. Avoid talking about potentially polarizing topics such as politics and religion. Asking about a person's family or history, without being overly personal, is generally accepted as polite conversation. Conversing about music, movies, sports, or non-political current events is probably okay, as long as you avoid passing judgment on another person's choices and interests.

Treat others as you wish to be treated. Project managers are role models. They set the tone. If team members are to treat one another with respect, project managers must demonstrate this respect through their behavior.

Keep the following guidelines in mind:

- Be aware of your feelings and attitudes regarding differences
- Recognize that, while these differences exist, human beings share many similarities
- Demonstrate respect for differences
- Deal with the implications of the "self-fulfilling" prophecy
- Recognize what it feels like to be "different"
- Avoid bias, prejudice, and discrimination

Work to mitigate language differences. When you have a language difference with someone, go the extra mile.

Here are tips to help overcome language difficulties:

- Respect the challenges that people whose native language is other than English may have
- Speak more slowly when communicating with the other person (and not more loudly)
- Monitor your use of language and keep it simple while avoiding slang

As with all communication, check to ensure the other person understands you. Simply saying, "Do you understand?" may not be enough. Ask the person to repeat your message to make sure it was understood. You may also want to observe the individual to help determine if the instructions you communicated were understood.

What to do:	**Other actions:**
☐ When communicating in situations with language barriers, simplify your message, maintain a steady volume, avoid slang, and be mindful of personal space and other aspects of body language.	☐ ______________________
☐ Be mindful of your beliefs and attitudes about people from different cultures or age groups.	☐ ______________________
☐ Learn about differences between you and people from other cultures and generational groups.	☐ ______________________

Remember

✓ Above all, be respectful, friendly, and patient when communicating with people whose native language is not the same as yours.

✓ Understanding individual differences can help you relate more effectively.

✓ If you are not sure about how a person wants to be addressed, topics they are comfortable discussing, or use of personal space, ask them.

✓ Just because people come from different generational groups does not mean they cannot work well together. Drawing on the different strengths of each group can produce synergy: the condition where the whole is greater than the sum of its parts.

Enhance Your Learning

Watch this humorous 3-minute clip from the TV show Modern Family by TheFunkypineapple. Granted, English is Gloria's second language; however, the clip shows that communication is a necessary skill no matter if the setting is professional or casual.

TheFunkypineapple. (2012). *The Many, Many, Many Mispronunciations of Gloria from Modern Family.*	Available at: https://www.youtube.com/watch?v=om9EhbBo1Yc

Reinforce Your Learning

Generations in the Workplace. What are the two most common generations in your workplace?

What communication challenges have you experienced or observed as members of one generation interact with those from another?

What could be done to improve understanding (or reduce challenges) between members of different generations?

What could you do to better communicate with someone from a different generation?

Enhancing Project Communication #5.8

Conquer Workday Communication Challenges

As we have discussed, to be a better communicator, you must overcome communication barriers and the challenges that can result from differences among people, such as cultural and generational differences.

Other types of challenges will also affect your ability to communicate. These are the typical problems project managers and teams encounter in the course of ordinary workdays. For example, it can be difficult to talk with someone who does not appear to be listening. This can happen for a number of reasons. They might be busy. And most of us are a little nervous when we must give a presentation to senior managers.

In this concept, we discuss everyday communication challenges and what you can do to communicate more effectively.

How to Manage Common Communication Challenges

In general, you will have better results if you plan and rehearse before what you expect will be a challenging interaction. Preparing and rehearsing your message may feel awkward and seem unnecessary, but it is helpful. As your skills improve with practice, you will need less planning and rehearsal. You can then count on your "natural" behavior to handle all but the most challenging situations.

What follows are problem situations you may encounter, and tips for responding so that you and the other person feel satisfied with the result.

Challenge: Interacting with a person who does not seem to be listening.

Model active listening behavior, which includes maintaining eye contact, leaning in toward the person, nodding, and using affirming sounds or words. The other person is likely to mirror your behavior, which will increase their listening bandwidth.

If you pick up nonverbal cues that the other person is not listening, stop talking. Even though it may be difficult to do, simply remain quiet until the other person speaks or otherwise signals an awareness of the silence.

If the other person interrupts before you have finished, politely let them know you have not finished and ask them to wait before responding.

Periodically, ask the other person to repeat back to you what they think you have said.

If nothing seems to work, ask the person if they are too busy to talk. For example, you may learn the person is waiting for an important call they do not want to miss. You can then set a time to talk when you both will be more comfortable. Just showing the other person you care about their time will go a long way toward resolving the problem.

Challenge: Communicating sensitive or critical information.

While situations and information vary, the following ideas may be helpful when you need to communicate about a sensitive subject.

Plan how to express the message while keeping in mind that people often react with at least some resistance or defensiveness. Among other reactions, this could take the form of anger, denial, assuming a victim role, or blaming you or others for the situation.

Make sure that you are calm and able to maintain emotional control. Do not overreact to the person's response.

Avoid being judgmental or accusatory.

Use a calm delivery tone with frequent pauses. Encourage the other person to share their ideas or concerns and any information they think you need to know.

Be clear about the issue or behavior you want to discuss. If you plan to ask the person to make some kind of a change, be prepared to offer suggestions for change, but do not impose them.

If the other person overreacts to your message, try using a clarifying question. A good one is, "What did you hear me say?" Their response will give you a chance to correct any misunderstanding and refocus on the issue.

Challenge: Receiving and processing information that conflicts with what you "know" or believe.

Focus on the information, especially the facts or evidence provided, and avoid personalizing the points of disagreement or conflict.

Do what you can to assume a calm, attentive attitude. As an example, if this started as an informal conversation and you are standing, suggest that you and the other person sit down in a private location and discuss the issues.

If you are "shocked" by the information, ask for time to reflect before you talk further.

Manage your paralanguage: keep your volume low, pace your speaking on the slower rather than the faster side (but not so slow as to be offensive), and avoid a sarcastic or challenging tone.

Do not be afraid to admit your mistakes, to listen to the input of others, and to change your mind in the face of new evidence or information.

Speaking to Groups

Most of us find it challenging to speak in front of a group, and especially when we give formal presentations. The following tips will help you meet this challenge.

Plan your message. Start by writing down the intent or purpose of your message. What do you want to accomplish? This will help you identify the words and style to use, along with any non-verbal support cues and paralanguage that may enhance the message. Write down the intent from the perspective of what you want the audience to feel, think, understand, or do as a result of the communication.

Consider your audience. What do you know about them? What do they know about the subject? Do they have any preconceived ideas about the subject or you? If so, will that help you, or does it present a barrier you must overcome?

Identify your key points. List the points you want to cover. Include a statement of your intention and supporting information listeners need to know to understand your message and find it credible. Also include questions you might ask to confirm listener understanding and the level of agreement.

Plan for audience response. Anticipate how your audience might respond. What questions are they likely to ask? How will you respond? What objections might they raise, and how will you explain or counter each? Writing down the answers to these questions can be an essential step in helping you gain listener agreement and commitment to your message.

Practice your delivery. Standing in front of a mirror, practice delivering your message using different combinations of words, non-verbal cues, and paralanguage. Listen to yourself to ensure you use the best words and appropriate paralanguage. Watch yourself in the mirror. Pay attention to your facial expressions, gestures, and body language and ensure they support and do not contradict your message.

What to do:	Other actions:
☐ Choose a communication challenge you want to handle better and identify ways to improve.	☐ ______________
☐ Identify situations or interactions where you want to maintain emotional control better.	☐ ______________
☐ Take adequate time to plan your key points and communication style, and to practice before giving your next presentation.	☐ ______________

Remember

- ✓ Before communicating with others, collect the facts, and then prepare your message.
- ✓ Do not be afraid to admit mistakes, to listen to the input of others, and to change your mind in the face of new information.
- ✓ Challenging communication situations require the use of well-honed skills in framing messages and listening actively.

Enhance Your Learning

Watch the following 8-minute video by Improvement Pill to learn more about how to improve your communication skills.

Improvement Pill. (2016). *Communication Skills - How to Improve Communication Skills - 7 Unique Tips!*	Available at: https://www.youtube.com/watch?v=

Reinforce Your Learning

Typical Communication Challenges. List some of the typical challenges you face in an ordinary workday.

For one or more of the challenges, ask your peers to help you identify the source of the problem. Is the problem with the sender and the message or with the receiver and the feedback? Or are barriers interfering with the communication process?

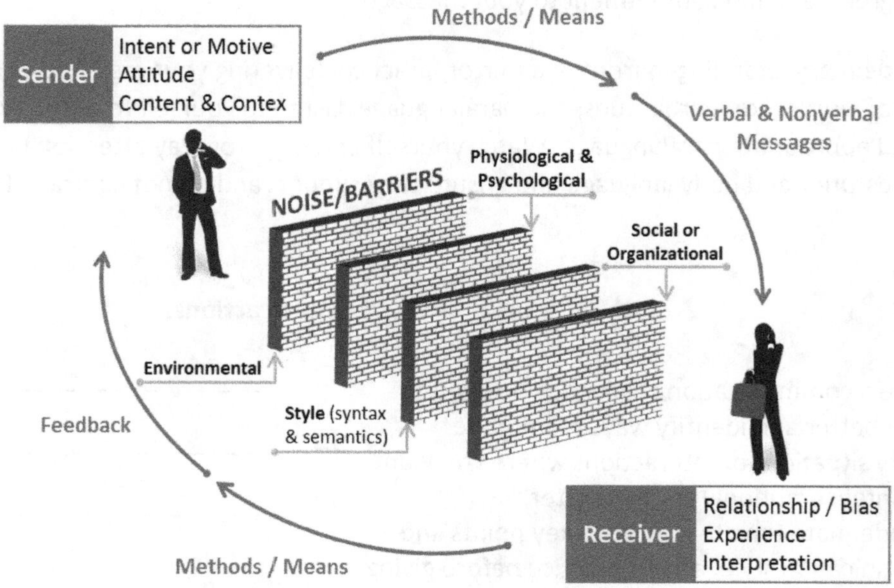

Is this problem a _____ Sender-Message _____ Listener-Feedback _____Barrier-Noise?

At what point in the communication process can you begin to correct this situation? _____ Preparation _____ Exchange _____ Both

Suggestions for communication improvement:

What is the message?

Who needs to be included and why?

What is your recommended medium?

Why did you choose this medium?

What communication method is best? For each of the scenarios that follow, briefly answer the questions in the space provided. For the recommended method question, enter the appropriate code, such as **A** for Personal Conversation.

Communication Scenarios

Communication methods: A. Personal Conversation. **B**. Personal Email. **C.** Personal Conversation + Email Follow up. **D**. Staff Meeting Discussion. **E.** Email to Staff. **F.** Staff Meeting Discussion + Email Follow up. **G.** Others?

1. Your field office has a lunchroom with a coffee maker. Recently, you have noticed coffee grounds and drips spilled on the countertop. Several non-coffee drinkers have complained.

What is the message? ______

Who needs to be included and why? ______

What is your recommended method? ______

Why did you choose this method? ______

2. John is a new client representative who just had his best month ever and helped to propel your team over their goals. His co-workers were impressed by his accomplishments.

What is the message? ______

Who needs to be included and why? ______

What is your recommended method? ______

Why did you choose this method? ______

3. Over the past weeks you have noticed a drop in your work unit's productivity on Friday afternoons. People tend to linger in the lunchroom to discuss their weekend plans—almost always with Chris.

What is the message? ______

Who needs to be included and why? ______

What is your recommended method? ______

Why did you choose this method? ______

4. Last week a safety audit revealed some people were not wearing their eye protection at all times. Incidents were reported throughout the shop, but especially in one work unit.

What is the message? ______

Who needs to be included and why? ______

What is your recommended method? ______

Why did you choose this method? ______

5. You run an estimating department that is behind schedule and missing quote bid dates. Your boss wants you to implement mandatory overtime to catch up.

What is the message? ______________________________

Who needs to be included and why? ______________________________

What is your recommended method? ______________________________

Why did you choose this method? ______________________________

6. Amy (a new employee) sent a text message to her coworker Mary this morning saying she is calling in sick today. Amy did not have your cell phone number, so she asked Mary to forward the message to you via email.

What is the message? ______________________________

Who needs to be included and why? ______________________________

What is your recommended method? ______________________________

Why did you choose this method? ______________________________

7. Your procurement team needs to follow a new, detailed procedure for entering orders. Although the new process is detailed, it is not difficult, but it is critical that all the steps are followed.

What is the message? ______________________________

Who needs to be included and why? ______________________________

What is your recommended method? ______________________________

Why did you choose this method? ______________________________

8. You were just copied on an email where two employees (one in your work unit, one in another) are having a dispute over a process change. You noticed one employee made a rude comment about the other.

What is the message? ______________________________

Who needs to be included and why? ______________________________

What is your recommended method? ______________________________

Why did you choose this method? ______________________________

Enhancing Project Communication #5.9

Use Email and Other Virtual Communication Effectively

There seems to be a paradox when it comes to communication in the modern world. In many ways we communicate more than ever before. For example, we use our cell phones for calls, text messages, and sharing on social media several times a day, and we use our computers at work for sending emails and video conferencing. Yet we seem to spend less time communicating face-to-face. While communication has become more pervasive, many people feel we have become less effective at communicating.

In our quest for speed, we sometimes seem to sacrifice genuine communication. As a project manager, you set the tone for how and when people use email and other communication methods in the workplace. You must guard against losing time and productivity because people do not use technology efficiently, while also ensuring they can communicate effectively.

In this concept, we discuss telephone calls, emails, texts, video chats, and conferencing, and recommend strategies for ensuring that you and your team use these tools effectively.

What Do You Want to Accomplish?

You must always **choose the best technology for the job**. For example, you would probably want to avoid emailing the person down the hall just because you do not like to see them face to face. On the other hand, you would use email if it is the best way to track your conversation, or you know the person is busy and unavailable for a chat.

As another example, technology allows us to communicate easily across space and time, but the way we use technology makes a difference. For example, it is easier to email a collaborator in Australia than to call them by phone. But if your goal is to forge a new relationship, the phone may be a better choice, at least in the beginning.

Telephone Calls

Next to face-to-face interaction, a phone call offers the most precise means of communication. You hear more and understand better by phone than email. However, you usually do not have a written record of the conversation.

When using the phone, be polite and clear. End conversations by summing up any action items, if applicable.

Respect time zone differences. This is true for all forms of virtual communication but vital for telephone interaction.

Emails

Use **descriptive subject lines** and mention if action is required in the subject line. This will make it more likely that your email is read.

Put action items first, so readers do not miss them if they do not read the entire message.

Never send a message in anger. Always take time to cool down before you send an email.

Use the Reply All button only when everyone included on the message needs to see your response. Most of the time, the Reply button is more appropriate.

Write in a way that is respectful of the receiver's time, well organized, and easy to read. For people you do not often write to, include pleasantries such as "Dear Cheryl" and sign your name.

Remember to add any attachments you mention and to provide context for any links and attachments. With concern over viruses, people often will not (and should not) open links or attachments unless they are described or identified.

Avoid sending jokes, tasteless messages, or non-work-related articles and links to people at work.

Emoticons (smiley faces and the like) can add a casual feel to a message and can help to convey feeling. But consider how well you know the receiver and whether you want to send an informal message.

Texts

Text only people who you know use text. Stop texting if the person asks you to stop.

Never text in anger.

Keep texts with coworkers work-related.

Keep texts brief and to the point. Use email or phone for lengthier messages.

Video Chat and Conferencing

With modern technologies, video chat and conferencing are now cost-effective. They support two or more locations communicating via two-way video and audio. Ensure an appropriate location, environment, and visual surroundings when you conduct these sessions.

What to do:	Other actions:
☐ Review, simplify, and correct grammar and spelling in emails before sending.	☐ ________________
☐ Jot down ideas and responses to consider before responding to emails.	☐ ________________
☐ Coach your team on the best practices for email.	☐ ________________

Remember

- ✓ Tell your team your expectations regarding communication technology. Ensure they understand how to use email, and what your policy is about personal use of email and other technologies during work.
- ✓ Establish communication guidelines about the type of medium to use.
- ✓ Project managers set the tone for the use of communication technologies in the workplace.

Enhance Your Learning

Watch the following 2-minute video to learn more about effective email communication.

Massey, K. (2011). *Effective email communication.*	Available at: https://www.youtube.com/watch?v=HF79m5KjuJg

Reinforce Your Learning

Communication Policies. Consider these kinds of communication technologies and answer the questions below.

Telephone Cellphone Email Tweeting Chat Texting Video Conferencing Other

What are your organization's communication policies for the above technologies?

Our email systems are flooded with unnecessary emails. When is it appropriate/inappropriate to copy (cc) others?

When is it appropriate/inappropriate to blind copy (bc) others?

Communicate Project and Job Information Efficiently

As a project manager, you will spend a good part of your day communicating project and job information. For example, you will respond to requests from clients and other stakeholders, discuss the project with your team, and keep everyone informed about project issues and status.

In this concept, we discuss the kinds of communication you will routinely do as part of managing your projects.

Communicating with Stakeholders

Effective communication takes thought, observation, and planning. As a first step it may be helpful to identify who you think are effective communicators and why. Examine how people are delivering messages now. Determine what communication technologies could or should be employed and what methods appear to be most effective.

In a previous chapter, we reviewed the **Stakeholder Communication Matrix**. It is a useful tool for managing communication with stakeholders. As shown in the example matrix below, stakeholders can be anyone involved in your project, such as clients, and administrative managers.

You use the matrix by identifying all the people you need to communicate with and listing them down the left column. Next, consider the kinds of information you must communicate, such as contracts and project plans. Then consider the method or medium of communication and frequency and mark the cells on the matrix accordingly.

Once you have a completed matrix, you can schedule the information you need to communicate, so nothing slips between the cracks.

Stakeholder Communication Matrix Example

Stakeholders \ Project Information	Contracts	Project Plan	Informal Updates	Design Submittal	Kick-off Monitoring	Progress Reports	Scope Changes	Meeting Minutes	Project Specific Info.	Invoices	Etc.
Directors											
Project Managers											
Client Project Managers											
Administration											
Discipline Managers											
QA Managers											
Regulatory Agency											
Etc.											

Communication Medium

1. Written Reports
2. Telephone
3. Web Page
4. Meeting
5. Email
6. Project Related Information
7. Other

Frequency:

D: Daily
W: Weekly
M: Monthly
S: Schedule
O: Occurrence

Communicating with Employees and Contractors

Ensure that employees and contractors who work on your project have copies of organizational policies and handbooks and that you are available to answer questions and assist them in finding what they need to know. For example, employees may have questions about benefits, and contractors may have questions about when they will be paid.

Project managers are responsible for ensuring that employees and contractors who work on the project have the following kinds of information:

- Expectations for their jobs (scope, budget, quality, and timelines)
- Work hours and time-off policies
- How performance is evaluated
- Policies for using work equipment, including email and phones
- Training requirements and opportunities
- Safety, health, and compliance rules
- Emergency procedures

The Way You Deliver Information Matters

When communicating **information, choose the right way to deliver your message**. If the situation is pressing, you might call a staff meeting, send an email, or send a message to all voicemail boxes. Make your decision based on urgency and the content of your message. Have a support structure in place if you are going to relay information that may be upsetting, so people will know how to ask questions.

Consider what recipients need to know. If you are communicating something that will negatively affect people, such as a workforce reduction, also express the positive side of the message. For example, let recipients know if the reduction is part of a plan to improve conditions and what they can expect in the future as a result.

Give your audience time to absorb the information. Be prepared to answer questions. Anticipate the most likely questions and have the answers ready or know where to direct people to get answers.

Get the timing right. Some messages should be communicated before events happen. Other times, the details might need to be addressed before you can communicate. Consider the timing of your message and make informed decisions about when, where, and how to communicate.

If **documentation** matters, letters, memos, and emails will be most effective. Keep the following tips in mind:

- Target your reader or audience
- Craft a short attention-getting header
- State the purpose in the first sentence
- Be clear, concise, and use simple language
- Avoid passive voice and jargon
- Try to limit the communication to a couple of paragraphs

What to do:	**Other actions:**
☐ Provide people with easy access to company policies and procedures.	☐ ______________________
☐ Give people copies of safety and emergency procedures and post them on bulletin boards.	☐ ______________________
☐ Identify the critical job information you are responsible for communicating.	☐ ______________________

Remember

✓ Communicating critical information can be challenging. Each situation is unique and must be considered carefully.

✓ Think about how sensitive information is likely to impact employees, how the timing works best, and the most effective way of disseminating the information.

✓ Anticipate questions and be prepared with appropriate responses.

Enhance Your Learning

The ability to communicate critical project information is an important skill. Review the following article by ProjectManager.com to learn about strategies.

Ray, S. (2017). *12 Effective Communication Strategies at Work.*	Available at: https://www.projectmanager.com/blog/12-effective-communication-strategies-work

Reinforce Your Learning

Critical Project and Job Information. Project team members, contractors, and staff will find a source for the information they need. Ensure that management is that source.

Note three examples of information you regularly communicate to your project team in the space on the next page. Some examples are meeting notes, task assignments, new scope documents, and updated project plans.

Examples of information you regularly communicate to your project team:

1.
2.
3.

What communication method is best? Select one of your examples and answer the questions below. For the method of delivery question, select the appropriate code from the list, for example, **A** for Personal Conversation.
Methods of Delivery: **A.** Personal Conversation **B.** Personal Email **C.** Personal Conversation + Email **D.** Staff Meeting **E.** Email to Staff **F.** Staff Meeting + Email **G.** Others.

What is the message?

Who needs to be included? Why?

What is your recommended method of delivery?

Why did you choose this method?

Summary

A Final Word: *Enhancing Project Communication*

This project management chapter focused on enhancing project communication and the process by which people exchange information with each other. We emphasized the various ways project managers and other people communicate, stressed that perception matters, and concluded with tips on delivering project and job information.

A few topics you have encountered in this chapter include:

- Identifying the key elements of the communication cycle.
- Recognizing the key causes of breakdown in interpersonal communication.
- Understanding the importance of nonverbal communication to transmit a message.
- Using active listening techniques that enhance understanding.
- Recognizing the various types of information that a project manager should consider communicating and to whom.

In the next Project Mastery chapter, we will focus on Developing Project Relationships.

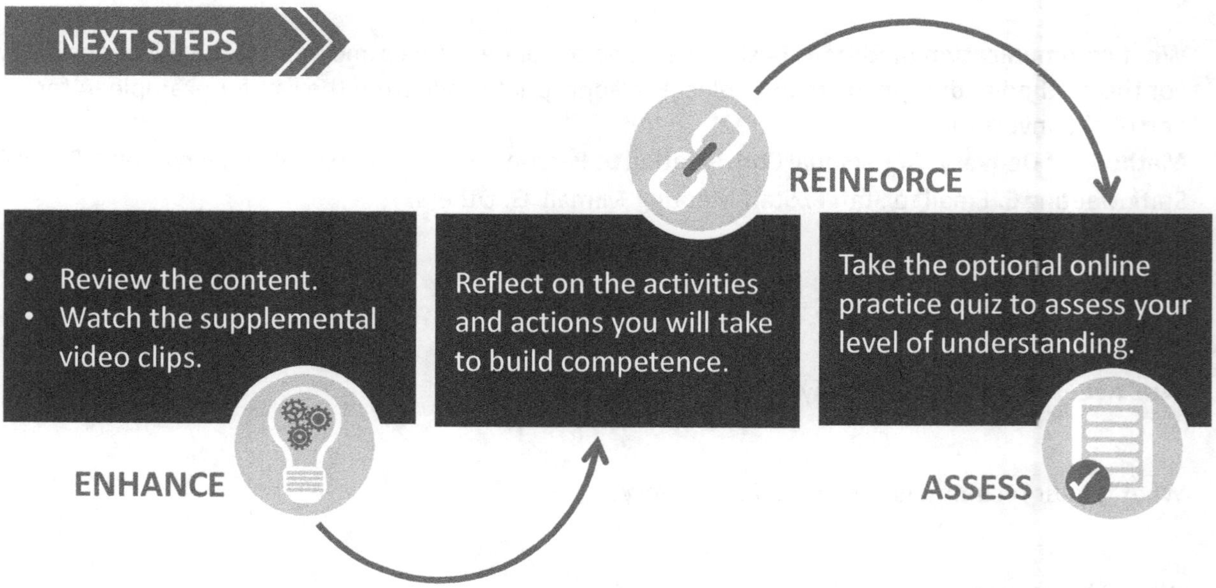

Enhancing Project Communication—Recap Checklist
Project management success depends on effective communication. You must communicate with sponsors, clients, and other stakeholders, and with your team, peers, and management. This course focuses on interpersonal communication, the process by which people use messages to exchange project information.
1. **Understand Communication as an Interpersonal Process** □ As you plan to interact with another person, be mindful of any environmental factors that might interfere with sending or receiving your message. □ Pay attention to how you communicate and for beliefs or behaviors that may cause misunderstandings. □ Make it clear to receivers that you want to hear their questions.
2. Appreciate How Perception Affects Communication □ Consider your perceptions and reflect on how others perceive you. □ Challenge your assumptions and listen attentively. □ Use the four-element model to improve your planning for interpersonal communication.
3. **Recognize Barriers and Why Communication Breaks Down** □ Hold conversations in environments free from noise. □ Consider the psychological conditions and social noise that might be present in your interaction with others and act to alleviate such barriers. □ Assess your communication style and consider your word choices. Could you use better words?
4. Be Aware of the Power of Nonverbal Communication □ Pay attention to your communication habits, especially your nonverbal behavior and paralanguage. □ Pay attention to nonverbal behavior and paralanguage of the people around you. □ Ask clarifying questions when you receive a mixed message.
5. Realize Listening is a Skill □ Examine and improve your nonverbal listening behavior. □ Practice listening for feelings, attitudes, and meaning. □ Commit to paying full attention to the speaker. Avoid interrupting or focusing on what you want to say.
6. **Use Active and Reflective Listening** □ Commit to asking questions to enhance your understanding of a speaker's message. □ If you are too rushed to listen fully, apologize and set a time when you can actively listen. □ Select one of the active listening techniques that you do not use and start using it.
7. **Consider Differences as You Communicate** □ When communicating in situations with language barriers, simplify your message, maintain a steady volume, avoid slang, and be mindful of personal space and other aspects of body language. □ Be mindful of your beliefs and attitudes about people from different cultures or age groups. □ Learn about differences between you and people from other cultures and generational groups.
8. **Conquer Workday Communication Challenges** □ Choose a communication challenge you want to handle better and identify ways to improve. □ Identify situations or interactions where you want to maintain emotional control better. □ Take adequate time to plan your key points and communication style, and to practice before giving your next presentation.
9. **Use Email and Other Virtual Communication Effectively** □ Review, simplify, and correct grammar and spelling in emails before sending. □ Jot down ideas and responses to consider before responding to emails. □ Coach your team on the best practices for email.
10. **Communicate Project and Job Information Efficiently** □ To prevent potential problems, provide people with easy access to policies and procedures. □ Give people copies of safety and emergency procedures and post them on bulletin boards. □ Identify the critical job information you are responsible for communicating.

Knowledge Review Test—Questions

Enhancing Project Communication

1. Communication is often enhanced when the sender _____ the receiver.
    A. speaks up to
    B. uses more physical movements
    C. talks slowly to
    D. shows concern for the perspective of

2. For a variety of reasons, problems can occur at both the sending and the receiving ends of a communication. Which of the following are problems that often occur?
    A. Assumption of understanding.
    B. Excessive information.
    C. Failure to ask for clarification.
    D. Stereotyping.
    E. All the answers are correct.

3. A "mixed message" is:
    A. When a person talks and uses a lot of body language at the same time.
    B. When a person uses a mixture of languages in conversation.
    C. When a person says one thing verbally and something different non-verbally.
    D. When a person uses visual pictures to help communicate what they are saying with words.
    E. None of the answers are correct.

4. Regarding perception in communication, which of the following is not accurate?
    A. We describe and see things in terms of our experience.
    B. We tend to find in a situation what we are looking for.
    C. All people in a meeting hear and understand what is said in the same way.
    D. We tend to simplify things we do not understand.

5. Communication breaks down for all but one of the following reasons:
    A. Senders assume that receivers understand what has been communicated.
    B. Information is not condensed.
    C. Differences in background and experience.
    D. Methodical presentation of ideas.

6. Common barriers to communication include all except:
    A. Physical or environmental noise.
    B. Physiological conditions.
    C. Well-crafted message.
    D. Emotional conditions.
    E. Unchecked body language.

7. Which of the following are considered non-verbal behaviors?
    A. Pauses, hesitations, sighs.
    B. Eye contact.
    C. Tone of voice.
    D. Gestures.
    E. All the answers are correct.

8. Studies have shown that which of the following has the biggest impact on communication:

A. Verbal message.
B. Tone of voice.
C. Body language.
D. None of the answers are correct.

9. When listening to an individual who has something important to discuss, it is important to:

A. Let the individual talk without interruption.
B. Talk to the individual as soon as possible.
C. Ask the individual to sit down.
D. Let the individual air his or her feelings first.
E. All the answers are correct.

10. The average speaking rate ranges from 120 to 200 words per minute, and the listening comprehension rate ranges from ___ to ___ words per minute.

A. 100 to 150
B. 250 to 350
C. 350 to 500
D. 600 to 1000

11. The existence of communication barriers is likely to cause:

A. The project to be on time.
B. Trust level to be enhanced.
C. Conflict to occur.
D. Senior management to be pleased.

12. A crucial part of active listening is the ability to ask questions to clarify or confirm your understanding of what a speaker said. All the following are benefits of asking questions, except:

A. Lets the speaker know you have been listening.
B. Helps you understand what the other person has said.
C. Allows the other person to clarify what they said.
D. Enables you to get your point across.
E. Encourages the other person to provide feedback.

13. Which of the following does not increase communication effectiveness?

A. Simplifying the content.
B. Giving details in order.
C. Using more than one communication channel.
D. Speaking very slowly.

14. Which of the following are tips to help overcome language difficulties:

A. Respect the challenges that people whose native language is other than English may have.
B. Speak more slowly when communicating with the other person (and not more loudly).
C. Monitor your use of language and keep it simple while avoiding slang.
D. All the answers are correct.

15. Typical problems project managers and teams encounter during ordinary workdays.

   A. Differences among people, such as cultural and generational differences.
   B. Difficult to talk with someone who is distracted and does not appear to be listening.
   C. People are too busy on the job.
   D. Presentations to senior managers can be nerve racking.
   E. All the answers are correct.

16. The following are all good tips for speaking in front of a group or giving a formal presentation, except:

   A. Plan your message.
   B. Consider your audience.
   C. Identify your key points.
   D. Plan for audience response.
   E. For the delivery, wing it.

17. Formal written communication with the customer is required when:

   A. The project is over budget.
   B. The customer requests additional work not covered in the contract.
   C. The project has schedule slippage.
   D. Defects are detected.

18. The following points are all good to consider about email communication, except:

   A. Review, simplify, and correct grammar and spelling in emails before sending.
   B. Jot down ideas and responses to consider before responding to emails.
   C. Writing in all caps is a good habit to follow.
   D. Coach your team on the best practices for email.

19. Project managers are responsible for ensuring that employees and contractors who work on the project have the following kinds of information. Which of the following are important to communicate?

   A. Expectations for their jobs (scope, budget, quality, and timelines).
   B. Work hours and time-off policies.
   C. How performance is evaluated.
   D. Safety, health, emergency procedures, and compliance rules.
   E. All the answers are correct.

20. If documentation matters, letters, memos, and emails will be most effective. Keep all of the following tips in mind, except:

   A. Target your reader or audience.
   B. Craft a short attention-getting header.
   C. State the purpose in the first sentence.
   D. Be clear, concise, and use simple language.
   E. Try to limit the communication to a couple of pages.

Enhancing Project Communication —Answer Sheet

1.	___ A	___ B	___ C	_X_ D	___ E
2.	___ A	___ B	___ C	___ D	_X_ E
3.	___ A	___ B	_X_ C	___ D	___ E
4.	___ A	___ B	_X_ C	___ D	___ E
5.	___ A	___ B	___ C	_X_ D	___ E
6.	___ A	___ B	_X_ C	___ D	___ E
7.	___ A	___ B	___ C	___ D	_X_ E
8.	___ A	___ B	_X_ C	___ D	___ E
9.	___ A	___ B	___ C	___ D	_X_ E
10.	___ A	___ B	___ C	_X_ D	___ E
11.	___ A	___ B	_X_ C	___ D	___ E
12.	___ A	___ B	___ C	_X_ D	___ E
13.	___ A	___ B	___ C	_X_ D	___ E
14.	___ A	___ B	___ C	_X_ D	___ E
15.	___ A	___ B	___ C	___ D	_X_ E
16.	___ A	___ B	___ C	___ D	_X_ E
17.	___ A	_X_ B	___ C	___ D	___ E
18.	___ A	___ B	_X_ C	___ D	___ E
19.	___ A	___ B	___ C	___ D	_X_ E
20.	___ A	___ B	___ C	___ D	_X_ E

Without continual growth and progress, such words as improvement, achievement, and success have no meaning.

—Benjamin Franklin

Chapter 6

Developing Project Relationship

Overview

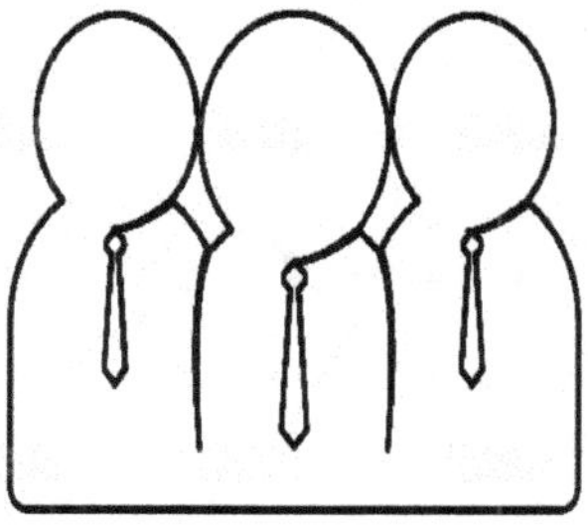

IT'S ALL ABOUT RELATIONSHIP

> ***"Whatever words we utter should be chosen with care for people will hear them and be influenced by them for good or ill."***
>
> —Buddha

Being a project manager requires that you continually work to improve your relationship skills. You must develop and maintain relationships with the leadership above you, with the people you lead, and with stakeholders, peers, customers, and suppliers.

One study of 500 employees found that **77 percent wanted to feel they contributed** to their manager and team, and **74** percent said they **wanted their manager to show some appreciation** (as cited in Business Management Daily, 2012).

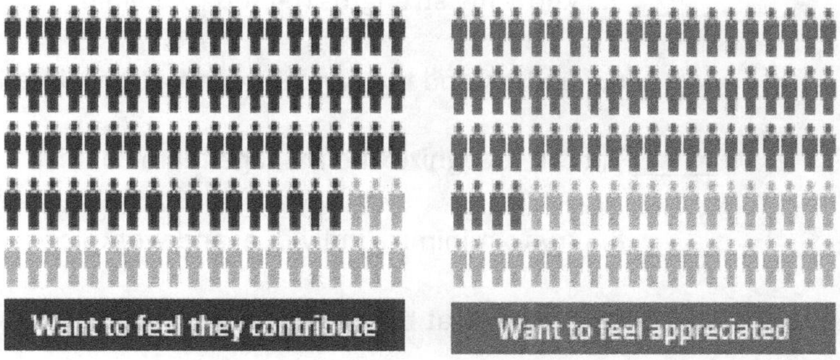

This chapter describes how to build and nurture project relationships. You will assess your interpersonal style and learn how your behavior and communication skills impact your effectiveness, and how developing collaborative relationships can make your project more productive.

After completing this chapter, you will know how to:

- Describe the characteristics and building blocks associated with high-quality work relationships
- Build trust with team members and stakeholders
- Explain what is meant by interpersonal style and how style differences can affect the quality of communication
- Identify four interpersonal styles and describe their characteristics
- Describe interpersonal conflict, typical causes, and methods of resolving

The competencies associated with this chapter include:

Interpersonal Relationship Building	Conflict Management	Influencing/Negotiating

Take Your Temperature for Developing Project Relationships

The following assessment will help you identify what you know about developing interpersonal project relationships and the areas where you can improve. With you and your organization in mind, read each statement and consider how much you agree with it. Use the numbers from 1 to 10, where **1** means you **strongly disagree** and **10** means you **strongly agree**.

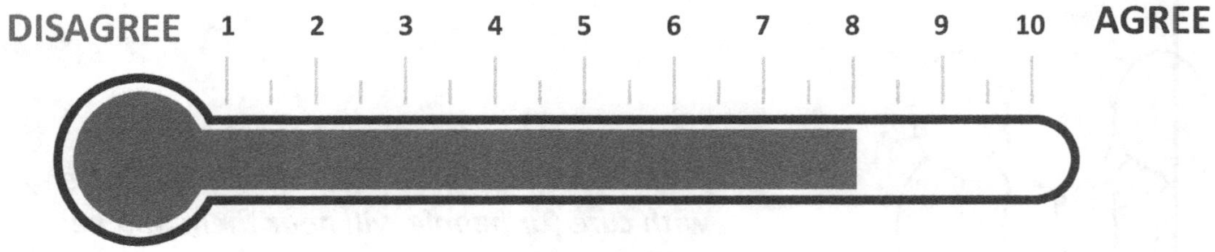

_____ 1. I can identify the characteristics of high-quality work relationships.

_____ 2. I am good at building trust with team members and stakeholders.

_____ 3. I am aware of my personal preferences and biases.

_____ 4. I know how to recognize my own and someone else's interpersonal style.

_____ 5. I know my strengths related to building relationships.

_____ 6. I understand the concept of emotional intelligence.

_____ 7. I can recognize and manage team conflict.

_____ 8. I try to minimize individual and workplace stress.

_____ 9. I am good at balancing inquiry and advocacy.

_____ 10. I use principled negotiation techniques.

_____ **Total**

Take a moment to reflect on the following before you move forward in the chapter.
In your work, what problems have you experienced or observed related to developing relationships at work?

Take a few minutes to also reflect on your self-assessment. List two or three areas you want to improve.

As you progress through this workbook consider what actions you can take to demonstrate competence in the following three competency areas:

1. **Interpersonal Relationship Building.** Considers and responds appropriately to the needs, feelings, and capabilities of others; seeks feedback and accurately assesses impact on others; provides helpful feedback; builds trust with others.

2. **Conflict Management.** Anticipates and seeks to resolve disagreements, complaints, and confrontations in a constructive manner.

3. **Influencing and Negotiating.** Keeps groups and individuals informed; appropriately uses negotiation, persuasion, and authority in working with others to achieve goals; builds productive networks.

Developing Project Relationships #6.1

Identify Characteristics of High-Quality Work Relationships

Work does not exist in a vacuum. You must build relationships with your project team, stakeholders, and associates. Regardless of your job title and responsibilities, you can serve as a role model for building and maintaining high-quality work relationships. You can take the initiative to develop productive relationships with everyone.

Relationships must be nurtured as well as built. This requires a good understanding of yourself and other people, and the ability to find areas of commonality when differences might reduce the quality of the relationship. It also requires that you act to make things better. High quality work relationships contribute significantly to a healthy work environment, which we define as an environment where individuals can experience job satisfaction and productivity, and better respond to workplace stress.

In this concept, we discuss what is meant by high-quality relationships and the attitudes that will help you develop such relationships.

Openness is Crucial in Building Relationships

Openness in communication is key to developing work relationships because it is essential for developing understanding.

First, you must be willing and able to communicate who you are, what your feelings are, and your honest regard for what other people contribute. Also, you must be willing and able to receive communication from other people; who they are, what their feelings are, and how they regard you.

When you understand other people, you are aware of and appreciate their needs, opinions, values, concerns, and expectations. The more effort you expend to achieve understanding, the more you benefit from the "loop effect" in interpersonal relationships. This occurs when others feel understood, valued, and accepted. They are then more inclined to consider your needs, opinions, values, concerns, and expectations.

Work Relationships are Interpersonal

According to *Merriam-Webster,* an *interpersonal relationship* is an association relating to or involving relations between people. The association may be between two or more people and may range in duration from brief to enduring. This association may develop from a variety of sources: kinship, neighborhood, social or recreational activities, regular business interactions, and, of course, places where people work together.

Effective work relationships require good communication skills. In high-quality work relationships, each person is responsible for setting a positive tone and for using sound communication practices. These are practices based on a fundamental understanding of how fragile interpersonal communication is. It is easy for messages to be misunderstood, for distractions to interfere with listening, and for untested assumptions to overshadow what another person says.

Knowing a bit about the "beyond work" lives of your coworkers will help you build relationships with them. This does not mean that you must develop friendships or pry into personal situations but being aware of someone's personal life helps establish rapport. A reasonable level of familiarity is sufficient. For example, you might discover that you both enjoy the same hobbies or activities outside work.

Members of high-quality work relationships embrace and value diversity. We differ from each other in many ways, such as physical and cultural attributes, religious beliefs and moral values, personality traits, styles of thinking and learning, competencies, and the goals we seek to achieve through our work. Working with people who come from different backgrounds and who have different skills and interests will promote a broader range of ideas and possibilities for new solutions and best practices.

Trust is a cornerstone of productive work relationships. People who work closely together do so from a base of mutual trust. It is not always easy to extend trust to people you do not know well or have not worked with before. Nor is it guaranteed that people will "automatically" extend trust to you. Trust must be built. Given today's increased dependence on collaboration and teamwork, you must make it a priority.

Shared responsibility for the relationship is an essential building block. In the famous book, *How to Win Friends and Influence People* (Carnegie, 2010), we learn that you cannot fake being interested in other people. To build relationships, you need to find ways to generate enthusiasm and interest in others. Practice listening with genuine curiosity; make the words of others matter to you.

Create the feeling of a team. Whether working as members of an official team, or as individual contributors in a workgroup, everyone must feel they are part of the team. People need to feel they are making a contribution to something important, and that others depend on them.

Show your team members and associates respect. This means using appropriate language, providing encouragement, giving support and constructive feedback, and acknowledging them as people.

What to do:	Other actions:
☐ Build a good rapport with each person on your project or in your work unit.	☐ ______________________
☐ Value diversity and encourage a wider range of ideas to create new solutions and best practices.	☐ ______________________
☐ Create the feeling of a team by showing other people you value their contributions.	☐ ______________________

Remember

- ✓ Be mindful of your behavior. Having good relationships does not occur by accident. If you act without thinking, you risk unwelcome consequences.
- ✓ Be conscious of the quality of your relationships and what you are doing to maintain them.
- ✓ Trust, respect, and effective communication are vital components of quality work relationships.
- ✓ High-quality work relationships are human relationships. Having at least a casually personal component helps us remember the human element. It is not just about getting the job done. It is also about connecting with others to do so. It is easier to extend care and support to someone you have taken the time to get to know.

Enhance Your Learning

Watch the following 3-minute video to learn more about building relationships in the workplace.

Halloran, A. (2009). *Workplace Communication – Three Steps to Building Relationships.*	Available at: https://www.youtube.com/watch?v=Hrd06zuiofI

Reinforce Your Learning

Work Relationships Self-Assessment. Think of three people with whom you work closely. For each one, write a few things you know about these people. Examples are where do they live, do they have a committed relationship, children or pets, anything about their hobbies, recreational activities, and your perception of them.

1.

2.

3.

Could you answer the questions about all three people? Could they answer the questions about you?

Developing Project Relationships #6.2

Build Trust with Team Members and Stakeholders

In project management, we start with the premise that a team exists because what needs to be done is too complex or of such scale that one person, or a series of individuals, cannot do it as effectively as a group working together. Teams achieve their objectives in the same measure as they are able to draw on the contributions of each member. This However, this teamwork—sharing information, problem-solving, and working together to complete tasks—relies on trusting relationships.

It has been said that trust takes years to build, seconds to break, and forever to repair. It is also true that trusting relationships do not just happen. People build them through communication and interaction. In general, the better the understanding between two or more people, the easier it will be for them to work together, resolve conflicts, reach agreements, and influence each other in productive ways.

In this concept, we discuss how trust affects projects and what you can do to increase trust levels in projects you manage.

Trust and Project Success

Trust allows people to focus on work with the confidence that others will follow through on their commitments. However, when people treat others in deceitful, condescending, intimidating, closed-minded, or other negative ways, trust is challenged or destroyed. Our sense of security is lost. When many work relationships lack trust, employees at all levels may also lose trust in the organization and its values.

With low trust, productivity is likely to decrease. Team members tend to gossip, miss deadlines, fail to bring their best thinking and creativity, and do not make an extra effort to get the job done. As team members become more and more disengaged, they may choose to find work in other projects or organizations.

Conversely, as you build trust in a team, not only will the team flourish, but your work output, sense of satisfaction, and level of commitment will grow. You and the project team will reap the benefits of an environment typified by high morale, where individuals take pride in their own and the group's achievements.

How to Increase Trust

Work side-by-side. Trust is built as people work with others and gain a better understanding of their skills, communication habits, and work styles.

To build trust in a new relationship, find opportunities for people to work together. This does not mean they must work together every day, but gaining direct experience is the best way to build trust.

Do not tolerate dishonest behavior. To promote trust in the workplace, make sure people know you will be honest and fair in dealing with them, and that you expect them to treat you and each other the same way.

Keep others informed and share information. Positively sharing helpful information helps build trust and shows respect. Some people will inevitably have work-relevant knowledge and information that others do not have. Encourage people to share with others. Withholding breeds distrust while sharing invites communication and collaboration.

Model this behavior by asking others for input as you work on a task or make a decision.

Give credit where credit is due. Take the time to acknowledge other people's accomplishments and support and encourage team members to do the same. Let people know you are aware of their contribution and that you appreciate it. Not giving credit when deserved, or worse, taking credit for something someone else did, will seriously erode trust.

Be consistent and forthright. You can lose trust when you act or speak in ways that cause others to see you as inconsistent, unpredictable, or unreliable. It is difficult to trust someone when you do not know which "version" of that person you might be dealing with, or when you are not sure you can believe what they say.

Demonstrate that you trust your team members and stakeholders. Give your team members and stakeholders every reason to believe you trust them. If, as sometimes happens, trust is lost, it should not be because you did not behave in a trustful way. Also, when you demonstrate trust, it is more likely that others will trust you.

What to do:	Other actions:
☐ Work side-by-side with your team and stakeholders to build relationships and trust.	☐ ____________
☐ Give credit where credit is due.	☐ ____________
☐ Be consistent, forthright, and keep others informed.	☐ ____________

Remember

- ✓ You show your team and stakeholders respect when you trust them to do their jobs and to behave in ways that align with the best interests of the project and organization. You also show respect when you work with them and share relevant information.
- ✓ Trust is a precious commodity, more difficult to earn than to lose, and difficult to regain once lost.

Enhance Your Learning

Watch the following 3-minute video to learn four tips for building trust, competence, connection, communication, and honesty.

Madsen, S. (2015). *How to Build Trust - Project Leadership Training.*	Available at: https://www.youtube.com/watch?v=afgzypHh8Oc

Reinforce Your Learning

Building Trust and Respect. Consider a time you worked closely on a task with one or more team members or stakeholders. Briefly describe the situation below and then respond to the other questions.

Situation (people involved, type of task, how long it took to do the work):

Did trust exist when you first started the task and/or at the end of the task?

Was trust, or the lack thereof, raised as an issue?

Consider that situation and the outcome. Did the job get done? If so, what did you do to contribute to the result? If the job did not get done, what could you have done differently?

Developing Project Relationships #6.3

Recognize Personal Preferences and Bias

We all have preferences. Some examples are the type of work we like, specific people we want to work with, and how we prefer to communicate.

In the workplace, you will encounter people with preferences different from your own. Productive work relationships require that you respect and try to understand the differences, and that you help others perform in concert with their preferences.

Unfortunately, acting on one's preferences can sometimes lead to favoritism or bias. For example, a preference to work with a certain kind of person and to see other people as less desirable will not lead to a work environment where people can be "the best they can be."

In this concept, we discuss the human tendency to have preferences and sometimes even bias, and how being aware of these tendencies will help you be a better project manager.

People Want to Contribute and Feel Appreciated

When asked about satisfaction in the workplace, workers at all levels had similar responses. If they liked their work and the people they worked with, they were satisfied (Mercanix, n.d.).

While this was an overall finding, each respondent had specific reasons for why they liked their jobs and co-workers. By getting to know your team and stakeholders, you can understand what aspects of the job they like and do not like and use that information to create a work climate that is more positive for everyone.

Most people prefer to work for a leader who is honest and fair and whom they respect and trust. One study of 500 employees found that **77 percent of employees** wanted to **feel they contributed** to their supervisor's team (and collaborated), and **74 percent said they simply wanted their manager to show appreciation** (as cited in BMD, 2012).

However, it is a mistake to assume that because someone is doing a specific job, that it is their preference. You will learn what people on your team need and what their preferences are if you ask. Once you know their preferences, you may find that switching things up even a little bit leads to a more energized work environment.

When it comes to your preferences, think about **what you like about your job or project** and what you **want out of it**. If you are not sure, devote a bit of time each day to consider what went well, what you enjoyed, and what you did not enjoy. Even remembering what you miss about a previous job or project can help clarify your preferences.

Once you understand your preferences, you might talk to your manager with the goal of doing more of the things you prefer and less of the things you do not like.

At the same time, be careful you do not let your preferences impact how you do your job and how you interact with others. There is a **fine line** between **preference** and **discrimination or bias.** It is natural to connect with some people more than others, but do not allow your preferences to exclude or demean those who you prefer less.

Communication Preferences Matter

People have deep-seated preferences for how they want to engage with others. The people on your team will not all communicate the same way, and it is doubtful you communicate the same in every relationship.

Pay attention to how your team members express themselves. For example, notice how they listen, if or how easily they speak up in a group, how directly they agree or disagree with someone, and whether they speak differently to peers and management. You will see that some people are direct, while others are more guarded. Some are reluctant to say anything that might be viewed as critical, while others are comfortable offering their opinions to anyone.

Observing and noting communication preferences, and being aware of your own, will help you deal with differences and avoid falling into the traps of negative bias or favoritism.

What to do:	Other actions:
☐ Identify what you like about your project or job and what you want out of it.	☐ ____________________
☐ Be observant and reflect on the preferences of your team members and associates.	☐ ____________________
☐ Do not let your preferences negatively affect how you treat people.	☐ ____________________

Remember

- ✓ Understanding your preferences for what you do, with whom you work, and how you communicate can help increase your satisfaction with your job and project.
- ✓ Understanding the preferences of your team and associates can help you prevent unnecessary breakdowns in communication and trust.
- ✓ One study of 500 employees found that 77 percent wanted to feel they contributed to their team, and 74 percent said they wanted their manager to show some appreciation (as cited in BMD, 2012).

Enhance Your Learning

Read the following article from Chron/Hearst to learn about the difference between discrimination and preference in the workplace.

Scott, S. (2021). *The Difference between Discrimination & Preference.*	Available at: http://smallbusiness.chron.com/difference-between-discrimination-preference-10338.html

Reinforce Your Learning

Awareness of Personal Preferences. List three things you can do to ensure you are not showing favoritism on the job.

1.

2.

3.

List three things you can do to learn more about the preferences of your team and associates.

1.

2.

3.

Developing Project Relationships #6.4

Be Aware of Social Styles in the Workplace

People have different **social** or **interpersonal styles**, which means individuals differ in how they interact with other people. Understanding style differences matters in the workplace because you will work with people of all styles, and you will need to communicate in ways that are productive and have a positive impact on the organization.

Being aware of social **styles** can help you in many situations. For example, you will better understand why someone acts or speaks in a certain way, why some people work well together while others do not, and why you find it difficult or easy to communicate with a particular person.

In this concept, we discuss social styles in the workplace and how you can use this information to understand better people. We will go into more detail about social style in the next concept and how knowing your style can help you be a more effective project manager.

What is Social Style?

Social style refers to the way an individual interacts with others. It is not exactly the same as our **personality style**, which is how we are motivated and think internally.

While there are several models of social style, most describe three main behaviors: assertiveness, responsiveness, and versatility.

Assertiveness describes the degree to which people take responsibility for influencing the thinking and actions of others. In other words, it is the extent to which an individual is direct and action-oriented in interacting with others.

Responsiveness describes the degree to which a person displays or controls their emotions while interacting with others.

Versatility describes the extent to which a person is adaptable or willing to change their behavior to accommodate the needs of others or the workplace.

When you consider social style, you can look at your team members and associates as individuals, or collectively, as a group.

Considering a person as an individual will help you understand why your interactions with the person are of the quality they are, why you may find points of irritation or frustration with each other, how best to ask the person for help or offer support, and what you need to share to influence the person.

Considering a group's style will help you understand the degree to which you feel included and accepted in the group, why group meetings are the way they are, where the group's momentum seems to be directed, and how effective the group is in functioning as a shared-responsibility team.

You can use your knowledge of a person's social style to help shape your communication with the person; this includes differences of opinion and conflicts.

Of course, **no person fits into a single square box**. We all have different aspects to our personalities, a certain amount of style flexibility, and different interpersonal strengths that shine through at certain times. Understanding others' styles and strengths can improve the quality of your relationships and can also reduce the extent to which you pigeon-hole people. That is, it can help you curtail what seems to be a natural human tendency to lock people into specific roles and therefore see them in a certain light.

Social Style Assessments

Many personal and **social styles assessments are available**. You can research them, take classes, and speak with your human resources department to determine if any are available in-house. Any of these assessments can help you become more aware of your social style and how to adapt it to maximize your relationships with other people.

The following are some of the most well-known assessments:

- Keirsey Temperament Sorter®
- DiSC®
- Interpersonal Style Profile
- Myers Briggs Type Indicator (MBTI)®

What to do:	Other actions:
☐ Review one of the social style assessments that are available.	☐ ______
☐ Encourage your team members and associates to complete the same assessment and share the results.	☐ ______
☐ Consider how you might use your knowledge of social styles to improve your relationships with team members and associates.	☐ ______

Remember

✓ Understanding the various social styles of others can help you get beyond points of irritation and increase your appreciation of the contributions of their styles.

✓ Many commercially available assessments can help individuals identify and better understand their social style and the styles of others. Examples include the Interpersonal Style Profile, Keirsey Temperament Sorter®, the DiSC®, and the Myers Briggs Type Indicator (MBTI) ® assessment. All of these can help you become more self-aware, and you may realize more about how to adapt your social style to better communicate with others.

Enhance Your Learning

Watch the following 5-minute video to learn more about social styles: *TRACOM's SOCIAL STYLE Model HD Version.*

Tracom Group. (2010). *TRACOM's SOCIAL STYLE Model HD Version.*	Available at: https://www.youtube.com/watch?v=wRBx8lkV-kQ

Reinforce Your Learning

Social Style Dimensions. Imagine you are organizing a team to work on a project, product design, or initiative. Consider the three basic social style dimensions of assertiveness, responsiveness, and versatility.

If you put three people on your team who all are highly assertive but low in versatility, what might happen?

Can you think of two people in your organization or an organization you are familiar with who might not work well together? Without disclosing their names, describe the characteristics that cause you to think these people might not work well together.

1.

2.

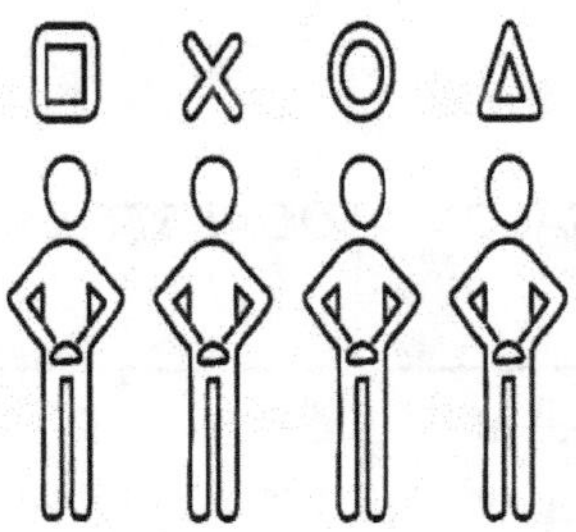

Developing Project Relationships #6.5

Know Your Social Style

Teams and organizations that respect individual differences and recognize the value in differences thrive. Such teams and organizations tend to be more innovative, productive, and responsive to change, and they tend to have people loyal and committed to organizational success.

Recall that social style speaks to the way individuals differ in how they interact with others. In this concept, we go into more detail about the differences in social styles and discuss how you can use this information to work more productively and to build truly collaborative teams.

The Four Main Types of Social Style

While many models for social styles are available, some listing up to 16 different categories, four main types have been used for decades. Understanding these four types will help you determine your dominant style.

Below, and in the illustration that follows, we refer to these four styles as I, II, III, and IV:

People with a **Type I** social style are interested in the big picture and results. They are not bothered by details and tend to be goal-oriented and practical. They are often competitive, quick to take charge, and very direct in their communications.

People with a **Type II** social style are most concerned with social acceptance. They want to be recognized for their accomplishments and valued. They tend to be spirited, enthusiastic, creative, good at influencing others, and thrive in times of change.

People with a **Type III** social style crave stability, harmony, and consistency. They are good listeners and team players, considerate of the needs of others, and prefer incremental, non-disruptive change.

People with a **Type IV** social style tend to be detail-oriented with have a strong need to be systematic and do things the right way. They value logic and accuracy, are disciplined and methodical in their work, and resist "change for the sake of change."

Four Main Types of Social Styles

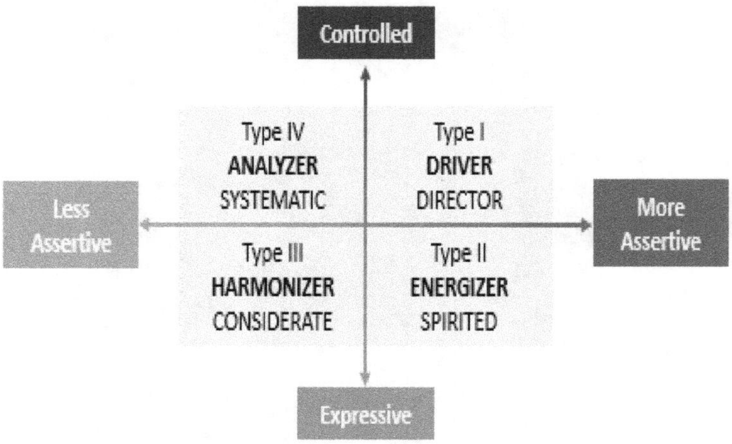

What is Your Social Style?

Consider what style or type you might be. You can use your knowledge of your own interpersonal or social style to evaluate your strengths and weaknesses. Think about how you communicate with other people. You might find that you can learn something by considering what is positive about each type.

Of course, as we mentioned before, no person fits into a single box. We all have different strengths and weaknesses that come through in certain situations.

Use Your Social Style to Benefit Your Project and Team

Understanding your strengths and areas for improvement can help you interact more productively with your team and associates. It will also make you more aware of those same strengths and weaknesses in other people, which can help you support and encourage their contributions and growth.

Additionally, consider how your strengths and weaknesses may affect the functioning of your project team. For example, if you are very much a type IV, would it benefit the team if you encourage the participation of people who are more type II? Or, for example, would you create a more energetic team if you reminded yourself to value the contribution of people with a type III style?

What to do:	Other actions:
☐ Focus on your social style and determine what you can do to build on your strengths.	☐ ________________
☐ Identify people who seem to represent each style and identify ways you can use the strengths of that style to enhance your team's success.	☐ ________________
☐ Identify at least one associate with whom opening-up communication might improve your relationship.	☐ ________________

Remember

- ✓ Social style refers to the way an individual interacts with others. It is not exactly the same as our personality style, which is how we are motivated and think internally.
- ✓ It is helpful to understand the various social styles of your team members and associates.
- ✓ Recognizing the social types of yourself and your team can help you build more effective teams.

Enhance Your Learning

Social or Interpersonal Styles Inventory. Click the first link below and take the free online Keirsey Temperament assessment and/or click the second link and download the free Centrestar Interpersonal Style Profile inventory:

https://profile.keirsey.com/#/b2c/assessment/start

https://www.centrestar.com/index.php?option=com_osdownloads&view=item&id=2

After completing the Keirsey Temperament assessment or the Centrestar inventory, respond to the questions in the "Reinforce Your Learning" activity on the next page.

Reinforce Your Learning

Social or Interpersonal Style Profile Inventory.

After completing the Interpersonal Styles inventory or the Keirsey Temperament assessment, reflect on the four styles or temperaments and identify people you know who fit each style. Then, with your style or temperament in mind, consider what you could do to enhance relationships with each type and note your ideas below.

Drivers/Guardians:

Harmonizers/Idealists:

Energizers/Artisans:

Analyzers/Rationals:

Social or Interpersonal Style: Questions for further reflection.

1. Why do organizations need your particular social style?
2. What if everyone in the organization were the same style as you?
3. How do you prefer to be treated?
4. How can leaders better apply knowledge of social styles to enhance organizational success?

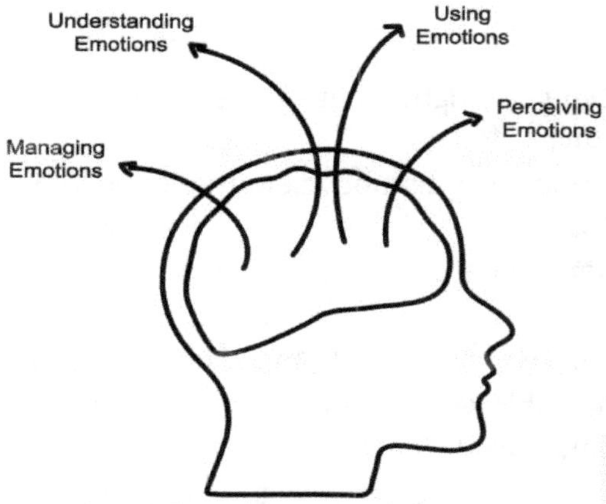

EMOTIONAL INTELLIGENCE

Understand Emotional Intelligence

"Our feelings are not there to be cast out or conquered. They're there to be engaged and expressed with imagination and intelligence."

—T.K. Coleman, author

Interpersonal communication in the workplace is complex, and many factors affect the quality of messages. By *quality*, we mean a message as understood by a receiver is the same message a sender intended to transmit.

Planning content and timing can help improve quality, but some people may appear to be gifted at getting their messages through. However, the reality is that most people who excel in interpersonal communication have developed their abilities through effort and practice. Excellent communicators have developed their emotional self-awareness and accuracy in reading the emotions of others, which are vital components of emotional intelligence.

In this concept, we describe emotional intelligence and discuss how you can improve the quality of your messages by honing your emotional intelligence skills.

Emotional Intelligence and Success

People with high emotional intelligence tend to be effective in developing productive and satisfying work relationships, and they tend to succeed in a variety of professional roles.

Research indicates that emotional intelligence may account for **58 percent of your success in life**, both professional and private. However, despite its importance, one study found that only **36 percent of people** studied were emotionally intelligent enough to recognize their emotional state (Van Camp, 2015).

In another study examining how emotional intelligence and IQ affect performance, the **lower IQ** person **outperformed** the higher IQ person **seven out of ten times** if the person with a lower IQ had **higher emotional intelligence** (Bradberry, 2014).

What is Emotional Intelligence?

Daniel Goleman, one of the chief contributors to our current understanding of emotional intelligence (EI), defines it as "the **capacity for recognizing our feelings and those of others**, for motivating ourselves, and for **managing emotions** well in ourselves and our relationships."

EI refers to abilities distinct from, but complementary to, academic intelligence, which is measured by IQ.

Daniel Goleman's EI model describes a wide array of competencies and skills associated with success in many professional roles, including leadership.

Goleman identifies **five areas involved in EI**:

- **Self-awareness**—knowing one's emotions, strengths, weaknesses, drives, values, and goals, and recognizing their effects in one's life
- **Self-regulation**—controlling or redirecting one's disruptive emotions and impulses and adapting to changing circumstances
- **Motivation**—emotional tendencies that facilitate achievement, such as achievement drive, commitment, and initiative
- **Empathy**—being aware of other people's feelings, needs, and concerns
- **Social skills**—adeptness at inducing desirable responses in others, through communicating, influencing, leading, and building bonds

Goleman describes a set of emotional competencies within each of these areas. Emotional competencies are not innate talents. They are learned capabilities you can develop, and doing so will improve your performance.

Competencies exhibited by emotionally intelligent people include the following:

1. Identify and name one's emotional states and understand the link between emotions, thoughts, and actions.
2. Manage one's emotional states to control emotions or to shift undesirable emotional states to more appropriate ones.
3. Tap into emotional states associated with a drive to achieve and be successful.
4. Read, be sensitive to, and influence other people's emotions.
5. Enter and sustain satisfactory interpersonal relationships.

In Goleman's model, **the emotional competencies build on one another in hierarchical order**. In other words, you must first learn to identify your emotions before you can manage them. Likewise, you must learn to identify and manage your emotions before you can tap into the drive-to-achieve.

The first three competencies, when applied to other people, lead to the fourth: read and influence other people's emotions in a positive way.

Finally, the first four competencies lead to the fifth, an increased ability to enter and sustain good relationships.

What to do:	**Other actions:**
☐ Ask a trusted friend how they think you display and regulate your emotions.	☐ ____________
☐ The next time you are on the verge of displaying a disruptive/extreme emotion, practice controlling it.	☐ ____________
☐ The next time a team member displays a disruptive/extreme emotion, consider how you might help them manage it.	☐ ____________

Remember

- ✓ You may not get a second chance to get your message across. More often than not, you may not even realize your message was not received as you intended. Being mindful before you speak is more effective than explaining your meaning.
- ✓ Emotional self-awareness is the foundation of emotional intelligence. Self-regulation, motivation, empathy, and social skills all extend from it.
- ✓ Emotional intelligence involves the personal competencies of recognizing and managing emotions and the social competencies of recognizing others' emotions and managing relationships.
- ✓ Emotional self-awareness means you can identify an emotion at the time you experience it and understand how the emotion affects your behavior.
- ✓ Emotions are natural, normal responses to events, and they greatly influence our thinking and behavior.

Enhance Your Learning

Watch the following 7- and 2-minute videos on the importance of emotional intelligence. Daniel Goleman explains that to be successful in a top profession a person needs, aside from a high IQ, the ability to be self-motivated and emotionally intelligent.

Goleman, D. (2015). *What it takes to be a great leader.*	Available at: https://www.youtube.com/watch?v=_QpUH7i7EhM

Goleman, D. (2021). *Emotional Intelligence Takeaways.*	Available at: https://www.youtube.com/watch?v=5_09ArEkFxY

Reinforce Your Learning

Emotional Intelligence Self-Assessment. Review each competency listed on the chart that follows and indicate if it is a strength, neutral, or a weakness by checking the corresponding box.

Then ask a trusted friend to review your completed chart and to tell you whether they agree or disagree with your assessments of yourself.

The chart is adapted from works of the Consortium for Research on Emotional Intelligence in Organizations.

EI Competency Self-Assessment	Your Self-Assessment			Friend's Assessment	
	Strength	Neutral	Weakness	Agree	Disagree
Self-Awareness					
Emotional-Awareness: Recognizing one's emotions and their effects					
Accurate Self-Assessment: Knowing one's strengths and limits					
Self-Confidence: Sureness about one's self-worth and capabilities					
Self-Regulation					
Self-Control: Managing emotions and impulses					
Trustworthiness: Maintaining honesty and integrity					
Conscientiousness: Taking responsibility for personal performance					
Adaptability: Flexibility in handling change					
Innovativeness: Being comfortable with and open to new ideas and information					
Self-Motivation					
Achievement Drive: Striving to improve or meet a standard of excellence					
Commitment: Aligning with the goals of the group or organization					
Initiative: Readiness to act on opportunities					
Optimism: Persistence in pursuing goals despite obstacles and setbacks					
Social Awareness					
Empathy: Sensing other's feelings and perspectives; taking an active interest in their concerns					
Service Orientation: Anticipating, recognizing, and meeting customer needs					
Developing Others: Sensing what others need to develop and helping to bolster their abilities					
Leveraging Diversity: Cultivating opportunities through diverse people					
Political Awareness: Reading a group's emotional currents and power relationships					
Social Skills					
Influence: Wielding effective tactics for persuasion					
Communication: Sending clear and convincing messages					
Leadership: Inspiring and guiding groups and people					
Change Catalyst: Initiating or managing change					
Conflict Management: Negotiating and resolving disagreement					
Building Bonds: Nurturing instrumental relationships					
Collaboration and Cooperation: Working with others toward shared goals					
Team Capabilities: Creating group synergy in pursuing collective goals					

Developing Project Relationships #6.7

Recognize and Manage Team Conflict

Conflict happens in many ways. It arises among project team members, employees and managers, and different work units. Conflict often occurs when people are problem solving and making decisions. For example, a few members of the team may see an issue one way, while others have a different approach.

The goal of conflict resolution is not to suppress, deny, or eliminate conflict, but rather to accept the fact that conflict will happen, and to develop effective strategies for taking advantage of its potential benefits. It is a positive thing that we do not all see things the same way. Otherwise, we would not discover new ways of doing things. The issue is to manage conflict so that it does not derail project work.

In this concept, we describe conflict resolution as a process with certain stages you must go through, how to support people as you work to resolve differences, and the typical ways people behave when they try to resolve conflict.

Be Aware that Conflict Resolution is a Process

Conflict resolution is a **process** with distinct stages involved in reaching a favorable conclusion. How well you handle each stage will impact the quality of the conflict resolution.

Stage	Description
Recognition of Conflict	A precipitating event occurs, and people have a sense of frustration or a feeling of being blocked by someone or something. With this awareness, we do some quick, preliminary thinking about the cause of the conflict, the odds of resolving it, and how we perceive the people involved.
Planning and Analysis	We more carefully define the issues surrounding the conflict, identify our needs and wants, consider the needs and wants of others, identify our preferred outcomes, and decide how to move toward resolution.
Interaction	We exchange information and discuss ideas about how to resolve the issue. This stage ends with some type of outcome—win-win, win-lose, lose-lose, stalemate, or a temporary holding pattern.
Follow-up	If the interaction produced an agreement, we check back with the people involved at a later time to ensure that the agreement is still in effect (the resolution worked). If the agreement did not work, we modify it as needed.

How to Use the Conflict Resolution Process

It may be tempting to skip one or more stages in the conflict management process, but that will not lead to resolution. You must follow all four stages of the process as you work with any kind of conflict. It's a mistake to stop at the **recognition** stage, deciding not to attempt resolution because you think the situation is not resolvable, or that it is too risky. This closes the door to the potentially positive ideas and behavioral choices that could come during the **planning and analysis** stage. Even a short time spent thinking clearly about the situation—how important it is, what are the likely consequences of interacting, what are the consequences of the conflict escalating—can give you insights into your best course of action.

The **interaction** stage, where people discuss the conflict, is what we tend to think of first when we think about interpersonal conflict. The purpose of this stage is to allow all parties involved to express their ideas. You will be most effective if you look for points of agreement and then work to manage any remaining differences.

Do not overlook the **follow-up** stage. Checking back with someone provides the opportunity to mention any "afterthoughts" or additional information to consider in going forward.

Differentiate between Supportive and Non-Supportive Behaviors

When trying to resolve a conflict, **speak and behave** in ways that create a climate supportive of a **favorable outcome**. Being supportive does not require you to agree to conditions that compromise the team or project. You do not have to agree to whatever someone wants, nor do you have to tolerate any undesirable behavior. Being supportive simply means that you create a communication climate that supports a positive resolution and reduces defensiveness or other counterproductive responses.

People become defensive when they feel they are being attacked, judged, or controlled. When we are treated in these ways, we become self-focused and entrenched in our positions. We build protective walls that interfere with our abilities to listen, reason, and be open to the ideas of others. In conflict resolution, it is much **easier to prevent defensive behavior** than it is to remove walls.

Conflict Resolution Styles

Most of us have habitual ways of behaving in conflict situations, and we may respond the same way in most or all conflict situations. A model developed by Kenneth Thomas and Ralph Kilmann defines **five styles**, which they refer to as **modes**. Their research supports the idea that most of us have a dominant style, which we tend to use even if the conflict would be better resolved using a different style (Thomas & Kilmann).

The Thomas-Kilmann model is based on two dimensions of behavior. One is **assertiveness**, the degree to which individuals put energy into achieving an outcome favorable to themselves. The other is **cooperativeness**, the degree to which individuals put energy into achieving an outcome favorable to the others.

As illustrated below, the five modes, or styles, reflect the degree to which each of the dimensions is activated during a conflict.

Modes of Conflict Resolution

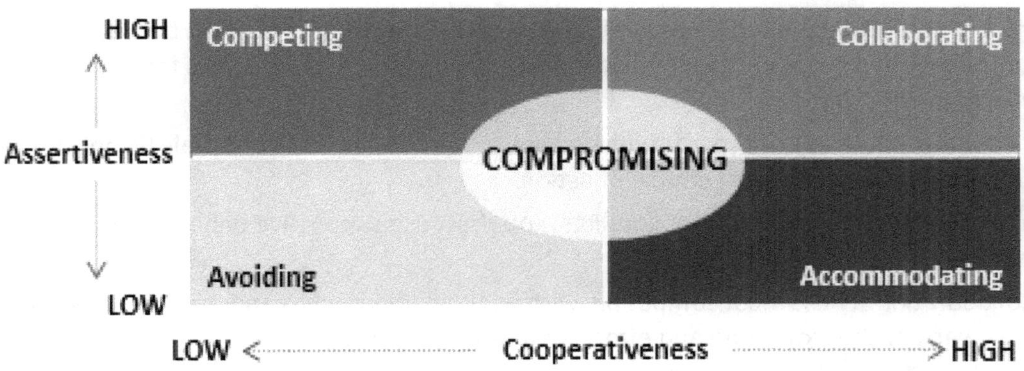

Adapted from Thomas and Kilmann

The Five Styles

- **Competing** is high in assertiveness and low in cooperativeness. A person using this style focuses on getting what they want, often drawing on power, persuasion, or rank, and they pay little or no attention to helping others achieve their wants.
- **Accommodating** is low in assertiveness and high in cooperativeness. This style is the opposite of the competing style. Someone using this style focuses on satisfying the concerns of others, with little or no attention to their own concerns.
- **Avoiding** is low in both assertiveness and cooperativeness. Someone using this style does nothing to satisfy their own concerns or the concerns of others. Instead, the behavior involves circumventing, postponing, or withdrawing from the conflict.
- **Collaborating** is high in both assertiveness and cooperativeness. This style is the opposite of the avoiding style. This person works to satisfy their concerns and the concerns of others. This style works to optimize gains and build win-win outcomes.
- **Compromising** is intermediate in both assertiveness and cooperativeness. Someone using this style works to achieve an acceptable resolution that minimizes losses to themselves and others.

The Thomas-Kilmann model is situational. In other words, none of the styles is inherently or universally the best. There are situations where each of the styles would be the most appropriate, and there are constraints and potential negative fallout associated with each style.

Even though it is a situational model—**the characteristics of the situation dictate which style is used**—people tend to have more comfort and practice using one or two styles. As part of planning for the interaction stage, it is helpful to think about what your dominant or preferred style is, and what you know about the typical behavior (style or styles) of the people involved in the conflict.

What to do:	Other actions:
☐ During the recognition stage, avoid the tendency to dismiss a conflict as not resolvable.	☐ ______________________
☐ Consider the consequences of interacting and not interacting to resolve a conflict.	☐ ______________________
☐ If you tend to avoid moving to the interaction stage more than you would like, consider how you can improve what you do in the planning and analysis stage.	☐ ______________________

Remember

- ✓ The goal of conflict resolution is not to suppress, deny, or eliminate conflict, but rather to accept the fact that there will be conflict and to develop effective strategies for realizing its potentially positive benefits.
- ✓ Even a small amount of time spent thinking about the situation and your realistic options can give you insights into your best course of action.
- ✓ Defensive behaviors do not resolve conflicts. You must use supportive behaviors and encourage others to do the same.
- ✓ The collaborating style is most compatible with win-win approaches and outcomes, but not all conflicts warrant the extra time and effort required for this style.
- ✓ Most people have one or two conflict resolution styles with which they are comfortable, and they tend to use these styles even when they do not fit the situation. Better results are achieved when you choose a style based on the nature of the conflict. Avoid staying with your preferred style just because it is comfortable.

Enhance Your Learning

Watch the following 3-minute video to learn more about the Thomas-Kilmann Conflict Mode approach:

Thibeaux, J. (2021). *Thomas-Kilmann Conflict Model Explained.*	Available at: https://www.youtube.com/watch?v=HxgSjnWSzf8

Reinforce Your Learning

Your Experiences.

List three types of interpersonal project conflicts you have experienced or observed. What conflict mode was or should have been employed?

1.

2.

3.

Reinforce Your Learning

Self-Assessment: Approaches to Conflict Assessment* The proverbs below describe some approaches to resolving conflict, as found in literature and traditional wisdom. Use the codes below to indicate how desirable each approach is to you by writing the appropriate number in the space preceding each proverb.

Use the numbers from **1** to **5**, where **1** means **completely undesirable** and **5** means **highly desirable**.

_____ 1. You scratch my back, and I'll scratch yours.

_____ 2. When people quarrel, the one who keeps most silent is most praiseworthy.

_____ 3. Soft words win hard hearts.

_____ 4. Come now and let us reason together.

_____ 5. It is easier to refrain than to retreat from a quarrel.

_____ 6. Better half a loaf than no bread at all.

_____ 7. A question must be decided by knowledge and not by numbers if it is to have the right decision.

_____ 8. When one hits you with a stone, hit back with a piece of cotton.

_____ 9. The arguments of the strongest always have the most weight.

_____ 10. By digging and digging, the truth is discovered.

_____ 11. Smooth words make smooth ways.

_____ 12. If you cannot make a person, think as you do, make the person do as you think.

_____ 13. The person who fights and runs away lives to fight another day.

_____ 14. Might overcomes right.

_____ 15. Tit for tat is fair play.

_____ 16. Seek till you find, and you will not lose your labor.

_____ 17. Kill your enemies with kindness.

_____ 18. A person loses least in a quarrel who keeps tongue in cheek.

_____ 19. One gift for another makes good friends.

_____ 20. Do not stir up a hornet's nest.

To score your assessment, transfer the number you recorded beside each of the 20 items above to the corresponding number below. Then total each column.

* Original source unknown

Avoiding	Accommodating	Compromising	Competing	Collaborating
Item 2	Item 3	Item 1	Item 9	Item 4
Item 5	Item 8	Item 6	Item 12	Item 7
Item 18	Item 11	Item 15	Item 13	Item 10
Item 20	Item 17	Item 19	Item 13	Item 16
TOTALS ____	____	____	____	____

Review your conflict style scores with the highest values indicating your preferences. Consider the Characteristics and Uses of Modes of Conflict Resolution that follows and reflect on what actions you might take to enhance your conflict resolution skills.

Characteristics of Modes of Conflict Resolution

Avoiding	• Best to avoid as much disagreement as possible. • Easier to refrain than to retreat from an argument. • Silence is golden. • See no evil, hear no evil, and speak no evil.
Accommodating	• Play down the difference and emphasize common interests. • Issues that might cause divisions or hurt feelings are not discussed. • Value the relationship more than the outcome of the issue.
Compromising	• Splitting the difference. • Bargaining. • Search for an intermediate position. • Concerned more for protection against loss than for realizing gain.
Competing	• Each conflict is seen as a win/lose situation. • Participants are antagonists—one person's victory must be the other person's loss. • Any tactic that will help one win is fair game. • Value "winning" the issue more than the relationship. • High ego involvement with the conflict outcome.
Collaborating	• Open exchange of information about the conflict or problem. • Differences are explored and worked through to reach a solution beneficial to all. • "Friction" is kept issue-oriented, not personality-oriented. • Only tactics that resolve the conflict without damaging the relationship are used. • High value placed on both the outcome and relationship.

Developing Project Relationships #6.8

Minimize Individual and Workplace Stress

We all experience stress in some form or another and working on projects can be stressful. Most of us try to avoid or reduce stress, but in some ways stress can be useful. It pushes us to accomplish tasks, forces us to leave certain situations, and moves us forward. But too much stress, or poorly managed stress, can wreak havoc in our lives.

A chronically stressful work environment is less productive. Excessive workplace stress costs organizations in a variety of ways (Lee, 1997). People lose time at work due to emotional and physical illnesses. They often become distracted, behave carelessly, and take short-cuts, increasing the risk of accidents. Stress can worsen interpersonal conflicts, which can lead to losses in productivity, litigation, and even physical harm. Stress can make it difficult for people to accept change, to think creatively and innovate, and to deal properly with upset coworkers and customers. The list goes on and on.

Managing workplace stress is vital to your project's success. You must manage your own stress, recognize stress in your project team, and take steps to help resolve mutual problems.

In this concept, we discuss how you can better manage stress in the workplace.

What is Stress?

Stress is the **body's physical, mental, and chemical reactions to events or circumstances that frighten, excite, confuse, endanger, or irritate**. It also defined as a combination of the bodily changes that prepare an individual to survive by **fight** or **flight**. It is the body's protective mechanism: a series of interrelated reactions that are triggered when something threatens our physical, mental, or emotional well-being.

Sources of stress, or stressors, include life-cycle changes, social pressures, physical pain, emotionally charged situations, and any number of specific health and living conditions. Typical project or job-related stressors include:

- Conflict with coworkers and authorities
- Job dissatisfaction
- Overwork
- Lack of support or appreciation
- Unclear job expectations
- Time pressures
- Monotonous or unchallenging work

Use the 4 A's to Manage Stress

You can use the **4 A's to manage stress**: avoid, alter, adapt, and accept (Ridpath, 2012). Use these ideas to help manage your stress, but also consider how you can use them to help your team deal with stress.

- **Avoid** stress when you can. Stay away from stressful situations when possible. Be proactive and deal with situations before they become problems.
- **Alter** a bad situation. You can sometimes change a situation or turn a negative into a positive. Look for opportunities to do that.
- **Adapt** when necessary. Sometimes you cannot change a stressful situation, but you can change your behavior to make your experience of it less stressful. Remember, life is only five percent what happens to you and 95 percent how you react to it.
- **Accept** what you cannot change. There is a saying: "God give me the strength to accept the things I cannot change, the courage to change the things I can, and the wisdom to know the difference." Sometimes, that is the best you can do.

One good way to avoid stress is not to create it. Here are some guidelines that will help.

- **Plan ahead**. For example, anticipate what tasks must be done as opposed to ones you can postpone or avoid.
- **Manage your time well**. Do not procrastinate. Try using "to do" lists and prioritizing what you must do versus what you would like to do if you have time.
- **Learn from past mistakes**. Be aware of stress when it occurs. Make changes to avoid the same problem from happening again.
- **Be proactive**. Deal with problems head-on without trying to pass them to others or letting them fester.
- **Change your expectations and attitudes.**

Strengthen Your Communication Skills

Communicating well can help everyone avoid and reduce stress, especially the kind of stress that comes from working with others. Strengthen your communication skills by doing something to improve them every day.

Do not bring your work home with you. Likewise, do your best not to bring outside stress into the workplace.

Learn to say no to tasks that will add unmanageable stress.

Understand the Stressors that Affect You and Your Team

Analyze the stressors that may be unique to you, and also analyze the stressors that may be affecting your team. Think of past situations that were stressful and determine if some of the same things may be happening now. Also, consider the following questions in terms of your behavior and your team's behavior:

- Can you identify one major part of the situation that was the primary stressor?
- What effects of the stress did you feel?
- How did you react?
- What parts of your reaction helped remove the stress, and what parts were not useful?
- If you could go through the situation again, what might you do differently?
- What did you learn about yourself that you can rely on in the future?

Organizational Culture and Stress

Organizational culture plays a role in the general level of stress, and the same is true for projects. If leadership is stressed and puts undue pressure on people, or conveys uncertainty to people, the stress that filters through the organization or project will be palpable.

Managing project and workplace stress must start at the top with the awareness that unrealistic expectations add to workplace stress. Besides managing stress in your team and project, another proactive step you can take is to keep your management informed if you sense that people are close to being "stressed out."

What to do:

- ☐ Identify your primary sources of stress at work. Identify your team's primary source of stress.
- ☐ Recognize the symptoms of stressful situations and work to eliminate or mitigate the root causes.
- ☐ Learn and practice strategies for managing stress both in yourself and in your team.

Other actions:

- ☐ ____________________
- ☐ ____________________
- ☐ ____________________

Remember

✓ Managing workplace stress matters. You need to be able to manage your stress and work with others to identify strategies to help keep stress at a manageable level for the team.

✓ You cannot change how other people behave, but you can control your behavior. If you react in ways that elevate your stress or lead you to doubt your competence, you will feel stress, and it will benefit you to find ways to change your reactions.

✓ Never express anger in harmful or hurtful ways. Raising your voice or sending angry emails or messages worsens the situation. Take time to cool off and think things through before responding.

✓ Maintain appropriate expectations and let people know you value open communication and cooperative relationships.

✓ Relationships make the world go around. Remind yourself, daily, if necessary, of the value and importance of having productive, collaborative, high-quality work relationships. This will help reduce your stress and that of everyone around you.

Enhance Your Learning

One way we reduce workplace stress is by managing communication. Watch the following 5-minute video to learn how you can use communication to change the direction of a negative interaction.

O'Connor, D. (2009). *4 Magic Phrases You Can Use to Respond to ANYTHING	Power Phrases for Work.*	Available at: https://www.youtube.com/watch?v=g5RknemM8Hw

Reinforce Your Learning

Stressful Situations. Below are some stressful situations that people face at work. Check all the ones you have experienced. If you think of others, add them to the list.

- ☐ 1. Dealing with people when they are overly emotional.
- ☐ 2. Having to give instructions repeatedly that you thought were clear.
- ☐ 3. Dealing with people who have a social style you find irritating.
- ☐ 4. Giving or receiving negative feedback on performance.
- ☐ 5. Dealing with an arrogant or narcissistic co-worker.
- ☐ 6. Administering or receiving discipline.
- ☐ 7. Dealing with resistance to your ideas or authority.
- ☐ 8. Denying or being denied a request or suggestion.
- ☐ 9. Enforcing or complying with an unpopular rule or change.
- ☐ 10. Observing an unethical or illegal activity.
- ☐ 11. Dealing with psychologically unstable individuals.
- ☐ 12. Giving or receiving positive feedback or praise.
- ☐ 13. Dealing with criticism.
- ☐ 14. Dealing with a workplace crisis.
- ☐ 15. Other:

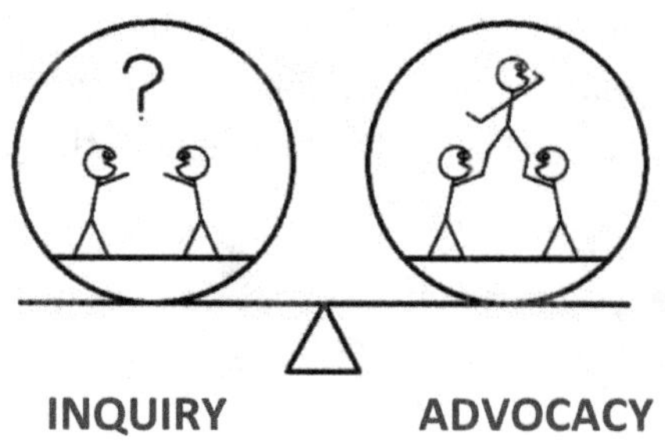

Developing Project Relationships #6.9

Balance Inquiry and Advocacy

Researchers have found some interesting facts about interpersonal communication. In some situations, the **words** you choose may constitute very little of the message others receive from your communication. Much of your message is conveyed through your **non-verbal behavior,** and much of what your receiver "hears" is rooted in their beliefs and past experiences.

Have you noticed what happens when you try to listen to another person without intending to respond? Perhaps not, because most of us tend to listen with a primary focus on our response. The result is we often do not hear what a speaker said.

A further impediment to understanding is that our listening is rarely pure, not like an audio recording. It is filtered through our perceptual screens. We often hear what we expect to hear, regardless of what the speaker said. We find it difficult (and, frankly, mostly unnecessary) to differentiate between what a person says and how we interpret what they said. This is because our own beliefs and past experiences affect how we hear, interpret, and respond.

In this concept, we discuss how you can improve your communication and make your discussions with people more productive by balancing inquiry and advocacy as you share ideas and problem solve with others.

The Ladder of Inference

William Isaacs, of the Dialogos Institute at the Massachusetts Institute of Technology, revisited a model initially developed by Chris Argyris that describes our thinking process as we experience and respond to events in our lives.

The model, referred to as the Ladder of Inference, is a seven-step process, as depicted in the illustration below. The model shows that as we move up the ladder, our beliefs affect what we infer about what we observe and, therefore, become part of how we experience our interactions with other people.

Ladder of Inference

I take ACTIONS based on my beliefs.

I adopt BELIEFS about the world.

I draw CONCLUSIONS.

I make ASSUMPTIONS based on the meanings I added.

I add MEANINGS (cultural and personal).

I select DATA from what I observe.

All the information in the world.

An example of this would be "hearing" only that which supports your argument. But the process is usually much more subtle. Your background influences the meanings you ascribe to what you hear, which in turn leads you to make assumptions. Your beliefs affect which data you select in the first place.

If you take the time to "walk" down the Ladder of Inference, you can learn a great deal about how your beliefs, assumptions, background, culture, and other influences that affect how you interpret what others say and your interactions with them.

Creating an Environment of Understanding

The Ladder of Inference is a useful tool for decision making, but also for improving understanding between people. As you interact with people, try walking down the ladder to gain a better understanding of how you and they think. Listen carefully to what people say and avoid quick interpretations.

- Listen for conclusions and beliefs, yours and theirs.
    - Do they jump to conclusions?
    - What conclusions are you making as you listen?
- Listen for directly observable data.
    - Can you form a picture in your mind of what they are saying?
    - Ask yourself: What led them to think as they do?
- Suspend your certainties and conclusions.
    - Do they act as if their conclusions are obvious?
    - Do you?
    - Are there other ways of seeing things?

Positive and Productive Discussions: Inquiry and Advocacy

To encourage positive discussions, you must first inquire and then advocate.

Inquiry involves engaging with other people and learning from them. In this mode, you are not judging, arguing, or trying to present your viewpoint. You are listening inquisitively.

This is also a time for observing your thoughts and checking your Ladder of Inference. During the inquiry phase, bring your assumptions forward to make them explicit so you and others can explore their meaning and impact.

Advocacy involves "selling" an idea or position or directing attention to certain facts you think are relevant. When in this mode, you evaluate ideas, narrow the field, and work toward consensus.

A **problem during many discussions** is that **participants move too quickly to advocacy without adequate inquiry.** This has the effect of stifling interaction and creative thinking and undermining trust.

Balance Inquiry and Advocacy

The key—both within yourself and in working with a team—is to balance inquiry and advocacy. You need both. In your efforts to develop self-knowledge, be aware of the intentions during inquiry and advocacy, and strive to balance the two. Then work to enable your group to do the same.

Consider the message you are sending. Then consider the non-verbal language you usually use. Check to ensure your non-verbal signals do not conflict with your message. **Do not send mixed messages** because that will confuse other people and short-change communication.

As previously discussed, consider how you can use non-verbal signals to **support your message**. For example, look at the person and stand firmly. This shows that you believe in your words.

What to do:	Other actions:
☐ When people state opinions, ask yourself what led them to think as they do.	☐ ____________
☐ Practice listening for conclusions, beliefs, and bias.	☐ ____________
☐ Identify communication role models and observe their body language. Note what you can do to improve your communication effectiveness.	☐ ____________

Remember

- ✓ Your verbal and non-verbal messages must align. Any incongruence will cause confusion and may even call your credibility into question.
- ✓ Be aware when you or others are jumping to conclusions.
- ✓ Walk others down the Ladder of Inference and find out what data you are operating from. Use unaggressive language such as "Can you help me understand your thinking here?" to encourage that assumptions be made explicit.
- ✓ Listen without resistance. Be aware of your filters as you listen to ideas. Respect differences.

Enhance Your Learning

For further information on the Ladder of Inference, watch this 4-minute video where Ed Muzio describes Chris Argyris' "Ladder of Influence" model and how you can use it to avoid making incorrect judgments.

Muzio, E. (2011). *The Ladder of Inference Creates Bad Judgment.*	Available at: http://www.youtube.com/watch?v=K9nFhs5W8o8

Reinforce Your Learning

Non-Verbal Cues. Assume that a project sponsor attends a project meeting where they sit down in a chair and turn to face the project team. The sponsor leans forward a little and puts his or her hands on their knees. You start discussing some of your ideas with the team. What do you think the non-verbal cues say about the project sponsor's interest in your ideas, and why do you believe this?

A. The sponsor is disinterested and might be thinking about something else as you explain your ideas.
B. The sponsor is reluctant to accept your idea and may even be a bit upset that you suggested it.
C. The sponsor is indicating openness to listening to your ideas, is giving you their attention, and may even be interested.
D. The sponsor feels they have other more important responsibilities to be concerned with at this time.
E. The sponsor has serious doubts about the value of the idea and its application to the work you are supposed to be doing.

Developing Project Relationships #6.10

Apply Principled Negotiation Techniques with Stakeholders

Knowing how and when to negotiate is an essential skill for project managers. Whether the negotiation involves the project management triangle (performance, cost, and time) or something else, having a basic understanding of negotiating techniques will serve you well.

Negotiating within your project team is different from negotiating with an outside person or organization. If you are negotiating with a supplier, for example, you will want what is best for your organization, so you may take a hard-core approach, "dig in your heels," and try to get the best deal.

But if you were, for example, managing a project team tasked with making a design decision and you and other people have strong opinions, would a hard approach be the best thing for your organization? Others on the team may have information you do not have, and it would benefit the team to spend some time in creative problem-solving discussions. You would achieve the best results by being open to ideas and approaches, and you would negotiate accordingly.

In this concept, we discuss how project managers can use principled bargaining to achieve the best results when negotiating with stakeholders and working with project teams.

Principled versus Positional Bargaining

Principled bargaining is more about finding a way to meet and determine an outcome that works for everyone. It gives everyone the opportunity to influence the result. It is usually done in a "win-win" mental framework with one objective being to protect the relationship.

Positional bargaining can be summed up as a type of negotiation where you stick to your position, very firmly, so you make the other person come to you for concessions. This is usually done in an "I win, and you lose" mental framework.

Negotiate for "Win-Win"

The following discussion about bargaining positions is adapted from the classic work of Fisher and Ury. It begins with the idea that every negotiation involves two aspects: **what you are negotiating** and **how you go about the negotiation**.

Most people tend to use positional negotiation. They tend to "dig-in" to their position, sometimes merging their identity with their negotiating position. This means that any attack on their position is taken personally. This type of bargaining causes people to become defensive, they can close down, and it can severely slow down, even halt, the negotiating process.

Another reason positional negotiation is often less likely to produce a positive outcome while keeping relationships intact is that this type of bargaining tends to overlook the underlying reasons people have for their positions. Positional bargaining usually assumes that one side will "win" and one side will "lose." A more modern view of negotiation is to pursue a win-win outcome, which is common to principled bargaining, and which will help you build and maintain relationships.

Positional bargaining tends to involve negative persuasion techniques such as threatening to walk out, taking excessive time to make a decision, or merely refusing any offers. This type of bargaining tends to build adversarial relationships and break down trust.

Principled bargaining is more concerned with **looking for a mutually beneficial solution**. It begins by looking at the needs of each party and how all needs can best be met. The goal is to have both parties reach an agreement that works for them, and so is mutually beneficial. This is the kind of interaction that you, as a project manager, want to have with your stakeholders. You want them to win—to get what they want—while you also win in the sense that the solution you agree to also works for the project (costs, resources, and schedules).

To conduct principled bargaining, focus on issues, and separate them from people so that challenging a position is not the same as challenging a person.

You also focus on what is in everyone's best interests, not just on their stated position. You look for options and try to meet the needs of everyone involved as best as possible.

Principled bargaining also applies to negotiations with your project team. No person should hold fast to their position solely for the sake of "saving face" or "winning." Instead, everyone must try to do what is best for the team and the organization.

What to do:	Other actions:
☐ Listen more and talk less while looking for win-win solutions.	☐ ______________
☐ Uncover the value proposition: what is of value to you and what is of value to others.	☐ ______________
☐ Work toward an outcome that works for everyone and is mutually beneficial.	☐ ______________

Remember

✓ Principled bargaining is a better choice over positional negotiation when working with your stakeholders and team.

✓ Effective negotiation strategies, such as principled bargaining, can strengthen relationships and build bridges to future negotiations.

Enhance Your Learning

Watch the following 16-minute TED Talk to learn a few points about effective negotiation. Consider how you can apply this information in your work:

Ury, W. (2010). *The walk from "no" to "yes."* TED.	Available at: https://www.youtube.com/watch?v=Hc6yi_FtoNo

Reinforce Your Learning

Apply Principled Negotiation and Conflict Resolution Strategies to a Team Conflict. Assume that a project-wide continuous improvement team headed by Paul Principle was asked to formulate procedures for improving safety, operational efficiencies, and reducing waste. The improvement team did its work, but some of the changes, which have been implemented, are controversial.

The "attitude" in one work unit is "we don't want to abide by the new procedures because we were not consulted," and "it cuts down our ability to meet our deadlines if we have to be looking over our shoulder to keep every little thing under control." Or, to put it another way, "we are here to get the job done on time and within budget, not to worry about a few workarounds or a little waste."

Seth Stressor, a foreman in a work unit resisting the new procedures, asked for a meeting with the continuous improvement team. The team's position is that its work is complete, and the procedures are now implemented. Paul Principle, however, is willing to meet with Seth to attempt to resolve the situation.

What are the issues for the work units that are resisting the new procedures? Does Paul Principle have enough information to understand the issues?

What is the commonality of interests in this situation?

Think back to the conflict resolution process described in Concept 7. Review the planning and analysis stage. What steps could Paul follow to help resolve the conflict? What would "win-win" look like?

Summary

Final Word: *Developing Project Relationships!*

You have just completed a chapter on Developing Project Relationships. Being a project manager requires that you continually work to improve your relationship skills. You must develop and maintain relationships with the leadership above you, the people you manage, and stakeholders, customers, suppliers, and peers.

A few topics you encountered in this chapter include:

- Describing the characteristics and building blocks associated with high-quality work relationships.
- Building trust with team members and with stakeholders
- Explaining what is meant by social styles and how style differences can affect the quality of communication.
- Naming four social styles and describing their characteristics.
- Describing conflict in project work, typical causes, and methods of resolving it.

In the next Project Mastery chapter, we will focus on Building High-Performance Project Teams.

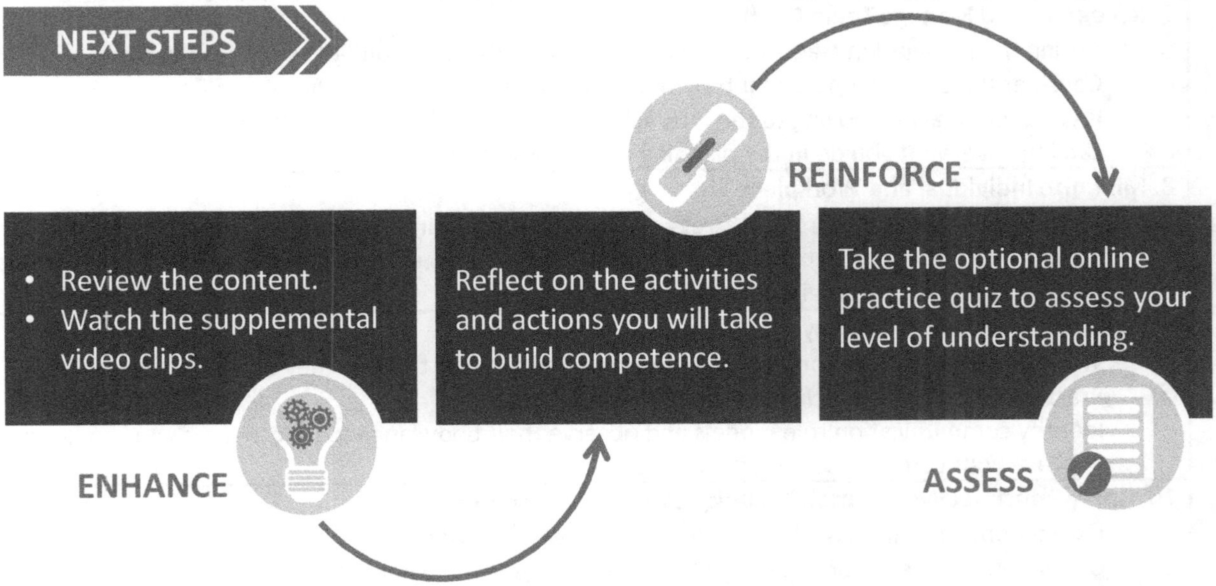

Developing Project Relationships—Recap Checklist Being a project manager requires that you continually work to improve your relationship skills. You must develop and maintain positive relationships with the leadership above you, with the people you lead, and with stakeholders, peers, customers, and suppliers.
1. **Identify Characteristics of High-Quality Work Relationships** ☐ Build a good rapport with each person on your project or in your work unit. ☐ Value diversity and encourage a wider range of ideas to create new solutions and best practices. ☐ Create the feeling of a team by showing other people you value their contributions.
2. **Build Trust with Team Members and Stakeholders** ☐ Work side-by-side with your team and stakeholders to build relationships and trust. ☐ Give credit where credit is due. ☐ Be consistent, forthright, and keep others informed.
3. **Recognize Personal Preferences and Bias** ☐ Identify what you like about your project or job and what you want out of it. ☐ Be observant and reflect on the preferences of your team members and associates. ☐ Do not let your preferences negatively affect how you treat people.
4. **Be Aware of Social Styles in the Workplace** ☐ Complete one of the social style assessments that are available. ☐ Encourage your team members and associates to complete the same assessment and share the results. ☐ Consider how you might use your knowledge of social styles to improve your relationships with team members and associates.
5. **Know Your Social Style** ☐ Focus on your social style and determine what you can do to build on your strengths. ☐ Identify people who seem to represent each style and identify ways you can use the strengths of that style to enhance your team's success. ☐ Identify one associate with whom opening up communication might improve your relationship.
6. **Understand Emotional Intelligence** ☐ Ask a trusted friend to describe how they think you display and regulate your emotions. ☐ The next time you are on the verge of displaying a disruptive/extreme emotion, practice controlling it. ☐ The next time a team member displays a disruptive/extreme emotion, consider how you might help them manage it.
7. **Recognize and Manage Team Conflict** ☐ During the recognition stage, avoid the tendency to dismiss a conflict as not resolvable. ☐ Consider the consequences of interacting and not interacting to resolve a conflict. ☐ If you tend to avoid moving to the interaction stage more than you would like, consider how you can improve what you do in the planning and analysis stage.
8. **Minimize Individual and Workplace Stress** ☐ Identify your primary sources of stress at work. Identify your team's primary source of stress. ☐ Recognize the symptoms of stressful situations and work to eliminate or mitigate the root causes. ☐ Learn and practice strategies for managing your stress.
9. **Balance Inquiry and Advocacy** ☐ When people state opinions, ask yourself what led them to think as they do. ☐ Practice listening for conclusions, beliefs, and bias. ☐ Identify communication role models and observe their body language. Note what you can do to improve your communication effectiveness.
10. **Apply Principled Negotiation Techniques with Stakeholders** ☐ Listen more and talk less while looking for win-win solutions. ☐ Uncover the value proposition: what is of value to you and what is of value to others. ☐ Work toward an outcome that works for everyone and is mutually beneficial.

Knowledge Review Test—Questions

Developing Project Relationships

1. Effective project relationships require the ability to ____________ effectively and be open to good communication.

   A. give orders
   B. communicate
   C. delegate
   D. All the answers are correct.

2. Which of the following is NOT true of creating effective project relationships

   A. Trust is a cornerstone of a good relationship between you and an associate.
   B. You should show your co-workers respect.
   C. You don't need to really be interested in an associate as long as you look interested.
   D. All the answers are true.

3. Which of the following is NOT true about generating trust from your stakeholders?

   A. To do so you must work with them every day.
   B. You should keep your stakeholders informed.
   C. Give credit where credit is due.
   D. Being inconsistent causes you to lose trust.

4. Remember that ___________ one's trust is much easier than _______ what has been lost.

   A. maintaining / re-earning
   B. re-earning / maintaining
   C. losing / maintaining
   D. maintaining / losing

5. One way that you can earn your boss's trust is:

   A. Telling them they can trust you.
   B. To do your job well.
   C. Spreading gossip about people your boss dislikes.
   D. All of the answers are good ways to earn trust.

6. One of the primary objectives of principled negotiations is to:

   A. Get the most from the other side.
   B. Protect the relationship.
   C. Be the winner.
   D. Define your objectives up front and stick with them.

7. If you really want to know the interests and preferences of your workers what can you do?

   A. Ask them.
   B. Insist that they fill out free personality and preference inventories.
   C. Ask your HR representative to find it in their application.
   D. Read their emails.

8. Which of the following is true?

    A. It costs more to keep an existing employee than to hire a new employee.
    B. It costs more to hire a new employee than to keep an existing employee.
    C. Training is too expensive to bother with.
    D. None of the answers are correct.

9. Stating that you only have one hour to make your plane is an example of what type of negotiation technique?

    A. Good guy, bad guy.
    B. Delay.
    C. Deadline.
    D. Extreme deadline.

10. Which of the following is NOT one of the four aspects of emotional intelligence?

    A. Self-management.
    B. Self-awareness.
    C. Self-regulation.
    D. Self-motivation.
    E. Self- reliance.

11. Which of the following NOT true?

    A. EI is arguably more important than IQ.
    B. EI is short for emotional intelligence
    C. EI is short for social intelligence
    D. EQ = EI = emotional intelligence

12. To keep your skills current and be relevant in our fast-paced world you have to be what?

    A. A lifelong learner.
    B. Kaizen.
    C. Listening to your supervisor.
    D. In training every day.

13. There are three dimensions or aspects of social style that are generally discussed:

    A. Assertiveness, reasonableness, and versatility.
    B. Assertiveness, responsiveness, and versatility.
    C. Acquiescence, reasonableness, and versatility.
    D. Anger management, retreating and value.

14. Knowing social styles can help you to do which of the following?

    A. Appreciate work assignments.
    B. Maintain schedules.
    C. Build teams.
    D. All the answers are correct.

15. The power of "threat" causes stress and is what source of power?

   A. Reward.
   B. Coercive.
   C. Legitimate.
   D. Expert.
   E. Connection.

16. The Emotional Intelligence model introduced by Daniel Goleman focuses on EI as a wide array of competencies and skills that drive leadership performance. Goleman's model outlines five main EI constructs:

   A. Self-awareness; Self-regulation; Social skill; Empathy; Motivation.
   B. Self-awareness; Sympathy; Social skill; Empathy; Motivation.
   C. Sympathy Self-awareness; Social skill; Empathy; Motivation.
   D. Self-awareness; Self-regulation; Sympathy; Empathy; Motivation.

17. Which of the following styles is best for resolving workplace conflicts?

   A. Competitive style.
   B. Collaborative style.
   C. Avoiding style.
   D. Accommodating style.

18. Which of the following are traits that a person can have which can make them naturally adept at resolving conflict?

   A. High emotional and social intelligence.
   B. Empathy.
   C. Excellent communications skills.
   D. All the answers are correct.

19. The best way to promote good relationships at work is what?

   A. Send out a memo.
   B. Put it on people's job descriptions.
   C. Assign everyone to a team.
   D. Be a positive role model.

20. Principled negotiating includes which of the following tactics?

   A. Focuses on interests rather than positions.
   B. Invents options for mutual gain.
   C. Insists on objective criteria.
   D. Negotiates on merits.
   E. All the answers are correct.

Developing Project Relationships — Answer Sheet

1.	___A	_X_ B	___C	___D	___E
2.	___A	___B	_X_ C	___D	___E
3.	_X_ A	___B	___C	___D	___E
4.	_X__A	___B	___C	___D	___E
5.	___A	_X__ B	___C	___D	___E
6.	___A	_X__ B	___C	___D	___E
7.	_X_ A	___B	___C	___D	___E
8.	___A	_X_ B	___C	___D	___E
9.	___A	___B	_X_ C	___D	___E
10.	___A	___B	___C	___D	_X_ E
11.	___A	___B	_X_ C	___D	___E
12.	_X_ A	___B	___C	___D	___E
13.	___A	_X_ B	___C	___D	___E
14.	___A	___B	___C	_X_ D	___E
15.	___A	_X_ B	___C	___D	___E
16.	_X_ A	___B	___C	___D	___E
17.	___A	_X_ B	___C	___D	___E
18.	___A	___B	___C	_X_ D	___E
19.	___A	___B	___C	_X_ D	___E
20.	___A	___B	___C	___D	_X_ E

Chapter 7

Building High-Performance Project Teams

Overview

"Individual commitment to a group effort — that is what makes a teamwork, a company work, a society work, a civilization work."

—Vince Lombardi

When we observe successful teams, certain facts emerge. For example, a **high-performance** team requires both an **established purpose** and **members committed** to the purpose. Another fact is success **depends on the support organizations give to their teams, and on the skills of team members** and their **ability to apply the skills**. Necessary skills are not limited to **technical expertise**. They include **interpersonal skills** and the ability to engage in **group problem solving**.

In this chapter, we discuss information and insights about teamwork and the responsibilities of project managers and team members. We also review the kind of support an organization must give a project team for it to succeed. The course covers the stages of team development, how to deal with team challenges, and what is needed to motivate and engage team members.

After you complete this chapter, you will know how to:

- Describe the skills and processes associated with project team success.
- Describe the support needed for project teams to function at high-performance levels.
- Define the necessary skills needed to lead project teams.
- Understand how to mitigate team dysfunctions and promote shared responsibility.
- Understand how to motivate and engage teams.

The competencies associated with this chapter include:

Teamwork and Cooperation	Interpersonal Relationship Building	Conflict Management

Take Your Temperature for Building High-Performance Project Teams

The following assessment will help you identify what you know about building high-performance project teams and the areas where you can improve. With you and your organization in mind, read each statement below, and consider how much you agree with it. Use the numbers from 1 to 10, where **1** means **strongly disagree**, and **10** means **strongly agree**.

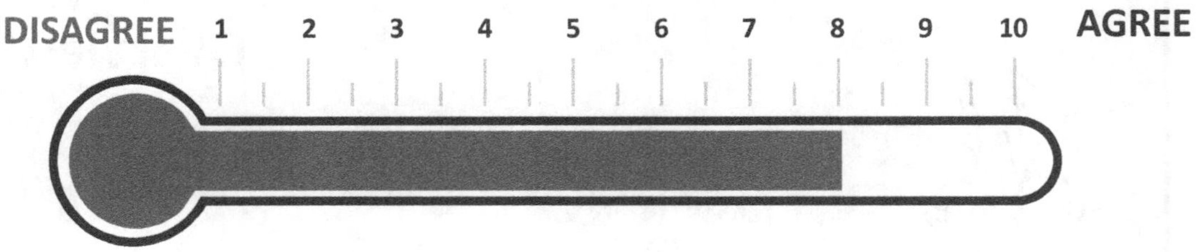

_____ 1. I can explain the characteristics and benefits of successful teams.

_____ 2. I know the key elements for structuring project teams.

_____ 3. I can describe the stages of team development.

_____ 4. I encourage communication openness, support, and cooperation.

_____ 5. I can deal with project team issues, concerns, and conflict.

_____ 6. I know how to enhance a team's problem-solving capabilities.

_____ 7. I understand and embrace the values that support team success.

_____ 8. I know how to review performance.

_____ 9. I can describe what motivation is and how to achieve it.

_____ 10. I have ideas about how to engage team members.

_____ **Total**

Take a moment to reflect on the following before you move forward in the chapter.
In your work, what problems have you experienced or observed related to building high performance project teams at work?

Take a few minutes to also reflect on your self-assessment. List two or three areas you want to improve.

As you progress through this workbook consider what actions you can take to demonstrate competence in the following three competency areas:

1. **Teamwork and Cooperation.** Demonstrates and fosters cooperation, communication, and consensus among individuals and groups.

2. **Interpersonal Relationship Building.** Considers and responds appropriately to the needs, feelings, and capabilities of others; seeks feedback and accurately assesses impact on others; provides helpful feedback; builds trust with others.

3. **Conflict Management.** Anticipates and seeks to resolve disagreements, complaints, and confrontations in a constructive manner.

Recognize the Characteristics and Benefits of Successful Teams

Project teams that succeed offer tremendous benefits to the organization, its customers, and its employees. These well-documented **benefits** include **higher productivity**, **decreased costs**, **increased quality**, **higher customer satisfaction**, **lower absenteeism**, **higher morale**, and **better-quality decisions**. This is just a sampling of the benefits associated with successfully functioning teams (Katzenbach, 1993).

The benefits to individual team members are also significant: work is less stressful, relationships are more positive and supportive, and team members feel empowered and encouraged to bring their best effort to the job.

But there is a catch. Not all teams succeed. Everyone has a story about a failed team experience. The failures are sometimes enough to stop organizations from using project teams and lead some managers to believe the payoff from teamwork is not worth the time and effort it takes to develop teams.

Team success or **failure** is not rooted in the nature of teams, nor is it merely a function of luck. It is **rooted in how teams are structured,** and the **value project managers and organizations place on their teams**.

In this concept, we discuss the various types of teams and team characteristics that make for success.

What is a Team?

There are different types of teams and different purposes for which teams are established. Some examples are leadership teams, functional teams, project teams, problem-solving teams, task forces, committees, self-directed teams, and virtual teams, to name a few.

No matter the type of team or its purpose, here are three components that make a team:

- The group has a shared goal, purpose, or reason for being.
- The structure is defined for housekeeping routines such as how meetings are conducted and record-keeping.
- The group establishes processes and procedures that all members use to communicate, make decisions, resolve conflict, and influence each other in positive ways.

Team Structure. Teams can be structured in a variety of ways, with the most common being cross-functional, permanent, and virtual.

- **Cross-functional teams.** A cross-functional team is one formed of individuals from different workgroups or units. They can be charged with completing a special project or solving a specific problem. Or they can have an ongoing job such as monitoring and making recommendations for employee safety. In some organizations, technical specialists can belong to more than one team at a time. They use their skills and knowledge where they are most needed rather than affiliating with a single organizational unit.
- **Permanent work teams.** A permanent work team is a group of people operating as a team. The team can be self-directed or have a manager or project coordinator as it works to perform its work responsibilities. In small businesses, the entire organization may be operating in a team structure. A group of managers may function as a team to provide uniformity in direction and coordination across an organization.
- **Virtual teams.** A virtual team may include team members from other organizations, and customers, vendors, and suppliers.

Characteristics of Effective Teams

The following are conditions that characterize a team, and they are necessary for a team to be effective:

- Clear goal with measurable objectives; achievement requires a team effort
- Sufficient time for team building, training, and group processes
- Supportive culture
- Knowledge and use of various problem-solving techniques

Depending on the type of team and the nature of the work, the following additional characteristics are often found:

- Members share a collective identity
- Compelling mission or purpose
- Team members have clearly defined roles
- Team members have the skills, knowledge, and other attributes required for their roles
- Effective leadership and coaching for both task completion and interpersonal process development
- Team tasks are interdependent, and team results require members to collaborate
- Team members have joint accountability for team successes and failures
- Problem-solving and decision-making processes are effective
- Cohesive unit with members who want it to succeed

What to do:	**Other actions:**
☐ Recognize there are significant payoffs to organizations when teams are appropriately used, structured, and supported.	☐ ____________
☐ Pay attention to the number, type, and success of the teams in your workplace.	☐ ____________
☐ Consider whether the teams you work on meet the characteristics of effective teams.	☐ ____________

Remember

- ✓ Effective teams get results because of their process, which is how they interact with each other and complete their work. Core processes include communication, decision making, conflict resolution, and influencing each other in positive ways.
- ✓ Key characteristics of successful teams include, among others, a clear goal with measurable objectives whose achievement requires a team effort, sufficient time for team building, and a supportive organizational culture.
- ✓ While team building is not easy, and the benefits are not guaranteed, evidence shows that successful teams offer significant benefits to organizations and individual members. Well-positioned and well-developed teams are worth the investment of time and effort.

Enhance Your Learning

Watch the following 3-minute video outlining the results of a three-year study that identified eight performance indicators great teams have in common.

Tanner, R. (2015). *Eight Characteristics of Effective Teams.*	Available at: https://www.youtube.com/watch?v=nSpsfXaL9A8

Watch the following 2-minute video on the Wisdom of Geese.

Hasrian, F. (2013). *Wisdom of Geese.*	Available at: https://www.youtube.com/watch?v=5rOg4WfNDfM

Reinforce Your Learning

Team Geese. The instinctive team participation and cooperation of migrating geese are worthy of note. You have probably seen geese flying in a v-formation during migration in the spring and fall. Scientists have discovered the reasons for the v-formation. When one bird flaps its wings, it creates an updraft for the bird immediately behind. By flying in this way, the geese significantly increase their flying range as compared to flying alone. There are other benefits, as well

By working in teams, geese (and people) can realize their goals more quickly and easily because they support each other. They get better results than when working alone. When a goose falls out of formation, it feels the drag of going it alone and quickly gets back into formation to take advantage of the support of the group.

Describe two or three other lessons you can glean from Team Geese.

1.

2.

3.

Characteristics of Effective Teams. Consider a team you work on and assess how well it matches the characteristics below. Check all the ones you think your team possesses. If you are not a member of a team now, use the checklist to assess a team you were part of in the past.

Team: ____

- ☐ The team had a compelling purpose.
- ☐ The team had shared leadership.
- ☐ The team had just enough structure.
- ☐ The team had full member engagement.
- ☐ The team embraced member differences.
- ☐ The team applied unexpected learning.
- ☐ The team worked to strengthen team member relationships.
- ☐ The team had great results.

Building High Performance Project Teams #7.2

Appreciate the Key Elements of Team Structure

Project teams, even self-directed ones, exist within the context of the larger organization and operate for a specific purpose and within guidelines. Teams can be both horizontal and vertical. In small businesses, the team structure may define the entire organization.

Organizations set parameters for team authority and autonomy, and organizations must also provide financial, developmental, and political support for the team. While the team structure and membership may shift as the team develops or conditions change, some fundamental elements of structure and process are established when the team is formed.

In this concept, we discuss the structural elements that lead to teams being effective.

Start with a Team Charter

Many organizations use a **team charter** for larger projects. The charter defines the team's **structure** and **authority** and ensures that team members and management are clear about what the team is expected to accomplish.

Think of a team *charter* as an **initial specification** of team composition and responsibilities. It typically provides the following kinds of information:

- Team purpose or mission
- Deliverables, usually what the team will produce, such as reports, models, new procedures, or a new product or service
- Team sponsor, the organizational member authorizing the team
- Team leader or a description of how leadership responsibilities will be handled
- Team facilitator (if external to the team)
- Team member names and, if it is a cross-functional team, their work units
- Resources available to the team, for example, their budget and information sources
- Constraints such as schedules and decision-making authority

Team Processes and Procedures

Another element of team structure involves the processes and associated procedures the team will use to complete its work.

Processes and procedures agreed to during the initial team formation should be considered flexible and subject to change, rather than permanent. As teams develop, it is common for members to make changes to their procedures, just as it is common for there to be changes in tasks and responsibilities.

During the formative stage, it is difficult to anticipate everything the team will need in order to do its best work. The most successful teams are those that learn and evolve as they work together.

The following questions about team procedures, tasks, and responsibilities are helpful to ask when you first structure a team (Johnson & Johnson, 2002).

Procedures:

- How will decisions be made?
- How will goals be set and met?
- How will tasks be accomplished?
- What procedures seem most appropriate?
- What problems might keep the team from functioning effectively?
- How can the group work out these problems?

Goals and tasks:

- What are the specific goals to be met?
- What are the maximum results possible?
- What are the strengths and resources of each member?
- How many ways can the goals be stated?

Factors to consider:

- What factors might aid or obstruct goal achievement?
- What are the situational "givens" that might impede progress on tasks, subtasks, interfaces, and outcomes?

Tasks:

- What tasks are necessary for goal achievement?
- How should tasks be broken down for individual assignments?
- What schedule, considering critical task interfaces, is reasonable?
- What alerts can we use to highlight problems and for lapses in the accountability of individual team members?

Evaluating the effectiveness of the team in goal achievement. How will the team know when the goals have been achieved, and how will the results be evaluated? The team should evaluate each task, interface, and partial product outcome and how ongoing corrections can be made.

Evaluating process (team effectiveness). How will the team know how effective they are in working together as a team? The team should conduct process evaluations periodically throughout the team assignments and hold a debriefing session after completing major elements or final products.

Team Roles and Responsibilities. Several roles are involved in a team, as illustrated below, although not all teams will need all these roles.

Team Roles

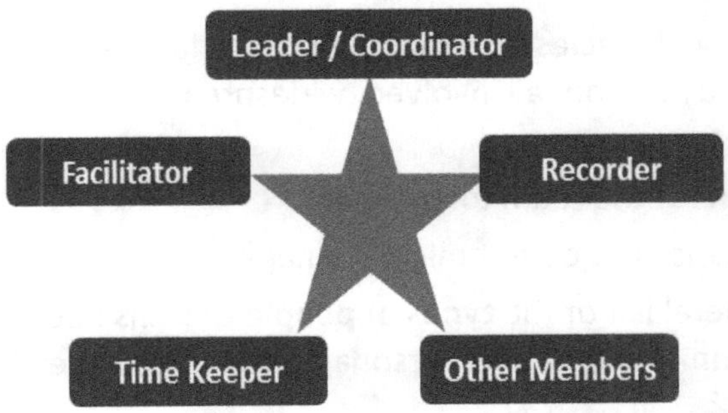

No matter what kind of project or team you are leading, the following **general guidelines** will help you establish roles and responsibilities for the members:

- Invite input from everyone on the team.
- Communicate with all team members, so each person knows the scope and everyone's involvement.
- Empower everyone in group sessions by allowing each member to shape and agree to their responsibilities.
- Ensure that schedules, roles and responsibility charts, key checkpoints, and workflow diagrams are agreed to and owned by the team. This might include, for example, target dates, quality standards, critical interfaces, authority limits, and monitoring procedures.

The **objective of developing the team roles and responsibilities** is to facilitate understanding of what each member is to contribute and also to foster their level of commitment. An essential element of empowerment is that each team member has the opportunity to shape and own what they will do. But the roles and responsibilities will only achieve these goals if they meet the following criteria:

- Expressed in terms that are understandable by the team member who will assume them.
- Recognized as that team member's responsibilities by the other team members.
- Committed to by the team member, and not only to do their part but to do so in a way that supports other team members.
- Appear in writing, clearly showing how each responsibility relates to team goals and other responsibilities, and the interdependencies of team members.
- Viewed as temporary responses to the team's current understanding of its needs and subject to change if demands change, projections were off, or initial plans didn't work out, or individual competencies need realignment to achieve the desired results.

The following guidelines are useful for assigning task responsibilities for each role:

- Establish task assignments with full team input.
- Clarify the tasks from the outset and share them with all team members so that each person knows the total scope, plus everyone else's involvement.
- Get agreement from team members in group sessions.
- Openly review, document, and post all information.

Management Support. Management can help **prepare teams for success** by:

- **Developing clear goals** for what you and the organization expect the team to accomplish.
- **Creating at least rudimentary criteria** for how the team's performance will be evaluated.
- **Clearing away obstacles** that the team might face, such as lack of uniform leadership, buy-in from all involved, widespread awareness of the team, and necessary resources.
- Leaving room for the team to **fine-tune.**
- Provide guidance but do not micro-manage.
- Create a general **list** of the **types of people or skills** needed on the team and consider technical skills, interpersonal skills, experience, and knowledge.

Team Structure Checklist . A Team Structure Checklist is a helpful tool for structuring a new team and is also useful for diagnosing difficulties in an established team.

Use the following checklist to help you assess how well you have set up a team structure and then later, as the project progresses, use it to assess how well the team structure is working.

- ☐ The role of leader is defined, and also the role of facilitator, if needed.
- ☐ The team's purpose is clear and is acceptable to the members.
- ☐ The tasks to be completed by the team have been identified.
- ☐ The team sponsor has been designated and can provide needed support.
- ☐ The team members appear to be clear about the team's decision-making authority and level of empowerment.
- ☐ A team leader has been designated, or leadership responsibilities have been assigned.
- ☐ Initial roles and responsibilities of team members have been defined.
- ☐ Members appear to have a basic understanding of the commitment they are making to the team.
- ☐ Initial procedures have been established for the team to communicate with the sponsor and other stakeholders, and to access external information, training, or coaching.
- ☐ Initial plans for the team to evaluate its results have been established.
- ☐ Team members appear to understand how the team's success will benefit the organization and its customers.

What to do:	Other actions:
☐ Ensure your team has a charter specifying the team's purpose, structure, and responsibilities.	☐ ____________________
☐ Complete the team structure checklist to identify what needs strengthening or clarifying on your team.	☐ ____________________
☐ Ensure team member roles and responsibilities are defined and understood by all.	☐ ____________________

Remember

- ✓ Teams function within the context of the larger organization and operate within parameters set by that organization. For larger projects, a team charter helps make these parameters clear.
- ✓ Teams need guidance, or imposed structure, in the beginning, but must have the latitude to change procedures, tasks, and even membership as the team develops.
- ✓ The roles and responsibilities of team members are best established with the full input of the team and must be understood and committed to by the member assuming them.

Enhance Your Learning

Read the following article that gives 10 tips for better teamwork.

Heathfield, S. (2021). *10 Tips for Better Teamwork.*	Available at: https://www.liveabout.com/tips-for-better-teamwork-1919225

Reinforce Your Learning

Team Structure Worksheet. Use the following worksheet to help establish a new team. Or, for an existing team, use the worksheet to assess how well the team is performing. If you have a team that is struggling, use the worksheet to help identify any structural problems. Use additional paper as needed.

1. What is the team's purpose?

2. What tasks must be completed by the team?

3. Who has the appropriate authority to sponsor the team?

4. Is it clear what decisions the team can and cannot make? Is the team's level of empowerment clear?

5. Has a leader been designated, or leadership responsibilities assigned?

6. Do the members appear to accept the team's purpose and have a basic understanding of the commitment they are making to the team?

7. How will the sponsor and other stakeholders be kept informed of team progress?

8. How will the team's need for external information, training, or coaching be met?

9. How will the team evaluate its results?

10. What are the benefits of membership on the team?

11. How will the team benefit the organization? Its customers?

12. Briefly describe any concerns or considerations regarding the project and how they may affect the team.

Teams Add Value: The Moon Explorer Problem. (Adapted from Hall & Watson)

Assume you are a member of a space crew originally scheduled to rendezvous with a mother ship on the lighted side of the moon. Due to mechanical difficulties, your ship was forced to land at a spot 200 miles from the rendezvous point. During re-entry and landing, much of the equipment aboard was damaged. Survival depends on reaching the mother ship, and so the most critical items from those available must be chosen for the 200-mile trip. Below are the 15 items left undamaged after landing. Your task is to rank-order them by importance in allowing your crew to reach the rendezvous point. Enter number 1 by the item you consider most important, number 2 by the second most important item, and so on through number 15, the least important item to your crew's survival. Record your rankings in the column headed Indiv. Rank (Individual Ranking). When you are finished, wait for further instructions.

	Item	Indiv. Rank	Team Rank		
____	a. Box of matches	______	______	______	______
____	b. Food concentrate	______	______	______	______
____	c. 50 feet of nylon rope	______	______	______	______
____	d. Parachute silk	______	______	______	______
____	e. Portable heating unit	______	______	______	______
____	f. Two .45 caliber pistols	______	______	______	______
____	g. One case dehydrated milk	______	______	______	______
____	h. Two 100 lb. tanks of oxygen	______	______	______	______
____	i. Stellar map (as seen from moon)	______	______	______	______
____	j. Life raft	______	______	______	______
____	k. Magnetic compass	______	______	______	______
____	l. 5 gallons of water	______	______	______	______
____	m. Signal flares	______	______	______	______
____	n. First-aid kit w/ injection needles	______	______	______	______
____	o. Solar powered FM receiver and transmitter	______	______	______	______
			Total Scores	______	______

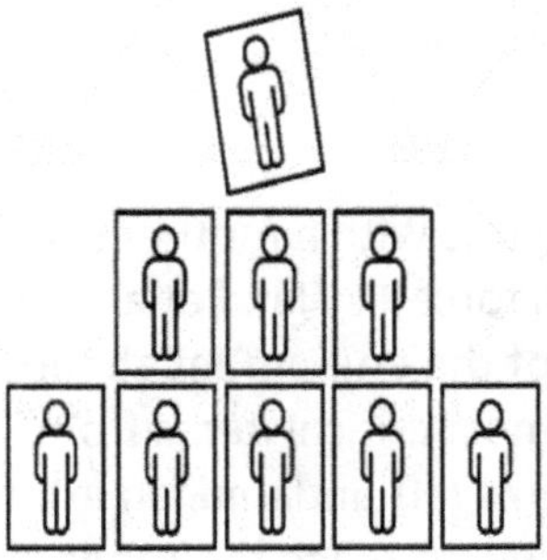

Building High Performance Project Teams | #7.3

Review the Stages and Challenges of Team Development

Project teams rarely, if ever, start at their peak performance level. They move to that level as members develop their ability to work together productively, creatively, and efficiently, while learning how to meet the many challenges they face.

To develop from a collection of individuals to a genuinely productive team, a group of people must move through a series of stages. The many models of team development may use different terminology, but all the models describe stages where the group must resolve a series of issues. When the issues at any one stage are resolved, the team moves to a higher level of performance, and so on.

In this concept, we discuss how groups go through stages where they learn how to funct on as a team. We also address typical challenges that many teams face at one time or another.

The Tuckman Model

A **popular model** based on the classic works of Bruce Tuckman, an educational psychologist, uses five stages: **form, storm, norm, perform and adjourn (or reform).**

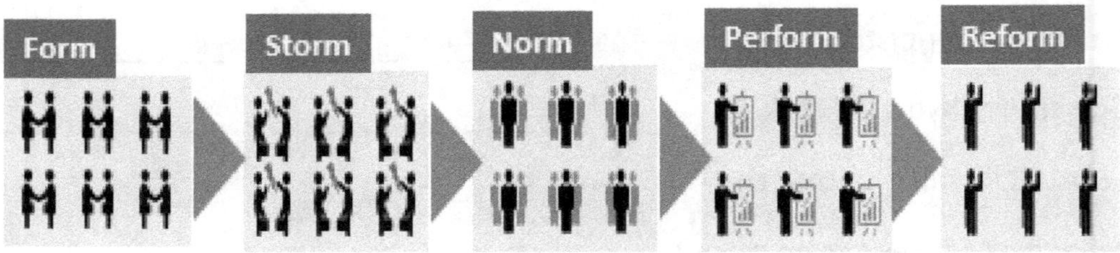

Form. As the name indicates, *form* is the first stage, when the team initially comes together. Team members get to know and trust one another, formulate goals, and establish roles and responsibilities. During this stage, some members are enthusiastic and positive, while others are uncertain or apprehensive.

Storm. This is a time of conflict as members gain insight into what being on the team may cost them in terms of personal freedom and control. As they discuss issues and responsibilities, disagreements will surface. At this stage, members have not developed sufficient trust in each other to work together productively, and they may start to question the organization's wisdom in setting up the team. Disagreements can be based on perceived differences in work habits, work ethics, and focus. Some members focus on minutia and seem to get lost in detail while others concentrate on settling broader issues.

Norm. This is the stage where team members "get serious" about structuring the team. They revisit and clarify their goals, tasks, processes/procedures, and responsibilities. At this stage, members have developed sufficient trust to subordinate their personal preferences to the preferences of the team. Participation is more open and authentic as members gain respect for the abilities and commitment of fellow members, the leader, and the sponsor.

Perform. At this stage, the team is cohesive. Members share a collective identity and purpose and proactively work for the team's goals. For the first time since its formation, the team is fully functioning with focused energy aimed at achieving the team's goals. This is the stage of **synergy.** This is different from earlier stages where some of the team's energy might have focused on issues other than **team achievement**: personal acceptance, developing relationships, establishing procedures, and the like.

Adjourn or Reform. When the team's work is complete, the team may be disbanded or reformed for another purpose with some changes in membership or leadership. This can be an emotional time for many members. Some may be sad to leave people who have now become friends or nervous about what lies ahead. Others may be excited to take on a new challenge.

Managing a Team as it Moves Through the Stages

In team development, there is no guideline for how long a team will stay in one stage before moving to the next. Further, teams do not necessarily proceed in a straight line through the stages. For example, a team in the **perform** stage may not stay there for the duration of its existence. Any number of changes—in membership, performance expectations, or leadership—may require that the team **reform** by moving back to **norm**, **storm**, or even the **form** stage before it can regain excellent performance. For example, as teams begin to raise new ideas and tasks during the norm stage, they can temporarily return to the storm stage until the differences are reconciled.

Teams need different types of support and direction at different stages of development. In the beginning, they need clear guidance and encouragement that they will become more self-directed as they work together. They may need help to clarify expectations and work through how they will function as a team. At subsequent stages, teams may need help in learning how to resolve interpersonal conflict, influence each other in positive ways, establish procedures, and refine team process.

Everyone—the project manager, team members, and functional managers—must be aware of this growth process, or they may have unrealistic expectations of the team and demand "too much, too soon." Managers may lose patience or conclude, prematurely, that the team has failed. Team members may become disenchanted and feel too much time is wasted, or they may be tired of dealing with "the same old things." Project managers, frustrated with an apparent lack of progress, can assume highly directive, authoritarian styles, which further discourages the members.

An excellent tool to help teams achieve—or return to—the perform stage is to periodically ask members to identify the team's stage of development. This will help individual members remember that the team must go through stages to become cohesive and productive. The group can then discuss what is needed to resolve any current difficulties and to move to the next level.

Team Challenges

As well as proceeding through developmental stages, certain conditions present challenges to teams as they strive to be successful.

Consider the following kinds of situations:

- Team members are new or membership changes often.
- Some people are unsure of or dissatisfied with their roles.
- Time pressures make it difficult for the team to adequately develop and use group processes, or threaten norms already established.
- Locations or workloads of members make it difficult for team members to have regular contact with each other, or with the leader.
- Uncertainty about the goals or tasks to be accomplished, or individual members are unsure of their task proficiency.
- Differences exist among team members: they simply do not see things the same way.

- Communication breaks down, and there is insufficient time to clarify, reflect, or correct misunderstandings, or the complexity of communication has been underestimated.
- "Taboo" ideas affect the team's efforts, but team members do not feel safe to discuss or explore them.
- Participation is not spread among team members—some dominate, some are only partially engaged, and some may drop out or may never have been team players.
- The team has a failure experience or team members share a perception of not measuring up to other teams.
- Too many changes happen to the team; examples are changes in team leadership, the tasks to be completed, team goals, and deadlines.
- Barriers to excellent teamwork exist; for example, unclear or inadequately delegated responsibility for decision making (level of empowerment), unnecessary approval required for team decisions (micro-management), and an organizationa climate that does not value new ideas or risk-taking.
- Insufficient resources allocated to ensure excellent teamwork. Examples are **the team did not receive** sufficient time to develop into the **perform** stage; not enough training to develop interpersonal and group interaction skills; schedule flexibility insufficient to accommodate teamwork, not enough access to information needed by the team, and insufficiently skilled, knowledgeable, and experienced team members.

What to do:	**Other actions:**
☐ Be mindful that teams must develop to become fully functioning and that development takes time.	☐ ____________________
☐ Manage the various conditions that slow down or derail team development.	☐ ____________________
☐ Periodically assess your team's stage of development.	☐ ____________________

Remember

✓ Even though all teams go through the **form, storm, norm, perform** and **reform** stages, how they do so is not entirely predictable or orderly. The time it takes a team to move from **form** to **perform** varies and is dependent upon the previous team experience of the members and the complexity of the team project or assignment. Also, teams often drop back to an earlier stage of development in response to changing conditions. This results in a temporary lowering of team performance.

✓ The storm stage is challenging and demanding, but essential as part of team development. Project managers, team members, and functional managers can give up on the team during this stage. However, if members are forewarned about this stage and know they must work through it, they will be well-positioned for authentic team participation.

✓ Many conditions present serious challenges to team success, including insufficient resources, insufficient skills to meet project needs, limited decision-making authority, and disruptive change.

Enhance Your Learning

Watch the following 3-minute video that explains the five stages of group development.

Muzio, E. (2010). *Tuckman's Model: Fight Right*!		Available at: https://www.youtube.com/watch?v=MNgzjYb02JM

Reinforce Your Learning

Moving to a Higher Level of Project Team Development. Based on your experience with current or past teams, describe actions or ideas that can help teams advance to the next stage.

Ideas for moving from **form to storm:**

Ideas for moving from **storm to norm**:

Ideas for moving **from norm to perform**:

Ideas for **adjourning** the team:

Encourage Positive Communication

To a high degree, a team's **success will be determined by how team members interact** with one another. Team success is related to how well a team communicates and how productively they pool their skills and knowledge to resolve issues and advance positive actions.

Team members may find it difficult to remain positive, open, and respectful for no other reason than they differ from each other in many ways. Differences in knowledge, skills, experience, interests, and personality can all create barriers to understanding. Even when team members may seek to appreciate their differences, productive communication can be challenging.

In this concept, we discuss **authentic participation** and give guidelines for how you can help team members communicate in ways that foster cooperation and collaboration.

What is Authentic Participation?

Positive team communication is open and balanced. Team members can present their viewpoints and listen to the views of teammates. All members sense that their ideas are valued and welcomed by others and that they can be honest and direct without fear of rejection or personal criticism. They can participate fully and authentically—respectfully, and without deceit or glossing over disagreements or difficulties.

The open and balanced atmosphere that is most positive for team communication applies to one-on-one conversations between members, as well as to meetings of the entire team.

For **authentic participation**, members need the following:

- Trust in the leader and other members
- Sense of team purpose or mission
- Knowledge about the goal or problem
- Skill in working with others in a group setting
- Value cooperation and collaboration instead of competition
- Separate disagreement from personal rejection, disapproval, or "put-downs"
- Willingness to share, follow, lead, and learn

Authentic Participation Questions for You and Your Team to Ask Yourselves

Understanding and working with personality differences **promotes positive communication** because personality disposes people to certain mind-sets or behaviors and thus influences interaction.

To help achieve **authentic participation**, you and each member of your team—on a focused, continuous basis—should ask and answer the following questions:

- How can I speak so that others will want to listen and understand my views?
- How can I listen so that I can understand and benefit from the views, information, and experience of others?
- How can I remain open to the knowledge and experience of others, and mesh that with my own to achieve growth, new ways of thinking, and solutions?
- How can we collaborate, with mutual influence, to come up with satisfying resolutions to our conflicts?
- How can I keep my focus, and that of other team members, on relevant issues, helpful discussions, constructive feedback, and breakthrough ideas?
- How can I ensure that the best idea, regardless of who has it, gets shared?

The questions may seem simple and obvious. Yet human realities attest to their difficulty. For example, we do not always listen well, especially when we disagree. We tend to be self-centered and limited by our own experience. We focus on our positions and wants before we consider those of other people and rely on our experience to make sense of things. We find it difficult to see things from someone else's perspective, and most of us think we are right, at least most of the time. So, ask the questions and encourage your team to do the same.

Effective Leaders Model Productive Communication

How you lead your team will significantly affect how members communicate. Leaders are role models for everyone on the team. As such, you will do well to review your communication behavior periodically. An honest assessment will help you uncover problems and make corrections before they have a chance to affect the project.

Here is an excellent question to ask yourself: Do I, as a leader, promote a positive atmosphere for communication?

Yes, if you:

- Help others express their ideas.
- Listen alertly and with understanding to what others say.
- Communicate your ideas well.
- Seek out other people's opinions and are sensitive to others' feelings.
- Are tolerant of opposing viewpoints.
- Encourage other people's ideas.
- Own your ideas and feelings.
- Support the team's use of effective process, tools, and techniques.

No, if you:

- Avoid conflict when it should be addressed.
- Are overly stubborn about your opinions.
- Are overly aggressive in fighting for your view of reality.
- Give in too quickly when opposed.
- Participate too much or too little in discussions.
- Deny other people ownership of their ideas.
- Provide feedback in destructive rather than constructive ways.
- Appear apathetic about the team, its goals, or its members.

Encourage Team Members to Support Positive Communication

As a leader, your communication behavior will affect the team, but unfortunately, you cannot build commitment to collaborative approaches without the support of the team. Each member has an impact on team performance and therefore has a responsibility to communicate in ways that enhance team performance.

Team members support positive communications when they:

- Seek consensus (from other members) on decisions, objectives, and plans.
- Openly share feelings, opinions, thoughts, and perceptions about problems.
- Participate and involve others in the decision-making process.
- Express genuine concern and support for other members.
- Use active listening and try to see others' viewpoints.
- Acknowledge and respect individual differences.
- Acknowledge and openly confront conflict.
- Use the ideas and suggestions of others.
- Encourage and appreciate feedback on their behavior.
- Do not engage in *win-lose* interactions with other team members.
- Resist dominating the discussions or forcing their opinions on others.

Encourage Constructive Behavior and Discourage Destructive Behavior

Some behavior relates to accomplishing a team's tasks while others focus on developing and maintaining group harmony and cohesiveness. These are referred to as *task* and *maintenance* behaviors, respectively.

Some of these behaviors are constructive, and others are destructive. Some examples are shown below.

Constructive Task Behavior

- **Initiating.** Proposing tasks, goals, objectives, problem definitions, and suggestions for solving problems.
- **Seeking information and feedback.** Requesting ideas, facts, and feelings about the subject discussed and the group reaction.
- **Giving information, suggestions, and feedback.** Participating in a discussion by offering one's information, insights, and feelings.
- **Clarifying and elaborating.** Asking questions to gain a full understanding of another member's point of view.
- **Confirming.** Summarizing one's understanding of another member's information or position to assure accuracy.
- **Coordinating.** Managing and sequencing the flow of ideas or information.
- **Evaluating use.** Testing suggestions or proposed actions for practical application; offering constructive criticism.
- **Summarizing and seeking consensus.** Periodically summarizing the progress of the discussion to assist the group in reaching a conclusion.

Constructive Maintenance Behavior

- **Active listening**. Using active listening techniques, which are reflecting, paraphrasing, summarizing, and acknowledging the other person is being heard and understood.
- **Encouraging and supporting.** Being friendly, warm, and inviting others to share their ideas, and responding positively to their input.
- **Gatekeeping.** Keeping communication channels open, suggesting procedures that encourage participation from all members, reminding members to "listen fully" before rejecting another person's idea, reminding people not to interrupt.
- **Managing conflict.** Reconciling disagreements and getting people to explore differences and strive for a positive resolution.
- **Climate testing.** Assessing the satisfaction of members with process and procedures; testing the readiness of the group to continue a discussion or reach closure; asking members to comment on "how things are going," and "what do we need to focus on."

Destructive Behavior

Some ways of communicating inhibit a free and open exchange of ideas and information and therefore interfere with authentic participation.

Destructive behaviors include:

- **Negative criticism.** Consistently playing the *Devil's Advocate*, finding reasons why things will not work; insisting on the worst-case scenario.
- **Sniping.** Making personal attacks on the motives, attitudes, or character of others.
- **Pulling rank.** Asserting one's power to dictate the course of the discussion or the decision; interrupting or preventing others from stating their ideas.
- **Deserting.** Withdrawing from active participation, becoming indifferent or aloof; depriving the group of one's input.
- **Horsing around.** Engaging in "sidebars," making comments to others, clowning and drawing attention to oneself, and disrupting discussions.
- **Sidetracking.** Introducing extraneous subjects, particularly to plead some special interest; distracting the group from the agenda.

Tools and Techniques to Help You Improve Team Communication

To improve the quality of communication in a team, you may find it helpful to use a **process facilitator.** A good facilitator can help the team improve its ability to communicate in positive and productive ways. A process facilitator is especially useful in the early life of a team but can also be helpful later when conflict arises, or difficult decisions must be made. A facilitator can help team members recognize and develop constructive ways of communicating and avoid destructive ones.

Training in group communication and reviewing examples of effective team interactions can help increase a team's self-awareness.

Having a team-building session in which members assess and discuss their personality types or interpersonal styles can be helpful as well. It allows the members to see their differences in positive and nonjudgmental ways and gives them insight into how to communicate better with each other.

What to do:	Other actions:
☐ Commit to asking yourself the authentic communication questions regularly.	☐ ____________
☐ If you sense that communication within a team is not open or balanced, respectfully raise these perceptions during an upcoming meeting and invite discussion.	☐ ____________
☐ Monitor task and maintenance behaviors: increase constructive behaviors and avoid destructive ones.	☐ ____________

Remember

✓ Open, respectful, shared, and balanced communication among team members is a necessary ingredient for success.

✓ Some people are easily influenced, coerced or otherwise encouraged not to give their opinion, either directly, for example by threat, or indirectly, for example, they are shy or insecure.

✓ The leader cannot build commitment to collaborative approaches without the support of the team. Keep in mind the team behaviors that support positive communications. Hold yourself and your team accountable for those behaviors.

Enhance Your Learning

Watch the following 3-minute video to review five tips for being an effective team member.

Yap, C. (2014). *Qualities of an Effective Team Member.*	Available at: https://www.youtube.com/watch?v=mf9FRXPHMUg

Reinforce Your Learning

Authentic Participation. Encouraging authentic participation involves helping everyone engage and share openly, respectfully, and honestly. With this in mind, answer the questions below.

How can I speak so that others will want to listen and understand?

How can I listen so that I can understand and benefit from the views, information, and experience of other people on the team?

How can I keep my focus, and that of other team members, on relevant issues, helpful discussions, constructive feedback, and breakthrough ideas?

Building High Performance Project Teams #7.5

Resolve Conflict and Promote Shared Responsibility

Project leaders use several methods to ensure that divergent and conflicting views are expressed and processed for the betterment of a team. They seek results that are more than the mere sum of individual efforts or a simple majority opinion. Productive leaders and teams aspire to reach *synergy*—the condition where "two plus two equals more than four."

Teams work diligently to reach decisions all members can support, even if there is no absolute agreement on everything. To manage the tension between different views, members must learn not to jump to conclusions or take sides too quickly. Team members must recognize the value of examining all aspects of an issue before reaching a conclusion. This does not mean that the best ideas are compromised. Members of effective teams monitor their processes to ensure they are not sacrificing deep thinking and real collaboration. They know that "giving in" is not the same as reaching consensus through understanding and deliberation.

In this concept, we discuss how you can lead your team to value diverse viewpoints and conflicting approaches, discover alternatives, and seek solutions by sparking creative thinking that leads to a strong sense of shared responsibility for decisions and results.

Teams Can Learn to Value Collaboration

Many organizations offer **team-building** exercises or expert facilitation for new groups to help members appreciate collaboration and understand that conflict is a natural and essential part of teamwork.

This type of support is especially useful when some people on the team have not experienced a successful team in the past, have limited knowledge of how to work collaboratively, or are accustomed to individual rather than shared responsibility.

Team members may also benefit from learning about **tools** such as the nominal group technique, force field analysis, and round-robin brainstorming. It may be beneficial to offer training to team members in these techniques, as well as in systematic problem solving and a structured process for resolving conflicts.

Project Managers Must Manage Conflict

Numerous types of conflict occur in team settings. Typical examples are conflict between two team members, conflict between one member and the rest of the team, conflict between different factions (several members in conflict with several other members), and conflict between the team and someone outside the team (a sponsor or other stakeholder from whom information, cooperation or approval is needed).

One common cause of conflict within a team is "non-team behavior" on the part of one or more team members. Teams depend on individual members to contribute to goal achievement, to follow structural and procedural guidelines, and to interact positively and productively with other members. When members do not do these things, they are engaging in "non-team behavior." Such behavior is destructive to team processes, and you must deal with it as a condition of membership.

However, for conflicts that arise around solutions to problems or decisions—conflicts of central importance to the team's work—you will get the best results by using a structured approach to resolving the issues and by working with team member perspectives.

Most team members must learn (or relearn) at least two fundamentals:

1. Disagreement is an essential part of teamwork. It is the authentic expression of differences among members that gives a team its potency of synergy.
2. When people care deeply about the team, its mission and members, they will have strong feelings and emotions, and this is especially so when they hold conflicting views, values, or needs. We cannot expect people to be authentic if all they are allowed to express is information or logic. Our best solutions come from an exhilarating blend of what makes sense and what feels right.

The following are guidelines that will help you develop the team's ability to engage in productive conflict:

- Acknowledge that the conflict exists.
- Define conflict issues in terms of shared needs rather than in terms of opposing viewpoints.
- Keep the focus on issues and not personalities.
- When generating options, remind people of the brainstorming ground rule: no criticizing statements by other people until all ideas are gathered.
- Encourage everyone to listen to different points of view.
- Get everyone involved.
- Do not stifle occasional outbursts of anger, but do not dwell on them either.
- Seek to understand all angles.
- Attack the issue, not one another.
- Develop an action plan.

Use a Process to Manage Conflict

The five dysfunctions of a team based on the popular book of the same name (Lencioni, 2002) are the absence of trust, fear of conflict, lack of commitment, avoidance of accountability, and inattention to results.

To deal with challenges and dysfunction, a team needs a clear mission, a goal and a timeline, and the resources to tackle the work assigned to the team. They also need a process for managing conflict.

A structured process for handling conflict, such as the one that follows, will help the team handle the intense feelings that often accompany conflict. Whatever structured approach you use will be most effective if you introduce the approach to the team and get their buy-in before you use the approach to tackle a problem or make a decision.

The following structure is one approach to helping a team resolve conflict:

- Invite each person to state their view briefly (no speech-making).
- Have neutral team members reflect areas of agreement and disagreement.
- Explore areas of disagreement for specific issues.
- Have opponents suggest modifications to their own and the other people's points of view.
- If blocked, ask opponents if they can accept the team's decision.
- If the team's decision cannot be accepted, the ultimate resolution will revert to the team sponsor.

What to do:	Other actions:
☐ Recognize that conflict, in various forms, is a natural and essential part of teamwork.	☐ ________________
☐ Follow a structured process for helping the team resolve conflict.	☐ ________________
☐ Teams with effective processes resolve conflicts and develop shared responsibility. Know how effective your team processes are.	☐ ________________

Remember

✓ When people care deeply about a team's mission and members, and when they participate authentically, they will sometimes disagree with each other. An absence of conflict, rather than being a good thing, might be a signal that people are either afraid to express their viewpoints or are apathetic about the team.

✓ To use their diversity to achieve higher performance, team members must keep their interactions focused on the team purpose and the performance issues at hand and not lapse into personalizing disagreements. Structured processes for problem solving and conflict resolution can help with this.

Enhance Your Learning

Watch the following 7-minute video to learn more about team dysfunctions, considering how you can apply this information to your work.

Lencioni, P. (2019). *5 Dysfunctions of Teams.*	Available at: https://www.youtube.com/watch?v=RoONBgHo_a8

Reinforce Your Learning

Five Dysfunctions of a Team. Below are five dysfunctions of a team based on the book of the same name by Patrick Lencioni. Consider each dysfunction and respond to the questions with an eye toward resolving conflict, building collaboration, and promoting shared responsibility in a team.

Dysfunction: Absence of trust
Goal: Team members are comfortable being vulnerable, open, and honest. How can team members or the leader instill healthy vulnerability in the team, and why does it matter?

Dysfunction: Fear of conflict
Goal: Members are comfortable engaging in healthy conflict. When is conflict among team members a good thing?

Dysfunction: Lack of commitment
Goal: Members are committed to the team. Consensus among the members is typically a good thing. When can consensus be a bad thing?

Dysfunction: Avoidance of accountability
Goal: Team members hold each other accountable, peer to peer. How do you hold team members accountable, especially when the person you are dealing with is a peer and friend?

Dysfunction: Inattention to results
Goal: Team members focus on team results rather than individual results. What do people usually focus on when there is a lack of attention to team results?

Building High Performance Project Teams #7.6

Enhance Team Problem-Solving Capabilities

The quality of team decisions, solutions to problems, and other products of team effort depend on the quality of the interactions among team members.

Significant research into group functioning has helped to sort out the more effective characteristics of interaction from the less effective, and to identify techniques to increase group effectiveness. This research has also identified the challenges and difficulties incurred when individuals work together as a group with a common purpose. Many of the problems grow out of the very interactions required for team success. While all groups are prone to certain patterns of interaction, each group has its own set of dynamics. Each group develops its own identity and standards of behavior.

In this concept, we discuss how to support and build your team's problem-solving capabilities and thereby improve its performance.

Content and Process

Content refers to the product, service, or result of a team's effort. It relates to the problems to be solved, and the decisions to be made (what we might call *the what*) as a team works on achieving its goal.

Process is separate from content. ***Process*** refers to the approaches, procedures, rules, group dynamics, and styles of interaction within the group (*the how*).

It is useful to keep these distinctions in mind as you work to improve your team's proficiency in solving problems. Keeping in mind whether you are working on "what" or "how" will help you and the team avoid "muddying the water" and reduce confusion. It will also help individuals focus their attention.

Problem Solving and Decision Making

A ***problem*** is a gap between something desired and what exists. ***Problem solving*** is the approach a group uses to identify a problem, collect relevant information, identify causes, and select, implement, and evaluate a solution. *Decision making* involves reaching a conclusion or making a choice.

Teams must make many decisions as they work on solving problems. Examples are how to define the problem, what information to collect, and who should be involved when and in what decisions.

Team members interact by exchanging task-relevant information, examining the meaning and implications of data, stimulating creative and critical thinking, challenging knowledge and assumptions, exploring novel ideas, and finally reaching agreement on actions or recommendations. These interactions reflect both content and process.

Use a Systematic Approach to Problem Solving

As we said, *problem solving* is the process used by an individual or group to identify a problem and to work through various steps to determine and implement a solution.

Team problem solving is most effective when members follow a *systematic process,* and when everyone focuses on the same step of the process at the same time. In other words, a team avoids chaos when everyone follows the same work steps to problem solve, and everyone works on the same step at the same time.

There are a number of different, but similar, problem-solving and decision-making processes. The following graphic depicts one of the most used problem-solving processes.

Six Step Problem-Solving Process

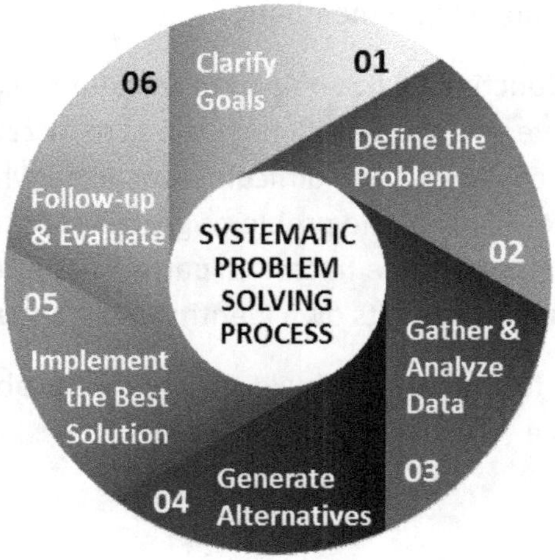

In addition to a systematic process, the following conditions also help groups improve their problem-solving capabilities:

- Access to more or better information
- Idea exchange (to stimulate thinking)
- Altered focus to help overcome a mindset
- Absence of hidden agendas (all members seek the group's best answer)
- Meeting climate supports cooperation over competition
- A facilitator who sees their role as stimulating creative thinking and encouraging members to air their ideas and make their own decisions

Barriers to Problem Solving

Obstacles to team problem solving include conditions that prevent the free expression of ideas. Typical conditions are majority rule, participation bias, failure to search for alternatives, failure to define or understand the problem, and adoption threshold factor, and groupthink, as discussed below.

Majority rule. If minority opinions are not solicited, members eventually suppress them. Not only do they not share them, but they often do not even think of them.

Participation bias. Some studies suggest that the individual who participates the most in a group discussion gets more of their ideas accepted than less participative members, regardless of the quality of the ideas.

Failure to search for alternatives. Most often, we make up our minds about something quickly, based on a few facts and casual thinking. Instead of using our phenomenal brainpower to help us evaluate ideas, we use it to prove that the "quick" decision we made is the right one!

Failure to define or understand a problem. It is not uncommon for a team to start coming up with "solutions" before they have agreed on what the problem is or what factors may have caused it.

Adoption threshold factor. Once an idea has passed a threshold of acceptability, subsequent ideas are not examined as carefully. Often, the first idea to cross this threshold is adopted, even though other, better ideas are available. In essence, we stop searching for them.

Pressure toward uniformity or groupthink. Groupthink is both an individual behavior and a group dynamic.

Groupthink refers to the fact that most people naturally **strive for agreement**, and some people will agree with a larger group simply to keep the peace. A person who believes one thing but is faced with people saying other things may simply remain silent. Their silence suggests agreement, and this allows the group to reach a conclusion contrary to what this person (and likely other people) believes or wants. While it is desirable to have cohesion within a group, groupthink can **shut down creativity** and healthy **debate**.

Teams can avoid groupthink by ensuring that the group is **open** and **invites divergent opinions,** especially from those who are silent or appear to disagree even though they do not express that verbally.

Sometimes breaking a larger group into smaller groups—even groups of two or three—can increase participation because it helps people feel more comfortable in expressing opinions.

Facilitators

Facilitators can help teams, especially newly formed ones, to become more comfortable with a systematic problem-solving process and with various techniques designed to help teams work through problems efficiently and effectively. Brainstorming is one of the techniques commonly used, as well as nominal group and force field analysis.

As well as helping the team focused and teaching techniques, facilitators can help keep discussions organized, and intervene to help the team manage disagreements or conflicts that emerge.

Encourage Creative Thinking

One of the benefits of bringing a group together to solve a problem is that the potential for creative solutions increases. As people interact, an idea offered by one person triggers someone else's thinking, and a new idea emerges.

Additionally, pooling their knowledge, experience, information, and insights produces solutions that, very often, would not be arrived at by an individual working alone. Creative thinking can lead to useful answers and often innovative and imaginative ones.

Being creative requires you to look at a situation from every angle and be flexible in the way you think. Some people mistakenly think they are not creative. The truth is, some people think more creatively, more often, than others do. But **creative thinking** is a **skill** that **can be learned** and improved upon with practice. It is a skill worth developing. In today's complex and quick-paced work organizations, creative thinking skills are in high demand.

Enhance Creative Thinking

Do some creative exercises as part of team building. Examples are playing games that force the team to come up with inventive answers, challenging members to try something in a new way, and asking members to spend time with people they do not generally interact with. All these exercises may help a team broaden their horizons.

Daydreaming or going on imaginary excursions is also a good exercise. For example, members could try envisioning the perspectives of other team members, thinking about how their minds work, and what is important to them.

Shape the environment to be conducive to creative thinking. Pick a setting that the team finds comfortable, both physically and psychologically. Remove anything that could be distracting. Choose a time when team members are not hungry, tired, or feeling stressed. Having a suitable environment will help all to focus on being creative.

With a focus on the problem and in a comfortable environment, try brainstorming. Start by clearing your mind, pushing aside any worries or troubles you and other team members may be experiencing. Then let your minds wander and use a brainstorming technique to stimulate idea exchange. Try the basic brainstorming steps described below.

Stay open-minded. A valid solution can be overlooked because people are not comfortable trying something new. Do not pass up a good idea just because it is somewhat unfamiliar.

Brainstorming

Whether you are brainstorming alone or with a team, here are **the basic steps.**

Generate as many ideas as possible. Invite people to shout out their ideas to the group. Assign a person to record the idea and record everything—even the ideas that sound crazy. Often, as you write them down, talk about them, or think them through, you will find those crazy ideas are not so crazy, or they lead to other ideas. So, encourage the team to keep thinking of ideas and list them all.

The following are the **rules of engagement.**

For group brainstorming, **every idea must be recorded** publicly and as offered—no editing or paraphrasing allowed.

No comments allowed (either positive or negative) as people offer ideas. The recorder may ask for the idea to be repeated if necessary to record it correctly.

After the idea generation phase, the group reviews, clarifies, consolidates, classifies, and evaluates the ideas. The purpose is to **reduce the list** and reach an **agreement** on the **best ideas** or a **course of action.**

What to do:	Other actions:
☐ Be mindful of thinking that restricts the free flow of ideas in team problem solving.	☐ ____________________
☐ Take five minutes and brainstorm ideas for solving a team problem you are facing.	☐ ____________________
☐ Consider using a facilitator to help the team follow a systematic problem-solving process.	☐ ____________________

Remember

- ✓ Teams solve problems by exchanging task-relevant information, examining the meaning and implications of data, stimulating creative and critical thinking, challenging knowledge and assumptions, exploring novel ideas, and reaching an agreement for actions or recommendations.
- ✓ Creativity is the ability to look at a problem and come up with a solution that is both effective and original. Some people think that creativity is beyond them, that it is an inherent talent they do not possess. This is a common misconception.
- ✓ Groupthink can interfere with the full examination of differing opinions. If you think any members of the team are in unspoken disagreement, encouraging them to express their views can only enhance the team's performance.

Enhance Your Learning

Watch the following 4-minute video about where ideas come from.

Johnson, S. (2010). *Where Good Ideas Come From.*	Available at: https://www.youtube.com/watch?v=NugRZGDbPFU

Reinforce Your Learning

Brainstorming and Categorizing Ideas. Consider a typical team problem you have experienced or observed, such as **keeping a team on track**. Take five minutes to list as many ideas as possible for solving the problem and record the ideas using the template below. Then take another five minutes to reduce your list by eliminating and merging ideas. Then categorize your ideas to help find a course of action.

Team Problem: ____________________

Category?	Ideas
	1.
	2.
	3.
	4.
	5.
	6.
	7.

Building High Performance Project Teams #7.7

Recognize Values That Support Team Success

A team's performance—its value to an organization—is directly related to the support, planning, preparation, development, and reinforcement the project manager and organization provide to the team.

High-performance teams can deliver remarkable results. But the results are not merely the sum of the skills and effort of the people involved. *Teams produce results because their capacity to perform at a high level has been developed.* Teams flourish when project managers and organizations value them and invest in their development.

Teamwork is more productive when it occurs in an environment that attracts people to teams, an environment where people take team responsibilities seriously, and appreciation is shown for teamwork.

In this concept, we discuss teams and organizational support and what you can do to make the value of teamwork part of your organization's culture.

Organizations Must Support Teamwork

Organizations establish project teams when there is a beginning and end to selected work or activities, and they are convinced the job is best suited to a team effort. But this is not enough to ensure that the teams will be successful.

Organizations with project teams that consistently produce results place a high priority on teams and teamwork. Such organizations fund **training** and support **team-building** activities and **reward team results**. These organizations also align human resource **policies and practices** to support a team-based environment focused on achieving results.

To realize the benefits teamwork can provide, organizations must proactively work to create and maintain a **culture of teamwork**, a culture that genuinely **believes two heads are better than one** and that consensus decision making will on-average produce the best results.

Organizations must think of project teams as "works in progress" and always be prepared to make accommodations as teams mature and conditions change.

Top management should ensure that sponsors, team leaders, and functional managers offer development and facilitation support when teams are formed, and when a team appears to be stuck.

Does your organization value teamwork? Does it reward people who recognize that planning, thinking, creating, making decisions, and taking action are more effective when people work together?

If you do not already know, find out the views of your top leadership. Are they *pro* or *con* when it comes to project teams? If *con*, consider how you might help advance a more positive viewpoint.

You will also do a better job of managing projects if you have a good understanding of your organization's history with project teams. Look at its past and current experiences, successes, and the impact of those successes. Look at and analyze the failures and understand the reasons for those failures. Also, learn what happens in response to productive project team performance.

Reinforce the Value of Teamwork in Your Organization

Examine your performance evaluation process. Does your current performance evaluation process properly reinforce the value of teamwork? Do not be surprised if you find that what counts are individual contributions, not team efforts.

Examine your organization's recognition and reward systems. Identify if your organization has rewards—and not penalties—for team contributions.

Check the current policies and procedures regarding funding for project team development, including skills education and "team-building" activities. Look for examples that demonstrate support and communicate these examples as indicators of the organization's commitment to supporting teams.

Review your organization's mission statement. Every organization needs a clear, **written mission statement** (Hull, 2013). A mission statement tells the world what the organization does, how they do it, for whom they do it, and **what value** they bring (Hull, 2013). The value of project teams should be addressed in the "how the organization does it" section of the mission. Check your organization's mission statement to see if it mentions teamwork.

Tell stories that demonstrate the fruits of collaboration. Every organization has their stories, although not all are positive. Repeat stories that reflect the success of project teams and how people feel about being part of successful teams.

If you do not know the stories, ask people who have been in the organization longer to share theirs. Sharing such stories helps build internal support for teams.

Find ways to publicize these stories at appropriate times. For example, on the anniversary of a successful project, you might ask your communications department to publish a story about the team in the organization's newsletter.

What to do:	**Other actions:**
☐ Research the views of your top leadership. Are they pro or con when it comes to project teams?	☐ ____________
☐ If the performance evaluation, reward, and recognition system you use do not encourage project teamwork, what changes can be made to recognize teams?	☐ ____________
☐ Be mindful of the cultural characteristics and resource support needed to make teams successful.	☐ ____________

Remember

- ✓ For project teams to succeed, they need dedicated support from all levels of management. This support must extend beyond stated values to include investing in team development.
- ✓ A culture of teamwork begins from the top, with leadership's commitment to teamwork and collaboration. Realistically, however, a culture of teamwork requires the engagement of everyone.
- ✓ Understand two heads are better than one, and consensus decision making will on-average produce the best results.

Enhance Your Learning

Watch the following 4-minute video that overviews the effect of shared values on a business team.

Beare, S. & McMillan, M. (2007). *The power of teamwork.*	Available at: https://www.youtube.com/watch?v=mR0_SK1K8xY&feature=related

Reinforce Your Learning

Team Process Checkup. Use the inventory below to consider how your project team is operating and use the results to help the team continue developing. Be thoughtful in your responses so the results will genuinely benefit the team. For each statement, circle the number you think is most descriptive of your project team **now** with **1** being **low/seldom** and **7** being **high/almost always.**

Project Team Name: ________________

1. Project team members understand group goals and are committed to them.	1 2 3 4 5 6 7
2. Project team members are friendly, concerned, and interested in each other.	1 2 3 4 5 6 7
3. Project team members acknowledge and confront conflict openly.	1 2 3 4 5 6 7
4. Project team members listen with understanding to each other.	1 2 3 4 5 6 7
5. Project team members include each other in the decision-making process.	1 2 3 4 5 6 7
6. Project team members recognize and respect individual differences and team member diversity.	1 2 3 4 5 6 7
7. Project team members contribute ideas and solutions to problems.	1 2 3 4 5 6 7
8. Project team members value the contributions and ideas of other members	1 2 3 4 5 6 7
9. Project team members recognize and reward team-supportive behavior.	1 2 3 4 5 6 7
10. Project team members seem to take pride in being part of the team.	1 2 3 4 5 6 7
11. Project team members show respect to each other, such as not interrupting when another is talking and avoiding personal attacks.	1 2 3 4 5 6 7
12. Our project team meetings are efficient and generally a good use of time.	1 2 3 4 5 6 7
13. Our project team's level of empowerment is clear and sufficient for where we are now.	1 2 3 4 5 6 7
14. Our project team has shown evidence of effective learning—about each others' areas of responsibility, the "big picture," and how to improve communications and processes.	1 2 3 4 5 6 7
15. Our project team makes most of its decisions using consensus.	1 2 3 4 5 6 7

Building High Performance Project Teams #7.8

Understand Performance Management

According to Armstrong and Baron (1998), the **goal** of **performance management** is to **increase** the **effectiveness of** the organization by **maximizing** the **capabilities of teams** and **individuals**. As such, *performance management* involves evaluating the performance, knowledge, skills, and abilities of teams and individual members, and comparing those with the mission and goals of the team and organization.

Performance management is an indispensable aspect of project management. Where you identify gaps between performance and the goals of the organization, you must provide support to equip the team and team members with the knowledge and skills that will help them improve their performance in order to close these gaps.

In this concept, we discuss performance management and what you can do to ensure you are making the best use of this tool.

Project Managers and Performance Management

Effective performance management takes place in a **positive work environment** that facilitates individual **engagement** and **empowerment** and supports the success of the team and team members. Studies find that employees want effective communication, involvement, and autonomy (Applegate, 2008). They want to feel empowered and engaged in their work.

An effective project manager in a positive work environment inspires individuals to set goals and motivates them to accomplish those goals. Consider the benefits that rewarding exceptional performance might have for a project. Conversely, also consider the impact a lack of follow-through on poor performance might have.

In meeting responsibilities for performance management, project leaders are involved in five key areas, each of which is described over the next few pages.

- Clarifying performance expectations
- Helping teams and individuals succeed
- Evaluating performance
- Rewarding performance
- Following through on poor performance

Clarify Performance Expectations

Ensure that everyone on the team knows what is expected of them in terms of the following:

- Tasks
- Job responsibilities
- Specific assignments
- Quality and quantity levels

- Time and cost parameters
- Resource use
- Quality of work such as the level of teamwork, open sharing of information, and participation in problem solving and decision making

The following **tools** will help you clarify performance expectations:

- Objectives
- Standards
- Position or job descriptions
- Models and samples
- Standard operating procedures or "best practices"
- Information updates
- Rules and policies
- Ongoing feedback about performance

Help Teams and Individuals Succeed

Effective performance management is based on open **communication** at all levels. Toward that end, you must communicate details about the system you use to manage performance, so people do not think this is "secret" information or a threat to them. Instead, work to have people see your performance evaluation system as constructive feedback and an opportunity for skills development.

Before conducting a performance appraisal or evaluation for a team or individuals, ensure that the following exist:

- A clear delineation of specific job duties, policies, and procedures
- A well-defined outline of how performance is measured
- A list and explanation of goals you expected people to achieve

Add Performance Management Tasks to Your Schedule

Informing individuals about their performance and evaluating their results against expectations requires that you work on performance management as a routine part of your job. In other words, your performance management responsibilities and opportunities to not begin and end with an employee's evaluation.

Plan your schedule so that you do the following:

- Observe performance and keep records throughout the life of the project
- Periodically conduct review sessions to compare progress against expectations
- Aligned with the organization's policies and procedures, conduct a formal appraisal (usually annual) "for the record"

Reward Performance

Making individuals aware that good performance counts and is valued by you and your organization will go a long way toward helping teams and individuals achieve their goals.

You can reward performance through all of the following:

- Salary increase
- Promotion or lateral transfer
- Special assignment
- Future development
- Special recognition

Follow-through on Poor Performance

Addressing poor performance with teams and individuals can be challenging but is vital to project success. Your goal is to work with people to point out performance issues, help them set improvement goals, and work with them as they make the necessary changes. You must do this in an atmosphere of fairness where you judge people against published expectations and where you help them identify the things they must change.

If improvement does not happen, you must continue the cycle of performance reviews, and in some cases, you may need to take some of the actions listed below. Depending on the problem and your organization's policies, poor performance may include any of the following:

- Lack of salary increases
- Demotions or no promotion
- Withdrawal of valued assignment
- Lack of special recognition
- Disciplinary action
- Dismissal

What to do:

- ☐ Identify the areas where team performance is not optimal.
- ☐ Explore opportunities for individuals to develop their skills within your team to the benefit of both.
- ☐ Inspire team members to set goals and ask for support when needed, and then provide it.

Other actions:

- ☐ ______________________
- ☐ ______________________
- ☐ ______________________

Remember

- ✓ The goal of performance management is to increase the effectiveness of teams and individuals.
- ✓ Effective performance management is based on open communication.
- ✓ Performance management is an ongoing process through which you recognize individual team member's strengths and accomplishments, as well as give constructive feedback for areas of improvement.
- ✓ Maintaining focus, adjusting conditions that foster good performance, providing feedback, and removing barriers to performance are ways of helping individuals succeed.

Enhance Your Learning

Review key terms and basic elements of a Human Performance Management system and what makes it work by viewing this 2-minute video.

Weatherburn, K. (2013). *Introduction to developing & managing performance management systems.*	Available at: https://www.youtube.com/watch?v=LC7bWCEtyHc

Reinforce Your Learning

Help Teams and Team Members Succeed. For a project position you are familiar with, rate how your organization helps people succeed. 5 = Excellent and 1 = Poor

_____ 1. Diagnosing non-performance

_____ 2. Providing ongoing feedback about performance

_____ 3. Coaching and counseling

_____ 4. Correcting performance problems

_____ 5. Administering positive discipline

_____ 6. Planning for improved performance

_____ 7. Reinforcing good performance

_____ 8. Providing opportunities for the development of needed skills

_____ 9. Training in new procedures, processes, and equipment

_____ 10. Communicating an effective career progression

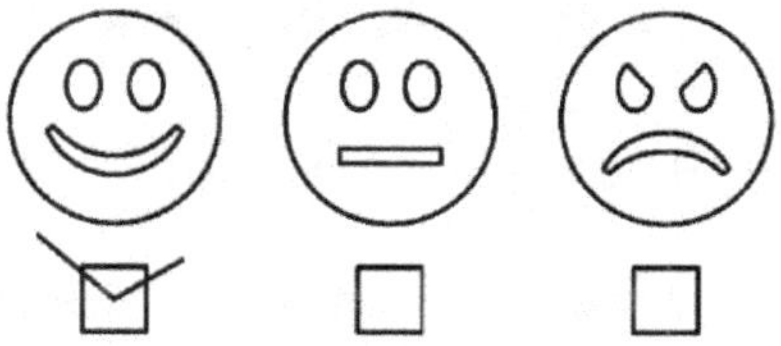

Understand Motivation

One survey found that 90 percent of employees are dissatisfied with their jobs, and dissatisfied employees are 11 times more likely to move to a new organization within 12 months (Cooper, 2012). Clearly, most organizations have room for improvement in motivating their employees.

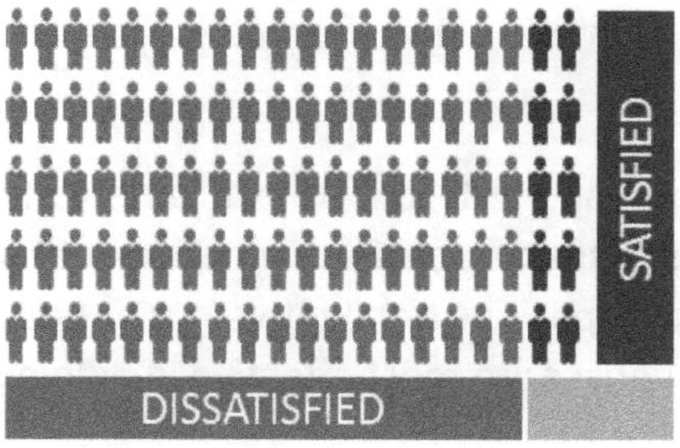

In this concept, we discuss motivation and what project managers do to help motivate teams and individuals.

What is Motivation?

To enhance team performance, project managers need to understand human motivation and how it affects work behavior and job satisfaction. Of particular importance is understanding the role a leader plays in creating a climate for positive performance.

Motivation is a significant driving force for performance. By definition, it pertains to motives. A **motive** is defined as "**something *within* the individual**, rather than without, that **incites** them to **action**." However, that something within the individual is highly influenced by the procedures, practices, and behaviors of others. Therefore, leaders must examine their processes, procedures, and behaviors to see whether they encourage, rather than discourage positive motivation.

> Motivate...to provide with a motive...something that causes a person to act.
>
> To motivate is to induce people to act in a desired manner.
>
> *Adapted from Webster Dictionary*

What Motivates People?

Whether an individual performs at a high level is determined by several factors, only some of which are related to the internal state of motivation. Several environmental factors affect behavioral choices, and thus the ultimate performance of any individual.

One model, developed by **Thomas Gilbert** in his landmark book, *Engineering Worthy Performance*, suggests six factors to examine when trying to understand and enhance performance.

As the chart below illustrates, some of these factors are environmental, and some pertain to the individual.

Environment	Information	Resources	Rewards and Incentives
Individual	Capacity and ability	Skills and knowledge	Motives and attitudes

Gilbert recommends starting with the environmental factors because they are more likely to produce results.

Leaders and other "behavior influencers" use external methods of influence—primarily through communication—and "reality structuring" to help people make the most appropriate behavioral choices. However, regardless of what an influencer does, the choice is still made by the individual.

Leaders are also human, so they respond within the context of the work environment as well. They choose to behave in ways that satisfy their internal needs and subjective interpretations of reality.

Managing Consequences

Although human motivation is complex and is described by a variety of theories, understanding the role of **behavioral consequences** in shaping performance is essential for project managers. Keep in mind that consequences can be positive or negative but notice the last rules in the list below. It states that positive consequences are most effective in helping to motivate people.

The following are **four basic rules** for managing consequences:

1. Reinforcement encourages repetition.
2. The faster the consequence, the stronger the effect.
3. Consequence importance varies from person to person (importance is subjective).
4. Positive consequences are more effective than negative ones.

Extrinsic and Intrinsic Rewards

Organizations have traditionally relied on the "reward-punishment" approach to try and manage performance. However, research has found that **extrinsic rewards**, apart from a proper paycheck, are fruitful mainly when tasks are mechanical or straightforward, and only over the short term.

Intrinsic rewards, those that arise from the individual's human needs, lead to more and better results when it comes to motivating people to perform.

Many theories of motivation are available, but the most useful in the workplace are humanistic. These theories suggest that leaders will have the most success by identifying the motivational needs of their team members and responding accordingly.

Well-known theories about human needs.

- Abraham Maslow – Hierarchy of Needs
- Frederick Herzberg – Two Factor Theory
- Douglas McGregor – Theory X – Theory Y
- Victor Vroom – Expectancy Theory
- John Adams – Equity Theory

Probably the most famous theory is **Abraham Maslow's Hierarchy of Needs**, which is illustrated in the chart below. It holds that people have levels of needs ranging from lower-level physiological needs to higher-level needs such as the need for accomplishment and growth. Subsequent levels of needs cannot be satisfied until satisfaction of the previous level is achieved.

Maslow's Hierarchy of Needs

Fredrick Herzberg's Two Factor Theory states that certain environmental and individual factors in the workplace cause job satisfaction, while a separate set of factors cause dissatisfaction. People are influenced by two sets of factors: motivation and hygiene, as illustrated in the chart below. According to Herzberg, individuals are not content with the satisfaction of lower-order needs at work; for example, those associated with minimum salary levels or safe and pleasant working conditions. Instead, individuals want to satisfy higher-level psychological needs having to do with achievement, recognition, responsibility, advancement, and the work itself.

Herzberg's Two Factor Theory

Motivator Factors	Hygiene Factors
• Achievement	• Pay and Benefits
• Recognition	• Company Policy and Administration
• Work Itself	• Relationships with Co-workers
• Responsibility	• Supervision
• Promotion	• Status
• Growth	• Job Security
	• Working Conditions
	• Personal Life

Douglas McGregor's model, Theory X and Theory Y, describes two contrasting models of workforce motivation. Theory X assumes individuals are inherently lazy and will avoid work. As a result, individuals need to be carefully managed, and comprehensive systems of controls developed. Theory Y assumes individuals are ambitious and self-motivated. As a result, given the right conditions, most individuals will want to do well at work.

McGregor's Theory X & Y

Theory X	Theory Y
Work avoiding	Work is natural
Need to control	Capable of self-direction
Avoid responsibility	Seek responsibility
Workers seek security	Can make good decisions

Victor Vroom's Expectancy Theory assumes that behavior results from conscious choices among alternatives whose purpose it is to maximize pleasure and minimize pain. Vroom states that an individual's performance is based on individual factors such as personality, skills, knowledge, experience, and abilities.

Vroom's Expectancy Theory

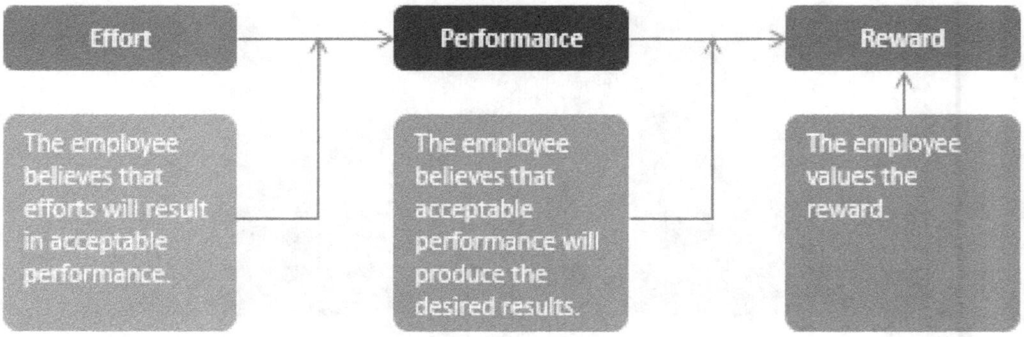

John Adams' Equity Theory explains relational satisfaction in terms of perceptions of fair and unfair distributions of resources, and how fairly a person is treated as compared to others. The theory proposes that individuals who perceive themselves as either under-rewarded or over-rewarded will experience distress, and this distress leads to efforts to restore equity.

Adams' Equity Theory

What to do:	Other actions:
☐ The most useful motivational theories in the workplace are humanistic.	☐ ______________
☐ Identify the needs of the individuals on your team and work to help them meet those needs.	☐ ______________
☐ Remember, the faster the consequence, the stronger the effect.	☐ ______________

Remember

- ✓ A motive is defined as "something *within* individuals that incites them to action."
- ✓ Motivation is the act or process of giving someone a reason for doing something.
- ✓ Getting to know your team members will help you in determining their individual motivational needs. Responding to those needs will increase individual engagement and performance, which will benefit the project, organization, and individual.
- ✓ Maslow's Hierarchy of Needs is probably the most well-known theory of motivation.

Enhance Your Learning

Watch this 3-minute video that explains theories developed by Abraham Maslow: Maslow's Hierarchy of Needs.

Sprouts. (2017). *Maslow's Hierarchy of Needs.*	Available at: https://www.youtube.com/watch?v=O-4ithG_07Q

Reinforce Your Learning

Well-Being and Satisfaction. What do the **rules of consequences** suggest that a project manager should do in response to the situations listed below?

Excellent performance.

Off-target performance.

Why are positive consequences more likely to be effective than negative ones?

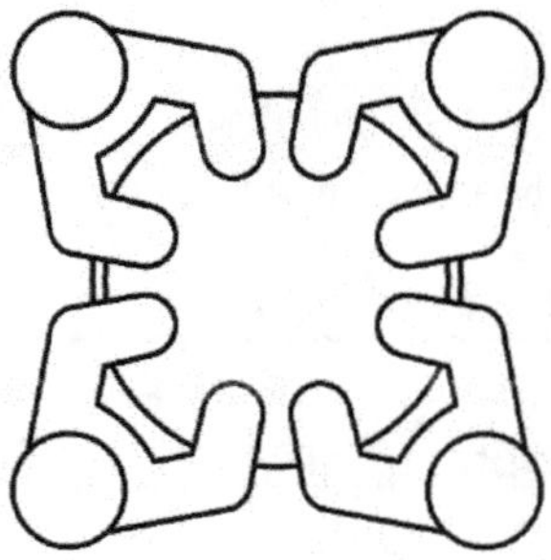

Building High Performance Project Teams | #7.10

Engage Team Members

> ***"Some quit and leave, others quit and stay. Their names are still on the payroll, but their hearts did not come to work. Commitment isn't going to show its face until you reconnect people to the organization."***
>
> —Price Pritchett

Organizations that actively promote employee **engagement** outperform organizations that do not. This fact was reported by the Gallup Business Journal, along with the following benefits:

37 % decrease in absenteeism

25-65 % decrease in turnover

48 % decrease in safety incidents

41 % decrease in quality defects

10 % increase in customer ratings

21 % increase in productivity

22 % increase in profitability

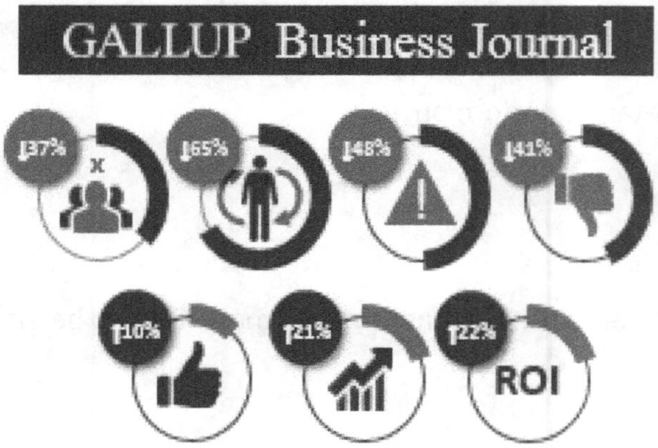

High-performance teams expect excellent performance from all team members. To meet this expectation, project managers and team members must be motivated and engaged.

In this concept, we discuss ways you can engage the people on your project and thus increase performance and productivity.

What is Engagement?

As a leader, it is your responsibility to develop a **positive work climate**, one that supports the success of your team. According to Cooper (*Forbes*, 2013), satisfied employees have a direct impact on the bottom line of an organization. Profits of the organization increase when employees like the organization.

Engagement in the workplace can be defined as follows:

- Level of enthusiasm toward and involvement in a task
- [The state in which] workers care about their work and its results
- Emotional commitment an employee has to the organization and its goals (Forbes.com, June 2012)

Promote Engagement and Empowerment

Leaders promote engagement in the following ways (Spiers, 2012):

- Spending time listening to employees and exploring individual needs, as well as career aspirations.
- Providing opportunities for employees to learn through education and training.
- Allocating roles and responsibilities appropriately.
- Treating employees fairly and equally.
- Motivating employees to meet or even exceed goals.
- Nurturing employee self-confidence and high morale.
- Offering recognition and rewards commensurate with the task performed.

Empowerment relates to giving employees autonomy and authority to use their knowledge, skills, and experience to help drive project work. A leader engages and empowers a team by giving them the power, skills, resources, and opportunities that encourage them to take initiatives, make decisions, solve problems, and develop innovative ideas.

Engaged and empowered team members exhibit increased job satisfaction and demonstrate greater competence in their work.

What to do:	Other actions:
☐ Ask your team members about their career aspirations.	☐ ______________________
☐ Ask individuals what they think their unique talents are.	☐ ______________________
☐ Give team members the autonomy and authority to use their talents to solve problems and contribute to the team's effectiveness.	☐ ______________________

Remember

✓ A culture of engagement and empowerment increases performance.

✓ A project manager's goal is to improve the performance of each individual in a way that aligns with project and organizational goals, and that also meets the needs of the individual.

✓ Engagement in the workplace can be defined as the level of enthusiasm and involvement in a task, or the state in which individuals care about their work and its results, or the emotional commitment an individual has to the project and its goals.

Enhance Your Learning

Watch the following 10-minute video that discusses the surprising truth about what motivates people by RSA Animate-Drive.

Pink, D. (2010). *The surprising truth about what motivates us.*	Available at: https://www.youtube.com/watch?v=bGRY8MHBHcY

Watch the following 18-minute TedTalk by Simon Sinek, who has a simple but powerful model for inspirational leadership all starting with a golden circle and the question "Why?"

Sinek, S. (2010). *How great leaders inspire action. TEDx.*	Available at: https://www.youtube.com/watch?v=bGRY8MHBHcY

Reinforce Your Learning

Positive Work Climate. Brainstorm ways to increase team member communication and engagement.

Communication: How can I get information that increases my understanding of the motivational needs of my team members and how engaged they are?

1.

2.

3.

4.

5.

6.

7.

9.

10.

Engagement: How can I encourage team members to "want" to engage and participate in tasks and activities and commit to assuming authority and responsibility?

1.

2.

3.

4.

5.

6.

7.

9.

10.

Summary

A Final Word: *Building High-Performance Project Teams*

Your team will perform as well as the value and support you and your organization provide to the team.

A few important topics you have covered in this chapter and workbook include:

- Listing and describing the skills and processes associated with project team success.
- Describing the support needed for project teams to function at high-performance levels.
- Defining the basic skills to lead teams and others on a project.
- Identifying ways to mitigate team dysfunction and promote shared responsibility.
- Describing methods to motivate and engage teams.

In the next Project Mastery chapter, we will focus on Leading and Managing Projects.

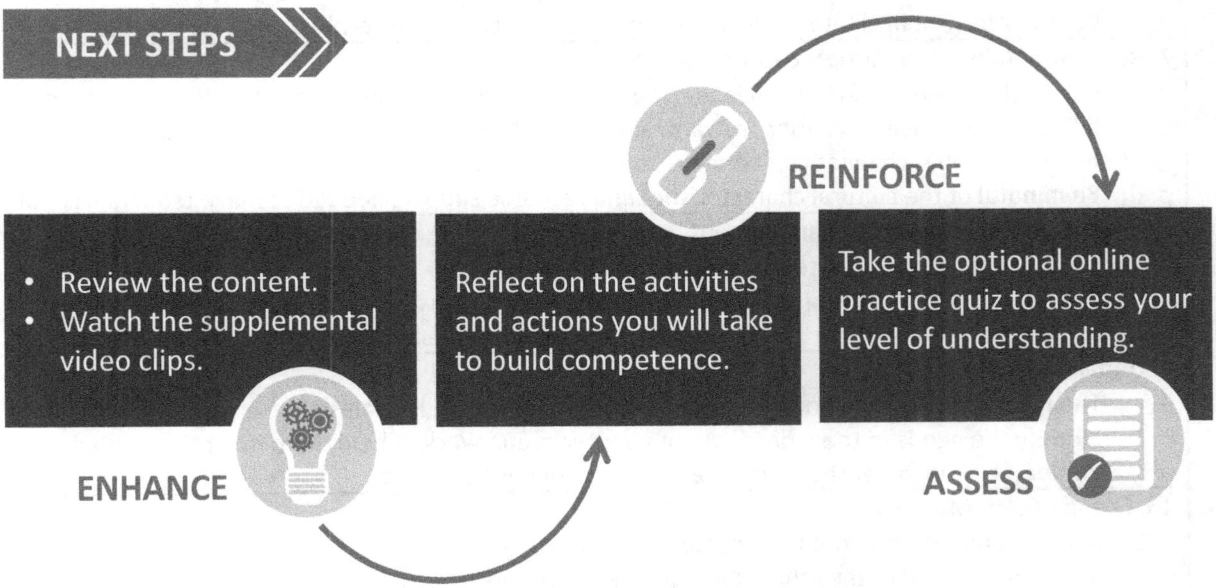

Building High-Performance Project Teams—Recap Checklist
Team success requires not only an established purpose but also members who are committed to the purpose. Success depends on the skills of team members and their ability to apply those skills within the team as they transition through the classic developmental stages of form, storm, norm, perform, and reform.
1. **Recognize the Characteristics and Benefits of Successful Teams** □ Recognize there are significant payoffs when teams are appropriately used, structured, and supported. □ Pay attention to the number, type, and success of teams in your workplace. □ Consider whether work teams you work on meet the characteristics of effective teams.
2. **Appreciate the Key Elements of Team Structure** □ Ensure your team has a charter specifying the team's purpose, structure, and responsibilities. □ Complete the team structure checklist to identify what needs strengthening or clarifying on your team. □ Ensure team member roles and responsibilities are defined and understood by all.
3. **Review the Stages and Challenges of Team Development** □ Be mindful that teams must develop to become fully functioning and that development takes time. □ Manage the various conditions that slow down or derail team development. □ Periodically assess your team's stage of development.
4. **Encourage Positive Communication** □ Commit to asking yourself the authentic communication questions regularly. □ If you sense that communication within a team is not open or balanced, respectfully raise these perceptions during an upcoming meeting and invite discussion. □ Monitor task and maintenance behaviors: increase constructive behaviors and avoid destructive ones.
5. **Resolve Conflict and Promote Shared Responsibility** □ Recognize that conflict, in various forms, is a natural and essential part of teamwork. □ Follow a structured process for helping the team resolve conflict. □ Teams with effective processes resolve conflicts and develop shared responsibility. Know how effective your team processes are.
6. **Enhance Team Problem-Solving Capabilities** □ Be mindful of thinking that restricts the free flow of ideas in team problem solving. □ Take five minutes and brainstorm ideas for solving a team problem you are facing. □ Consider using a facilitator to help the team follow a systematic problem-solving process.
7. **Recognize Values that Support Team Success** □ Research the views of your top leadership. Are they pro or con when it comes to project teams? □ If the performance evaluation, reward, and recognition system do not encourage project teamwork, what changes can be made to recognize teams? □ Be mindful of the cultural characteristics and resource support needed to make teams successful.
8. **Understand Performance Management** □ Identify the areas where team performance is not optimal. □ Explore opportunities for individuals to develop their skills within your team to benefit both. □ Inspire team members to set goals and ask for support when needed, and then provide it.
9. **Understand Motivation** □ The most useful motivational theories in the workplace are humanistic. □ Identify the needs of the individuals on your team and work to help them meet those needs. □ Remember, the faster the consequence, the stronger the effect.
10. **Engage Team Members** □ Ask team members about their career aspirations. □ Ask individuals what they think their unique talents are. □ Give team members the autonomy and authority to use their talents to solve problems and contribute to the team's effectiveness.

Knowledge Review Test—Questions

Building High-Performance Project Teams

1. Which of the following would not be a way for your organization to help teams flourish?
    A. Ensuring that there are rewards for team contributions.
    B. Investing in team development.
    C. Aligning human resource policies and practices to ensure compatibility with a team-based environment.
    D. Focusing on individual contributions to a project.
    E. None of the answers are correct.

2. Which of the following statements regarding a team charter is false?
    A. The team sponsor is typically responsible for developing the charter.
    B. The charter can be renegotiated to adjust to changing conditions.
    C. Team charters should include clearly stated, specific and measurable performance expectations.
    D. The charter should be a written document and typically given to key team members.
    E. The charter can act as a tool for team self-evaluation.

3. Which of the following could be considered a reason that teams succeed?
    A. Once a team is formed, success is only a matter of time.
    B. Team success is basically a function of luck.
    C. Team success is rooted in organizational support and development of teams.
    D. All the answers are correct.

4. The duties of a team sponsor would best be described as:
    A. Directing the team.
    B. Establishing the team, allocating resources, and approving results.
    C. Facilitating the actions of a team.
    D. Micromanaging the various duties of a team.
    E. Instructing team members to rely on direct supervision.

5. In the early stages of development, a team would most likely need help doing which of the following?
    A. Learning how to resolve interpersonal conflicts.
    B. Influencing each other in positive ways.
    C. Establishing procedures.
    D. Refining team processes.
    E. Learning how to function as a team.

6. Active listening would be an example of which type of skill?
    A. Technical Skill.
    B. Interpersonal Skill.
    C. Group Skill.
    D. Analytical Skill.
    E. Directive Skill.

7. Which of the following actions would help you determine the desired state of your team?
    A. Examining the past work experience of potential team members.
    B. Having potential team members do a self-assessment of their skills.
    C. Having a skilled facilitator interview potential team members.
    D. Reviewing the team's purpose, performance expectations, and progress.
    E. None of the answers are correct.

8. Which of the following differences is especially important to overcome when trying to promote positive communication within a team?
    A. Differences in knowledge.
    B. Differences in skills.
    C. Differences in personality.
    D. Differences in experience.
    E. Differences in interests.

9. Which of the following actions would help you collaborate with other team members?
    A. Encouraging all members to contribute to the discussion.
    B. Learning to quickly decide which side of an issue you agree with.
    C. Compromising ideas for the sake of reaching an agreement.
    D. Sticking rigidly to your own position until you win.

10. Which of the following is not a common factor that contributes to team success?
    A. The roles of the team members are well-defined and well-executed.
    B. There is balanced, responsible participation.
    C. The team has well-defined methods for resolving conflicts.
    D. The team is aware it will go through an evolution to be productive.
    E. None of the answers are correct.

11. It is important to have a good understanding of your organization's history with teams. Which of the following needs to be acknowledged?
    A. Look at past successes.
    B. Look at past failures.
    C. Consider why past teams have failed.
    D. Consider what your organization has done in response to good experiences.
    E. All the answers are correct.

12. Which of the following is NOT a step in developing a team charter?
    A. Launch the team by involving the team members in charter development.
    B. Spell out all responsibilities and actions.
    C. Ensure that team members understand that the terms of the charter can be renegotiated should conditions change.
    D. Do not include performance expectations until the assessment part of the process.
    E. Ensure the charter is a written document, given to every team member.

13. Which of the following is NOT an example of how a team can benefit an organization?
    A. Increased productivity.
    B. Decreased cost.
    C. Increased absenteeism.
    D. Increased quality.
    E. Increased customer satisfaction.

14. When considering team composition, whose role is at the operational level, and involves being "facilitative" rather than "directive"?
    A. Self.
    B. Functional manager.
    C. Team leader.
    D. Facilitative leader.
    E. Sponsor.

15. What is a common result when project managers may have unrealistic expectations for team results, demanding "too much, too soon"?
    A. The team feels motivated and delivers.
    B. The team members all quit.
    C. The team is doomed to failure.
    D. The team can lose patience or become disenchanted.
    E. All the answers may be correct.

16. Project managers should manage conflict by:
    A. Informing the sponsor of conflicts.
    B. Having a team meeting and tell your team members how you have handled such conflicts in the past.
    C. Anticipating and seeking to resolve disagreements, complaints, and confrontations in a constructive manner.
    D. Ignoring the problem.
    E. None of the answers are correct.

17. When assessing the skills and development of your team, what two aspects should you compare?
    A. Interpersonal skills and intrapersonal skills.
    B. Analytical state and metamorphosis potential.
    C. Current state and desired state.
    D. Past, present and future abilities.
    E. Communications versus vulnerability.

18. Which of the following is not true of human motivation?
    A. The faster the consequence, the stronger the effect.
    B. Reinforcement encourages repetition.
    C. The importance of consequences varies from person to person.
    D. Money has an important impact on long-term performance.
    E. Positive consequences are more effective than negative ones.

19. Which of the following is NOT a tool to encourage members to contribute to group discussions and problem solving?
    A. Pressure tactics and belittlement.
    B. Nominal group technique.
    C. Force field analysis.
    D. Round robin brainstorming.
    E. All the answers are correct.

20. As a project manager, you should promote teamwork and cooperation by:
    A. Hiring entertainers to keep conversations light and lively.
    B. Designating one of the team members to monitor team member output and cooperation.
    C. Securing an outside facilitator to run all project meetings.
    D. Demonstrating and fostering cooperation, communication, and consensus among individuals and groups.
    E. All the answers are correct.

Building High Performance Project Teams — Answer Sheet

1.	____A	___B	___C	_X_D	___E
2.	_X_A	___B	___C	___D	___E
3.	___A	___B	_X_C	___D	___E
4.	___A	_X_B	___C	___D	___E
5.	___A	___B	___C	___D	_X_E
6.	___A	_X_B	___C	___D	___E
7.	___A	___B	___C	_X_D	___E
8.	___A	___B	_X_C	___D	___E
9.	_X_A	___B	___C	___D	___E
10.	___A	___B	___C	___D	_X_E
11.	___A	___B	___C	___D	_X_E
12.	___A	___B	___C	_X_D	___E
13.	___A	___B	_X_C	___D	___E
14.	___A	___B	_X_C	___D	___E
15.	___A	___B	___C	___D	_X_E
16.	___A	___B	_X_C	___D	___E
17.	___A	___B	_X_C	___D	___E
18.	___A	___B	___C	_X_D	___E
19.	_X_A	___B	___C	___D	___E
20.	___A	___B	___C	_X_D	___E

Chapter 8

Leading and Managing Projects

Overview

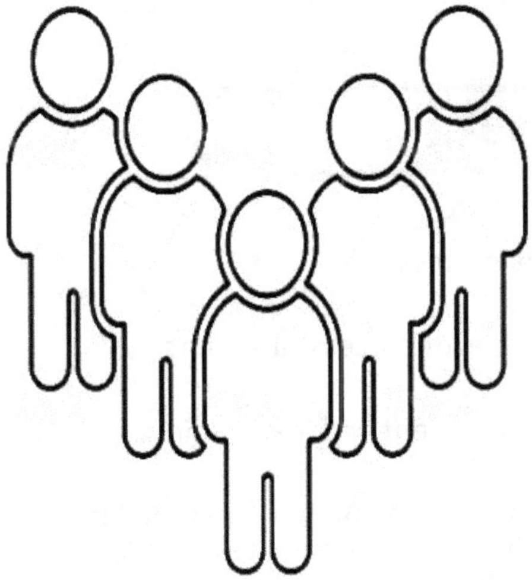

"If your actions inspire others to dream more, learn more, do more and become more, you are a leader."

—Vincent Van Gogh

Leadership and management are challenges even on straightforward projects, and the issues increase exponentially with complex projects. To improve your chances of success, you are wise to hone your proficiencies before you find yourself on the firing line.

This chapter covers essential tools and approaches for leading projects of any size and complexity. The chapter also discusses relevant best professional practices.

After you complete this chapter, you will know how to:

- Understand the relationship between management and leadership.
- Describe the basic approaches of leadership that you can apply to a wide range of projects.
- List and describe the key leadership skills and competencies associated with project success.
- Plan the development of your leadership competencies in areas of importance
- Describe how to coach and develop yourself.

The competencies associated with this chapter include:

Leadership and Coaching	**Interpersonal Relationship Building**	**Managing a Diverse Workforce**

Take Your Temperature for Leading and Managing Projects

The following assessment will help you identify what you know about leading and managing projects and the areas where you can improve. With you and your organization in mind, read each statement below, and consider how much you agree. Use the numbers from 1 to 10, where **1** means you **strongly disagree** and **10** means you **strongly agree**.

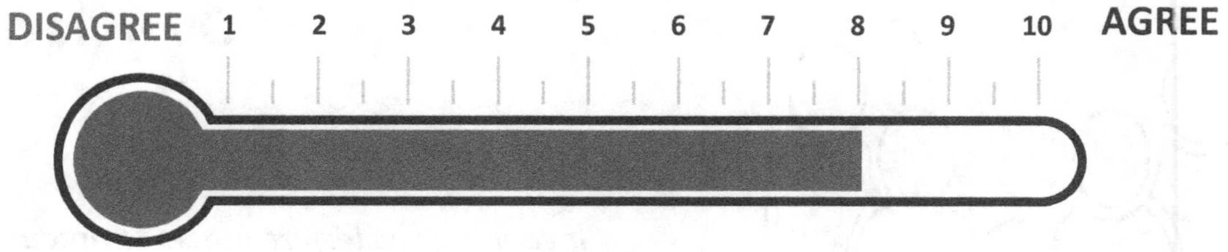

_____ 1. I can differentiate between effective and ineffective leaders.

_____ 2. I can identify the key characteristic and traits of successful leaders.

_____ 3. I can explain the difference between management and leadership.

_____ 4. I can distinguish between transactional and transformational leadership.

_____ 5. I am aware of the power and influence approach to leadership.

_____ 6. I can explain what is referred to as the behavioral approach to leadership.

_____ 7. I can list personality traits and four interpersonal styles.

_____ 8. I can describe the leadership cycle and its stages.

_____ 9. I know how to build my leadership competence.

_____ 10. I am good at coaching others and developing myself.

_____ **Total**

Take a moment to reflect on the following before you move forward in the chapter.
In your work, what problems have you experienced or observed related to leading and managing projects?

____

____

Take a few minutes to also reflect on your self-assessment. List two or three areas you want to improve.

____

____

As you progress through this chapter consider what actions you can take to demonstrate competence in the following three competency areas:

1. Leadership and Coaching. Models and encourages high standards of ethical behavior; adapts leadership styles to situations and people; empowers, motivates, guides, and coaches.

2. Interpersonal Relationship Building. Considers and responds appropriately to the needs, feelings, and capabilities of others; seeks feedback and accurately assesses impact on others; provides helpful feedback; builds trust with others.

3. Managing a Diverse Workforce. Recognizes the value of cultural, ethnic, gender, and other individual differences; provides employment and development opportunities for a diverse workforce.

Observe the Qualities Effective Leaders Share

You probably can think of leaders you felt were effective, and some who were not. You may recall a leader who inspired you to serve your team and clients better. You most likely can also remember managers who made you want to quit.

While no human is perfect, and no one is an ideal example of a "good leader" every single moment of every day, the best leaders share many similarities. They may not share the same backgrounds or have the same life experience, but they share many of the same qualities.

In this concept, we discuss the traits that effective leaders share, and that can help you become a better leader.

Traits Successful Leaders Share

As *Inc*. magazine says, we know good leadership when we see it. But "good" leadership is not just something you can recognize in others. Becoming aware of the various traits good leaders share can also help you develop your leadership skills.

To be an effective leader, you must have **confidence**. It is imperative. You must look as though you know what you are doing and that you are confident of success. To be clear, you need to be confident and exude confidence, the latter being perhaps the more essential.

You must be **decisive**, which is a trait that requires confidence. You must know how to assess a situation, consider the evidence, and make a decision.

You must be **ethical**. You must behave in a way that is ethical, both morally and according to the culture of your organization. You must inspire and demand ethical behavior in those you lead. You must also demonstrate integrity, which is an effort to be fair and honest in your dealings with others (*Economy*, 2013).

As we have discussed, you must have excellent **communication skills**.

You must be optimistic and help build the confidence of those around you (*Economy*, 2013).

You must have **charisma;** in other words, be outgoing, friendly, and persuasive (at least when necessary).

You must be **caring**. You must care about the project, about the people who work for you, and about clients and other stakeholders. You demonstrate caring by how you treat people, communicate, and behave. You manage with compassion and see people as more than the bottom line. This must be genuine. Attempts to "pretend to care" are easy to see through and detract from integrity and effectiveness.

You value **teamwork**.

You know how to **listen**.

You are **collaborative** in that you know when, and how, to solicit input from others.

Celebrating success also matters. This means you must know how to praise and reward your team's achievements and encourage them to continue performing because they feel good about working for you.

What to do:	Other actions:
☐ Select a leader you admire and list the specific things you admire.	☐ ____________
☐ Identify your most effective leadership traits.	☐ ____________
☐ Ask a person you trust to identify your leadership weaknesses.	☐ ____________

Remember

✓ If there were just one skill of an effective leader, it would probably be good communication. Notice in the discussion above that many traits are tied to effective communication, such as integrity, collaboration, confidence, teamwork, and decisiveness.

Enhance Your Learning

Watch this 12-minute Ted Talk by Simon Sinek to learn what makes great leaders.

Sinek, S. (2014). *Why good leaders make you feel safe.*	Available at: http://www.ted.com/talks/simon_sinek_why_good_leaders_make_you_feel_safe

Reinforce Your Learning

Most Admired Leader. Consider a leader you admire. This could be someone you know, like a previous boss, or a famous leader like a military general. Write down at least three but as many as five things you admire about the leader. Then, reflect on how you can exemplify those things in your leadership.

Your Best Leadership. Think of a time when you believe you did your best as a leader. Describe three or four things you did that enabled you to be your best.

What are the primary lessons you learned from that experience?

Leading and Managing Projects #8.2

Identify the Key Competencies of Successful Leaders

Communicating effectively, valuing collaboration and teamwork, and being ethical are a few of the key competencies all leaders share. But there are many more.

Competencies may be defined as the combination of knowledge, skills, attitudes, and behaviors that are observable and measurable. For project leaders they are necessary to perform a task or to understand what is needed. For example, in an engineering organization, a core competency for a leader might be the ability to understand a specific design process or method. While technical skills such as this will vary according to the organization, other competencies do not change. For example, as a leader, you must be proactive in identifying problems and solutions, though not all problems will be the same in all organizations.

In this concept, we discuss the leadership competencies that project managers must develop to deliver successful projects.

Authority

One **component** of successful leadership is **authority**; the **power inherent** in a leadership position. The level of authority, and the expectations for how you use the authority, will vary by the type of leadership position. For example, as a project manager, you have less authority than a vice president, but you do have jurisdiction over the work of your project and project team.

The authority for your project and team is given to you based on your job description. Even so, if you fail to execute that authority, you will most likely experience problems. For example, if two groups on your team have a conflict about how work should be done, you will need to help them resolve their differences and create a work plan. If you find it challenging to deal with conflict and fail to do so, the groups may "assume" the authority you are dodging and continue their battle while the project work suffers.

As a leader, you also have **perceived authority.** It is the aspect of your **power** that is **recognized** and **respected** by the people you lead. If you appreciate your position, your project, and your team, and proactively lead them to be successful, your team will **perceive** your authority.

Team Building

One **competency** you must have to be an effective leader is the knowledge and ability to leverage team strengths and **build teams** in a way that helps everyone achieve their work goals. You must also support people in developing their skills.

As a **competent** leader, you need the necessary **technical knowledge** to do, or at least understand, the work that takes place under your management.

An effective leader is trained in leadership principles. They know the key **components, competencies,** and **characteristics** of leadership. Using this knowledge, they continually strive to improve.

As an effective leader, you:

- **Observe** and watch for signs of distress in those you lead, and anticipate problems
- Work to develop your **communication** skills and those of your team
- Know how to **set** achievable and measurable **goals** for you and your team
- Plan and **prioritize** what you do with your time
- **Manage conflict** between yourself and others and among those you lead
- **Problem solve** by analyzing situations, considering all relevant information, and then taking action

What to do:	**Other actions:**
☐ Be mindful of what authority you have and do not have when making decisions.	☐ ____________________
☐ Know your technical strengths and who can help you in your areas of weakness.	☐ ____________________
☐ Identify the leadership competencies you need to be successful in your job.	☐ ____________________

Remember

✓ Effective leaders are made, not born. Sure, some people are born with more leadership abilities than others. However, with the right training and effort, almost anyone can become an effective leader.

Enhance Your Learning

Read the following article to learn more about the leadership skills of successful project managers from AboutLeaders.

Marsh, F. (2022). *10 Great Leadership Skills of Project Management.*	Available at: https://aboutleaders.com/10-great-leadership-skills-of-project-management/

Reinforce Your Learning

Leadership Potential. Review the statements below. Indicate whether you **agree** or **disagree** by placing an **A** or **D** in the space provided.

_____ 1. Leaders are born, not made.

_____ 2. I tend to treat peers well when they listen and do what I want.

_____ 3. Good leaders depend on their followers as much as on themselves.

_____ 4. I always include the reason when I ask employees to do a task.

_____ 5. Good leaders will achieve their objectives at any cost.

_____ 6. A key to good leadership is being consistent in how one leads.

_____ 7. If justified, I would recommend a colleague for promotion above me.

_____ 8. If my team failed to achieve an objective because of one team member's failure, I would explain it to the sponsor.

Leading and Managing Projects #8.3

Differentiate Between Management and Leadership

If you search online for "management versus leadership," you most likely will find quotes about how managers have subordinates and leaders have followers, and so forth. But what does this mean? What are the significant differences between managers and leaders?

In this concept, we discuss the differences between management and leadership, but also that these roles are complementary. Project managers must shift back and forth between the management and leadership roles based on project and organizational needs.

Technical, Human, and Conceptual Leadership

Leadership involves three broad skills areas: technical, human, and conceptual. These skills apply in varying degrees of importance to everyone who leads others.

As you can see in the diagram below, technical skills are essential at the lower levels of leadership, while conceptual skills are essential at higher levels.

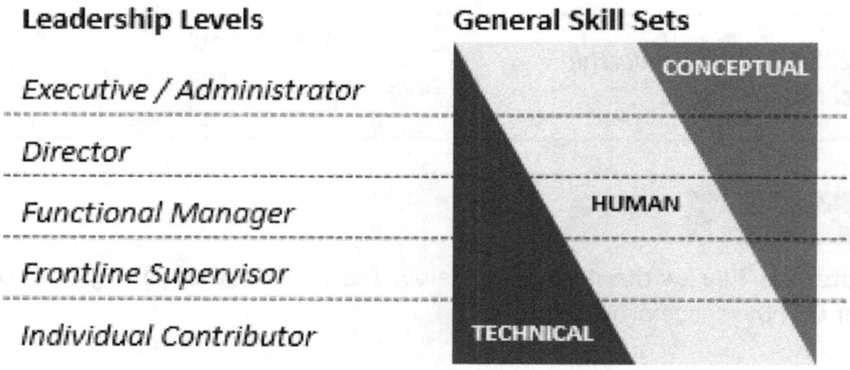

Management and Leadership

A **manager's** job tends to be more **transactional**, whereas a **leader's** job tends to be more **transformational**. In addition to managing, leaders must also **inspire** and **motivate**. (Warren Bennis).

Management and leadership go hand in hand. They are linked and complementary. Project managers shift between manager and leader countless times over the course of a project, even many times during a day. They may find themselves one moment serving as a leader, other times supervising directly, and sometimes managing people or processes.

No matter the title, in today's world, most leaders, including those who manage projects and operational processes, must also have leadership skills.

The chart below summarizes the difference in emphasis between these roles:

Manager	Leader
Do things right	Do the right things
Strives for efficiency	Strives for effectiveness
Administer	Innovate
Focus on systems and structure	Focus on people
Rely on control	Rely on trust
Organize and staff	Align people with a direction
Emphasize tactics	Emphasize philosophy, core values, and shared values
Have a short-term view	Have a long-term view
Ask "how" and "when"	Ask "what" and "why"
Accept the status quo	Challenge the status quo
Focus on the present	Focus on the future
Bottom-line orientation	Focus on the horizon
Develop detailed steps and timetables	Develop visions and strategies
Seek predictability and order	Seek change
Avoid risks	Take risks
Motivate people to comply with standards	Inspire people to change
Use position-to-position influence	Use person-to-person influence
Require others to comply	Inspire others to follow
Operate within organizational rules, policies, and procedures	Initiate appropriate changes to organizational rules, policies, and procedures
Are given a position	Take initiative and lead

The Project Manager's Role as Manager

As a project manager, you fulfill a management role. The **four classical functions** of **management** are **planning**, **organizing**, **directing**, and **controlling**.

In the past, organizations and managers were guided by the following philosophy assumptions:

- Workers, if not controlled, will act irresponsibly.
- Workers are motivated primarily by external factors.
- Management can exert unilateral control over the quality and quantity of work.
- Only management makes critical decisions.
- Continuity of production/operations must be maintained.

Contemporary assumptions about management in many successful organizations are now based on the following ideas:

- Workers can exert self-control and behave responsibly.
- Workers are motivated by both external and internal factors.
- Managers maintain awareness and knowledge by helping others develop.
- Workers are able and willing to make decisions.
- People must commit to continuous improvement.

The Project Manager's Role as a Leader

As a project manager, recognize you **play many roles**, and the roles change as situations unfold. They include leader, teacher, coach, counselor, listener, decision-maker, organizer, coordinator, climate builder, information provider, and process change agent.

You wear **many hats**, serving as a worker one moment, a role model the next, a manager the next, and often **you will be called to lead**. To prepare for meeting these opportunities, familiarize yourself with your job description and talk to your senior leaders about what they **expect** of you.

A project manager must be knowledgeable about the areas and positions they lead. This does not mean you know everything, but that you understand what the people you lead do and the skills needed for their jobs. Being aware of the skills will help you be more confident, **connect with and care about those you lead,** and encourage others to place their confidence in you.

Realize that as a leader, you are a **role model** for your team, and you **lead by example**.

At times you will need to inspire people to grow, and so you will serve as a **coach**. You must also be a cheerleader who encourages others to meet challenges.

As a leader, you must **uphold** your organization's **ethics**, vision, and mission, and help build a positive and productive culture.

You must also have solid communication skills, be able to communicate with your team effectively, and **generate urgency to achieve results**.

As an effective leader, you must be open to change and **new ideas**.

As a good leader, you remain calm and **maintain self-control**. You work to **have fun** and **energize others**.

What to do:	**Other actions:**
☐ Connect with and care about those you lead.	☐ ______________________
☐ Lead by example and remember that your role as a project manager is to plan, organize, direct, and control work. Your role as a leader is to inspire and motivate.	☐ ______________________
☐ Work to have fun and energize others.	☐ ______________________

Remember

✓ Strive to embody the best characteristics of both managers and leaders.

Enhance Your Learning

Watch the following 45-minute TEDx Talk by Simon Sinek to learn more about receiving feedback. Consider how you can apply this information in your work.

Sinek, S. (2013). *Why Leaders Eat Last.*	Available at: https://www.youtube.com/watch?v=ReRcHdeUG9Y

Reinforce Your Learning

Test Your Knowledge. Draw a line from each of the actions below to either management or leadership.

MANAGEMENT **LEADERSHIP**

enforce deadline

empower

encourage collaborations

sign paychecks

create budgets

model behavior

coach

plan resources

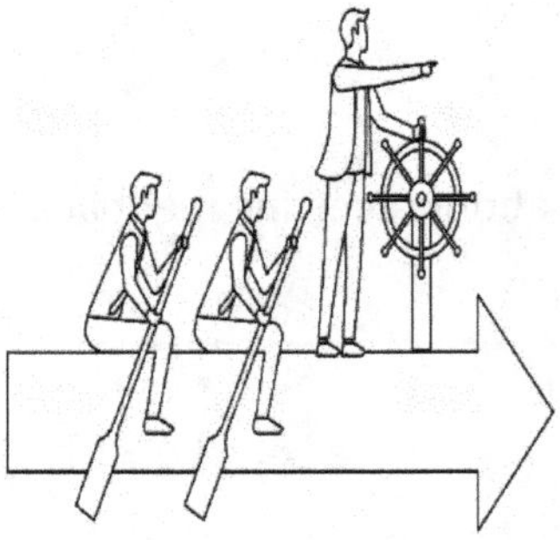

Leading and Managing Projects #8.4

Distinguish Transactional from Transformational Leadership

Change is part of running an organization. Leaders and organizations must continually work to meet the day-to-day operational challenges of the marketplace while also planning for how they will transform the organization to meet future goals.

Project leadership is no different. You must manage your current projects for success, while at the same time planning how you will improve your processes and capabilities for delivering future projects required by the organization's changing goals and strategies.

In this concept, we discuss transactional and transformational leadership and how project managers must use both sets of skills, and also that you will increase your effectiveness by understanding more about how you think and lead.

Mental Models

We all have mental models or paradigms that provide **structure** for our thinking. The models help us sift through all the information we encounter by showing us what to pay attention to, how to arrange what we pay attention to, how to interpret data, and how to make decisions.

Our unconscious assumptions guide us all, and they can be especially useful when they help us structure and act in unfamiliar situations.

However, our mental models can also impede our ability to understand and respond to situations. For example, perceiving people with different mental models as confused, unpredictable, irrational, immoral, or unethical ways can lead to misunderstandings and conflict.

To be effective, project managers need to continually review their mental models to ensure the models are helping them be proactive and productive in meeting current and future project challenges.

Transactional and Transformational Leadership

The following chart provides examples of **transactional** and **transformational** leadership:

Transactional Behaviors	Transformational Behaviors
Supervise others	Plan for the future
Evaluate proposed projects	Remain current with emerging trends
Integrate conflicting perspectives	Communicate a sense for the future
Manage performance	Foster commitment
Focus on results	Emphasize organizational values
Solve problems	Challenge people with new goals
Influence operational decisions	Create a sense of excitement or urgency
Manage the efficiency of operations	Inspire people to take action

Transactional behaviors typically occur at the **foreman** or **supervisory level** of leadership, whereas transformational leadership often happens at higher levels. But as a project manager, you will face situations where you will need to, for example, plan for future changes and challenge people with new goals.

To move **from being a transactional leader to** a more **transformational** leader requires that you are **open** and self-aware; it can involve some risk.

Understand How You Lead

Three ways to learn more about yourself and how you lead are self-assessment, self-reflection, and soliciting authentic feedback.

One way to solicit feedback is to consider ideas related to the **Johari Window,** which focuses on how people communicate.

As shown below, the Johari Window is a model that provides a visual way to think about self-knowledge. It is organized around an exchange between "me" and "you."

Johari Window		
	Known to Self	**Unknown to Self**
Known to Others	Open Area	Blind Area
Unknown to Others	Hidden Area	Unknown Area

The ***open area*** is what we both know about me and openly share.

The ***hidden area*** is what I hide from you about myself.

The ***blind area*** is what you know about me, what you keep from me, what you observe about me, or think or feel about me, of which I am unaware.

The ***unknown area*** is part of me, from my past, that neither of us knows about, at least not consciously.

The more we can increase the parts of ourselves that are known to self and others, the higher our potential for building effective relationships, both at home and in the workplace.

The most effective, evolved leaders expand their open area, creating a picture that looks like this:

Johari Window		
	Known to Self	**Unknown to Self**
Known to Others	Open Area	Blind Area
Unknown to Others	Hidden Area	Unknown Area

When leaders make their reasoning and thinking **apparent** to others, they **build trust** over time. As a result, others are more willing to give them the benefit of the doubt during tough times or when the leader cannot share information.

Becoming more open means becoming **more transparent**, which means showing people more of your thinking, more of the things you are wrestling with, more about your objectives, and your likes and dislikes.

Have the leaders you trusted and respected been open about themselves? Did they let you in on **their preferences**, **biases**, strengths, and weaknesses?

Moving from operating in a more transactional way to a more transformational way has the potential **payoff of greater trust** and **better communication**.

What to do:	**Other actions:**
☐ Move from being a transactional leader to a transformational one by being more open.	☐ ________________
☐ Reflect on your preferences, biases, strengths, and weaknesses.	☐ ________________
☐ Be more transparent. Show more of your thinking and the issues you wrestle with.	☐ ________________

Remember

- ✓ As a project leader, you will often need to change your behavior and move from being a transactional leader to a transformational one.
- ✓ Open communication is vital for leaders.
- ✓ Being more open involves some risk. But the potential payoffs are greater trust and better communication.

Enhance Your Learning

Watch the following 4-minute videos to learn more about transactional versus transformational leadership and the Johari Window.

StaubLeadership. (2012). *Transactional vs. Transformational Leadership.*	Available at: https://www.youtube.com/watch?v=SqOjtO-g6QI
Blake, B. (204). *The Johari Window Model.*	Available at: https://www.youtube.com/watch?v=BWii4Tx3GJk

Reinforce Your Learning

Transactional versus Transformational Leadership. What observations can you make about what you do as a transactional leader and as a transformational leader? Using the lists below, check the leadership behaviors that apply to you.

Transactional Behaviors	Transformational Behaviors
☐ Supervise others	☐ Plan for the future
☐ Evaluate proposed projects	☐ Remain current with emerging trends
☐ Integrate conflicting perspectives	☐ Communicate a sense for the future
☐ Manage performance	☐ Foster commitment
☐ Focus on results	☐ Emphasize organizational values
☐ Solve problems	☐ Challenge people with new goals
☐ Influence operational decisions	☐ Create a sense of excitement or urgency
☐ Manage the efficiency of operations	☐ Inspire people to take action

Leading and Managing Projects #8.5

Be Aware of the Power and Influence Approach to Leadership

Theories about leadership flourish. They explore what leadership is, why it works and does not work, how people become better leaders, and what characteristics are natural, and what characteristics people can develop.

The power and influence approach to leadership is one such theory. It states that leaders have power and influence given their job title, but there are other sources as well.

In this concept, we discuss the kinds of power and influence available to project managers and how you can use them in your projects.

Types of Power

Power is a fascinating thing. They say that absolute power corrupts absolutely, and when a person has a taste of power, they want more. Power makes people feel stronger, better, and important. But power itself is not a bad thing. It is when we use power to take advantage of others or to deny them access because of who they are that power leads to problems.

Everyone has some type of power. As individuals in the workplace, we each influence others, at least to some extent. Where does power or influence come from? Why do you have it, and how do you use it?

One popular theory is an expansion of the classic work of French and Raven (1959), who looked at power as deriving from various sources, as discussed below.

- **Legitimate power** is the power of position or title. Within the legitimate range of your responsibilities, you can ask people to comply with your requests because your organization has given you this right. Being a team leader gives you an element of control over people but does not guarantee they will comply.
- **Reward power** is the power that comes from your ability to provide rewards people value, such as signing timecards, influencing raises, or recognizing their ideas and contributions.
- **Expert power** is the power of expert knowledge. You can influence others to support your ideas and causes because they respect your knowledge, skills, and understanding. They go along because they trust your expertise.
- **Referent power** is the power of personal appeal. You influence others to support you because they like you, want to be like you, want your respect, or do not want to cause you problems. This power comes from your personality, charisma, or reputation for accomplishments.

- **Information power** leads others to comply with your requests because you have access to information they need, and they cannot access it on their own.
- **Connection power** is the power of "who you know." Your power derives from connections with influential people who may have information, resources, or capabilities that are needed.

Power and Transactional Leadership

Transactional leadership, which we discussed in a previous concept, is another type of power. People do what you tell them because you have power and influence over them, and usually, because your name is written in just the right spot on the organizational chart.

Power and Leading by Example

You may be surprised to realize that role modeling or leading by example is also a type of power. When you act as a positive role model, your team is likely to emulate your behavior.

What to do:	Other actions:
☐ Identify what types of power you possess and the sources of your power.	☐ ____________________
☐ Determine your main characteristics related to power and influence.	☐ ____________________
☐ Determine what power you have gained or lost in the past year.	☐ ____________________

Remember

✓ Everyone has power.

✓ Power is not in itself negative. The issue is how a person exerts that power.

✓ Power and influence are necessary aspects of leadership. You cannot lead if you do not have power and influence. However, people who depend on their power and influence alone are not the best leaders.

Enhance Your Learning

Watch the following one-minute videos that illustrates one approach to leading others.

Magia3e. (2010). *Legitimate Power - Big Bang Theory style.*	Available at: https://www.youtube.com/watch?v=v4LIADSF93c
Wonstolen, L. (2014). *Big Bang Expert Power.*	Available at: https://www.youtube.com/watch?v=5thkrGlq-M0

Reinforce Your Learning

Bases of Power. For the types of power listed below, give an example of how you have seen the power used or abused.

Legitimate power:

Reward power:

Coercive power:

Expert power:

Referent power:

Information power:

Connection power:

What power source has the most approval or the most value in your organization? For example, does your organization value people with technical skills the most? One way to analyze how power works is to consider the power sources associated with the most influential people in your organization.

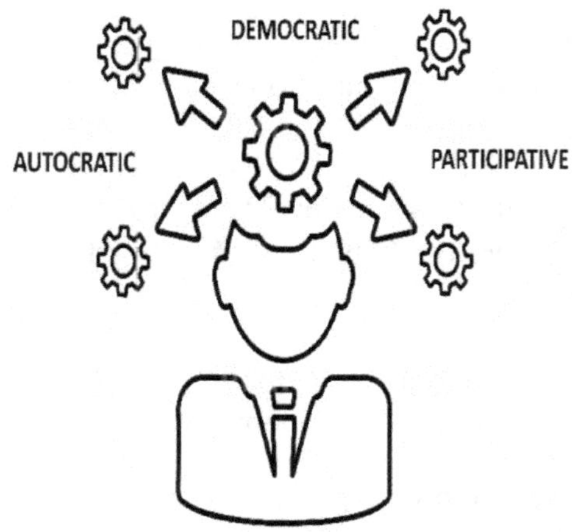

Leading and Managing Projects #8.6

Be Mindful of the Behavioral Approach to Leadership

According to the behavioral theory of leadership, you are what you do. Actions matter.

This approach supports a wide variety of behaviors, and in the past people have discussed whether leaders are born or made. In recent times, most people agree that with desire and effort most of us can learn how to be effective leaders.

In this concept, we describe various behavioral leadership theories and how you can use this information to better manage your project teams.

Behavioral Leadership Theories and Studies

The following are noteworthy theories and studies related to the behavioral approach:

- Kurt Lewin's autocratic, democratic, and Laissez-faire behavioral styles
- McGregor's Theory X (passivity and resistance) and Theory Y (motivation and desire)
- University of Michigan Study: employee-centered and production-centered leadership
- Ohio State University Study
- Blake and Mouton's "Managerial Grid"

Autocratic, Democratic, and Participatory Leadership Styles

In the 1930s, researcher **Kurt Lewin** postulated three behavioral leadership styles, autocratic, democratic, and participatory. Today, we can identify more than just these three, but they are still relevant (Cherry, n.d.).

Autocratic leaders make decisions on their own, quickly, without seeking input from others. In most cases, autocratic leadership is considered to be one of the least effective approaches. The most successful decisions usually require input from multiple sources. However, in a few situations, autocratic leadership can be decisive and effective (Bass & Bass, as cited in Cherry, n.d.). As an example, this style works effectively in the military.

Democratic leaders seek input from others, including team members, to make decisions. They seek agreement and approval; however, an agreement is not a requirement.

Laissez-faire or participatory leaders let employees make their own decisions with little interference from the leader. While this style can work well with highly motivated, dedicated employees, it can lead to problems when employees are less motivated.

Theory X and Theory Y

Douglas McGregor was an early theorist who postulated two types of organizational leadership: Theory X and Theory Y.

Theory X assumes that people are passive and resistant to organizational needs. Therefore, an authoritarian or ***autocratic*** management style is required to ensure the fulfillment of the leader's objectives.

Theory Y assumes that people are motivated and desire responsibility. Therefore, they have a say in decisions that influence them, and a more ***democratic*** management style is evident.

Theory Z is a catchall for various theories of human motivation built on Douglas McGregor's work. It assumes that people are loyal to their jobs and organizations. Therefore, a more ***participative*** management style with shared leadership is characteristic.

Blake and Mouton's Managerial Grid

Two landmark studies reported by Hunt (1984) advanced the behavior approach to leadership. These were the University of Michigan and Ohio State University studies.

The Michigan study looked at employee-centered and production-centered leadership. A popular leadership-training model emerged from this research called the Blake and Mouton's Managerial Grid.

As illustrated below, the model is a grid with **concern for production** as the x-axis and **concern for people** as the y-axis. Each axis ranges from 1 (Low) to 9 (High).

Blake Mouton Managerial Grid

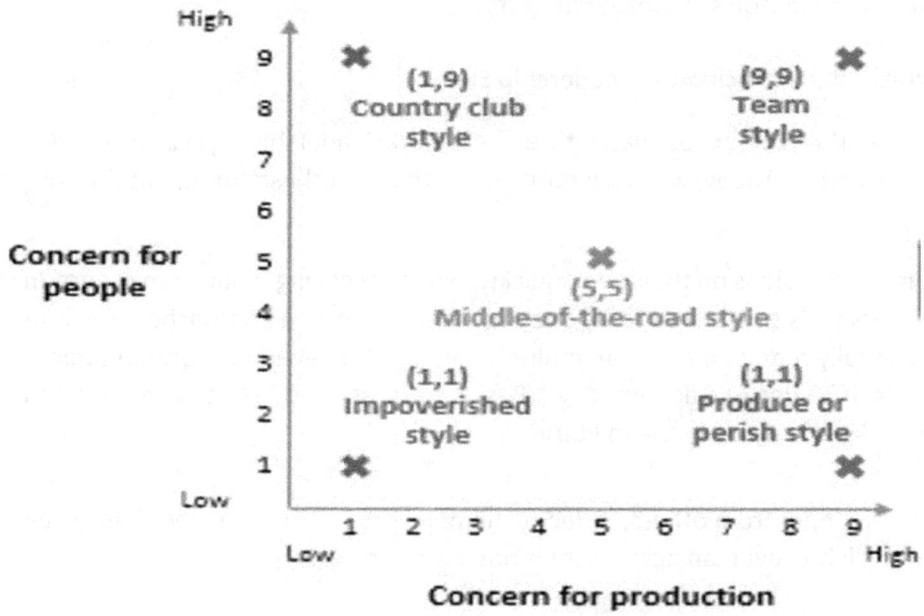

The ***impoverished style*** refers to managers who have little concern for production or people. They do just enough to get by, trying to stay below the radar.

The ***country club-style*** refers to managers who feel that by making employees happy, results will follow. However, while satisfying employees is a worthy goal, it is not usually sufficient to achieve productivity.

The ***produce or perish style*** of management is just the opposite, where the emphasis is on results, with little concern for employee well-being or satisfaction. This style of management cannot be sustained for long.

The ***middle-of-the-road style*** of management tries to balance company needs with employee needs, a concept that makes sense in most situations.

The ***team style*** focuses on people and productivity, emphasizing cooperation to get the job done in an atmosphere that values employees.

Emulating Effective Leadership Behavior

The behavioral approach emphasizes what leaders do on the job. When it comes to improving your leadership ability, one good practice is to observe what leaders you admire do and then model their behavior. For example, you might observe that a manager you admire makes eye contact, has a firm handshake, and begins conversations by saying something positive.

Emulating such behaviors can help you become a better leader. Although keep in mind that people are different, and to be successful, you must be authentic. Learn from your observations and adopt behavior that works for you.

What to do:	Other actions:
☐ Realize the behavioral theory of leadership asserts that people can learn to be effective leaders.	☐ ______________________
☐ Observe leaders you admire, watch their specific behaviors, and consider how you can be like them.	☐ ______________________
☐ Determine your behavioral style of leadership.	☐ ______________________

Remember

- ✓ The behavior approach emphasizes what managers and leaders do on the job. It includes a wide variety of possible behaviors.
- ✓ Most people agree that with desire and effort most of us can learn how to be effective leaders.

Enhance Your Learning

Watch the following 3-minute video that illustrates Blake and Mouton's styles. After watching the video, where would you place yourself on Blake and Mouton's grid?

MindToolsVideos. (2018). *The Blake Mouton Managerial Grid: Balancing People's Well-Being with Productivity.*	Available at: https://www.youtube.com/watch?v=lSvVdEKq9io

Reinforce Your Learning

Exercise in Management Values.* Enter the number that describes your level of agreement with the statements below.

3 – Frequently agree

2 – Occasionally agree

1 – Rarely agree

1. _____ Subordinates should be loyal and support the boss even when they disagree with work methods or decisions.

2. _____ Employees should be aware of organizational problems and, when appropriate, make recommendations about them.

3. _____ Everyone on a team must agree to make it work.

4. _____ Employers should evaluate employees on their ability to make unique contributions to the organization and not on whether they do or do not conform to authority structures/company tradition.

5. _____ "Management makes it happen."

6. _____ Subordinates at all levels have the right and responsibility to question and challenge authority when they believe management is not acting in the best interest of the organization.

7. _____ The chain of command is the official communication channel in any organization, and it should always be used if competitiveness and suspicion are to be discouraged.

8. _____ A good boss not only tolerates disagreements, but they encourage and create them.

9. _____ It is difficult to be an effective manager when subordinates keep criticizing you.

10. _____It is as much the responsibility of the managed as of the manager to create and maintain effective relationships.

Score Your Answers:

Score your answers to determine if your leadership style tends to be more autocratic, democratic, or participatory.

1. Sum of all even numbered items __________
2. Sum of all odd numbered items __________
3. Subtract line 2 from line 1 (Show + or - sign) __________

Plot your score on the scale below to illustrate your preferred behavioral style of leadership.

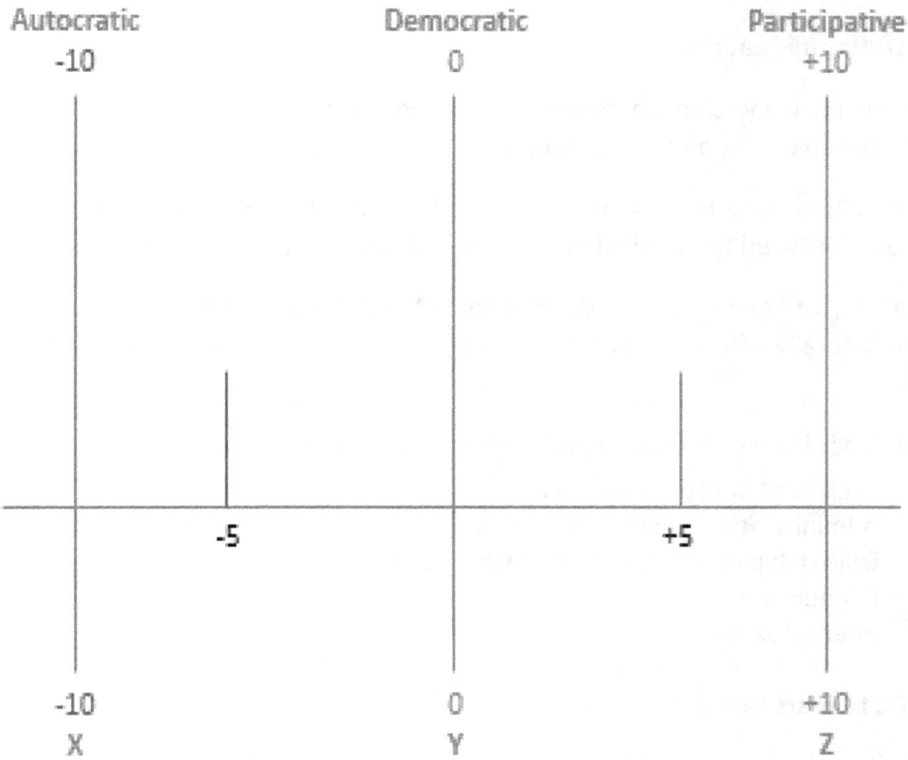

* Original source of activity unknown.

Appreciate the Personality Traits Approach to Leadership

As we mentioned, some people believe that good leaders are simply born. This idea comes from the personality traits approach to leadership. It emphasizes the personal attributes of leaders and suggests that certain traits, often those which are innate, are what make a person a good leader.

There is certainly value in the idea. But realize that no one can possess only positive leadership traits. Situations vary, and thus different leadership traits will serve one well in different situations (Brown, n.d.).

In this concept, we review the personality approach to leadership and the traits that effective leaders share.

Personality Traits and Leadership

The trait approach to leadership emphasizes the personal attributes of leaders and examines the relationship between quality and leader success.

According to this theory, leaders can be defined by their personality traits. Early researchers in leadership theory believed this applied only to the traits people are born with.

The theory has evolved to incorporate the idea that effective **leadership traits** are not limited to those who are naturally gifted. Most people now agree that these traits can be **learned**, **practiced**, and **developed**.

Recent research has focused on specific skill sets, including the following:

- Social and interpersonal skills
- Administrative and technical skills
- Leadership effectiveness and task motivation
- Friendliness
- Intellectual skills

What Traits Do Leaders Need?

Personality traits include such things as empathy, integrity, calmness, self-directedness, and assertiveness.

As mentioned previously, leadership trait theory has considered the idea that some people are born to be leaders. Certainly, we see this in military history, but today we know that by observing and emulating born leaders, we can all learn to be more successful leaders.

If you think about trait theory in practice, somebody like Robert Kennedy would be an example of a natural leader. He was inspiring, friendly, and outgoing—all great traits in a leader. Yet, other great leaders lacked these same traits. Abraham Lincoln, considered one of the greatest U.S. presidents, was introverted, solemn, and introspective by nature (Fowler, 2011). These examples show that commonly held “traits” associated with this theory are not always required of great leaders.

The value in trait theory comes from recognizing that many great leaders share common traits. Consider the positive traits that make a great leader. Identify the traits you feel to be the most valuable in your leadership situation and work to develop those in yourself.

Another challenge to trait theory is that many people possess qualities commonly associated with great leaders, yet they are not leaders. Again, we see that simply having certain traits does not make a person a leader. The desire to lead and to improve yourself is equally important.

Managerial Motivation and Skill Sets

Earlier research focused on personality traits. More recently, research has focused on managerial motivation and specific skills. They include social and interpersonal, administrative, technical, and intellectual skills, along with leadership effectiveness, achievement, social nearness, friendliness, support of group tasks, and task motivation and application. Social style refers to the way an individual interacts with others. It is related to our personality style, which is how we are motivated and think internally.

Several commercially available social style assessments can help you identify your interpersonal style. Examples are the Centrestar Interpersonal Style Profile™, Keirsey Temperament Sorter®, the DISC®, and the Myers Briggs Type Indicator (MBTI)® assessments. All these assessments can help you become aware of your preferences and, in turn, indicate how you can adapt your leadership style for maximum impact in your environment.

Although many models for assessing social styles are available, most can be reduced to four main categories, as illustrated in the graphic that follows.

Centrestar Interpersonal Style Profile™

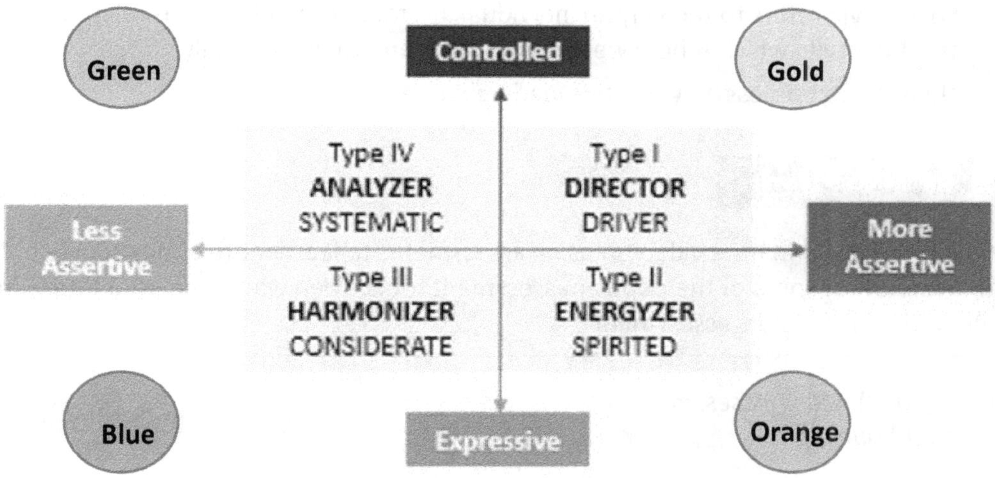

- **Type I Driver (Gold)** leaders are interested in the big picture and results. They are not bothered by details and tend to be goal-oriented and practical. They are often competitive and prone to be difficult as they can be bossy, self-involved, impatient, and direct.

- **Type II Energizer (Orange)** leaders are most concerned with social acceptance. They want to be recognized for their accomplishments and valued. They tend to be spirited, enthusiastic, and able to accept change, but also are good at influencing others.

- **Type III Harmonizer (Blue)** leaders crave stability, harmony, and consistency. They are resistant to change but tend to be good team players, considerate, and great listeners.

- **Type IV Analyzer (Green)** leaders tend to be detail-oriented, and while not competitive, they have a strong need to be systematic and right. They value logic and accuracy above everything else, are disciplined, and can be controlling of others.

Use your knowledge of your personality to evaluate your strengths and weaknesses. Consider what personality type you might be and how you communicate.

Also consider how people in the other types communicate. You might find you can learn something from looking at the differences.

Again, no person fits into a single box. We all have aspects to our personalities, different strengths and weaknesses that come through at certain times.

No matter what else you do, understanding your strengths and potential areas for improvement can help you to become a more effective leader.

What to do:	Other actions:
☐ Complete one of the available personality assessments.	☐ ____________________
☐ Encourage your co-workers to complete the same inventory and share the results.	☐ ____________________
☐ Consider how you might use your knowledge of personality in your job.	☐ ____________________

Remember

- ✓ The personality traits approach to leadership emphasizes the personal attributes of leaders.
- ✓ Social style refers to the way an individual interacts with others. It is related to our personality traits or style, which is how we are motivated and think internally.
- ✓ No one can possess only positive leadership traits.

Enhance Your Learning

Watch the following 5-minute video on using assessments to learn more about yourself. The video gives an overview of some of the popular assessment tools. Then watch the 3-minute video to learn more about the True Colors assessment .

Leduc, S. (2009). *Using Assessments to Learn about Yourself.*	Available at: http://www.youtube.com/watch?v=4YwmaHiT7b8
True Colors International. (2021). *What is True Colors.*	Available at: https://www.youtube.com/watch?v=XsM1wZ5WT8Q&t=32s

Reinforce Your Learning

Trait Approach. Consider what traits you feel you were born with that help to make you an effective leader. List them below.

Now consider: Were you really born with these traits? Or did you build them through experience? Plan how you can continue to build your skills in this module and beyond.

Understand the Leadership Cycle Approach

The leadership cycle approach focuses less on the leader and more on the employees, processes, and environment in which the leader practices.

The approach is situational in that leaders assess employee needs and then provide instruction, guided development, coaching, and monitoring as needed. As time goes on and the organization grows, the cycle begins again as the leader and employees respond to a changing situation.

The **leadership cycle approach** is about changing or adapting your leadership style in response to specific tasks in the context of a work environment. It requires that you understand the development level of the people you manage and provide what they need to be successful in their jobs.

In this concept, we discuss the leadership cycle approach and how you can use it to help develop individuals and project teams.

What is the Leadership Cycle Model™?

Leaders adapt their style to work situations with the assumption that people **have the ability and willingness** to perform even though they may lack certain skills or be **unaware of the need to develop.** The cycle is also **continuous**.

As illustrated in the graphic that follows, the **Centrestar Leadership Cycle Model™** has five phases: observe and assess, show and explain, guide development, coach and enable, and monitor independence.

The leadership cycle is ongoing and repetitive as workers and teams advance, and as the work environment changes.

Centrestar Leadership Cycle Model™

1. Leader assesses how workers understand a job.
2. Leader explain what workers need to know.
3. Leader guides development with instruction.
4. Workers engage fully in job with leader support.
5. Workers are independent and leader monitors.

Phase 1 Observe and Assess. Workers or teams in this phase are typically new to a task, or they are experiencing changes in their jobs due to reorganization or restructuring of their responsibilities. The leader observes the work to assess the workers' or team's level of development and awareness of what is needed for them to reach competent independence.

Phase 2 Show and Explain. Workers or teams in this phase are typically inexperienced, lacking knowledge, skills, and confidence. They may not know how unskilled they are, so the leader may need to show or explain how much they need to learn. The leader needs to encourage, perhaps even push, people to accomplish tasks.

Phase 3 Guide Development. In this phase, the leaders offer instruction and direction and attempt to get buy-in as people gain skills and competence. Workers are typically motivated but still lack the skills to complete the job effectively without guidance from the leader.

Phase 4 Coach and Enable. People are typically engaged in developing skills but still lack full competence and confidence to complete the job. They need coaching and support from the leader as well as an environment in which to practice.

Phase 5 Monitor Independence. People typically perform tasks independently and effectively because they now have skills, commitment, and confidence in their abilities. However, leaders must monitor performance because the work environment will constantly change. As new levels of skill and knowledge become necessary, the cycle begins again.

Examples of the Leadership Cycle Model™

Consider the cycle of leadership as it might apply in learning how to swim. In Phase 1, a good swimming instructor (leader) would observe and assess your skill at swimming and then fine-tune the training to fit your needs.

In Phase 2, the instructor would explain why it is important to learn how to swim, and demonstrate the skills needed.

In Phase 3, the instructor would ask you to jump into the shallow end, and there help you acclimate to the water and provide focused instruction and tools such as kickboards.

In Phase 4, the instructor would support you while you practice your new skills, let you gradually move to the deep end, and coach you on proper swimming techniques.

Finally, in Phase 5, the instructor would exit the water and let you swim independently while monitoring your progress and advising that you should not swim alone. Should you want to advance your skills beyond the novice level, the process cycle would continue.

Another example of the five phases is a situation where you, as a project manager, have an experienced team that is adept and reliable when you give them a goal and a timeline. You expect the team to complete work without much direction. But, if a new member joins the team or if the team is restructured, the situation would change. Then you might need to adapt your leadership by becoming involved and working through the five phases again.

Other Situational Models

Other theories and models that focus less on the leader and more on workers, processes, and the environment include:

- Gradual Release of Responsibility Model suggested by Vygotsky (1930s), Burch (1970s), Pearson and Gallagher (1980s)
- Fiedler Contingency Model developed by Fiedler (1960s)
- Situational Leadership Theory advanced by Hershey and Blanchard (1970s)

Gradual Release of Responsibility (GRR) Model. This is a structured method of teaching framed around a process of releasing responsibility from the teacher to the learner who gradually develops the knowledge and skills to work independently.

While Pearson and Gallagher (1983) coined the phrase "gradual release of responsibility" to describe this dynamic in the classroom, it was Russian educational theorist Lev Vygotsky's concept of the Zone of Proximal Development (1930s) and Bruner's idea of instructional scaffolding (1960s) that suggested the four-component model used by many teachers today.

The four components include direct instruction, guided instruction, collaborative learning, and independent work.

A related model developed by Noel Burch in the 1970s is called the Four Stages of Competence. It includes the following stages: unconscious incompetence, conscious incompetence, conscious competence, unconscious competence.

Examples of the Leadership Cycle Model™

Consider the cycle of leadership as it might apply in learning how to swim. In Phase 1, a good swimming instructor (leader) would observe and assess your skill at swimming and then fine-tune the training to fit your needs.

In Phase 2, the instructor would explain why it is important to learn how to swim, and demonstrate the skills needed.

In Phase 3, the instructor would ask you to jump into the shallow end, and there help you acclimate to the water and provide focused instruction and tools such as kickboards.

In Phase 4, the instructor would support you while you practice your new skills, let you gradually move to the deep end, and coach you on proper swimming techniques.

Finally, in Phase 5, the instructor would exit the water and let you swim independently while monitoring your progress and advising that you should not swim alone. Should you want to advance your skills beyond the novice level, the process cycle would continue.

Another example of the five phases is a situation where you, as a project manager, have an experienced team that is adept and reliable when you give them a goal and a timeline. You expect the team to complete work without much direction. But, if a new member joins the team or if the team is restructured, the situation would change. Then you might need to adapt your leadership by becoming involved and working through the five phases again.

Fiedler Contingency Model. Fred Fiedler, a scientist who studied the personality and characteristics of leaders, developed a leadership model in the 1960s. It states that there is no one best style of leadership. Instead, a leader's effectiveness is situational. This is the result of two factors—leadership style and situational control.

Situational Leadership Theory. In the 1970s, Hersey and Blanchard characterized leadership style in terms of the task behavior and relationship behavior leaders provide to their followers.

The authors later developed models using situational leadership theory. Hersey developed the Situational Leadership Model characterized by four behavior types: telling, selling, participating, and delegating. Blanchard developed the Situational Leadership II Model, also characterized by four behaviors: directing, coaching, supporting, and delegating.

What to do:	Other actions:
☐ Identify the phase you have experience leading **most effectively** and **least effectively**.	☐ ____________
☐ Assess one of your employees using the leadership cycle approach.	☐ ____________
☐ Develop a strategy based on the leadership cycle approach to use with this employee.	☐ ____________

Remember

✓ The Leadership Cycle Approach is about adapting your leadership style to fit a specific job in the context of the work environment, and understanding what workers need for successful performance.

✓ The Leadership Cycle Model has five phases: observe and assess, show and explain, guide development, coach and enable, and monitor independence.

✓ The leadership cycle process is ongoing and repetitive as workers or teams advance, and as the work environment changes.

Enhance Your Learning

Watch this 7-minute video to learn more about situational leadership theory.

Can-Indian Channel. (2018). *MBA 101 Leadership, Situational Leadership Contingency Model.*	Available at: https://www.youtube.com/watch?v=H1SiGs-SVdA

Reinforce Your Learning

Apply the Leadership Cycle Approach. Review the leadership **Strategy Development Guide** that follows.

After that, read the four case scenarios that follow and assess each from a leadership phase perspective (Phase 1, 2, 3, 4, or 5). Then develop strategies you would employ if faced with each situation. Use the **Strategy Development Guide below** to help develop your strategies. Be prepared to explain and/or conduct a role-play to demonstrate how you would handle each situation.

Leadership Cycle Approach: *Strategy Development Guide*

Phase	Employee Characteristics	Leader Behavior
1. Observe & Assess	• New to the job or task • Can be a new employee or a current employee whose job responsibilities or task completion processes have changed. • May lack ability to solve problems that occur relative to the task • May be unable to initiate action or take responsibility without supervisor's direction	• Provide description of job or task • Ask questions to see how well the employee understands the nature of the job or task • Ask questions (or check records of past performance) to see if the employee has skills or experience with similar tasks • Decide where to start building the employee's skills and confidence needed for successful performance
2. Show & Explain	• Inexperienced or low experience in the task • Insufficient knowledge or understanding of how to do the task • Lacks ability to solve problems related to the task • Low confidence in his/her ability to perform the task. Appears reluctant to initiate action or take responsibility • Will perform below levels or standards established for experienced workers • "Don't know what they don't know"	• Closely supervise • Teach employee how to perform task • Give clear, simple instructions and directions • Keep focus on the basics of the task and how to perform successfully • Encourage or push employee to do what the leader says to do
3. Guide Development	• Some experience but not skilled enough to complete task or solve task-related problems without supervision • Has confidence in his/her ability to perform (even though lacking all the skills) • Is motivated to work on his/her own--want less direct supervision—but will still perform at a lower level than fully competent employees • May "give up" easily when faced with a problem	• Continue giving direction and instruction • Ask questions to see how well the employee understands where the task fits and why it is important • Go beyond the basics, give reasons and information to help employee see why things are done as they are • Allow the employee a little more independence—within bounds of safe operations—but monitor fairly closely
4. Coach & Enable	• Skills needed for successful performance are improving, but employee still needs some supervision, direction and instruction • Even though the skill levels are increasing, employee lacks confidence in the ability to operate on his/her own • Employee may ask the supervisor to explain things (again) or to watch as he/she completes the task • May need a lot of positive feedback and encouragement from the supervisor • Generally meets established performance levels and standards • Able to solve most problems related to the task	• Maintain regular contact with the employee, to provide support or help if he/she gets stuck • Respond to employee's questions by first asking, in a positive tone, "what do you think?" or "what would you do?" • Coach employee as needed • Remind employee of how much he/she has learned and how well he/she is doing • "Counsel" employee if he/she loses confidence or reverts to a lower level of performance • Point out the next higher level task or responsibility (if there is one related to the task at hand) and encourage the employee to stretch toward that task
5. Monitor Independence	• Able to work independently and effectively • Appears to be committed to doing a good job and takes pride in his/her work • May ask for opportunities to develop further skills or take on added responsibilities • Fully meets or exceeds established performance levels or standards • Willingly takes responsibility and meets deadlines • Has a positive work attitude • Persists in completing work in spite of problems or set-backs	• Keep in touch with employee to maintain a positive work relationship (providing direction and encouragement is no longer needed) • Take care to avoid micro-managing employee • Look for ways to involve the employee in helping less- experienced co-workers develop their skills and confidence • Monitor the environment and technology to anticipate changes in skills needed to perform successfully • Inform employee of upcoming changes and how those will affect the employee's tasks or responsibilities

Leadership Scenario A: Terri Tentative

Terri has worked at CS Contracting for ten years. Her performance on the job, technical knowledge, understanding of job requirements, and ability to solve problems related to the job are all well above average.

Yet, Terri constantly seeks support from her peers as well as you, her project manager. She seems unable to deal with everyday issues that occur in the workplace. You must constantly be Terri's "mother," making sure details get taken care of so Terri is not overwhelmed. You have observed her being emotional on several occasions.

Terri does not seem to tolerate change, especially regarding details. As a result, Terri can be a drain on your time. Terri seems to lack self-esteem. She seems unsure of her capabilities and unable to sort out the details from the "big picture."

Assess Terri's situation from a leadership cycle perspective (Phase 1, 2, 3, 4, or 5), and develop strategies you would employ if you faced this situation. Be prepared to conduct a role-play to demonstrate how you would handle the situation.

Leadership Scenario B: Larry Learner

Larry has been with CS Contracting for a considerable time and was recently promoted. He did well during training and seemed to be learning his new job quickly. However, after training was complete, Larry's performance declined. Because Larry held his previous position so long, it had become easy for him. In that situation, his project manager could leave him alone and know that he would always succeed.

Larry senses he is not performing well in his new position, and you have recently seen him become withdrawn.

Assess Larry's situation from a leadership cycle perspective (Phase 1, 2, 3, 4, or 5) and develop strategies you would employ if you faced this situation. Be prepared to conduct a role-play to demonstrate how you would handle the situation.

Leadership Scenario C: Ina Inconsistent

Ina has worked at CS Contracting for two years. She progressed in her job maturity to the point where she is one of the quickest and most accurate workers. Her overall job performance has been good, but she seems to decline in performance during certain times, especially around the holidays.

You have observed she has not been consistent with following through on projects during these times, and when she does come through, she is often behind schedule.

Recently, she did not meet the schedule deadline for a new product launch. Besides catching flak from the administration, you are worried about whether you can count on Ina.

Assess Ina's situation from a leadership cycle perspective (Phase 1, 2, 3, 4, or 5) and develop strategies you would employ if you faced this situation. Be prepared to conduct a role-play to demonstrate how you would handle the situation.

____

____

____

Leadership Scenario D: Fred Fixable

Fred came to CS Contracting after he was let go due to a cutback at another organization. Fred has five years until he is eligible for retirement.

Fred has high energy levels. Over the years he has become knowledgeable technically, is task oriented, and decisive. He comes to your team appearing to be arrogant about his expertise. He also seems to distrust management and to exhibit a negative attitude.

Fred works with clients and recently you heard him on the phone telling a client they should listen to him because he has "more experience than they do."

Assess Fred's situation from a leadership cycle perspective (Phase 1, 2, 3, 4, or 5) and develop strategies you would employ if you faced this situation. Be prepared to conduct a role-play to demonstrate how you would handle the situation.

____

____

____

Leading and Managing Projects #8.9

Assess and Build Your Leadership Competence

As previously discussed in Concept 2, leadership competencies such as communication, team building, and conflict management are not the only knowledge and skills required for you to lead effectively. Attitudes and behaviors that drive you to use your knowledge and skills in meeting—or exceeding—your job standards also matter. For example, critical thinking is a competency. The ability to calmly analyze a problem and develop a solution is a vital ability. Being flexible and innovative are examples as well.

The most **effective organizations align** their business models with the **competencies** they deem to be **important**. For example, Apple and Google both pride themselves, as organizations, on being innovative. So, innovation is an important competency for leaders in those organizations.

However, before we dig deeper into the concept of competencies, we must first stress the importance of core values and beliefs.

Core Values

Core values and beliefs play a role in defining "who" your organization is and what it expects from its employees. The values vary based on many factors, for example, industry, location, founders, employees, and more.

What do you value in the workplace, and how important are the following to you?

- Honesty, **respect**, human dignity, people, fun
- Meaningful work, **integrity**, employee development, organizational learning, agility
- Work ethic, positive attitude, **responsibility**, delivering value, drive for results
- Quality products and services, effective and efficient operations, innovation, **discovery**
- Safe working conditions, environmental protection, customer-focused **excellence**
- Contribution to society, **community,** teamwork, ethics, and transparency

Keep in mind as we discuss the types of competencies you need to build as you develop your project management skills that core values and beliefs are at the heart of leadership and competency development.

Competencies Matter

Competencies are (ThinkWise, 2007) the measurable and observable **knowledge, skills, attitudes,** and **behaviors (KSABs)** critical to successful job performance. Competencies are **driven by** a leader's **personality** traits, **motives, attitudes** and so forth (ThinkWise, 2007).

In competency-based analysis, the **people** who perform a particular job **are the focus** of the analysis. The **performance** of **exemplary workers** is **analyzed,** and the job is thus defined in terms of how the exemplar utilizes the knowledge, skills, behaviors, and attitudes necessary to perform to the established standards. In this approach, the focus is on the employee and what they need to excel in a job.

A competency can be described as "an **enduring characteristic** of a person that **predicts behavior** across many workplace situations" (ThinkWise, 2007, p.1). For example, a leader who is a competent communicator knows how to provide direction and listen to feedback effectively.

What is defined as a desired or required **competency can vary** from organization to organization, and some competencies are specific to a particular organization. For example, a leader may need to be technologically competent and able to communicate effectively with a virtual, globally distributed workforce in a certain organization, but not all organizations require this particular competency. However, some competencies are necessary for all leaders. For example, all leaders need to be competent communicators.

The point here is that leader competencies must **align** with the mission, vision, culture, and **needs** of the **organization**. This is why many companies have competency-based terminology in their mission and vision statements as well as their job descriptions, such as "innovation, creativity, and customer service."

Competency Models and Approaches

Many leadership studies have been done from a military and government perspective. The United States Office of Personnel Management (OPM) developed a research-based conceptual model that included skills, knowledge, attitudes, and behaviors. This model incorporated twenty-two leadership competency dimensions identified as important for effective performance at leadership levels by the U.S. Government and was verified in a Federal-wide survey of supervisors, managers, and executives. This group represented the largest Federal occupational study ever conducted. The foundation underpinning much of OPM's work was a thirty-year research effort on twenty-six behavioral dimensions that had been found to correlate significantly with ratings of management potential and with subsequent achievement for American Telephone and Telegraph Company executives.

OPM's work was adapted for use in a research project (Donahue, 1996) that focused on the leadership perceptions of practicing engineers at different organizational levels. The research measured the importance of leadership skills for future success.

To build on this research an expanded Leadership and Professional Development Competency Inventory and Framework called the *CentreStar Leadership Competency Framework*, was developed by Dr. Wesley Donahue. It combines extensive research in both the public and private sectors.

The Leadership and Professional Development Framework (2011) that follows includes thirty-five competency dimensions. They are clustered into five overarching content domains listed below, with core values at the heart:

- Professional Competence
- Supervisory/Management
- Resource Management
- Technical Acumen
- Organizational Leadership

CentreStar Leadership Competency Framework

Your Role as a Project Leader

As we discussed, competencies are measurable and observable knowledge, skills, abilities, and behaviors (KSABs) critical to successful job performance with core values underpinning them.

In your role as a project leader, competencies refer to the KSABs your organization hired you to perform.

Your project leader role requires that you have many competencies, some technical and others termed "soft skills," such as the ability to communicate effectively.

As people progress to different leadership levels or positions, certain competencies become more important, as shown in the graphic that follows (Donahue, 2020).

TYPICAL COMPETENCIES by LEADERSHIP LEVEL

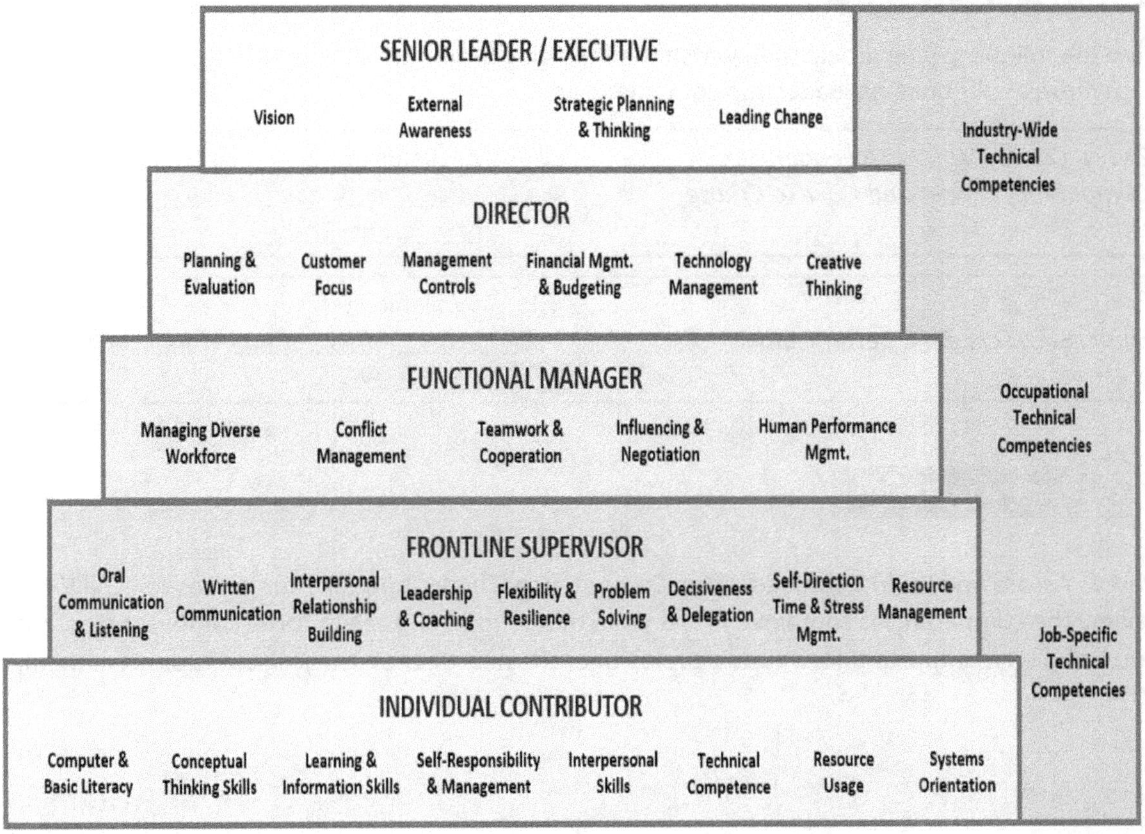

As a project manager, in addition to your technical competencies, you are typically responsible for the supervisory and functional manager competencies, and at times you will also need some of the competencies associated with higher management levels.

What to do:

- ☐ Identify three competencies you plan to develop over the next year.
- ☐ Select resources (people, books, work assignments) to help you develop the competencies.
- ☐ Search for exemplary performers for each competency you identified and solicit advice.

Other actions:

- ☐ ______________________
- ☐ ______________________
- ☐ ______________________

Remember

✓ Competencies are measurable and observable knowledge, skills, attitudes, and behaviors (KSABs) critical to successful job performance.

✓ The premise of competency-based leadership is leader competencies (and employee/volunteer competencies) must match the competencies valued by the organization. Alignment is paramount.

Enhance Your Learning

View the following blog article then watch the following 5-minute video (which is also embedded) to learn more about building leadership competencies.

Zavvy. (2022). *What Is A Leadership Competency Model And How To Create One?*	Available at: https://www.zavvy.io/blog/leadership-competency-model

Sinek, S. (2022). *Be a Better Leader.*	Available at: https://www.youtube.com/watch?v=sNdkwaW_LwQ

Reinforce Your Learning

Part 1. Assess and Build Your leadership Competence. Thinking about your job responsibilities, review the competencies contained in the table that follows and select **three competencies** that you believe are important for you to develop over the next year on the job. Circle or mark them.

COMPETENCIES			
1. Computer and Basic Literacy	9. Interpersonal Skills	17. Conflict Management	25. Technology Management
2. Technical Competence	10. Oral Communication	18. Teamwork and Cooperation	26. Creative Thinking
3. Resource Usage	11. Written Communication	19. Influencing and Negotiating	27 Vision
4. Resource Management	12. Leadership and Coaching	20. Customer Focus	28. External Awareness
5. Understand Systems	13. Flexibility and Resilience	21. Interpersonal Relationship Building	29. Strategic Thinking and Planning
6. Conceptual Thinking	14. Problem Solving	22. Human Performance Management	30. Management Controls
7. Learning and Information	15. Decisiveness	23. Planning and Evaluation	31. Diverse Workforce
8. Self-Responsibility and Management	16. Self-Direction	24. Financial Management and Budgeting	32. Leading Charge

Part 2. Feed-Forward—Learning from Peers. If practical, in groups of 3 to 5, this activity should take no more than 12-15 minutes to complete. Divide the time allotted equally among the number of members and assign a timekeeper. For your allotted time segment, you will listen to development suggestions from your other group members for your **three competencies** and document the suggestions in the workbook page that follows. During their allotted time segments, you will provide development suggestions for their selected competency areas.

Rules of engagement:

- Let go of the past
- Listen to suggestions without judging
- Document as much as you can
- Learn as much as you can
- Help as much as you can

If you cannot participate as a group, ask three colleagues that you believe are good leaders for their feedback.

Competency: ______________________________

____

____

____

____

____

Competency: ______________________________

____

____

____

____

____

Competency: ______________________________

____

____

____

____

____

Coach Others and Develop Yourself

Authentic leadership is a catchphrase we hear in business almost every day. It is the idea that people want to work with managers who are enthusiastic, engaged, and good at motivating others. Today's workers do not just want to be told what to do. They want to be part of something they believe in, and that must come through leadership.

In many cases, project team members do not directly report to project managers. This requires that project managers rely more on persuasiveness than on directives. It requires that project managers are creative in exciting the team about the opportunities associated with the project, and in making the project a priority.

In this concept, we discuss how you can engage and motivate team members by encouraging a culture of teamwork through mentoring and coaching as well as keeping an eye on developing yourself.

Authentic Participation

As a project manager, you are responsible for creating an atmosphere where participation is more than lip service.

To start, spend some time reviewing the environment your team is functioning in to determine if people have what they need, which includes the following:

- Trust in the leader and other members.
- Understanding of the problem or work goal.
- Skills in working with others in a group setting.
- Understand the value of cooperation instead of competition.
- Ability to separate disagreements from personal rejection or disapproval.
- Clear sense of team purpose or mission.

Culture of Teamwork

Creating effective teams requires creating a **culture of teamwork** where people can participate authentically. This is a culture that believes two heads are better than one and that **consensus decision making** will—on-average—produce better results.

Start by considering how you lead. Do you value teamwork? Do you develop and reward people who recognize that planning, thinking, creating, making decisions, and acting are more productive when people work together? Encouraging and rewarding teamwork is the foundation for creating teams that function well.

Coaching

Training, performance feedback, and coaching can significantly affect team success. Authentic coaching convinces people of their ability to do the job.

While traditional training can increase productivity by over 20 percent, training and coaching combined can result in nearly a 90 percent increase in productivity (as cited in TCS, 2012).

An authentic leader has faith that with the proper training and support, people will excel. Coaching involves providing training, support, and constructive feedback as team members carry out their responsibilities.

How to Conduct a Coaching Session

An effective coaching session begins with you as the project leader clearly understanding how you want to coach a team member. Observe the individual's behavior and collect data until you are sure of how you want to support the team member.

Also review any evaluations or notes from prior coaching sessions. Prepare to discuss anything of importance such as previous goals or task lists. Anticipate possible excuses and prepare to handle and respond to them.

Then pick a time and place and plan for the session that will be comfortable for you and the person you want to coach and set an agenda for the session.

The following is a discussion of the steps to follow in the session.

1. Describe the performance, behavior, or issue in a way that is non-confrontational.

The first step in solving a problem or issue is getting the individual to see the situation. You need to do so without making the individual feel defensive or attacked.

Begin by asking the person how they think things are going, and if they see any problems. If they do not, you can discuss your project goals and what problems you see in achieving the goals. Then present some objective evidence about how the person needs to do something differently. Explain your expectations for their performance with clear, specific examples of what they are doing that is not supporting project goals, and then explain the behavior you want to see.

2. Ask for the employee's assistance in solving the problem or issue you are addressing.

Just telling someone there is a problem will accomplish little. You must get the team member to agree there is an issue, agree on what that issue is, and agree to work together to improve things. Ask the individual to help find solutions. This will help the team member "buy in" to solving the problem.

3. Brainstorm causes or contributing factors for the performance, behavior, or issue.

Work with the individual to come up with potential causes of the problem. Ask questions, probe into their situation. Help them find out what is leading to this problem.

4. Identify possible solutions.

Once you have identified some sources of the problem, work with the individual in a similar manner to brainstorm solutions. Ask them questions about what solutions they envision, get them to agree to certain solutions, or come up with better ones together.

5. Agree on an action plan.

Once you have a list of solutions get the individual to agree that these solutions are valuable and doable. Set specific goals that are measurable, achievable, realistic, and time-based.

End the session with a new list of goals or next steps, as well as plans for following up. Get them in writing and ensure that both you and the employee have a copy.

6. Set a follow-up meeting.

Arrange a meeting where you will follow up with the individual to see how their action plan is proceeding.

Mentoring

Mentoring happens when an experienced person provides guidance and support in a variety of ways to a developing employee. Mentoring is a way to help introduce a person to the workings of the organization and assisting with professional development. A mentor must be willing to share their expertise and not be threatened by the person's future success in the organization.

Develop Yourself

Your chosen style of leadership will impact the authentic participation, attitudes, and productivity of those you lead.

Start by considering the various leadership theories and decide what aspects of those approaches you can apply to your work.

Harvard researcher, Jim Collins, in his classic book *Good to Great* suggests five levels of leadership (2005):

> **Level 1: Highly Capable Individual**. This person is a highly skilled, experienced worker, a valued employee. These people are often considered for management positions.
>
> **Level 2: Contributing Team Member**. This person has the skills and knowledge to be productive on teams.
>
> **Level 3: Competent Manager**. This person has learned to effectively organize teams and groups around objectives.
>
> **Level 4: Effective Leader**. This is the level that most top leaders fall into, where they can organize their employees to meet performance objectives.
>
> **Level 5: Great Leader.** This is the highest level of leadership, on which all the abilities of the previous four levels are met, and the leader also possesses the right blend of humility and emotional intelligence to be highly effective.

To become a level 5 leader, you must value yourself, recognize your strengths, and work to develop the traits of a great leader. You also need to **show humility**, which means giving credit to the team when things go well and taking responsibility when things go awry. You need to know how to solicit input and to ask for help when necessary. You also must find the right people to support your growth and development.

What to do:	Other actions:
☐ Practice coaching and giving credit to your team and others.	☐ ______________________
☐ Provide opportunities for team members to develop their skills.	☐ ______________________
☐ Solicit input from knowledgeable people to support your growth and competency development.	☐ ______________________

Remember

- ✓ The most important way to continually developing your leadership ability is to do just that—continually develop.
- ✓ Recognize that you always have room to improve. Reflect on your successes and plan for further success, remembering that your leadership style may need to be altered depending on the situation, your organization, and those you are leading.

Enhance Your Learning

Watch the following 14-minute video to learn more about developing your skills by the London Business School.

Birkinshaw, J. (2013). *Two tips for developing good management skills.*	Available at: https://www.youtube.com/watch?v=ILcgCEbsT08

Review Leadership Competency guidebook Building Leadership Competence available on Amazon.

Reinforce Your Learning

Competency Action Plan. Select one of the three competencies that you identified as an area of development in the last Concept. Plan and track your development activities by completing the **Leadership Competency Action Plan Worksheet** that follows. When completed, repeat the process for the other two competencies that you identified.

A blank worksheet template is included at the end of this workbook module as Appendix Part A. Copy and use the worksheets as you progress.

Leadership Competency Action Plan Worksheet

Competency: ____________________

Briefly describe how improvement in this competency will help you achieve important results or better meet your job responsibilities.

List courses, books, or independent study opportunities that could help you develop this competency.

Find one or more people who could help you, either as a role model or source of information. Write any questions you would like to ask each person.

WHAT SPECIFIC STEPS WILL YOU TAKE?	START	DATE FINISHED

Summary

A Final Word: *Leading and Managing Projects*

You have just completed a chapter on Leading and Managing Projects. Now that you have studied the basic approaches to leadership, reflect on what resonates with you. Focus on building the knowledge, skills, attitudes, and behaviors (competencies) you want to develop or exhibit. This is how you build your leadership abilities and a style that works for you.

A few important topics you have encountered in this chapter include:

- Understanding the relationship between management and leadership.
- Describing the basic approaches of leadership that can be applied to a wide range of projects.
- Listing and describing the key leadership skills and competencies associated with project success.
- Identifying and describing how to coach and handle typical project leadership situations.
- Building your leadership competence.

In the next Project Mastery chapter, we will focus on the business side of project management with Controlling Time, Money, and Risk.

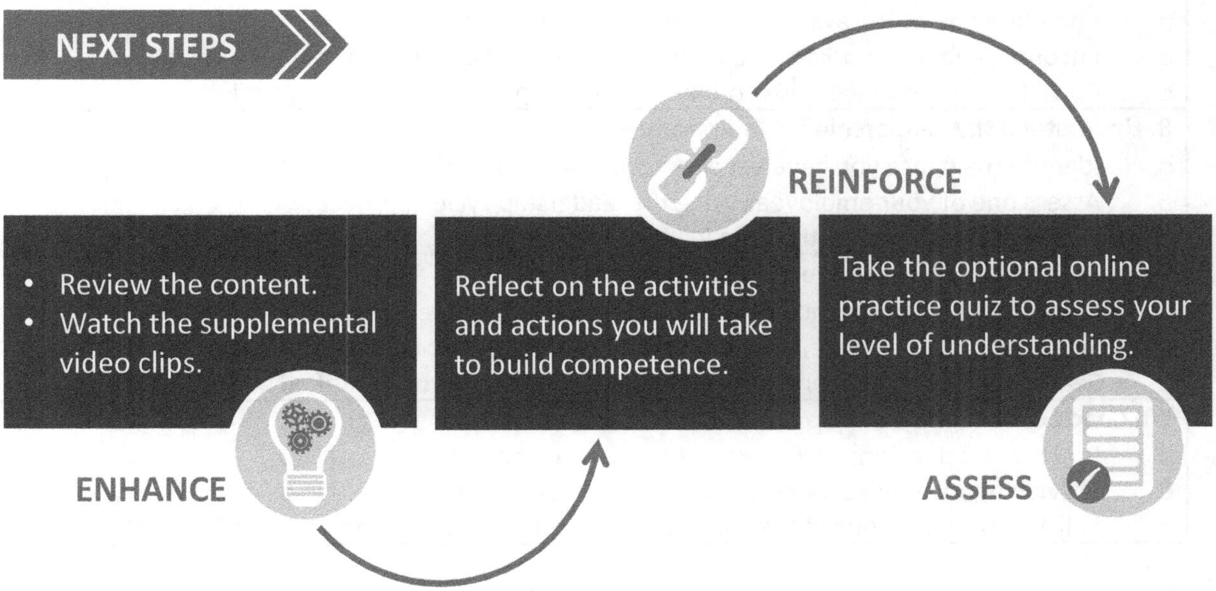

Leading and Managing Projects—Recap Checklist Many studies have tried to identify the differences between exemplary performers as compared to average performers. No matter what the specific differences, leadership has a significant impact on results.
1. Observe the Qualities Effective Leaders Share □ Select a leader you admire and list the specific things you admire. □ Identify your most effective leadership traits. □ Ask a person you trust to identify your leadership weaknesses.
2. Identify the Key Competencies of Successful Leaders □ Be mindful of what authority you have and do not have when you make decisions. □ Know your technical strengths and who can help you in your areas of weakness. □ Identify the leadership competencies you need to be successful in your job.
3. Differentiate Between Management and Leadership □ Connect with and care about those you lead. □ Lead by example and remember the role of a project manager is to plan, organize, direct, and control work. Your role as a leader is to inspire and motivate. □ Work to have fun and energize others.
4. Distinguish Transactional from Transformational Leadership □ Move from being a transactional leader to a transformational one by being more open. □ Reflect on your preferences, biases, strengths, and weaknesses. □ Be more transparent. Show more of your thinking and the issues you wrestle with.
5. Be Aware of the Power and Influence Approach to Leadership □ Identify what types of power you possess and the sources of your power. □ Determine your main characteristics related to power and influence. □ Determine what power you have gained or lost in the past year.
6. Be Mindful of the Behavioral Approach to Leadership □ Realize the behavioral theory of leadership asserts that people can learn to be effective leaders. □ Observe leaders you admire, watch their specific behaviors, and consider how you can be like them. □ Determine your behavioral style of leadership.
7. Appreciate the Personality Traits Approach to Leadership □ Complete one of the available personality assessments. □ Encourage your coworkers to complete the same inventory and share the results. □ Consider how you might use your knowledge of personality style in your job.
8. Understand the Leadership Cycle Approach □ Identify the phase you have experience leading most effectively and least effectively. □ Assess one of your employees using the leadership cycle approach. □ Develop a strategy based on the leadership cycle approach to use with this employee.
9. Assess and Build Your Leadership Competence □ Identify three competencies you plan to develop over the next year. □ Select resources (people, books, work assignments) to help you develop the competencies. □ Search for exemplary performers for each competency you identified and solicit advice.
10. Coach Others and Develop Yourself □ Practice coaching and giving credit to your team and others. □ Provide opportunities for team members to develop their skills. □ Solicit input from knowledgeable people to support your growth and development.

Knowledge Review Test—Questions

Leading and Managing Projects

1. Project managers and leaders can contribute to poor performance by a team member because:
   - A. They do not adapt to the readiness level of the team member.
   - B. They over-supervise.
   - C. They under-supervise.
   - D. All the answers are correct.
2. Project managers, in addition to the organization, can be held liable for all except:
   - A. Violating anti-discrimination laws.
   - B. Ignoring safety issues.
   - C. Enforcing OSHA rules.
   - D. Accepting bribes from contractors.
3. Which of the following is a competency of a successful leader?
   - A. Keys to the building.
   - B. Power.
   - C. Knowing how to leverage team strengths.
   - D. All the answers are correct.
   - E. None of the answers are correct.
4. Which of the following questions should not be asked of a candidate in an employment interview?
   - A. Have you ever been convicted of a crime?
   - B. Can you show proof of your eligibility to work in the U.S.?
   - C. When did you graduate?
   - D. Are you able to perform the essential functions of this job with or without reasonable accommodation?
5. Leadership involves three broad skills areas except:
   - A. Technical.
   - B. Perceptual.
   - C. Human.
   - D. Conceptual.
6. All of the following management functions can be used to control team members as well as develop them except:
   - A. Discipline.
   - B. Delegation.
   - C. Facilitation.
   - D. Training.
7. It is critical that both leadership and management be exercised in project management. One way of defining management is the process of using project resources to achieve specific objectives through the functions of planning, organizing, implementing, and controlling, whereas leadership involves:
   - A. Using power to influence others.
   - B. Establishing direction, aligning people, and motivating and inspiring others.
   - C. Using charismatic power to motivate others.
   - D. Getting the right resource to the right place at the right time and in the right numbers.

8. As a project leader, you will often need to change your behavior and move from being a transactional leader to what type of leader?

   A. Energizing leader.
   B. Harmonizing leader.
   C. Transformational leader.
   D. Directive leader.
   E. Analytical leader.

9. All the following are types of power derived from a project manager's position except:

   A. Reward.
   B. Discipline.
   C. Formal.
   D. Expert.

10. Which of the following factors "disempower" team members?

   A. Confusion.
   B. Lack of trust.
   C. Not being listened to.
   D. Lack of time to solve problems.
   E. All the answers are correct.

11. According to behavioral leadership theory the following styles are identified, except:

   A. Participative.
   B. Democratic.
   C. Conciliatory.
   D. Autocratic.

12. Military leaders are often examples of what type of leadership?

   A. Autocratic leaders.
   B. Democratic leaders.
   C. Laissez-faire leaders.
   D. Lewin leaders.

13. What is the best way to make yourself a better leader?

   A. Identify what type of leadership style is natural to you and work only in that style.
   B. Use various theories to craft a leadership style that works best for you and your organization.
   C. Exactly emulate another leader whom you admire.
   D. None of the answers are correct.

14. The management theory that states that people are motivated and desire responsibility and can direct their own efforts is:

   A. McGregor's Theory Y.
   B. Maslow's Theory.
   C. Blake and Mouton's Theory.
   D. Centrestar's Leadership Cycle Theory.

15. What type of leadership is less about the leader himself, or herself, and more about the people, situation, and circumstances?

   A. Telling leadership.
   B. Trait theory.
   C. Directed leadership theory of development
   D. Leadership cycle approach.

16. If you have a new worker with low job maturity, according to the Leadership Cycle Model, what is the most effective approach?

   A. Selling/persuading/coaching.
   B. Supporting/participating.
   C. Delegating/mentoring.
   D. Show and explain.

17. Which of the following is NOT true of the behaviors demonstrated in a competency-based approach to leadership?

   A. These traits must be inborn; they cannot be taught.
   B. You must demonstrate an understanding, knowledge, and ability to engage in behaviors that positively impact your job.
   C. Competencies must have the ability to be further improved with training.
   D. A competence may be defined as knowledge, skills, attitudes, and behaviors that are observable and measurable.

18. All the following are responsibilities of a project manager except:

   A. Maintain confidentiality of stakeholders.
   B. Make sure that conflicts of interests do not compromise the legitime interests of the organization.
   C. Determine the legality of the organization's procedures.
   D. Provide accurate and honest representations in budget estimates.

19. Which of the following is not a benefit of performance coaching?

   A. Coaching helps team members build their skills and competencies.
   B. Coaching enables the project manager to handle psychological issues employees experience.
   C. Coaching allows for greater delegation, giving the project manager more time to devote to other value-adding activities or responsibilities.
   D. Coaching enables sharing of leadership responsibilities.
   E. Coaching increases the likelihood of tasks being completed with quality and creativity.

20. What is the most important aspect of being an effective leader?

   A. Continually developing your leadership ability.
   B. Choosing a specific leadership style and sticking with it.
   C. Delegating all your work to others.
   D. Being an overbearing project manager.

Leading and Managing Projects — Answer Sheet

1.	___ A	___ B	___ C	_X_ D	___ E
2.	___ A	___ B	_X_ C	___ D	___ E
3.	___ A	___ B	_X_ C	___ D	___ E
4.	___ A	___ B	_X_ C	___ D	___ E
5.	___ A	_X_ B	___ C	___ D	___ E
6.	___ A	___ B	_X_ C	___ D	___ E
7.	___ A	_X_ B	___ C	___ D	___ E
8.	___ A	___ B	_X_ C	___ D	___ E
9.	___ A	___ B	___ C	_X_ D	___ E
10.	___ A	___ B	___ C	___ D	_X_ E
11.	___ A	___ B	_X_ C	___ D	___ E
12.	_X_ A	___ B	___ C	___ D	___ E
13.	___ A	_X_ B	___ C	___ D	___ E
14.	_X_ A	___ B	___ C	___ D	___ E
15.	___ A	___ B	___ C	_X_ D	___ E
16.	___ A	___ B	___ C	_X_ D	___ E
17.	_X_ A	___ B	___ C	___ D	___ E
18.	___ A	___ B	_X_ C	___ D	___ E
19.	___ A	_X_ B	___ C	___ D	___ E
20.	_X_ A	___ B	___ C	___ D	___ E

Part III: Resources

Part III: **Resources** focuses on the business management side of project management including controlling time, money, and risk.

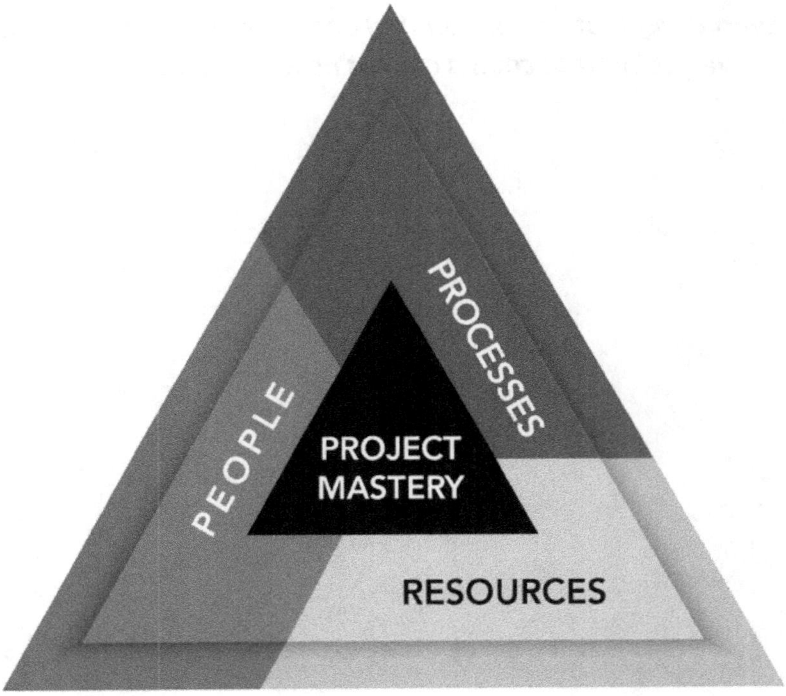

RESOURCES

9. Identifying and Mitigating Project Risk
10. Controlling Project Costs and Budgets
11. Managing Time and Project Meetings
12. Completing and Closing Projects

Everything that can be counted does not necessarily count;
everything that counts cannot necessarily be counted.

—Albert Einstein

Chapter 9

Identifying and Mitigating Project Risk

Overview

> ***"All courses of action are risky, so prudence is not in avoiding danger (it's impossible) but calculating risk and acting decisively. Make mistakes of ambition and not mistakes of sloth. Develop the strength to do bold things, not the strength to suffer."***
>
> —Niccolo Machiavelli, The Prince

Risk is a significant element of any project and can be identified during all phases of a project's life cycle. The original scope, cost, and schedule baselines all take risk into account, including the perceived risks categorized by probability and impact. Project managers must find ways to identify and manage risks and develop contingency plans as if the risks were to occur.

This chapter focuses on tools and techniques project managers use for risk planning and management. After completing this chapter, you will know how to:

- Take a proactive approach to analyzing project risk and planning for it
- Discuss industry models of risk management
- Identify and prioritize risk factors
- Apply risk analysis and management planning concepts to a project
- Recognize the importance of making ethical project decisions

The competencies associated with this chapter include:

Resource Management	Understands Systems	Management Controls

Take Your Temperature for Identifying and Mitigating Project Risk

The following assessment will help you identify what you know about identifying and mitigating project risk and the areas where you can improve. With you and your organization in mind, read each statement below, and consider how much you agree with it. Write a number from 1 to 10 for each item, where **1** means you **strongly disagree,** and **10** means you **strongly agree**.

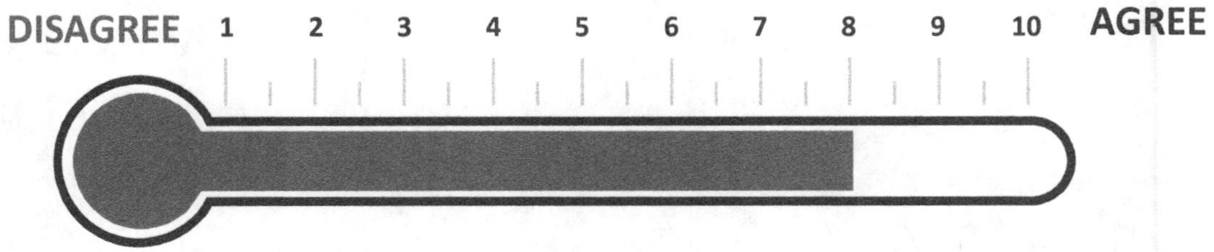

_____ 1. I design contracts and agreements to avoid risk and litigation.

_____ 2. I avoid risk by maintaining a safe working environment.

_____ 3. I know the importance of moving uncertainties to quantifiable risks.

_____ 4. I am skilled at planning for project risk.

_____ 5. I can identify and document project risk.

_____ 6. I can analyze project risk factors.

_____ 7. I can quantify project risk factors.

_____ 8. I understand risk and ethics and the importance of risk and ethics spotting.

_____ 9. I recognize the value of making ethical project decisions.

_____ 10. I expose potentially detrimental risks and unethical behaviors.

_____ **Total**

Take a moment to reflect on the following before you move forward in the chapter.
In your work, what problems have you experienced or observed related to identifying and mitigating project risk at work?

Take a few minutes to also reflect on your self-assessment. List two or three areas you want to improve.

As you progress through this chapter, consider what actions you can take to demonstrate competence in the following three competency areas:

1. Resource Management. Demonstrates awareness of technical resources; knows how to apply resources to achieve desired outcomes.

2. Understands Systems. Grasps complex interrelationships and interdependencies.

3. Management Controls. Ensures the integrity of the organization's processes; promotes ethical and effective practices.

Identifying and Mitigating Project Risk #9.1

Design Agreements to Avoid Risk and Litigation

A **contract** is a **legally enforceable agreement** that sets forth the obligations of two parties to each other. If either party to a contract breaches the obligations, it can face legal sanctions.

Avoid risk, legal issues, and litigation before a project starts by taking the time to think through the elements of a contract or formal agreement before signing it. As the famous quote from Benjamin Franklin says, "An ounce of prevention is worth a pound of cure."

In the construction industry, a well-written professional service contract is one that outlines a precise understanding of the duties and responsibilities of each party to the process, the client, the design professionals, and the contractors. This leaves as little room as possible for misinterpretation.

No matter the type of project, your goal is the same as in the construction industry; create agreements that leave little room for misinterpretation. You can often avoid or minimize your organization's exposure to costly lawsuits by including appropriate legal clauses in the planning stages. Here, as in many areas of an enterprise, knowledge is power.

In this concept, we discuss issues for your consideration and recommend you consult your organization's legal team when you are in doubt about anything in an agreement.

Oral and Written Agreements

A **verbal contract** is one that is based on the spoken word; a **written contract** is in writing. Both may be binding and thus enforceable. It is more difficult to prove the precise terms of oral agreements. The best approach for project managers is to get an agreement in writing at the earliest opportunity, and except in rare cases, before doing the work.

Binding Contracts Have Five Elements

An enforceable contract must have **five specific elements** for it to be *binding*, that is, to place the parties under a legal obligation to perform their contractual duties. You must pay careful attention to each of these five items when you **negotiate** and form a contract.

The five elements are as follows:

- An **agreement** requires an offer and an acceptance. An offer is one party's tender or proposal to perform an act or service in return for some form of consideration from the other party. Acceptance is the approval or favorable reception of the offer by the party to whom the offer is made.
- There must be **consideration** that each party gives; something of value. Consideration is the inducement term in the contract: a mutual promise to do or give something that constitutes an exchange of value to bind the contract. For example, a construction contractor promises to build a house, and the client promises to pay the contractor.

- Finally, **legal form** makes a contract enforceable because the terms of the agreement are neither illegal nor against public policy. Most contracts are enforced according to their terms, but a payment clause with a high-interest rate might be unenforceable. In some jurisdictions, only the offending clause is struck; in others, the entire contract might be voided. To prevent this from happening, contracts often contain a Severability and Survival clause.
- A contract must be executed by **competent parties**: namely, parties of legal age, mentally competent, and when representing business entities, they have the authority to bind or make commitments for those entities. The company must be qualified to do business in the state or country in accordance with state or national law.
- A contract must also have a **legal purpose**. An agreement to undertake illegal activity is not enforceable in court. For example, a contract to design a building that contains known and intentional violations of building codes or ADA requirements would fall into this category.

Thoughtful Planning is Essential

A well-planned project scope and schedule are essential to a successful project. **Careful project planning is the best defense against litigation**.

Regardless of the industry or type of project, and regardless of a project's size or complexity, successful project managers keep the project's objectives in mind. This *must* include serious consideration of the legal problems that could crop up as a result of a project.

Are you aware of the many legal clauses you should consider as you plan a project? Do you know how to incorporate them into a contract when necessary?

Countless situations can lead to legal action being taken against you, your organization, and subcontractors. Here are some examples:

- **Changed conditions**. Even with careful planning, what happens when actual project conditions vary from those expected? For example, what happens when the scope of work varies from the agreement, and the work must still be done?
- **Job site safety issues are paramount**. What happens when injury or something more serious occurs? Is your organization protected? What happens if your organization has "deep pockets?" Surprise! You may find yourself in court, even if you yourself are removed from any direct responsibility for the situation.
- **Billing and payment**. What happens when a client will not pay? Is there a legal clause in place? Are the terms and conditions clearly defined?
- What kind of claims can be made when **a delay in the project** occurs? Sometimes delays are beyond anyone's control, including careful planners. What happens in that situation?
- **Environmental and health hazards no one expected.** What happens if the project encounters a toxic substance, or one is discovered? Are you prepared to deal with a hazard no one can predict?
- **Limitation of Liability clause.** Do you know what it is and where to employ it?

What to do:	Other actions:
☐ Review the five elements of a contract that make it binding.	☐ ______________
☐ Consider situations that could go wrong and lead to legal action being taken against you, your organization, and subcontractors.	☐ ______________
☐ Plan to conduct a risk analysis and planning audit to avoid litigation.	☐ ______________

Remember

- ✓ Upfront planning can save on costs down the line.
- ✓ Get legal help if you are not sure about the legality of an agreement.
- ✓ Risk analysis and planning audits are proactively performed to ensure that what could go wrong with a project has been recognized and that actions were or will be undertaken to prevent them.

Enhance Your Learning

Read the following article about legal risk in project planning: Project management legal issues.

Bright Hub PM. (2022). *Project management legal issues: Watching out for Project Management Legal Issues.*	Available at: https://www.brighthubpm.com/risk-management/63360-project-management-legal-issues-are-you-at-risk/

Reinforce Your Learning

Plan to Avoid Messy Litigation.

Consider and describe a project situation where planning ahead made everything go better.

____

____

____

Can you think of a time when a lack of project planning caused problems? Please describe.

____

____

____

Identifying and Mitigating Project Risk #9.2

Avoid Risk by Maintaining a Safe Working Environment

Consider this statistic: **80 out of every 100 accidents are the fault of a person involved** in the incident, according to the Pennsylvania Department of Labor and Industry. Unsafe acts cause four times as many accidents and injuries as unsafe conditions. Results of another study by the DuPont Company (Ward, 2005) show that **96 percent of accidents and illnesses are caused by unsafe acts and 4 percent unsafe conditions**. In many cases, accidents could have been avoided by promoting and maintaining a safe work environment.

When it comes to managing projects and supervising the work of others, one task stands out above the rest. You might think this is to ensure that the organizational goals are met or to get the project done on time and on budget. While these are essential parts of a project manager's job, the most crucial part of the job is to promote and maintain a safe working environment.

Safety is, most certainly, the responsibility of every person in an organization. But ensuring the day-to-day review and implementation of safety procedures usually falls on a project manager, foreman, or supervisor.

In this concept, we discuss the things you need to consider in keeping the project team and work environment safe.

An Unsafe Work Environment is a Risk

An unsafe work environment is a severe risk in any organization and potentially damaging in many ways. Consider the following:

- Human life is always the most critical consideration in any project or organization
- Lack of safety increases injuries, absenteeism, and costs
- Safety issues threaten public relations
- Safety issues damage employee morale
- Safety problems are financially costly

You must make your team aware that safety is a serious concern and that your organization expects all safety protocols to be enforced. Make it known that it is not acceptable for anyone to ignore an unsafe situation. For example, anyone who sees an empty box of latex gloves or earplugs has a responsibility to replace them, and anyone who notices a cabinet blocking a fire extinguisher must report it. While you are ultimately responsible for safety, you must ensure that your entire team knows safety is their job as well.

Project Managers are Responsible for Safety

Project managers are responsible for ensuring that resources promoting safety are used. As a manager, you must ensure that any safety materials, such as gloves, glasses, hard hats, eye protection, eyewashes, safety showers, decontamination units, safety manuals, and more, are readily available and easy to access, and that each person is trained in their use.

It is vital that you, as a project manager, **model safe behavior**. It is also essential that your associates know you expect them to engage in safe acts in the workplace. In fact, if you ignore unsafe behaviors, fail to enforce safety protocols, or do not take timely action when you see any unsafe behavior or condition, you are condoning dangerous behavior.

One study found that workers were more likely to engage in safety behaviors in the workplace if they had a good relationship with their direct supervisor (Michael, Guo, Wiedenbeck, & Ray, 2006). The researchers asserted, "Organizations should foster positive social exchange between their employees and supervisors and enhance the leadership qualities of supervisors to help reduce workplace injuries" (Michael, Guo, Wiedenbeck, & Ray, 2006).

The same study also found that job satisfaction played a role in safety behaviors (Michael, Guo, Wiedenbeck, & Ray, 2006). People who were more satisfied with their job were more likely to follow safety protocols and practice safe behaviors in the workplace.

As a project manager, you can impact workplace behaviors in several ways:

- Providing positive reinforcement by observing positive behaviors and providing feedback (WorkSafeBC, 2008)
- Modeling safe behavior
- Evaluating safety performance during performance appraisals (WorkSafeBC, 2008)
- Intervening immediately when you see unsafe conditions or behavior (WorkSafeBC, 2008)
- Creating an environment that makes employees satisfied with their work

You must view safety as a system. Just as your organization establishes a goal, does research, develops a plan, implements it, and then follows up on the results, you must do the same with your safety procedures.

As you evaluate your existing procedures, ask the following questions (WorkSafeBC, 2008):

- How can you verify that your workers or associates are trained in safety?
- How can you best communicate safety information to your team?
- When did you last update your safety procedures? How often should they be updated?
- What process do you use to identify, analyze, and control risk?
- Do you have an accident investigation process?
- Are workers and associates aware of emergency protocols? Have they practiced them?

Make Team Members Aware of their Safety Responsibilities

Ensure that your team knows they also have a responsibility towards safety. Specifically, they should (WorkSafeBC, 2008):

- Follow established safety procedures and protocols
- Take reasonable care to protect themselves and those around them
- Comply with any applicable government or OSHA regulations
- Report any unsafe conditions or behaviors

As a project manager, you must respond quickly to any unsafe situation or incident. However, it is more important and much better practice to be proactive, actively seeking out unsafe behaviors, conditions, and hazards so you can resolve them before they cause a problem.

What to do:	Other actions:
☐ Enforce safety protocols, or you could be guilty of condoning unsafe behavior.	☐ ______________________
☐ Model safe behavior.	☐ ______________________
☐ Actively seek unsafe behaviors and hazards and resolve them before they cause problems.	☐ ______________________

Remember

- ✓ Unsafe acts cause four times as many accidents and injuries as unsafe conditions, and 80 out of every 100 accidents are the fault of a person involved in the incident.
- ✓ Results of a study by the DuPont Company (Ward, n.d.) indicated that 96 percent of accidents are caused by unsafe acts and 4 percent unsafe conditions.
- ✓ You, as a project manager, are responsible for reviewing safety procedures and ensuring that protocols are put in place and followed.
- ✓ If new workers or associates are assigned to you or if you are moving someone to a new job, verify they are adequately trained in safety procedures for the position.
- ✓ Regularly review your workplace to ensure safety signage is in place, safety equipment is present, and safety procedures are being followed.

Enhance Your Learning

Review the following website by the National Institutes of Health that outlines six safety responsibilities for supervisors.

National Institutes of Health. (2022). *6 Safety Responsibilities for Supervisors.*	Available at: https://www.ors.od.nih.gov/sr/dohs/HealthAndWellness/Pages/Safety-Responsibilities-for-Supervisors.aspx

Reinforce Your Learning

Observe Workplace Behaviors. Take a few minutes and walk around your workplace. Observe what people are doing with a focus on inefficiencies, waste, potential unsafe acts, and unsafe conditions. List possible improvements below or on a separate sheet.

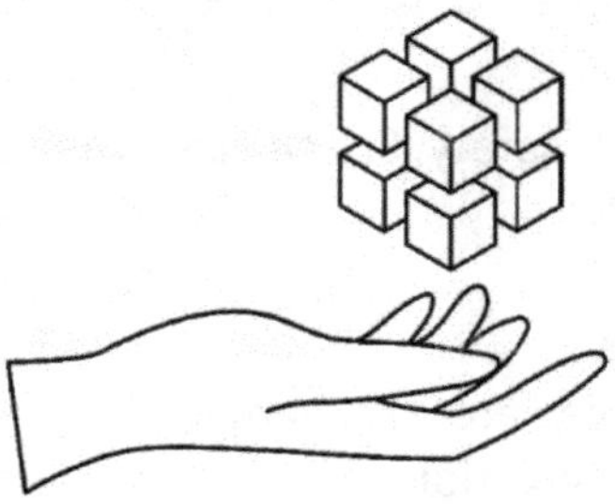

Identifying and Mitigating Project Risk #9.3

Differentiate Between Uncertainty and Risk

Uncertainty refers to situations or projects involving unknown or imperfect information or variables. The Project Management Institute, Inc. (PMI®) defines **project risk** as, "An uncertain event or condition that, if it occurs, has a positive or negative effect on a project's objectives."

Project management is filled with risk and uncertainty. Proactively identifying areas of uncertainty, gathering sufficient information to reduce the uncertainty, establishing risk probabilities, and then managing risk are responsibilities shared by all team members.

Risk management starts early in the project planning phase, should be subject to ongoing scrutiny throughout project implementation, and will ultimately lessen unexpected events by leaving as little to chance as possible.

In this concept, we discuss how project managers must gather information and work to transform uncertainties into quantifiable project risks.

Uncertainty

Uncertainty in the context of risk management is that point in a project when you or your team lacks sufficient information or experience to determine where potential problems, barriers, or variables to the project could cost time or money.

Risk Avoidance

Your challenge is to use available expertise and gather more information, so you can move from uncertainty to identifying potential pitfalls and evaluating them for possible loss to the organization, and from there to calculating a probability of occurrence.

A good example is what happened in the Interstate 99 project. From a cost-benefit perspective, that project had many questions: How much information is required to make a decision to predict sub-surface conditions? When do you drill and how many borings? Again, the goal was to move from a stage of uncertainty to one where the team could calculate potential risks and make decisions.

The opening of Interstate 99 in Centre County, Pennsylvania, was delayed nearly four years by the discovery of iron pyrite during the construction of the highway. This type of pyrite is no problem when it stays underground. However, this highly reactive material became an environmental nightmare when it was exposed to air and water, which caused it to leach sulfuric acid into a nearby stream and groundwater.

If additional test borings had been completed in the planning phase, the uncertainty of the geological substructure might have been identified and appropriate risk migration measures taken. Instead, the nearly four years of embarrassing remediation is referred to as "The $100 million dollar mistake on Skytop Mountain." The personal and emotional cost may have felt even greater to residents of the area, who saw well water go bad, and the value of their homes plunge.

Trying to move from uncertainties to risk probabilities should occur when you initially decide whether to pursue a project. You should consider those items that represent risks to the project, including identification and analysis of both internal and external risks.

Uncertainty versus Risk

In summary, project managers and teams deal with uncertainty and risk. Risk relates to events that can be quantified but not controlled. Uncertainty refers to that which is unknown. Risk is dealt with by the application of decision science; uncertainty is dealt with by gathering additional knowledge.

A risk-based decision is generally made based on expected value or an averaging of the outcome over time. An uncertainty-based decision is usually made based on whether an adverse outcome is acceptable or catastrophic, such as the failure of a dam, collapse of a building, or long-delay and environmental and human impact of an interstate's completion.

What to do:	Other actions:
☐ Understand the benefits of moving uncertainties to quantifiable risks	☐ ______________________
☐ Gather sufficient information to move uncertainties to quantifiable risks	☐ ______________________
☐ Apply risk-based decision making and basic probability concepts to project variables.	☐ ______________________

Remember

- ✓ Risk relates to events that can be quantified but not controlled, and it is dealt with by the application of decision science.
- ✓ Uncertainty relates to that which is unknown, and it is dealt with by gathering additional knowledge.
- ✓ One of the jobs of project managers is to facilitate the gathering of sufficient information needed to move uncertainties to quantifiable project risks.

Enhance Your Learning

Watch the following 5-minute video which describes uncertainty and risk from a business perspective.

Bizconsesh. (2018). *Risk vs. Uncertainty..*	Available at: https://www.youtube.com/watch?v=aX6Cr6enozw

Reinforce Your Learning

Uncertainty to Quantifiable Project Risk. Sometimes we pursue projects without a great deal of thought given to uncertainties, and the effort that may be required to move them to quantifiable project risks. Consider an example you experienced or observed related to project uncertainty where sufficient information related to the uncertainty was not gathered. Use the space below or another sheet of paper to describe how this affected the project.

Identifying and Mitigating Project Risk #9.4

Plan for Project Risk

Proper **identification** and **characterization** of project **risks** are two of the most significant activities associated with successful projects. Missing a risk completely, underestimating its impact, or underestimating the probability it will occur can all create problems for any project.

In this concept, we discuss the inputs, tools, and techniques for beginning the process of identifying and characterizing project risks. In the concepts that follow, we will cover the steps for developing a project risk plan, which includes evaluating the potential severity of the risks, and examining options for risk elimination or mitigation.

Inputs for risk planning and management include the following kinds of information:

- Project management plan
- Schedule management plan
- Project charter
- Stakeholder register
- Enterprise environmental factors
- Organizational process assets
- Resource calendars

The following are **tools** and **techniques** used to identify and characterize risk:

- Brainstorming
- Documentation review
- Interviewing
- Data and analytical techniques
- Expert judgment
- Meetings

Factors to Consider in Risk Planning

Analytical methods employed in risk planning depend on the size and complexity of a project, the **risk appetite** of the **sponsor or client,** and the nature of the strategic risk to the organization. These vary among projects, organizations, and sponsors/clients.

You will need to know your sponsor or client to determine their **tolerance for risk**. Some elect to **purchase insurance** as a method to **deflect** risk. The less risk tolerance they have, the more risk vetting you must do. This element could be added to a stakeholder analysis matrix and a stakeholder register.

Depending on how directly a project is associated with one or more strategic objectives, the project may impact other projects that also support strategic objectives. You must be in tune with your organization's strategic objectives, and how a project supports those objectives, to calculate risk. The project manager is typically charged with leading risk planning and facilitating team involvement to proactively focus on risk management throughout the project's life cycle.

The Risk Planning Process

No matter how well you assess risk before you accept a project or how well you plan for risk at the beginning of a project, you will discover that such pre-planning is not sufficient alone. Even the best-planned project will be susceptible to random events. You manage risks by identifying them in advance, communicating common pitfalls, assessing possible impact, and developing possible mitigation strategies to address them. However, you must also be prepared to facilitate and manage the process as information develops.

The following discussion describes typical elements to consider in risk planning and gives several tips for each.

Identify Risks. The ability to recognize that an adverse event or risk potentially exists can be a hurdle. Here are tips that will help you and the team identify risks:

- Provide the team with checklists of concerns or issues from past lessons learned to alert them to potential problems on the current project.
- Assign responsibility to team members to scrutinize the work breakdown structure and challenge the assumptions made on each item. Ask questions such as, Do we have the resources, facilities, and processability to accomplish the task?
- Review the schedule network. Are the dependencies between tasks clearly shown? What tasks form the critical path? Are these receiving priority attention?
- Ask an independent expert to look at your project plan and suggest things you may have missed.
- Invite the team to be candid about unspoken issues that may be causing them anxiety. Ask people: What is making you lose sleep? Opening this dialog may help identify the most likely issues and rally the team.

Communicate Risks. A common pitfall is to simply list adverse events or risks generically, such as shipping delays, test failure, and late payment. This may be adequate for a few experts, but the rest of the team may be confused or unaware that the risk is something that should concern them. Management, likewise, may presume one meaning when you might have a different problem to solve.

Here are tips that will help you and the team communicate effectively:

- Build a list of adverse events or risk statements and preserve them in a document, spreadsheet, or database file.
- Group similar items and identify an owner for each.
- Establish tools that will enhance team communication on the progress of these adverse events or risks.

Assess Risks. Next, assess or quantify each risk by determining the essential factors of the risk. The team's input will be valuable. However, the project manager must ultimately take responsibility.

Here are the essential criteria to consider:

1. What is the importance or severity of the potential risk or adverse event to the project and client?
2. What is the likelihood or probability that the risk or adverse event will occur?
3. What prevention or detection measures can we take to mitigate the risk or adverse event, and how effective will the measures be in achieving project success?

Mitigate Risks. Your ability to act on risks will require forethought and planning. If you need to be convinced of this, remember that a plan is the result of a thought process that employs judgment to make decisions well in advance. Where there is a risk, sooner or later, you will need to make a decision. You can choose the most effective action plan when you have sufficient time.

A successful plan does not rely on a single approach. Your risk management strategy must assess the severity of the risk to your project, the likelihood or probability that it might occur, and the effectiveness of prevention and detection measures.

Action Plans. After the assessment, your next step is to define the appropriate actions and contingency plans to mitigate each risk.

In a well-planned strategy, the team should have identified risks or adverse events, quantified them, defined appropriate mitigation plans, and have a clear understanding of assigned responsibility.

What to do:	Other actions:
☐ Identify and categorize the severity of possible risks.	☐ ______________
☐ Determine sponsor/client/stakeholder tolerance for risk.	☐ ______________
☐ Determine if purchasing insurance is appropriate to deflect selected risks.	☐ ______________

Remember

✓ Proper identification and categorization of risks are significant activities associated with successful projects.

✓ Missing a risk completely, underestimating its impact, or underestimating the probability it will occur can result in problems.

✓ Analytical techniques employed in planning for risk depend mainly on the size and complexity of a project, the risk appetite of the sponsor/client, and the nature of the risk to the organization.

Enhance Your Learning

Review the following article by T. Morphy of stakeholdermap.com to learn more about common project risks.

Morphy, T. (2008). *20 Common Project Risks - Example Risk Register.*	Available at: https://www.stakeholdermap.com/risk/register-common-project-risks.html

Reinforce Your Learning

Common Project Risks. As a first step in building a database of project risks, discuss the types of risks you have experienced with projects you managed or observed. List the risks that appear in more than one project and give an example of potential problems.

Technical risk—it may not work:

Schedule risk—it may not get done on time:

Budget risk—it may not get done within budget:

Contractual risk—it may expose us to excess liability:

Business risk—it may expose us to adverse publicity:

Other risk:

Identifying and Mitigating Project Risk #9.5

Document Project Risk

As discussed earlier in the course, it is often possible to avoid or minimize your organization's exposure to lawsuits by including appropriate legal language in a project's planning phase. As we said, **careful project planning is the best defense against litigation.**

Successful project managers recognize that planning is the key to success, and planning includes documenting all potential risks, so you can track and communicate risk information with stakeholders and the team throughout the project.

In this concept, we discuss two tools for documenting and communicating risk information— the probability and impact matrix, and the risk register.

The effect of risk on a project's schedule, cost, and quality creates an array of options. In a typical project, the most likely options are expected to occur. However, this is sometimes wishful thinking. To better understand the complexities, the probability and impact of the risks occurring are examined.

Probability and Impact Matrix

For large complex projects, project managers often complete a probability and impact matrix like the example that follows. The matrix shows the correlation between the probability of a risk occurring, and the impact it would have on a project if it did.

In the matrix, the best situation is to have nothing in the red blocks. For anything in the red blocks, you will need to develop a strong mitigation plan to minimize or eliminate the chances of that risk occurring.

Probability and Impact Matrix Example

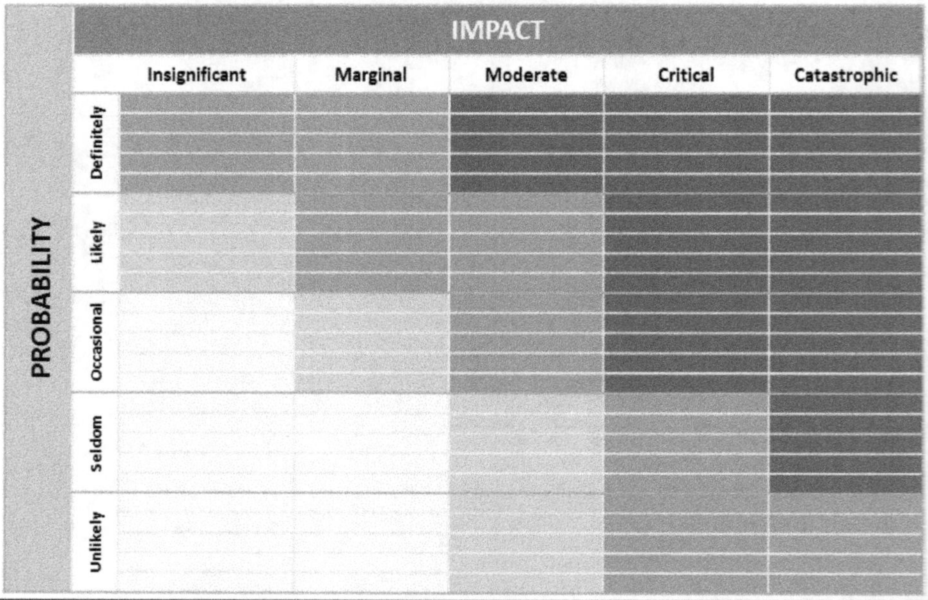

Risk Register

One of the most useful tools for tracking and documenting risk in any project is a risk register, as illustrated below. You use it to log all the risks you can identify during the planning phase of the project and to define or characterize their probability and potential impact on the project if they were to occur.

To develop the register, have the team brainstorm all the risks they can imagine for the project. Then rank each risk by how critical it is to the project. Criticality is determined by the probability of the risk occurring together with the impact on the project if it does.

In our example register, the first item "Resources not available" has a High probability of occurring and will have a High impact if it does. This is a risk you must mitigate (minimize the probability of occurring) and develop a strong contingency plan for action should it happen.

It is advisable to define the probability and the impact during the planning phase so you will not have to think through this information later when you might be dealing with a problem. Spend your time addressing the most critical risks first.

Risk Register Example

Ref. #	Risk Description	Probability	Impact	Other
1	Resources not available	H	H	
2				
3				

You may want to include a "categorization" column on the risk register. The information you track in this column can be what makes the most sense for your project or organization. For example, you could track whether the risk is internal or external, or something to do with technology.

Categorizing risks enables you to construct a risk breakdown structure (RBS), as illustrated below. In this example, the categorization is the role that identified the risk.

The layout of the register depends on what best conveys your message to the client/stakeholders. The key is for stakeholders to see what risks have been identified, so they can validate that all significant risks are accounted for or point out that some risks are missing. Do not forget that we live in a litigious society.

Risk Breakdown Structure (RBS) Example

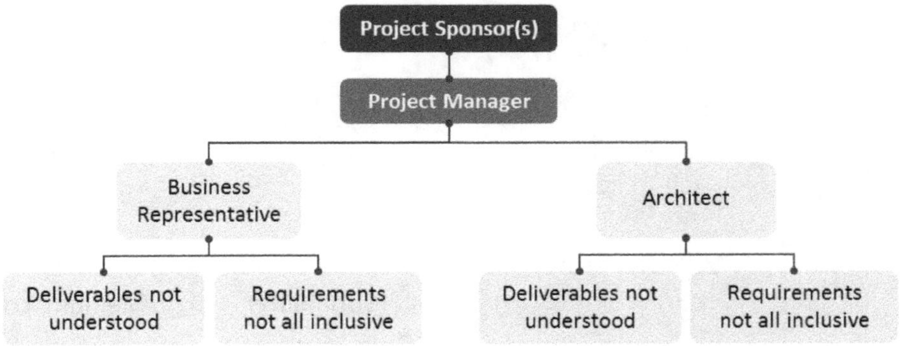

What to do:	Other actions:
☐ Develop a risk register for your project to ensure that all risks are accounted for (logged) and categorized.	☐ ____________
☐ Review the risk register with key stakeholders to ensure all aspects of risk are included.	☐ ____________
☐ Discuss mitigation and contingency plans for the most critical risks.	☐ ____________

Remember

- ✓ Tools such as the risk register and risk breakdown structure help you identify, categorize, and plan for project risks.
- ✓ The project team should develop mitigation and contingency plans for the most critical risks.
- ✓ Risk assessment continues throughout the life of a project. As the project changes, the risks also change.

Enhance Your Learning

View the following 10-minute video by Ally Business Developers about risk management.

Bender, M. (2013). *Project Risk Management – Simplified!*	Available at: https://youtu.be/O6sVZ5ZwST4

Reinforce Your Learning

Risk Register. Use the risk register that follows to log the risks associated with a typical project, such as planning a celebratory event. Based on your experience and brainstorming with others, list all the risks you can think of, then assess each risk's probability and impact.

Risk Register

Project Name: ______________________________

Ref. #	Risk Description	Probability*	Impact*	Other
1				
2				
3				
4				
5				
6				
7				

* **Probability:** High, Medium, Low

* **Impact:** High, Medium, Low

Identifying and Mitigating Project Risk #9.6

Conduct Risk Analysis and Planning Audit (RAPA) Part 1

RAPA Part 1

Project Name:

	A.	B.	C.	D.	E.	F.	G.
Ref. #	Key Project Activities	Potential Adverse Event(s)	Potential Cause of Adverse Event(s)	Potential Effects of Adverse Event(s) on the Project and Customer	Severity (1-5)	Estimated Impact in Time and Money	Likelihood (1-5)

Think of a risk analysis and planning audit (RAPA) as an ongoing process throughout your project. The objective of the process is to identify potential problems—and identify them as early as possible—and **not** pass them on.

In this concept, we discuss the first part of the risk analysis and planning audit process by showing you how you can refine and expand the information you have gathered about project risk. This will help you plan for the risk should it occur, which we will cover in the next concept (RAPA Part 2).

The RAPA Process

In the previous concept, you learned how to identify and log possible risks for the project using a risk register. You now know the initial identification of risks should take place in the planning stage of a project and be documented in the risk register.

Now you want to refine and expand the risk information to clarify what characteristics, requirements, or specifications are important to the sponsor/client/stakeholders and organization. You will want to identify the project requirements you must meet.

Guidelines for a project risk analysis and planning audit includes completing and updating your **risk register** and then completing a **risk analysis and planning audit (RAPA).**

RAPA has two parts:

- Identifying and quantifying **project risk factors (RAPA-Part 1)**
- Developing an appropriate **risk response plan (RAPA-Part 2)**

RAPA Part 1

The following is an example of how you document risk factors in RAPA Part 1:

Risk Analysis and Planning Audit Part 1 Example

	A.	B.	C.	D.	E.	F.	G.
Ref. #	Key Project Activities	Potential Adverse Events	Potential Cause of Adverse Event(s)	Potential Effects of Adverse Event(s) on the Project and Customer	Severity (1-5)	Estimated Impact in Time and Money	Likelihood (1-5)
1	Resources not available.						
2							
n							

A. **Key Project Activities:** Enter the project activities. Make the descriptions as clear and concise as possible. Be sure to include all the activities. If the Key Project Activities you identify include multiple activities and the activities differ significantly, it is desirable to list the activities separately.

B. **Potential Adverse Event(s):** An adverse event is a design error or change in the project which prevents it from being completed on time and within budget. Typical adverse events for a simple project might include workers not showing up, materials delayed, mechanical system problems, and delayed start date. The adverse event is typically expressed in physical terms, not in terms of what the client might experience.

C. **Potential Cause of Adverse Event(s).** Identify possible causes (root causes) of the potential adverse events with the objective of potential prevention strategies to minimize the occurrence.

D. **Potential Effects of Adverse Event(s) on the Project and Customer**: The potential effect is what the customer or client might experience as a result of the adverse event. List all conceivable effects, including violations of government regulations. Group together any effects that are similar or have equal severity.

E. **Severity:** A subjective estimate or judgment of how severe the customer or client will perceive the impact of the adverse event or how it might impact the project. For discussion purposes, the scale of 1 to 5, shown below, is used.

Rating	Severity or Impact Criteria
1	**Negligible severity.** It is unreasonable to think that this effect would impact the project's completion. The client will probably not notice the adverse event.
2	**Mild severity.** The effect would cause only a slight annoyance. The client will not notice any project performance change. Correction can be done before completion of project.
3	**Moderate severity.** The client will notice and tolerate adverse event. Correction is inexpensive with little project delay.
4	**High severity.** The client will notice an intolerable effect. Delays are expensive. Areas possibly subject to government regulations or safety hazards.
5	**Very high severity.** Dangerous effect may occur without warning. Nonconformance to laws.

F. **Estimated Impact in Time and Money:** An estimate of the impact in terms of the magnitude of time and money associated with the potential adverse effect.

G. **Likelihood:** A subjective estimate or judgment of the probability that the cause and, therefore, the adverse event will occur. For discussion purposes, the scale of 1 to 5, shown below, is used.

Rating	Likelihood or Probability Criteria
1	It is highly unlikely that an adverse event will occur.
2	The adverse event would occur only under rare circumstances.
3	Adverse event is somewhat likely.
4	Adverse event is very likely.
5	Almost certain that an adverse event will occur.

What to do:

- ☐ Update your risk register so it is current for all possible risks to your project.
- ☐ Identify and clarify possible adverse effects and quantify their severity.
- ☐ Determine the likelihood that adverse events will occur.

Other actions:

- ☐ ____________________
- ☐ ____________________
- ☐ ____________________

Remember

- ✓ A risk analysis and planning audit is an ongoing process throughout your project.
- ✓ The initial step is to identify and log possible risks in your risk register.
- ✓ Completing a risk analysis and planning audit helps to identify potential problems early on, and not pass them on.

Enhance Your Learning

Review the project risk functional probability and impact matrix template by Bright Hub PM.

Bright Hub PM. (2022). *Project Risk Matrix Template - Probability and Impact.*	Available at: https://www.brighthubpm.com/templates-forms/104889-using-a-probability-and-impact-matrix-for-project-risk-management/

Download a template containing the Risk Analysis and Planning Audit: RPA – Part 1.

Centrestar. (2022). *Risk Analysis and Planning Audit: RAPA-Part 1.*	Available at: https://app.box.com/s/7um2qchru95j2y06v3z1apkmc7nz1t4y

Reinforce Your Learning

Risk Analysis and Planning Audit Part 1. Continue work on a typical project, such as planning a celebratory event. Refer to your risk register from the previous concept and complete Part 1 of the Risk Analysis and Planning Audit (RAPA) worksheet that follows.

RAPA Part 1

Project Name: ______________________________

	A.	B.	C.	D.	E.	F.	G.
Ref. #	Key Project Activities	Potential Adverse Events	Potential Cause of Adverse Event(s)	Potential Effects of Adverse Event(s) on the Project and Customer	Severity (1-5)	Estimated Impact in Time and Money	Likelihood (1-5)
1							
2							
3							
4							
5							
6							
7							
8							
9							
10							
11							
12							
13							
14							
15							

Conduct Risk Analysis and Planning Audit (RAPA) Part 2

RAPA Part 2

Project Name:

	H.	I.	J.	K.	L.	M.
Ref. #	Recommended Actions to Eliminate or Enhance Prevention or Detection	Effectiveness (1-5)	Risk Priority Number (RPN)	Responsibility for Action Accepted by	What Actions Can You Take if It Does Occur?	Action Priority (A-E)

Now that you have identified and logged all potential risks associated with your project, you can begin the process of planning what to do if the risks should occur as the project progresses.

In this concept, we discuss how to plan for risk. We will do so by building on the work you did in Part 1 of your risk analysis and planning audit.

Tools and Techniques for Risk Response Planning

The following are some of the tools and techniques you can use when planning your response to project risks:

- Acquire needed goods or services from sources outside the immediate project.
- Develop contingency plans, which involves defining steps to take if an identified risk event should occur.
- Avoid risk events by changing the planned design or approach.
- Obtain insurance or bonding to deal with selected categories of risk.

Developing a Risk Response Plan

Risk response planning involves defining the steps for responding to the risks you have identified for a project. Responses typically fall into one of three categories:

- Avoidance
- Mitigation
- Acceptance

Avoidance refers to eliminating a specific threat, usually by removing the cause. The project team cannot eliminate all risk, but specific risks can often be eliminated.

You can **mitigate** project risk in two ways. You can reduce the expected monetary value of a risk by reducing the probability of its occurrence, for example, by using technology to lessen the probability that the project will not work. You can also reduce the cost of the risk, for example, by buying insurance.

Acceptance refers to accepting the consequences of a risk. Acceptance can be active; for example, by developing a contingency plan to execute should the risk event occur. Or acceptance can be passive; for example, by accepting a lower profit if some activities overrun.

RAPA Part 2

The illustration below is an example of the RAPA Part 2 process. Part 2 focuses on how you will respond to each risk you have identified.

Risk Analysis and Planning Audit Part 2

	H.	I.	J.	K.	L.	M.
Ref. #	Recommended Actions to Eliminate or Enhance Prevention or Detection	Effectivenes s (1-5)	Risk Priority Number (RPN)	Responsibility for Action Accepted by	What Actions Can You Take If It Does Occur?	Action Priority (A-E)
1	Authorize Overtime					

H. **Recommended Actions to Eliminate or Enhance Prevention or Detection.** Enter recommended actions for reducing the severity and likelihood that an adverse event will occur, or for improving the effectiveness of the prevention or detection methods.

I. **Effectiveness**. A subjective estimate or judgment of how effectively the prevention or detection measure will eliminate potential adverse events. It quantifies judgment of the soundness of a design feature or inspection procedure, the validity of a test, the sufficiency of a final test, and so on. For discussion purposes, the scale of 1 to 5, shown below, is used.

Rating	Effectiveness Criteria
1	The prevention or detection measure is foolproof. There is no chance that the adverse event might still occur.
2	The likelihood of the adverse event occurring is low.
3	Moderate likelihood. The prevention or detection method is effective only sometimes; the cause occurs repeatedly.
4	High likelihood. The prevention or detection method is effective only sometime; the adverse event occurs repeatedly.
5	Very high likelihood. The prevention or detection measure is ineffective. The adverse event will occur.

J. **Risk Priority Number**. The Risk Priority Number (RPN) is the product of the severity, likelihood, and effectiveness ratings. For example:

Severity		Likelihood		Effectiveness		RPN
4	x	5	x	3	=	60

Every adverse event has an RPN. Through rank-ordering the RPNs, it is possible to determine which adverse event is most critical and assign it a priority for corrective action.

K. **Responsibility for Action Accepted by**. Enter the name of the individual on the project team who has the authority to implement the corrective action.

L. **What Actions Can You Take If It Does Occur?** Identify specific actions you can take if the event does occur.

M. **Action Priority**. Assign priorities to the recommended corrective action by considering the RPN and other factors. Give high priority to corrective actions associated with adverse events of high severity. Below, for discussion purposes, is one method of classification.

R.P.N.	Priority Classification	Description
100-125	A	Action must be taken or implemented before project begins or immediately.
75-99	B	Action must be initiated or resolved or implemented within a week.
50-74	C	Action must be initiated or resolved or implemented within a month.
25-49	D	Action may or may not be pursued.
1-24	E	Action will not be pursued.

Risk Analysis and Planning Continues Throughout a Project

We have said that you must plan for risk at the beginning of a project. But we also stressed that as changes to the project occur, you must repeat the basic cycle of identify, quantify, and determine appropriate response actions. You may become aware of risks at any point throughout the lifecycle of a project and often when changes to the project occur. A risk analysis and planning audit (RAPA) ensures that project risks are recognized and that actions are planned to manage the risks.

What to do:

- ☐ As appropriate in your project, act to avoid, mitigate, and accept risks.
- ☐ Estimate how effectively the prevention or detection measures you plan will eliminate potential adverse events.
- ☐ Assign appropriate individuals the responsibility to implement corrective actions.

Other actions:

- ☐ ____________________
- ☐ ____________________
- ☐ ____________________

Remember

- ✓ Responses to risks typically fall into one of three categories: avoidance, mitigation, or acceptance.
- ✓ Some of the tools and techniques in risk response planning are acquiring needed goods or services from outside sources, contingency planning, avoiding risk events, and obtaining insurance.
- ✓ When project changes occur, the basic cycle of identify, quantify, and determine appropriate response actions is repeated.
- ✓ Adverse events may occur at any point throughout the lifecycle of a project.
- ✓ A risk analysis and planning audit (RAPA) is performed to ensure that risks are recognized, and actions planned to manage them.

Enhance Your Learning

Watch the following 4-minute video about qualitative and quantitative risk analysis.

Fremouw, B. (2014). *Project Management Concept #2: Qualitative Risk Analysis vs Quantitative Risk Analysis.*	Available at: https://www.youtube.com/watch?time_continue=3&v=hx-r0sdp2wk&feature=emb_logo

Download a template containing the Risk Analysis and Planning Audit: RPA – Part 2.

Centrestar. (2022). *Risk Analysis and Planning Audit: RAPA-Part 2.*	Available at: https://app.box.com/file/686380339807?s=toqq4jur65zqpze3p7ifjciwxqum3ejj

Reinforce Your Learning

Risk Analysis and Planning Audit: Part 2. Continue work on your celebratory event project. Refer to your Risk Analysis and Planning Audit (RAPA) Part 1 exercise and complete Part 2 using the worksheet that follows.

RAPA Part 2

Project Name: ______________________________

	H.	I.	J.	K.	L.	M.
Ref. #	Recommended Actions to Eliminate or Enhance Prevention or Detection	Effectiveness (1-5)	Risk Priority Number (RPN)	Responsibility for Action Accepted by	What Actions Can You Take If It Does Occur?	Action Priority (A-E)
1						
2						
3						
4						
5						
6						
7						
8						
9						
10						
11						
12						
13						
14						
15						

Describe Risk and Ethics Spotting

Ethics are a **set of morals**, **principles**, **values**, and **norms** that shape what is good or bad conduct. Different sets of ethics can relate to different areas of your life. For example, ***business or professional ethics*** are the standards of behavior you follow in the business world.

Personal ethics are the beliefs that shape the decisions you make for yourself. Personal ethics can come from internal factors such as religious beliefs and morals, from external factors such as your family values, the environment you grew up in, your culture, and so on.

Often, when making decisions at work, you need to merge your business and personal ethics, looking for commonalities between them and following both as a single set.

Also, be aware that **ethics** and **legality** are not necessarily the same thing. Laws serve as the bare minimum of ethics, and it is possible to do something legal that is not ethical. For example, it may be legal to promote your own interests by influencing the actions of your organization, but this would be unethical.

In this concept, we discuss ethics and how your ethical beliefs may affect your ability to effectively manage projects and represent your organization.

What is at Risk?

What is at risk in ethical situations is your personal **reputation**, your organization's status, and ultimately, the **public's trust** in you and your organization's ability to do the right thing. What is also at risk is your ability to successfully manage projects. It is often not enough to do the right thing. Project managers, organizational leaders, and public officials must avoid even the appearance of impropriety.

Ethical Choices

Ethics is a set of standards that guide our behavior, both as individuals and as project managers and members of organizations. The ethical principles for this discussion are simple standards of right and wrong that we learn as children, such as being honest and fair and treating others with respect.

As a project manager, you must make many decisions. To be effective, you will need to consider how your ethics affect your decision making. For example, how do you **spot project risk** or **possible ethical dilemmas**? How is your **decision-making ability** impacted by your beliefs?

Ethical "Do Nots"

The following are guidelines that project managers follow to help them avoid ethical problems in day-to-day project work:

- Do not use your position to seek personal gain. Examples of seeking personal gain include:
  - Soliciting gifts
  - Making decisions for your financial benefit
  - Using inside information gained through your job to benefit you or your family or friends
  - Using work time or property such as a phone or car for personal reasons
  - Using your official position and accepting compensation to endorse a product or service
- Do not exceed your authority or make promises.

Avoid even the appearance of ethical violations. Take the extra step of making sure your actions could not be unethical. Consider how your actions would read on the front page of the newspaper.

Spotting Risk and Common Ethical Dilemmas

Interpersonal Issues. This sort of ethical dilemma involves a person or group unfairly treating others. This can take several forms. *Discrimination* is the tendency to treat someone differently based upon race, gender, age, or religion. *Sexual harassment* involves unwanted sexual advances or behavior, which is not only unethical, but also illegal.

Ethical problems can also arise because of performance appraisals. The person being evaluated may wish to hide their performance problems. Or the person giving the appraisal may not treat each employee or associate in the same manner.

Safety Issues. An organization may want to keep costs low by cutting corners when it comes to safety. However, it is both unethical and illegal for an organization to force people to work in an unsafe environment.

Conflicts of interest. This ethical dilemma occurs when a project manager has outside interests that could be furthered by influencing the actions of the organization. A project manager may not work in the best interests of the organization; they are focused on personal gain.

Quality Issues. Sometimes organizations want to make their products or services appear better than they are. They may use false advertising or exaggerating or outright lying about a product or service. Organizations may also make a product or perform a service that is unsafe in some way because it is cheaper to do so.

Confidentiality. It is unethical to divulge customer or client information to outside parties without the customer's or client's consent. It is also unethical to share certain organizational information with others, such as details on unreleased projects, products, services, inside trading tips, and non-public information.

Resource issues. Some people are tempted to misuse project and organizational resources. This can be as small as taking home a ream of paper or other office supplies, but it can also involve more significant issues such as misrepresenting your hours or embezzling money. It is also possible to skew data to make you look more productive. All these actions are unethical and can hurt an organization.

What to do:	Other actions:
☐ In your own words, define ethics and how ethics impact decision making. ☐ Explain how something can be legal but not ethical, and vice versa. ☐ Cite at least one potentially detrimental risk or ethical challenge you experienced, spotted, or observed in the workplace.	☐ ________________ ☐ ________________ ☐ ________________

Remember

✓ Ethics are a set of morals, principles, values, or norms that shape what is good or bad conduct.

✓ *Ethics* and *legality* are not necessarily the same thing. Laws serve as the bare minimum of ethics. It is possible to do something legal but not ethical.

✓ Common ethical dilemmas include interpersonal issues, safety issues, conflicts of interest, quality issues, confidentiality, and resource issues.

Enhance Your Learning

Review and download the following project management code of ethics: *Code of Ethics & Professional Conduct.*

Project Management Institute. (2022). *Code of Ethics & Professional Conduct.*	Available at: https://www.pmi.org/about/ethics/code

Reinforce Your Learning

Spotting Potentially Detrimental Risk and Ethical Challenges. Write down at least one potentially detrimental risk or ethical challenge that have you experienced, spotted, or observed in the workplace and describe how it affected a project or work situation.

____

____

____

____

____

____

____

____

Identifying and Mitigating Project Risk #9.9

Recognize the Value of Ethical Project Decisions

One approach to identifying and mitigating risk is to view it as an ethical decision-making process.

An **ethical decision-making** process is essentially a **problem-solving process;** gather the facts, examine the issues, determine who will be affected, generate solutions, decide, and then reflect to ensure you are comfortable with the decision and that it is ethically sound.

In this concept, we discuss ethical decision making as it applies to project management.

Making Ethical Decisions is a Process

Gather facts. Before you can decide, you need to know the facts involved in the situation. Ask questions and gather information before jumping to a conclusion or deciding something is true based on your beliefs about a situation.

Examine the issues. Once you know the facts, examine the issues at hand. For example, ask what is causing a dilemma. Sometimes it is helpful to discuss your thoughts with other project managers or professional associates to help you gain perspective.

Determine who will be affected. When deciding, examine who will be touched by the result, both directly and indirectly. Start by examining the effect on those closest to the problem, which might be your project, the team, your department, or the organization. Then you can move outward and review the situation more broadly. For example, ask how a community, industry, or larger society may be impacted.

Generate solutions. Brainstorm as many answers as you can and then narrow the list to the solutions most likely to help the situation. Keep in mind any laws or codes of ethics you must uphold and consider the consequences of each potential solution. Try to choose a solution with the least number of negative consequences or lower risk of negative results. Consider the integrity of your decision. Is your choice something you would be proud of? How would you feel if your decision were made public?

Reflect on the situation. Once you have decided, give yourself a bit of time to reflect on your choice before you act. Make sure you feel comfortable with what you have decided and double-check to ensure the decision is ethically sound.

Case Scenario—NASA Shuttle Disasters

On January 28, 1986, NASA made the decision to launch the Space Shuttle Challenger, despite reservations voiced by some engineers as to whether the O-rings would function properly in the cold temperatures projected for the morning of the launch. Just seconds after launch, the O-rings failed, and the Shuttle exploded, killing the crew and shocking the nation (Snarski, 2012).

While many factors contributed to the loss, communication played a substantial role. When preparing to launch the Challenger, there was concern among several Morton Thiokol engineers that the O-rings used on the solid rocket boosters had shown a potential for failure when tested during cold temperatures (Rogers Commission, 1986). Various engineers sent multiple memos for years before the launch, warning of the potential problem. Reports show that the memos were mostly dismissed as speculation or not routed to the correct person (Rogers Commission). The situation illustrates two problems: the unwillingness of management to listen to their engineers, and the inability of the engineers to push a critical point to the forefront (Snarski, 2012). Sadly, the cultural problems that lead to the Challenger disaster were still causing problems almost 20 years later.

Following is a summary of an ethics and communication issue in NASA that led to two shuttle tragedies. While the subject matter may or may not directly apply to your daily work, the ethical concepts do.

The Headline

According to an article by MSNBC, "NASA's habit of relaxing safety standards to meet financial and time constraints set the stage for the Feb. 1 [2003] loss of the Space Shuttle Columbia and its seven astronauts, investigators said Tuesday. They warned that the agency's 'broken safety culture' would lead to tragedy again unless fundamental changes are made... the board's report laid out the cultural factors behind NASA's failings. It said NASA mission managers fell into the habit of accepting some flaws in the Shuttle system and tended to ignore or not recognize that these problems could foreshadow catastrophe. This was an 'echo' of some root causes of the Challenger accident, the board said. 'These repeating patterns mean that flawed practices embedded in NASA's organizational system continued for 20 years and made substantial contributions to both accidents,' the report said."

The Story

The night before the Challenger Shuttle launch in 1986, several Morton Thiokol engineers became gravely concerned about the potential malfunction of an O-ring in cold temperatures (Snarski, 2012). After escalating their concerns, they participated in a telephone conference between Morton Thiokol and Marshall Space Flight Center management (Rogers Commission). During this first conference, the engineers explained the potential for the O-rings to fail under cold conditions. They did not have hard numbers, and, according to some records, perhaps did not get their point across.

Shortly after the Shuttle disaster, President Regan formed the Presidential Commission, also called the Rogers Commission, to investigate. During the Rogers Commission investigation, participants in the telephone conference from Marshall Space Flight Center were asked, "So as early as after that first afternoon conference at 5:45, it appeared that Thiokol was basically saying delay. Is that right?" (Rogers Commission). One member replied, "That is the way it came across to me." But another person, attending the same conference, replied, "I did not perceive it that way. I perceived that they were raising some questions and issues which required looking into by all the right parties, but I did not perceive it as a recommendation to delay" (Rogers Commission).

The Rogers Commission learned that later the same day a second telephone conference was held (Snarski, 2012). For this conference, the Morton Thiokol engineers presented charts and data illustrating O-ring performance in various temperatures. Some management questioned the data, and the engineers admitted that it was incomplete and that the O-rings had not been tested specifically at the anticipated flight temperature. However, the engineers, as reported in the Commission review, specifically advised that the lift-off should be delayed. In the end, management decided not to delay the launch based on the following two facts: The O-ring in question was a secondary O-ring, and the testing data was not confirmed. Upon launch, the primary seal failed, as did the secondary seal, resulting in the complete loss of the Shuttle Challenger.

Upon completion of its investigation, the Rogers Commission found that "Marshall Space Flight Center project managers, because of a tendency at Marshall to management isolation, failed to provide full and timely information bearing on the safety of flight 51-L to other vital elements of Shuttle program management" (Rogers Commission). Also, it was found that communication is not just about sending or even about receiving. It is also about the communications climate, and the willingness of participants to listen to those whose opinions they solicit. Thus, the Rogers Commission further stated that "The nation's reliance on the Shuttle as its principal space launch capability created a relentless pressure on NASA to increase the flight rate. Such reliance on a single launch capability should be avoided in the future." Essentially, NASA was working in a political and business climate that left little room for action contrary to that which was planned (Snarski, 2012).

The Impact

There is no one person, or even one group, to blame for the Shuttle disaster. Instead, there were various incidents of inadequate communications—communications that were either unclear, misrouted, unsupported, or untimely. There was also a lack of receptiveness on the part of management. A climate existed that made postponing the launch a negative decision, and thus closed the minds of the Marshall Space Flight Center management to various issues presented. This is an effective, if sad, reminder that communication is a two-way street; a communicator must disseminate clear, concise, understandable information, and receivers must be open to what they are being told (Snarski, 2012).

Seventeen years and 88 Shuttle launches later, on February 1, 2003, tragedy struck again as the Columbia Shuttle flight 107 broke up on re-entry into the Earth's atmosphere due to damage sustained upon launch. During the launch of the Columbia, a piece of foam insulation broke free from the main propellant tank, striking the leading edge of the left wing and damaging the tile, part of the Shuttle's thermal protection system. Upon re-entry, the heat build-up caused the destruction of the Shuttle (Snarski, 2012).

The Columbia Accident Investigation Board (CAIB) embarked on a large-scale investigation to determine the cause of the disaster and make recommendations for the future. The Board reported that "The NASA organizational culture [the basic values, norms, beliefs, and practices] had as much to do with this accident as the foam." Specifically, the Board felt that after the Cold War ended in the 1980s, NASA lost much of its funding, and thus to stay profitable, had to increase its launches while decreasing its costs. This, reportedly, led to an atmosphere that left room for mistakes, such as Columbia (Snarski, 2012).

According to CAIB, concerns over the foam and tile had been expressed before the Columbia launch. However, like the Challenger disaster, the communications on the issue were not timely or specific, and management, given the political climate of NASA, was not willing to listen. The CAIB agreed with the Rogers Commission that failures in communications, incomplete and misleading information, and poor decisions allowed these launches when they should not have (Snarski, 2012).

As the Rogers Commission concluded, "The Commission applauds NASA's spectacular achievements of the past and anticipates impressive achievements to come. The findings and recommendations presented in this report are intended to contribute to the future NASA successes that the nation both expects and requires." Sadly, this cultural problem tarnished the reputation of the United States space program and continued to endanger lives. This situation did draw attention to the much-needed changes in the communication policies of the program and illustrated that effective communication and an ethical, open corporate culture can, in fact, be life-altering (Snarski, 2012).

What to do:	**Other actions:**
☐ Review the Code of Ethics & Professional Conduct–PMI that you previously downloaded.	☐ ______________________
☐ Identify parts of this code that could apply to situations like the Challenger disaster.	☐ ______________________
☐ Consider how adopting this code could help guide your decision making.	☐ ______________________

Remember

✓ The ethical decision-making process is essentially a problem-solving process: gather the facts, examine the issues, determine who will be affected, generate solutions, decide, and reflect to ensure the decision is ethically sound.

Enhance Your Learning

Read the following list of eight ethical dilemmas in project management developed by PMI.

Project Management Institute. (2008). *Ethical project management.*	Available at: https://www.pmi.org/learning/library/ethical-dilemmas-project-management-7084

Reinforce Your Learning

NASA—Code of Ethics. Locate the *Code of Ethics & Professional Conduct–PMI* you downloaded in the previous exercise. Consider the Challenger disaster and identify two or three specific parts of the code that could apply. Consider how the code could have guided the actions of the engineers and managers early on when they first questioned the temperature sensitivity of the O-rings. Consider also how the code could have guided the entire group as they discussed the situation the night before the launch. Consider what you would have done if you were one of the engineers, and then if you were one of the project managers. Consider how the code could help NASA in future projects. List some of your thoughts below.

Identifying and Mitigating Project Risk #9.10

Expose Potentially Detrimental Risk and Unethical Behavior

The ethical actions of leaders directly affect the health of their organizations. Many organizations promote ethical conduct by developing compliance programs that seek to define and highlight the benefits of ethical conduct.

Ethics are sometimes viewed as a soft issue for executives, but there are **concrete steps** that **organizations can take** to promote and **sustain an ethical workplace environment**. In organizations where senior leaders demonstrate ethical leadership, workers are less likely to observe unethical conduct and are more likely to seek advice on ethical issues. Organizations with ethics programs are more likely to report misconduct, to believe that violators are held accountable, and to feel less pressure to violate organizational standards of conduct (Barry & Lunday, 2004).

In this concept, we discuss how recognizing and exposing potentially detrimental or dangerous risk and unethical behavior is an essential component of a project manager's code of ethics and professional conduct. Such actions are vital to organizations because project managers link labor, contractors, shareholders, suppliers, and clients.

Guidelines for Exposing Detrimental Risk and Unethical Behavior

In the course of your work as a project manager, you may uncover potentially detrimental risks or unethical or illegal activity. Exposing such activities is referred to as whistleblowing, and it can be stressful. But if the actions you discover are harmful to others or the organization, it is in everyone's best interest for them to be exposed.

It is illegal for an organization to fire you for "blowing the whistle," but you should consider the situation carefully when deciding what you will do.

Consider the consequences of exposing the risk or unethical behavior. If you expose the behavior, you may damage your relationships with your coworkers or associates, and morale may be affected. However, if *not* exposing the potentially dangerous risk or unethical behavior will ultimately lead to others being hurt, you must expose the situation.

Consider your options. For small risks and unethical behavior, for example, a single employee falsifying a timesheet, it is likely that the matter can be solved in-house. However, for widespread corruption, it may be necessary to go outside the group or organization.

Make sure you have your facts straight. Document your findings, ensuring that any accusations you make are verifiable. Falsely accusing a coworker, associate, or an organization of unethical behavior can tarnish reputations, and it will not reflect well on you and your career.

Follow the proper chain of command. Your first step should be to report a problem to your manager unless your manager is involved in the problem. If you are not successful when speaking with your manager, go to that person's manager. You may also want to speak with other internal authorities such as the organization's safety or ethics department or the department that the situation or ethical problem affects most. Only go to upper management or publicly expose the problem after all other methods of reporting the problem have been exhausted.

When reporting a dangerous situation or unethical behavior, remain calm. Be as specific as possible with your facts and avoid placing blame. Your goal is to report the situation for management's review.

Ethical Decision Making

Ethical decision making requires being aware of your own and your organization's ethical values and applying them when necessary. It involves being sensitive to the impact of your actions and being able to evaluate complex, ambiguous, and incomplete facts.

The following are three components of ethical decision making:

- Commitment
- Consciousness
- Competency

Ethical commitment (or motivation) involves demonstrating a desire to act ethically and to do the right thing, especially when ethics imposes financial, social, or psychological costs. A crisis or emergency tests ethical commitment. You need to be clear about your own ethical values and understand your organization's ethical standards.

Ethical consciousness (or awareness) involves seeing and understanding the ethical implications of our behavior and expressing our values in our daily lives. Understand that people's perceptions are their reality. Understanding and adhering to ethical guidelines helps everyone understand and do the right thing, and especially in situations where the ethical lines may seem fuzzy.

Ethical competency (or skill) involves being competent in making decisions, which include:

- Evaluation—the ability to collect and evaluate relevant facts and knowing when to stop collecting facts and to make decisions based on incomplete or ambiguous information.
- Creativity—the capacity to develop resources for accomplishing goals in ways that avoid or minimize ethical problems.
- Prediction—the ability to foresee consequences and assess the likelihood or risk that persons will be helped or harmed by an action.

What to do:	Other actions:
☐ Be clear about your own ethical values and understand your organization's ethical standards.	☐ ____________________
☐ Understand that people's perceptions are often their reality.	☐ ____________________
☐ Collect facts and understand consequences as part of your decision-making process.	☐ ____________________

Remember

- ✓ The ethical actions of leaders directly affect the direction and health of organizations.
- ✓ People in organizations with ethics programs are more likely to report misconduct.
- ✓ Ethical decision making requires being aware of your own and your organization's ethical values and applying them when necessary.

Enhance Your Learning

Watch this 11-minute video reporting 25 biggest corporate scandals.

List 25. (2013). *25 Biggest Corporate Scandals Ever*	Available at: https://www.youtube.com/watch?v=LrJaOOGJLW0

Reinforce Your Learning

Exposing Detrimental Risk and Unethical Behavior. Based on your experience or observation, describe at least one example of where unethical behavior was, or should have been, exposed, and describe the situation.

Summary

A Final Word: *Identifying and Mitigating Project Risk!*

You have just completed a chapter on *Identifying and Mitigating Project Risk.* Information you reviewed includes:

- Developing a proactive approach to analyze project risk and plan for it.
- Reviewing industry models of risk management.
- Identifying and prioritize risk factors.
- Applying risk analysis and management planning concepts to a project
- Recognizing the importance of making ethical project decisions.

In the next Project Mastery chapter, we will focus on controlling project costs and budgets.

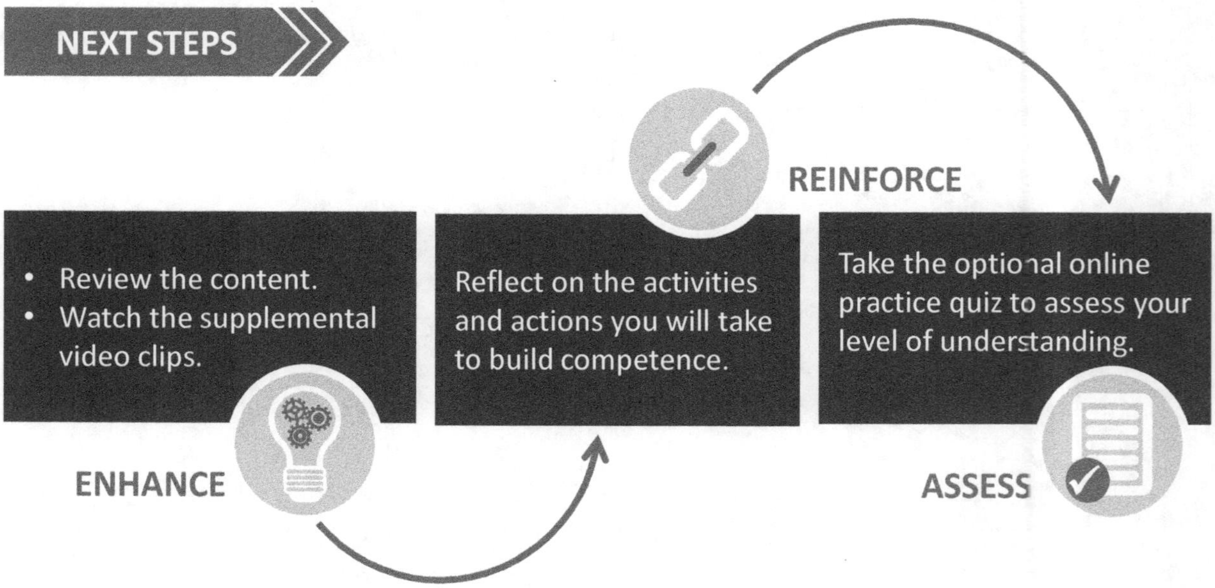

Identifying and Mitigating Project Risk—Recap Checklist Risk is a significant element of any project and can be identified during all phases of the project life cycle. The original scope, cost, and schedule baselines all take risk into account, including the perceived risks categorized by probability and impact. Also important is exposing potentially detrimental risk and unethical behavior.
1. **Design Agreements to Avoid Risk and Litigation** □ Review the five elements of a contract that make it binding. □ Consider situations that could go wrong and lead to legal action being taken against you, your organization, and subcontractors. □ Plan ahead to conduct a risk analysis and planning audit to avoid litigation.
2. **Avoid Risk by Maintaining a Safe Working Environment** □ Enforce safety protocols, or you could be guilty of condoning unsafe behavior. □ Model safe behavior. □ Actively seek unsafe behaviors and hazards and resolve them before they cause problems.
3. **Differentiate Between Uncertainty and Risk** □ Understand the benefits of moving uncertainties to quantifiable risks. □ Gather sufficient information to move uncertainties to quantifiable risks. □ Apply risk-based decision making and basic probability concepts to project variables.
4. **Plan for Project Risk** □ Identify and categorize the severity of possible risks. □ Determine sponsor/client/stakeholder tolerance for risk. □ Determine if purchasing insurance is appropriate to deflect selected risks.
5. **Document Project Risk** □ Develop a risk register for your project to ensure all risks are accounted for (logged) and categorized. □ Review the risk register with key stakeholders to ensure all aspects of risk are included. □ Discuss mitigation and contingency plans for the most critical risks.
6. **Conduct Risk Analysis and Planning Audit (RAPA) Part 1** □ Update your risk register, so it is current for all possible risks to your project. □ Identify and clarify possible adverse effects and quantify their severity. □ Determine the likelihood that adverse events will occur.
7. **Conduct Risk Analysis and Planning Audit (RAPA) Part 2** □ As appropriate for your project, act to avoid, mitigate, and accept project risks. □ Estimate how effectively the prevention or detection measures you plan will eliminate potential adverse events. □ Assign appropriate individuals the responsibility to implement corrective actions.
8. **Describe Risk and Ethics Spotting** □ In your own words, define ethics and how a person's ethics impact their decision making. □ Explain how something can be legal but not ethical, and vice versa. □ Cite at least one potentially detrimental risk or ethical challenge you experienced, spotted, or observed in the workplace.
9. **Recognize the Value of Ethical Project Decisions** □ Locate and review the *Code of Ethics & Professional Conduct–PMI* you previously downloaded. □ Identify parts of this code that could apply to situations like the Challenger disaster. □ Consider how adopting this code could help guide your decision making.
10. **Expose Potentially Detrimental Risk and Unethical Behavior** □ Be clear about your own ethical values and understand your organization's ethical standards. □ Understand that people's perceptions are often their reality. □ Collect facts and understand consequences as part of your decision-making process.

Knowledge Review Test—Questions

Identifying and Mitigating Project Risk

1. Risk identification outputs include the following except:
   A. Decision trees.
   B. Inputs to other processes.
   C. Risk symptoms.
   D. Potential risk events.

2. To be effective, the risk management process should:
   A. Be applied primarily during the concept and closeout phases and to some extent during the implementation and planning phases.
   B. Be applied throughout the project and at all levels of system decomposition and project organization.
   C. Include an assembly of certain stakeholders to identify risks and develop mitigation strategies.
   D. Focus on those risks that senior management finds most critical.

3. An unsafe work environment is a severe risk in any organization and potentially damaging in many ways. Which of the following are considerations?
   A. Human life is always the most critical consideration in any project or organization.
   B. Lack of safety increases injuries, absenteeism, and costs.
   C. Safety issues threaten public relations.
   D. Safety issues damage employee morale.
   E. All the answers are correct.

4. One of the first requirements for effective risk management is:
   A. Clear visibility of the information needed for decision making.
   B. Ownership of the risks that are identified.
   C. Appointment of the project manager early in the process to manage the identified risks.
   D. Project team members who are trained in risk and understand its causes to help construct and implement risk mitigation strategies.

5. Responses to risk threats include all the following except:
   A. Avoidance.
   B. Acceptance.
   C. Mitigation.
   D. Rejection.

6. Project managers and teams deal with uncertainty and risk. Which of the following are true?
   A. Risk relates to events that can be quantified but not controlled.
   B. Risk is dealt with by the application of decision science.
   C. Uncertainty is dealt with by gathering additional knowledge.
   D. All are true.

7. Risk response development is intended to:
   A. Create steps to identify project risks.
   B. Formulate strategies for dealing with adverse events.
   C. Construct a list of previous risks.
   D. Develop measurements to quantify project risks.

8. Categories of risk response are:

A. Technical, marketing, financial, and human.
B. Identification, quantification, response development, and response control.
C. Avoidance, mitigation, and acceptance.
D. Avoidance, retention, control, and deflection.

9. You have been assigned as project manager for a new project. There appears to be many risks on this project, but no one has evaluated them to assess the range of possible outcomes. What needs to be done first?

A. Risk identification.
B. Risk quantification.
C. Risk response planning.
D. Risk monitoring and control.

10. Which of the following is NOT true about project risk?

A. Tools such as the risk register and risk breakdown structure help you identify, categorize, and plan for project risks.
B. The project team should develop mitigation and contingency plans for the most critical risks.
C. Successful project managers recognize that planning is the key to success.
D. Risk assessment continues throughout the life of a project. As the project changes, the risks also change.

11. RAPA is an acronym for:

A. A musical chorus characterized by drumbeats.
B. Risk Assessment Problem Analysis.
C. Rhapsody and Paradise Association.
D. Risk Analysis and Planning Audit.

12. You as a new project manager have assembled your project team, identified 39 risks on the project, determined what would trigger the risks, ranked them on a risk matrix, tested their assumptions, and measured the precision of the data used. Your team is moving forward through the risk management process. What have you forgotten?

A. Simulation.
B. Risk mitigation.
C. Overall risk ranking for the project.
D. Involvement of other stakeholders.

13. Which of the following are true about project risk planning?

A. A risk analysis and planning audit is an ongoing process throughout your project.
B. The initial step is to identify and log possible risks in your risk register.
C. Completing a risk analysis and planning audit helps to identify potential problems early on, and not pass them on.
D. All are true.

14. Which of the following is NOT true?

    A. Responses to risks typically fall into one of three categories: severe, likely, or abnormal.
    B. Some of the tools and techniques in risk response planning are acquiring needed goods or services from outside sources, contingency planning, avoiding risk events, and obtaining insurance.
    C. When project changes occur, the basic cycle of identify, quantify, and determine appropriate response actions is repeated.
    D. Adverse events may occur at any point throughout the lifecycle of a project.

15. Risk tolerances are evaluated to help:

    A. The project manager to gauge the project risk.
    B. The project team rank the project risks.
    C. The team determine the schedule.
    D. Other project managers learn from mistakes.

16. Buying an insurance policy is an example of:

    A. Risk mitigation.
    B. Risk acceptance.
    C. Risk avoidance.
    D. Risk transfer.

17. Which of the following is an attribute or feature of the Delphi technique?

    A. Top-down approach.
    B. Organization development process.
    C. Subject matter expert opinions.
    D. Simulation.

18. A Monte Carlo technique is used to:

    A. Estimate an activity's duration.
    B. Obtain an indication of the risk involved in the project.
    C. Determine the possible schedule issues associated with a project.
    D. Identify possible material quality defects before installation.

19. Which of the following is NOT true?

    A. The ethical actions of leaders directly affect the direction and health of organizations.
    B. People in organizations with ethics programs are more likely to report misconduct.
    C. People's perceptions are often not their reality.
    D. Ethical decision making requires being aware of your own and your organization's ethical values and applying them when necessary.

20. Ethical decision making requires:

    A. Knowing the other part's motives.
    B. Being aware of your own and your organization's ethical values and applying them when necessary.
    C. Accepting the impact of your actions and decisions.
    D. Taking a bottoms-up approach to evaluate complex, ambiguous, and incomplete facts.

Identifying and Mitigating Project Risk – Answer Sheet

1.	_X_ A	___ B	___ C	___ D	___ E
2.	___ A	_X_ B	___ C	___ D	___ E
3.	___ A	___ B	___ C	___ D	_X_ E
4.	_X_ A	___ B	___ C	___ D	___ E
5.	___ A	___ B	___ C	_X_ D	___ E
6.	___ A	___ B	___ C	_X_ D	___ E
7.	___ A	___ B	___ C	_X_ D	___ E
8.	___ A	___ B	_X_ C	___ D	___ E
9.	_X_ A	___ B	___ C	___ D	___ E
10.	___ A	___ B	_X_ C	___ D	___ E
11.	___ A	___ B	___ C	_X_ D	___ E
12.	___ A	___ B	___ C	_X_ D	___ E
13.	___ A	___ B	___ C	_X_ D	___ E
14.	_X_ A	___ B	___ C	___ D	___ E
15.	___ A	_X_ B	___ C	___ D	___ E
16.	___ A	___ B	___ C	_X_ D	___ E
17.	___ A	___ B	_X_ C	___ D	___ E
18.	___ A	_X_ B	___ C	___ D	___ E
19.	___ A	___ B	_X_ C	___ D	___ E
20.	___ A	_X_ B	___ C	___ D	___ E

Balancing the budget is like going to heaven. Everybody wants to do it, but nobody wants to do what you have to do to get there.

—Phil Gramm

Chapter 10

Controlling Project Costs and Budgets

Overview

> ***"'Know your numbers' is a fundamental precept of business."***
>
> —Bill Gates

Everything in business is bound by money. One of management's critical jobs is to manage revenues and expenditures, and they do this mainly through budgetary planning and control. Even amazing projects must have budgets, strive to be cost-effective, and justify expenses. Project managers help develop and manage such budgets. Yet reportedly, **less than** a **third of all projects** were **completed on time** and **on budget** over the past year, and **33 percent of projects fail** because of a lack of involvement from senior management (Capterra, 2015).

The goal of creating and presenting a project budget is to get it approved so the project can move forward. Thus, understanding how to do a budget is an essential part of every project manager's job. Even for leaders without a formal accounting background, knowing how to justify costs is vital. This course walks the non-accountant professional through the budgeting process.

After completing the course, you will know how to:

- List and describe several typical types of budgets
- Explain the general process of budgeting
- Provide examples of costs and benefits
- Discuss standard financial analysis tools and techniques
- Describe how budgets and earned value analysis can help control projects

The development of the competencies and behaviors associated with this course include:

Financial Management and Budgeting	Management Controls	Resource Management

Take Your Temperature for Controlling Project Costs and Budgets

The following assessment will help you identify what you know about controlling project costs and budgets and the areas where you can improve. With you and your organization in mind, read each statement below, and consider how much you agree. Use the numbers from 1 to 10, where **1** means you **strongly disagree** and **10** means you **strongly agree**.

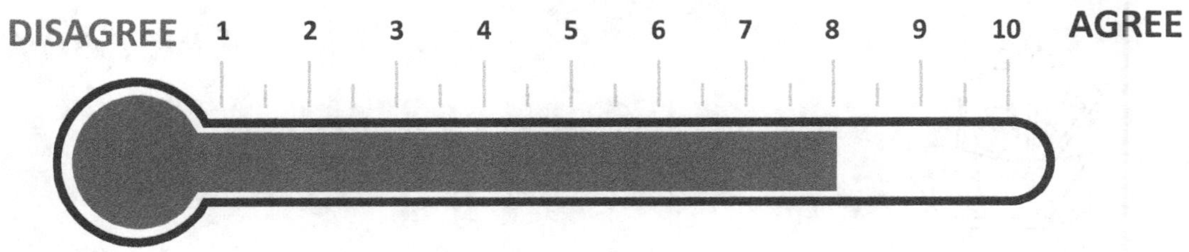

_____ 1. I can explain budget types and how they relate to projects.

_____ 2. I can describe budgeting principles and components.

_____ 3. I am good at forecasting project budgets.

_____ 4. I can estimate project costs.

_____ 5. I do a good job of conceptualizing project benefits.

_____ 6. I can do a break-even analysis.

_____ 7. I understand present value concepts.

_____ 8. I know how to perform an alternative choice analysis.

_____ 9. I am skilled at using budgets for project monitoring and control.

_____ 10. I understand the use of budgets and earned value analysis for project control.

_____ **Total**

Take a moment to reflect on the following before you move forward in the chapter.
In your work, what problems have you experienced or observed related to project costs and budgets?

Take a few minutes to also reflect on your self-assessment. List two or three areas you want to improve.

As you progress through this chapter consider what actions you can take to demonstrate competence in the following three competency areas:

1. **Financial Management and Budgeting.** Understands budget process; prepares and justifies budgets; monitors expenses; manages profit/loss ratios as appropriate.

2. **Management Controls.** Ensures the integrity of the organization's processes; promotes ethical and effective practices.

3. **Resource Management. D**emonstrates awareness of technical resources; knows how to apply resources to achieve desired outcomes.

Controlling Project Costs and Budgets #10.1

Understand How Organizations Use Budgets

Budgeting is a **necessary activity** for **every organization** and **project**. The budgeting process refers to how an organization goes about creating and using the various types of budgets they need. A small organization with a few employees might need just a brief operating budget when they get started. But as projects ramp up and the organization grows, various other budgets might be required.

As an organization grows, managers in the various departments are responsible for their budgets, and they develop their budgets using guidelines from management. Then, all budgets are brought together to complete the organization-wide budget.

The budgeting process is based on discipline and planning (Lotich, 2014). The development of a budget is not arbitrary. It is driven by the organization's vision (what the organization wants to accomplish) and strategic plan (the steps the organization will take to achieve the vision) (Lotich, 2014). When an organization maintains a steady focus on its strategic plan, it can decide where to invest money and work to keep money away from areas that do not support the vision (Lotich, 2014).

In this concept, we give an overview of how the budgeting process works and describe the types of budgets organizations use.

Financial Accounting and Managerial Accounting

To understand how budgeting works in an organization, you need to know a bit about accounting. The two types of overarching accounting in organizations are financial accounting and managerial accounting.

Financial accounting refers to the creation of information for **external** stakeholders, such as shareholders or creditors (Ahmed, 2014). For these people to make decisions about lending money or investing, they need financial information, generally at a high level. This information is usually provided in documents such as balance sheets and cash flow statements.

Managerial accounting provides information to be **used by managers inside an organization**.

Both managerial and financial accounting use the same underlying numbers, but each with a different focus. Financial accounting tends to report the past; it shows what has been happening with an organization financially (Ahmed, 2014).

The type of managerial accounting that project managers and functional managers use is more focused on the future. Managers seek to predict future earnings and fund projects that will turn those predictions into reality (Ahmed, 2014).

Financial accounting is concerned almost exclusively with actual, objective numbers— indicators of past performance. Managerial accounting will often include numbers that are educated guesses.

Budgets Reflect Organizational Goals

Managers use the budgeting process to detail the financial plan that supports the organization's goals. Whether simple or complex, budgets are statements of an organization's anticipated income and expenditures.

Managers budget for the entire organization and the divisions, departments, and projects within the organization.

You can think of a budget as a **quantified plan of action**. Managers create a budget by taking known financial data and then merging it with projections (for sales, income, competition, and so on) to create a financial plan that describes what will be spent and where.

Budgets are **educated guesses** about an organization's expectations regarding various opportunities. They answer questions regarding money, such as: **How much? From what sources? To whom? By when? For what?**

Operating Budgets and Capital Budgets

Two general categories of budgets are operating budgets and capital budgets. They support an organization's operating plans and capital plans.

The *operating budget* is about the present, and details how the organization's daily operations will be handled for the year.

The *capital budget* is longer term. It supports the organization's future plan (expected revenue and growth), how investments will be made, what future projects will be financed, and so forth. A capital budget typically covers five to ten years.

Budgeting Periods

A *budgeting period* is the time a specific budget covers. For example, most organizations create an annual operating budget. A project budget may have a budgeting period that is a month, a quarter, or six months—whatever time the project is expected to last.

Budgets are often referred to as long term or short term. A budget of a year or less is typically a short-term budget.

A budget of more than a year is often considered a long-term budget. They can extend to two, three, four, and even five years or more. However, managers find it challenging to create a budget that extends beyond two years because it is difficult to estimate market trends and what will be happening that far into the future (Demand Media, n.d.). That does not mean, however, that one should never create a long-term budget. A long-term budget can help create a framework and keep people on track and accountable (Demand Media, n.d.). But managers often re-evaluate long-term budgets as time moves on.

Fixed and Flexible Budgets

Budgets can be fixed or flexible.

A *fixed budget* is the most straight forward and easiest to prepare. Organizations use fixed budgets when income and expenditures are relatively stable and known. These budgets are used by organizations that are relatively stable in terms of money going out and coming in. They have a disadvantage in that they do not consider changes in the market or economy.

Organizations often use a *flexible budget* when they have variable income and expenditures, where things are more uncertain. Flexible budgets have the advantage of allowing flexibility in spending, but they are complicated to create because they require multiple budgets that allow for variances in income and spending.

Responsibility Budgeting

Responsibility accounting or responsibility budgeting is a kind of budgeting where managers are responsible for budgeting their own departments. Managers create budgets for their areas of responsibility, track the actual money in and out, compare the actual against the planned budget, and report performance to their management, typically monthly or quarterly.

Types of Budgets

As discussed, organizations prepare budgets for various reasons. Some budgets are for planning, whereas others can be for monitoring, motivating, facilitating communication, coordinating activities, allocating resources, and so forth.

For example, short-term budgets are mostly about planning, while long-term budgets can be for planning, motivating, and more. Responsibility budgets can be about planning, motivating, communicating, and coordinating activities. Every budget can serve multiple purposes.

Managers also prepare budgets for specific functional purposes.

- A **sales budget** lists the expected units sold and dollar value within a specific budget period. This budget often forms the bases of other budgets and thus must be detailed and accurate.
- An **operating budget** looks at revenues generated and fixed expenses.
- A **materials budget** lists the quantity and costs of raw materials that must be purchased during a specified period to meet project requirements or production within that same period or in a specified future period.
- A **direct-labor budget** is similar to the direct materials budget except that it looks at the required labor and costs to ensure project requirements or production are accomplished during a specified period.
- An **overhead budget** considers what costs will be incurred other than direct material costs and direct labor costs within a specific period.
- A **program budget** is for a specific program or project within an organization.
- A **cash budget** is a list of the expected inflow and outflow of cash within a specific period. It typically includes disbursements, receipts, travel expenses, cash surplus, and cash deficits.

Revenue Centers, Cost Centers, and Profit Centers

At the organizational level, budgets typically report costs by revenue centers, cost centers, and profit centers.

Revenue centers are areas of an organization that directly bring in money. For example, in an engineering firm, project managers bring in revenue from the services they provide to clients; or in a university setting, faculty bring in revenue by teaching courses.

Cost centers are areas of an organization that cost money to operate but do not directly bring in revenue. For example, in an engineering firm, while the project managers directly bring in revenue, the design engineers may not, and thus may be considered cost centers. Or in a university setting, faculty directly bring in revenue while advisors do not, and thus advisors might be considered cost centers.

Profit centers are areas that both spend money and earn money, but where the actual profit (more earned than spent) is likely to come from in an organization. A profit center needs to be managed so that income exceeds expenses.

What to do:	Other actions:
☐ Recognize that managerial budgets include educated guess about expectations, and answer questions such as how much money, from what sources, to whom, by when, and for what?	☐ ______________
☐ Identify the types of budgets you use or will use.	☐ ______________
☐ Describe the difference between a capital budget and an operating budget.	☐ ______________

Remember

- ✓ Two main types of accounting are financial accounting and managerial accounting. Financial accounting tends to focus on past performance. Managerial accounting tends to look toward the future and provides information for functional managers and project managers.
- ✓ Organizations typically use three main types of financial statements: income statements, balance sheets, and cash flow statements.

Enhance Your Learning

Watch the following 19-minute video that describes budget concepts and types of budgets. Consider how you can apply this information in your work.

Cholayil, S. (2013). *Budget Concept And Types.*	Available at: https://www.youtube.com/watch?v=nsesYlcp2BE

Reinforce Your Learning

Characteristics of Good Budgets. Below are some features of an effective budget. Check those that apply to the budgets you use. Then, list any questions you have about budgets.

- ☐ Promotes planning
- ☐ Communicates what to expect
- ☐ Links objectives to resources
- ☐ Improves decision making
- ☐ Fosters careful study
- ☐ Creates an opportunity to think
- ☐ Provides an early warning
- ☐ Identifies weak areas, problems, and threats
- ☐ Aids interdepartmental coordination and collaboration
- ☐ Flexible and computerized
- ☐ Provides trend data over time
- ☐ Indicates spending trends over time
- ☐ Other

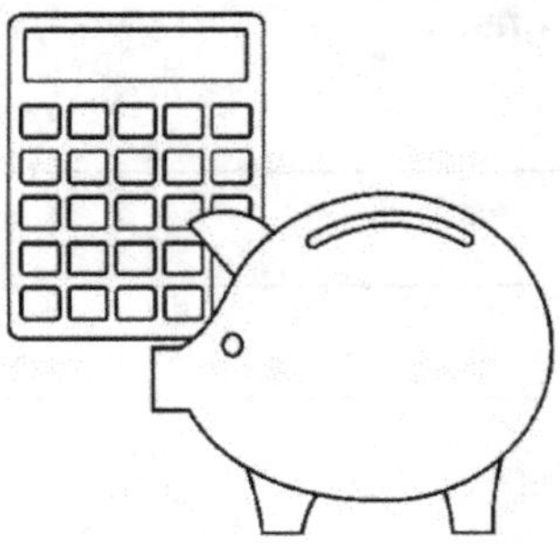

Controlling Project Costs and Budgets #10.2

Understand How Organizations Track and Report Money

Organizations must track the money they bring in, what they spend and want to spend, and balance it all out in the end. If an organization does not balance its budget, it must borrow money, which inhibits future cash flow, and repayment must be planned in future budgets. The point is, budgeting is vital in every organization, all the time, and at every level, including the project level. In this concept, we describe the basic tools and techniques organizations use to track and report money.

Chart of Accounts and General Ledger

A General Ledger is organized by ledger accounts. There is one ledger account for each account in the **Chart of Accounts**.

The General Journal is used to record daily transactions that occur. This is usually done electronically; but of course, could be written manually in a journal. The entries maintain the **General Ledger** record of the money coming in and going out. These include the following:

- Assets (cash, real estate, equipment)
- Liabilities (outstanding loans)
- Stockholder equity
- Operating revenue (sales or incoming fees)
- Operating expenses (rent or mortgage, employee pay, insurance)
- Non-operating revenue and gains (investment income)
- Non-operating expenses and losses (losses on disposal of equipment)

A **Chart of Accounts** is a list of all accounts (money in and out), even if they have a zero balance and are not recorded on the General Ledger. For example, some accounts in the chart of accounts for a services firm might be:

- Assets
    - Cash
    - Accounts Receivable
    - Equipment
- Liabilities
    - Salaries
    - Benefits
    - Advertising
    - Accounts Payable
- Owner Equity

Balance Sheet

A Balance Sheet is part of the General Ledger. The balance sheet lists all the assets and liabilities that carry over from year to year. It includes stockholder or owner shares in the organization, as depicted by the following equation:

Balance Sheet Equation

Assets = Liabilities + Owner Equity

Bookkeeping Conventions

In double-entry bookkeeping, **debits** and **credits** are entries made in account ledgers to record changes in value resulting from business transactions.

Two rules apply, as illustrated below:

- Assets and Expenses increase with a debit and decrease with a credit
- Liability, Equity, and Revenue accounts increase with a credit and decrease with a debit

BALANCE SHEET ACCOUNTS			
Assets		**Liabilities & Owner Equity**	
DEBIT	CREDIT	DEBIT	CREDIT
Increases	Decreases	Decreases	Increases

JOURNAL ENTRY				
Date	**Description**	**Ref**	**DEBIT**	**CREDIT**
xx	**Supplies** *(asset increase)*		**1000**	
	Accounts Payable *(liability increase)*			**1000**
	Supplies purchased on account			

BALANCE SHEET ACCOUNTS			
Supplies = Assets		**Accounts Payable = Liability**	
DEBIT	CREDIT	DEBIT	CREDIT
1000			1000

The following is a business example.

XYZ organization recorded the following transactions for a new venture:

1. Owner, YYZ ORG, invested $18,500 in a new business venture.
2. The organization borrowed $10,000 from the bank.
3. New equipment costing $12,000 was purchased for cash.
4. A payment of $1,000 was made on the bank loan.
5. The owner withdrew $500 from the business for personal use.
6. Office supplies were purchased on credit for $20.

Cash = Assets	
DEBIT	CREDIT
18,500 (1)	
10,000 (2)	
	12,000 (3)
	1,000 (4)
	500 (5)

Accounts Payable = Liability	
DEBIT	CREDIT
	20 (6)

Loans Payable = Liability	
DEBIT	CREDIT
	10,000 (2)
1,000 (4)	

Supplies = Assets	
DEBIT	CREDIT
20 (6)	

XYZ Corp = Equity	
DEBIT	CREDIT
	18,500 (1)
500 (5)	

Equipment = Assets	
DEBIT	CREDIT
12,000 (3)	

Assets		=	Liabilities		+	Owner Equity
Cash	15,000		Accounts Payable	20		XYZ ORG 18,000
Equipment	12,000		Loan Payable	9,000		
Office Supplies	20					

Balance Sheet for XYZ ORG's new venture:

Total Assets $27, 020 = Total Liabilities $9,020 + Owner Equity $18,000

Income Statement

The **Income Statement** is also part of the General Ledger. Unlike the balance sheet, it does not carry over from year to year, but rather starts over each year. The income statement includes all sources of income and expenses (operating revenue, operating expenses, non-operating revenue, and non-operating expenses), as depicted by the following equation:

Income Statement Equation

Revenue – Expense = Net Income

Cash Flow Statement

A Cash Flow Statement reports how much cash was generated and spent during a specific period.

What to do:

- ☐ Locate and review the Chart of Accounts for your organization or project or create one.
- ☐ Identify how daily transactions are recorded in your organization for your project.
- ☐ Describe how the General Ledger and Journal entries relate to a Balance Sheet.

Other actions:

- ☐ ______________________
- ☐ ______________________
- ☐ ______________________

Remember

✓ Budgets should support your organization's vision and mission (what your organization hopes to accomplish) and strategic plan (the steps your organization will take to achieve its goals).

Enhance Your Learning

Watch the following 4-minute video to learn more about managing the financial aspects of a project and consider how you can apply this information.

Whitt, J. (2012). *4 Project Financial Management Tips–Project Management Finance.*	Available at: https://www.youtube.com/watch?v=5DybRTTvrmY

Reinforce Your Learning

Start-up Budget and Chart of Accounts. A Chart of Accounts is a tool to organize and track expenses. Any project or business venture has specific accounts within their chart that correspond to their assets and liabilities. For a new business, you may need to estimate start-up activities and costs. You would consider the start-up activities as a separate project for which you would develop a chart of accounts and budget.

Start-up costs are expenses you will incur before your business opens or is operational. They vary according to the type of venture or product line. The worksheet below will help you assess your financial needs, so your venture is not undercapitalized. Add or subtract budget items and assign budget categories to establish your estimated start-up costs. This example of a **start-up budget** is for a shop with 1,000 sq. ft. of operating space.

Project: *Wes's Shop*

Tasks/Activities	Cost	Budget Categories
❖ Build-out/remodeling	10,000	Construction
❖ Furniture/fixtures	10,000	Fixtures
❖ Equipment	10,000	Equipment
❖ Installation (Fixtures, equipment)	5,000	Fixtures
❖ Information technology	5,000	Equipment
❖ Initial inventory	15,000	Inventory
❖ Supplies	3,000	Supplies
❖ Advertising and promotion	5,000	Marketing
❖ Insurance	2,000	Insurance
❖ Legal/professional services	5,000	Professional
❖ Misc./Other:	10,000	Misc.
Total Start-Up or Project Costs:	**$ 80,000**	

Your Start-up Budget and Chart of Accounts.

Revisiting the project of planning a retirement or other celebratory event, try your hand at developing a budget and chart of account. Start by listing the tasks needed to complete the project, estimate the cost of each task, and assign each cost a budget category.

If you plan on starting a small business or a new product line within an organization (not a single product or service) the same process would apply. In such cases, if available, you would use the organization's existing chart of account categories and structure.

Your Project (business venture or product line): ____________________

Tasks/Activities	$ Cost	Budget Categories
❖		
❖		
❖		
❖		
❖		
❖		
❖		
❖		
❖		
❖		
❖		
❖		

Total Start-Up or Project Costs: ________

Controlling Project Costs and Budgets #10.3

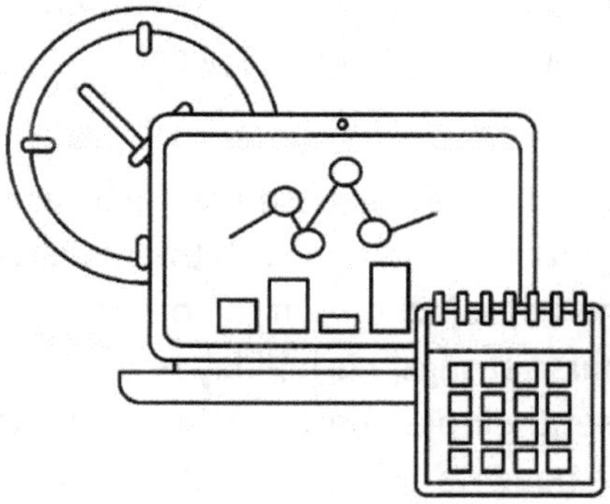

Appreciate How Budgeting Relates to Time

By their nature, budgets reflect what is happening in an organization, what it costs to produce a product or service, and the money the organization receives. Thus, budgets reflect organizational activity over time.

Budgets are also involved with time in that when managers create budgets, they must estimate future costs and profits, and they often do this using past numbers as a guide. As we all know, conditions can change, and not all estimates prove to be accurate.

In terms of accounting practices, how organizations account for costs and revenues can vary, and how they account for the loss in value of their physical assets can also vary. These accounting differences result in budget numbers being reported at different points in time.

In this concept, we discuss the factors most often involved with budgets and time, to help you better understand what you must consider when you plan your project budgets.

Cash Basis Accounting and Accrual Accounting

Organizations and project managers use two basic types of accounting: cash basis and accrual.

Cash basis accounting does *not* mean that everything was paid in cash. Rather, it means that expenses and income are **reported at the time they happen**, when money changes hands, or moves from account to account.

For example, if an organization receives a rent payment with cash basis accounting, the organization adds that rent to the ledger when payment is received. In other words, not annually, before the rent is due, but each month, as payment is made. This type of accounting works best for smaller organizations that receive most of their payments in cash at the time service is rendered or a product delivered, who maintain little or no inventory, and who do not carry long-term customer accounts.

Accrual-based accounting is a bit different in that things are **recorded when they are ordered** or **agreed upon**, even if payment is not made at that time.

For example, if a customer uses a credit card to buy a product, the cash method would not recognize the payment until the money clears the credit card company and is received. The accrual method, on the other hand, recognizes the value of the credit card payment. Thus, accrual accounting is more expensive, more difficult, but gives a more accurate picture of how the organization is really doing.

Optimistic, Pessimistic, and Most Likely Budget Scenarios

Managers use optimistic, pessimistic, and most likely budget scenarios. When they develop their budgets, managers often do not know precisely what specific income streams will produce, and thus the numbers must be estimates.

For example, ABC Company just created a new product they are marketing. Based on their research, they believe that they can sell 10,000 units in the first quarter. They hope that with the holidays coming they may sell 13,000 units; at least that is their optimistic projection. On the other hand, some naysayers are predicting consumers will tighten their purse strings this holiday season. ABC fears they may not even make 10,000 units in sales, so they set their pessimistic estimate at 6,000 units. Knowing these three sets of numbers allows for flexible budgets and contingency plans, as we will discuss in Concept 10.

Depreciation

Depreciation can have two meanings in accounting. *Depreciation* refers to an asset losing its value over time. *Depreciation* also refers to listing the cost or value of an asset over time as a fractional piece of the full value each year.

How organizations *depreciate* assets can impact tax payments as well as profit statements.

Budget Assumptions

Budget assumptions are **educated guesses** about budget line items; things one reasonably believes will hold true. For example, if an organization earned $10,000 last year in profits, having grown by 10 percent each year for the past three years—and other known factors remain mostly the same—it would be reasonable to make a budget assumption that profits will be $11,000 this year.

Responsibility Matrix

A responsibility matrix is a term for a list that shows what **person (manager)** in an organization is **responsible for what** aspect of the project or organization, and perhaps for what budget or line items within a budget.

What to do:	Other actions:
☐ Describe the difference between cash basis accounting and accrual accounting.	☐ ____________
☐ Cite where developing optimistic, pessimistic, and most likely budget scenarios would be appropriate.	☐ ____________
☐ Define depreciation and cite an example.	☐ ____________

Remember

✓ Every project your organization takes on should relate to a specific organizational goal. Then, each project must have a budget associated with it, a budget that fits into the larger organizational budget.

Enhance Your Learning

Watch the following 2-minute video that overviews the initial steps in developing a project budget.

Ray, B. (2009). *Making a Budget: Developing a Project Budget.*	Available at: https://www.youtube.com/watch?v=DQYEDejl66Y

Reinforce Your Learning

Start-up Budget Timelines. Part of budgeting is managing timelines by developing realistic expectations and a schedule. Consider your start-up project from the previous concept (or another of your choice) and prepare a start-up budget schedule using the template that follows. Place the $ Cost amount you estimated in the $ Cost column and segment as appropriate in the appropriate monthly column or columns. Then add the total in each column to determine your total and monthly start-up cash flow.

As an example, the following start-up budget is for the small shop previously introduced in the last course module.

Project or Business Venture: *Wes's Shop*

Budget Task/Activity	***$ Cost ($ 000)***	***1***	***2***	***3***	***4***	***5***	***6***	***7***	***8***	***9***	***10***	***11***	***12***
Build-out/remodeling	**10**	2	2	2	2	2							
Furniture/fixtures	**10**			2	2	2	4						
Equipment	**10**			2	2	2	2	2					
Installation (Fixtures, and equipment)	**5**			1	1	1	1	1					
Information technology	**5**				1	1	1	1	1				
Initial inventory	**15**						2	8	5				
Supplies	**3**						1	1	1				
Advertising and promotion	**5**							1	2	2			
Insurance	**2**	2											
Legal/professional services	**5**	5											
Misc./Other	**10**	1	1	1	1	1	1	1	1	1	1		
TOTAL	**80**	**10**	**3**	**8**	**9**	**9**	**12**	**15**	**10**	**3**	**1**		

Budget Timelines: Your Project or Business Venture: ____________________

Start-up Budget and Monthly Start-up Cash Flow Schedule

Budget Task/Activity	***$Cost ($000)***	***1***	***2***	***3***	***4***	***5***	***6***	***7***	***8***	***9***	***10***	***11***	***12***
TOTAL													

Controlling Project Costs and Budgets #10.4

Understand Budget Forecasting

According to Dictionary.com, *forecast* means to predict a future condition or occurrence; to plan or arrange beforehand; foresight in planning.

A forecast is an educated guess about the future. Thinking ahead and forecasting budget line items is a necessary part of project planning. A budget forecast helps managers, including the project manager, plan (for example) production volume, inventory and materials needs, labor hours and workers required, cash needed on hand, and financing needs. When you know where an organization is heading, you can make informed decisions about staffing, purchasing, pricing, and more (Ray, n.d.).

In this concept, we describe how budget forecasting works so you may better understand how you will plan the costs for your project, and what your company can expect to gain from the project.

How Managers Forecast

All managers plan for the future. Their budgets reflect what the managers expect to happen in terms of the money they will spend and what the organization will earn as a result.

Some people think of budgeting as a guessing game. Others view it as something uncanny, an ability to see the future. Budgeting is the **application of logic** and **common sense in forecasting** what will most likely happen.

Forecasting uses known facts about the past and the present to project future conditions rationally. While no amount of forecasting or preparation can guarantee success, when you create a well-researched budget forecast, you increase your chances of growing the organization (Ray, n.d.).

The simplest way to forecast future organizational performance is to consider recent performance (Ashe-Edmunds, 2015). Managers typically do this by creating a master budget that provides a "one-time snapshot" of how the manager thinks the organization will do over the year coming up (Ashe-Edmunds, 2015). Managers evaluate each line item in the budget to better understand how the cash is flowing (Ray, n.d.). Looking at records from the previous one to five years can lead to the most accurate forecasts (Ray, n.d.).

Of course, recent organizational performance is not the only consideration in creating the master budget. Managers must also consider the industry, relevant consumer trends, and other experiences (Ashe-Edmunds, 2015).

Further, the master budget is not prepared by a single manager living in a vacuum. Instead, the manager responsible for creating the budget talks to other relevant managers in the organization to get their perspectives on factors that could cause the future to be different from the past (Ashe-Edmunds, 2015).

Budgeting Terminology

Before moving on, here are some budgeting terms you will want to understand: pro forma, revenue, and profit.

Pro forma, a term commonly heard in accounting, means **projected** or predicted budget line items, not including things that are considered unusual or exceptions. Typically, in business, pro forma financial statements are **prepared in advance** of a planned transaction, such as a merger, an acquisition, a new capital investment, a change in capital structure, the incurrence of new debt, or issuance of stock.

Revenue refers to how much money an organization brings in from all sources before expenses are subtracted.

Profit refers to how much was earned after all costs and expenses are deducted from the revenue.

A revenue budget tracks all income received from every source: sales, investments, donations, shareholders, and so on.

Not-for-profit organizations sometimes refer to profit as discretionary income.

Qualitative and Quantitative Forecasting

Budget forecasting involves both qualitative and quantitative considerations.

Qualitative forecasting tends to use techniques that are more subjective than quantitative techniques. That is, they are more open to interpretation and rely on the experience of the person creating the budget. Managers use qualitative techniques when there is little quantitative information available, such as when an organization, industry, or product is new. Qualitative methods are varied but can include the Delphi technique, user surveys, market research, and more.

Quantitative forecasting techniques are more numbers-focused and tend to be more objective than qualitative forecasting. These often include projecting the past into the future but can also include competitor analysis.

What to do:	**Other actions:**
☐ Explain how forecasting is making an educated guess and applying logic and common sense.	☐ ____________
☐ Cite examples of where qualitative and quantitative budgeting approaches might be applied.	☐ ____________
☐ Develop a pro forma or projected operating budget forecast for your project or a project you are familiar with.	☐ ____________

Remember

- ✓ Budgeting is the application of logic and common sense in forecasting what will most likely happen in the future.
- ✓ In budgeting, managers like to look at past experience and present conditions, using what they know from the past to inform their budget for the future. However, for new ventures or projects, this can be difficult. Thus, forecasting based on industry and competitor analysis is sometimes necessary, and managers reevaluate budgets as they gather data.

Enhance Your Learning

Watch the following 5-minute video to hear a summary of the difference between budgeting and forecasting.

Accenture Academy. (2016). *Spotlight on the Difference between Budgeting and Forecasting.*	Available at: https://www.youtube.com/watch?v=FL2NcGYQ_Vc

Reinforce Your Learning

Initial Pro forma Operating Budget. Assume the start-up project or business venture you identified previously was successful, and you are looking ahead at your first year of operation. Develop a Pro forma or projected operating budget forecast and monthly cash flow schedule for the next budget year using the template that follows.

As an example, the following Pro forma Operating Budget and monthly Cash Flow Schedule is for the small shop in our previous example.

Project or Business Venture: *Wes's Shop*

Initial Pro forma Operating Budget and Monthly Cash Flow Schedule Example

Budget Item	***Total Cost ($000)***	***1***	***2***	***3***	***4***	***5***	***6***	***7***	***8***	***9***	***10***	***11***	***12***
Revenue													
Sales & Services	**230**	**30**	**10**	**10**	**10**	**10**	**10**	**10**	**10**	**20**	**35**	**40**	**50**
Expenses													
Accounting & Legal	**4**	1			1			1			1		
Advertising	**5**	1			1			1			2		
Insurance	**2**			1						1			
Maintenance & Repairs	**2**			1						1			
Materials & Inventory	**121**	6	6	10	10	6	6	6	10	20	20	15	6
Payroll Exp. & Benefits	**15**	2	1	1	1	1	1	1	1	1	1	2	2
Rent	**12**	1	1	1	1	1	1	1	1	1	1	1	1
Salaries and Wages	**55**	6	4	4	4	4	4	4	4	4	5	6	6
Supplies	**2**	1									1		
Taxes & Licenses	**1**							1					
Telephone & Internet	**2**	.1	.1	.2	.1	.1	.2	.2	.2	.2	.2	.2	.2
Travel	**2**			1					1				
Utilities	**2**	.3	.1	.1	.1	.1	.1	.1	.1	.2	.2	.3	.3
Misc	**6**		1		1		1		1		1		1
Sub-Total Exp.	**231**	**18.4**	**13.2**	**19.3**	**19.2**	**12.2**	**13.3**	**15.3**	**18.3**	**28.4**	**32.4**	**24.5**	**16.5**
Net Income	**14**	**11.6**	(3.2)	(9.3)	(9.2)	(2.2)	(3.3)	(5.3)	(8.3)	(8.4)	**2.6**	**15.5**	**33.5**

Your Project or Business Venture: ______________________________

Initial Pro forma Operating Budget and Monthly Cash Flow Schedule

Budget Item	***Total Cost ($000)***	***1***	***2***	***3***	***4***	***5***	***6***	***7***	***8***	***9***	***10***	***11***	***12***
Revenue													
Expenses													
Sub-Total Exp.													
Net Income													

Controlling Project Costs and Budgets #10.5

Estimate Project Costs

Budgeting basically involves cost accounting and forecasting. **Cost accounting** is the process of **accounting for expenditures**, or what has been spent. As we discussed, forecasting means making educated guesses about future income and expenses based on knowledge of past income and expenses, coupled with knowledge of the industry, economy, and so forth.

In this concept, we discuss the various types of project costs and how to estimate them.

Estimating Costs

While estimating costs may seem straightforward—for example, adding up what you will spend—many concepts and considerations are involved. Estimating costs in a reasonably effective way is imperative if your budget is to be accurate and enforceable.

Some of the tools and techniques used for estimating we touched on previously included:

- Expert judgment
- Published estimating data
- Historical or analogous project estimating
- Bottom-up estimating
- Probabilistic approaches

Types of Costs

The following is a discussion of various types of costs included in budgets.

Product or service costs are used in accrual accounting where the cost to create a product or perform a service is not listed on the balance sheet until the period in which the product or service is sold.

Product or service costs include direct labor costs, direct materials costs, and manufacturing overhead costs.

Period costs are costs that occur during the specific period but are not directly tied to income in that period. These are often fixed expenses such as rent, or even labor expenses like sales commissions.

Materials costs are costs that can be directly connected to a saleable product or service, for example, the metal used to make screws that are sold, or the books used in a training program.

Equipment costs are the costs of equipment used in manufacturing, office work, and throughout the organization.

Indirect costs are those necessary for the organization, but that cannot be directly tied back to an income stream.

Labor costs can be both direct and indirect; those directly involved in creating a sellable product or service and those that are not directly responsible for revenue but are necessary.

Selling costs are the necessary costs to sell products, advertise, and inform customers that a product or service is available.

Administrative costs are all costs associated with the overall management of the organization, such as executive pay, travel, and secretarial expenses.

Fixed and Variable Costs

Another way of looking at costs is as **fixed** or **variable**, or a **mix of the two**.

Fixed costs do not change much over time. For example, alarm system monitoring would be a relatively fixed cost, as would building insurance, rent expense, and depreciation expense.

Variable costs are subject to change over time. For example, the number of books used in a training program, health insurance costs on a per-person basis, or snow removal services might be variable depending on if there is a heavy snow season or a light snow season.

A budget ceiling can be one way to manage variable costs. The manager determines the maximum amount they are willing to spend on a specific line item and in the overall budget.

Fixed cost items and variable items have a push-pull relationship. For example, if a variable item cost rises, other variable items are usually adjusted. But sometimes fixed costs may need to be re-evaluated. Of course, given that fixed costs are, well, fixed, often nothing substantial can be done. Thus, variable items may need to be reviewed. For example, if a supplier raises materials costs, and no lower-priced alternative can be found, marketing may need to be slashed to make up the difference.

Mixed costs are a mix of fixed and variable costs. For example, consider the cost of running a large printing machine. Costs such as ink, labor, and paper might be fixed, but other costs might change with time and with the age of the equipment and maintenance needs. Another example has to do with management. One supervisor may be able to manage the work of up to five people, but if the organization hires a sixth person, it must also hire another manager.

Controllable and Uncontrollable Costs

Another difference between costs is that management can control some costs, but others are uncontrollable.

Controllable costs are those which **a manager has direct control** over. **Uncontrollable costs** are those **costs a manager does not have control over.** For example, controllable costs might include labor (because the manager decides how many people will work how many hours) and materials (when a manager negotiates the prices). Examples of uncontrollable costs are rent and taxes.

Activity-Based Costing

Activity-based costing is a somewhat complicated yet a logical way of accounting for product and service costs (AC, n.d.b). In activity-based costing, **overhead** production **costs** are **assigned to the product or service activity** they support. For example, the lighting bill for a company's manufacturing plant might be separated from the lighting bill for the office, and each bill allocated to a specific department.

Activity-based costing has become more important recently as manufacturing overhead costs have increased. Manufacturing overhead costs "no longer correlate with the productive machine hours or direct labor hours," diversity in new products and in consumer demand has grown, and "some products are produced in large batches, while others are produced in small batches" (AC, n.d.b).

Standard costing refers to **planned or anticipated costs.** These are often used in manufacturing accounting for inventories and production. When actual costs differ from the standard costs, variances are reported." (AC, n.d.c).

What to do:	Other actions:
☐ Describe how estimating costs is part of accounting for expenditures.	☐ ____________
☐ Describe the fixed and variable costs for your project or a project you are familiar with.	☐ ____________
☐ Estimate the budget costs for a training program applicable to your job or position.	☐ ____________

Remember

✓ When estimating costs, be aware that many types of costs can be included in a budget. For example, product costs, period costs, materials costs, labor costs, and more.

Enhance Your Learning

Watch the following 12-minute video on estimating costs that describes two reasons why cost estimation is important for project managers. First, it is extremely hard to estimate project costs well. Second, poor cost estimates are often the biggest reason for project budget over-runs.

Clayton, M. (2020). *How to Estimate Project Costs: A Method for Cost Estimation.*	Available at: https://www.youtube.com/watch?v=YQ2Wi3Jh3X0

Reinforce Your Learning

Training Program Costs. Estimate the budget costs for a training program applicable to your project. For simplicity, assume that the program is one day in length. Use the budget template that follows or create one.

Training Program: ______________________________

Program Dates						
Instructor						
# sessions						
Notes		**# people**			**Total hours**	
			Expenses			
Promotion	Printing	$		**Special features**		
	Advertising	$		Room rental	$	
	Promo	$		AV	$	
		TOTAL	$	Other	$	
Mailing	Mailing	$		Room/Food	$	
	Postage	$			**TOTAL**	$
	Other	$		**Food**		
		TOTAL	$	Breaks/meals	$	
Salary/Wages	University	$			$	
	Benefits	$			$	
	Non U	$			**TOTAL**	$
	Other	$		**Miscellaneous**		
		TOTAL	$		$	
Travel	University	$			$	
	Non U	$			$	
	Other	$			$	
		TOTAL	$		**TOTAL**	$
Instructional Items				**TOTAL DIRECT COSTS**		$
	Copies	$				
	Office sup.	$				
	Ed. Sup.	$		**ADMIN EXPENSES**		$
	Freight	$				
	Other	$				
		TOTAL	$	**GRAND TOTAL EXPENSES**		$

Controlling Project Costs and Budgets #10.6

Evaluate Project Benefits

The purpose of a budget is to detail how money will be spent to accomplish organizational goals. This is particularly true of project budgets because a project supports a specific goal.

Merely listing what will be spent and specifying line item after line item of budgeting numbers is not sufficient. A reasonable budget considers the **benefits** to be **derived** from specific expenditures—such as employee safety, increased revenues, decreased costs, and so on.

In this concept, we discuss the kinds of benefits that might come to the organization from a project and how to reflect benefits in a project budget.

Capacity Planning

Capacity planning is essential in budgeting. Too many organizations do not link their revenue and expense numbers to their operational capabilities (AT, 2015). For example, a manager might assume that sales will double in the coming quarter but not factor in higher commissions or sales force compensation (AT, 2015). Perhaps they think their sales staff will just be amazingly effective. But the truth is, capacity must be considered, as well as related costs such as training, expansion, additional marketing, and the like.

When doing capacity planning, consider the following kinds of questions (AT, 2015):

- Manufacturing center and equipment. Is the facility nearing capacity? Do changes need to happen to accommodate anticipated growth?
- Staffing. Are more people needed? Higher fees? Training?
- Product development. Have all products and any changes been adequately tested? What are the failure rates? Are new designs being rushed to production?
- Overhead positions. Are there administrative or overhead positions that must be filled to accommodate growth (positions such as marketing, accounting staff, receptionists, and so forth)?
- Will the benefits outweigh all these factors and costs?

Cost Reduction

Cost reduction refers to efforts to trim costs and expenditures. Sometimes the best way to achieve financial benefit and increase profits is simply to cut costs. As Benjamin Franklin said, "A penny saved is a penny earned."

Cost-cutting can be done as part of the budgeting process. However, cuts need to be evidence-based (not just emotional) and realistic.

You can also reduce costs by encouraging employees to look for cost-cutting measures or hiring an outside analyst to find cost reduction areas. Once any such changes are implemented, new budgets can use the reduced costs going forward.

Employee Safety

Employee safety is an area where money invested in reducing employee injury provides financial benefits later. Employee injuries cost the U.S. billions of dollars every year (OSHA, n.d.). These include medical costs, insurance rates, and loss of work. According to OSHA (n.d.), "Employers that implement effective safety and health management systems may expect to significantly reduce injuries and illnesses and reduce the costs associated with these injuries and illnesses, including workers' compensation payments, medical expenses, and lost productivity. In addition, employers often find that process and other changes made to improve workplace safety and health may result in significant improvements to their organization's productivity and profitability."

- One 2012 study found that employees who conducted safety inspections could save up to $3,555,000 in injury claims and costs related to employee injury and illness (OSHA, n.d.).
- Money spent on employee safety should be compared to the money saved, and the net value considered.

Employee Satisfaction

Talent development efforts to improve employee morale and turnover are also benefits to consider when budgeting projects. Employees with low morale are more likely to lose work time by wasting time on the job or using sick days. Additionally, it is more expensive to hire a new employee to replace one who leaves than it is to keep an employee happy.

When budgeting for employee costs, team building activities, added training, and so forth, keep in mind the costs of recruiting, selecting and training a new person, and productivity lost as they acclimate.

You can measure, from both a qualitative and quantitative standpoint, the positive benefits or impact that training and development programs can have on organizational performance.

You can review the **qualitative improvement** in organizational performance before and after the program for the following areas:

- Morale levels
- Quality of communication
- Work-related stress/climate
- Job satisfaction
- Contribution to furthering core values
- Employee retention
- Customer satisfaction

You can review the **quantitative improvement** in organizational performance expected or realized from conducting the training and development program as follows:

Annual Cost Savings: $ _______ Increased speed/output, less scrap, less rework, improved energy efficiency, less work-in-process inventory, improved inventory turns, less new hire training/orientation, lower absenteeism, less overtime, lower insurance/workers' compensation, and so on.

Cost Avoidance: $ ________Fewer customer complaints, fewer returns/allowances, fewer accidents/claims, avoid compliance fines, prevent workplace violence, prevent harassment, better work climate, prevent wrongful discharge, retain talent, and so on.

Increased Revenue: $ _______New product/service sales, service add-ons, achieve higher prices, improve sales or increase funding, favorable purchase terms, achieve earlier customer purchases, and so on.

What to do:	Other actions:
☐ Consider a project you are familiar with and identify the benefits to be derived from the project, such as employee safety, increased capacity, or decreased costs.	☐ ______________
☐ Consider the benefits of conducting business safely and ethically.	☐ ______________
☐ Estimate the expected benefits for a training program applicable to your job or position.	☐ ______________

Remember

✓ Projects must be budgeted to achieve the organization's vision and goals.
✓ Budgets must consider the benefits to be derived from a project.

Enhance Your Learning

Watch the following 4 ½-minute video about performing a cost-benefit analysis.

Conservation Strategy Fund. (2014). *Intro to Cost-Benefit Analysis.*	Available at: https://www.youtube.com/watch?v=7tdKkeNCIPE&list=PLBfu1mD9hk64sgOIH_nUEsndUzACDe-4Y

Reinforce Your Learning

Training Program Impact. For your training program in the previous learning activity, identify the anticipated qualitative and quantitative benefits using the format below, or create a format of your own.

Qualitative Benefits:

Observable Improvement in Organizational Performance

	before program					*after program*				
Qualitative Measures:	Poor ➔ Excellent					Poor ➔ Excellent				
	1	2	3	4	5	1	2	3	4	5
Morale Levels	☐	☐	☐	☐	☐	☐	☐	☐	☐	☐
Quality of Communication	☐	☐	☐	☐	☐	☐	☐	☐	☐	☐
Work-related Stress / Climate	☐	☐	☐	☐	☐	☐	☐	☐	☐	☐
Job Satisfaction	☐	☐	☐	☐	☐	☐	☐	☐	☐	☐
Contribution to Furthering Core Values	☐	☐	☐	☐	☐	☐	☐	☐	☐	☐
Employee Retention	☐	☐	☐	☐	☐	☐	☐	☐	☐	☐
Customer Satisfaction	☐	☐	☐	☐	☐	☐	☐	☐	☐	☐

Quantitative Benefits:

Bottom Line Impact that is Anticipated / Expected from the Program Conducted

Please check (✓) Anticipated ($) Impact

- **Annual Cost Savings: $ _______ or**

  (Increased speed/output, less scrap, less rework, improved energy efficiency, less work-in-process inventory, improved inventory turns, less new hire training/orientation, lower absenteeism, less overtime, lower insurance/worker comp., etc.)

$0	$20,000	$40,000	$60,000	$80,000
☐	☐	☐	☐	☐
$100,000	$200,000	$300,000	$400,000	$500,000
☐	☐	☐	☐	☐
$600,000	$700,000	$800,000	$900,000	$1,000,000
☐	☐	☐	☐	☐
More than $1,000,000	$			

- **Cost Avoidance: $ ________ or**

  (Less customer complaints, less returns/allowances, fewer accidents/claims, avoid compliance fines, prevent workplace violence, prevent harassment, better work climate, prevent wrongful discharge, retain talent, etc.)

$0	$20,000	$40,000	$60,000	$80,000
☐	☐	☐	☐	☐
$100,000	$200,000	$300,000	$400,000	$500,000
☐	☐	☐	☐	☐
$600,000	$700,000	$800,000	$900,000	$1,000,000
☐	☐	☐	☐	☐
More than $1,000,000	$			

- **Increased Revenue: $ _______ or**

  (New product/service sales, service add-ons, achieve higher prices, improve sales or increase funding, favorable purchase terms, achieve earlier customer purchases, etc.)

$0	$20,000	$40,000	$60,000	$80,000
☐	☐	☐	☐	☐
$100,000	$200,000	$300,000	$400,000	$500,000
☐	☐	☐	☐	☐
$600,000	$700,000	$800,000	$900,000	$1,000,000
☐	☐	☐	☐	☐
More than $1,000,000	$			

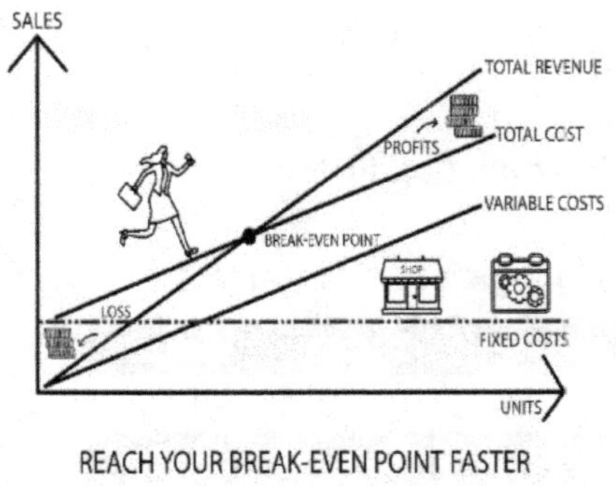

Controlling Project Costs and Budgets #10.7

Analyze the Break-Even Point

As we have discussed, managers must balance budgets to ensure the organization does not have more money going out than coming in. That is, essentially, the point of a budget.

You can do a break-even analysis to determine how many products or services must be sold for the organization to be on the cusp of profitability, neither losing money nor making it.

In this concept, we discuss the process of balancing a budget and review how to do a break-even analysis for your projects.

What is a Break-Even Point?

An organization is said to **break even** when **total expenses** match **total earnings.**

When expenses exceed earnings, we say an organization is "in the red," meaning there are red numbers (which denote negatives) in the final tally column.

An organization is "in the black" when earnings exceed expenses, hence the name Black Friday for the day after Thanksgiving, a day when many retailers move into "the black" for the first time in the year.

As you can see, **revenue** is an essential number in break-even analysis. Revenue is calculated by taking the selling price of an item or service times the quantity or units of sales. This number does not consider costs or expenses. *Revenue* refers to the money brought into an organization.

Costs, or **expenses**, are also crucial in break-even analysis. These typically contain both a variable and fixed element for a specified period.

Break-Even Point Example

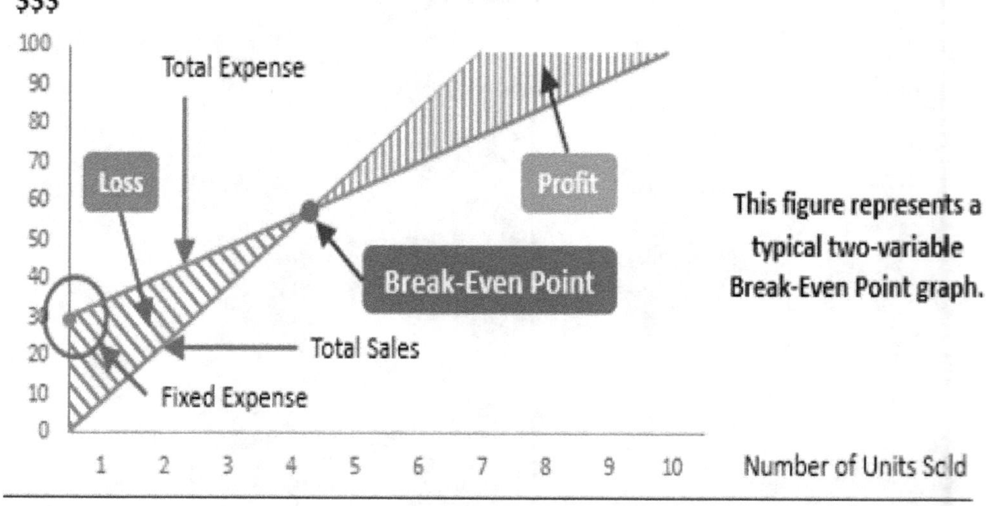

This figure represents a typical two-variable Break-Even Point graph.

A break-even analysis is typically a simple exercise. It provides a mental discipline that can help managers develop real-world solutions and help ensure that ideas and plans are attainable.

Cost-Volume-Profit Analysis (CVP)

CVP analysis is an essential tool for project managers, and it relates to break-even analysis.

Doing a CVP analysis will help you in understanding how variables such as cost, volume, and profit work together to impact the organization (Rehman, 2014). Specifically, CVP looks at the interaction of the following five numbers (Rehman, 2014):

- Prices of products or services
- Volume or level of activity
- Per unit variable cost
- Total fixed cost
- Mix of product sold

CVP analysis can help a project manager decide, for example, what products to manufacture, what pricing structure to use, what marketing strategy to use, and so forth.

Contribution Margin

The contribution margin shows the critical relationship among cost, volume, and profit. The *contribution margin* is the excess of sales revenues ***over*** variable costs.

The contribution margin concept is useful in business planning. It provides insight into the profit potential of a firm.

For example, the income statement below is prepared in contribution margin format:

Sales	$1,000,000
Variable Costs	600,000
Contribution Margin	400,000
Fixed Costs	300,000
Income from Operations	$ 100,000

In this example, the **contribution margin** of $400,000 is available to cover the fixed costs of $300,000. Once the fixed costs are covered, any remaining amount adds directly to the operating income of the firm. Think of the fixed costs as a bucket and the contribution margin as water filling the bucket. After the bucket is full, the overflow represents operating income (income from operations). Up until the point of overflow, however, the contribution margin contributes to fixed costs (filling the bucket).

The contribution margin helps a project leader or manager understand how much a particular product or service (or even a person) is contributing to the overall profit margin. It follows a simple formula:

Product Revenue – Product Variable Costs = Contribution Margin

A **contribution margin** can also be expressed as a percentage. A project manager might use this formula to evaluate different products the organization sells. They may find, for example, that products A, B, and C each have a contribution margin of around 30 percent, that is, each one is contributing about 30 percent to organizational profits. But product D is only making about 10 percent of profits. Thus, they may decide to drop or modify product D.

In manufacturing or sales, you can use a simple formula to determine how many units must be sold to break even. It requires knowing your fixed costs per period, the variable cost per unit, and the selling price per unit.

The formula is as follows:

Break-even point = Fixed Costs/(Unit Selling Price – Variable Costs)

or

Break-even point = Fixed Costs/Contribution Margin

As we discussed, the **break-even point** is the level of operations at which a business's revenues and expenses are equal. The business will neither realize an operating income nor incur an operating loss. The break-even point is useful in business planning, especially when expanding or decreasing operations.

For example, assume the fixed costs for the XYZ organization are estimated to be \$90,000. The unit selling price, unit variable cost, and unit contribution margin are as follows:

Unit selling price	\$25
Less Unit variable cost	<u>\$15</u>
Unit contribution margin	\$10

The break-even point is 9,000 units, computed by using the following equation:

Break-even Point (units) = Fixed Costs/Unit Contribution Margin

The break-even point is affected by changes in the fixed costs, unit variable costs, and unit selling price.

What to do:	Other actions:
☐ Understand that the break-even point refers to total expenses matching total earnings.	☐ ______________________
☐ Calculate the break-even point for an initiative with fixed costs of \$2000, variable costs of \$50 per unit sold, and a selling price of \$80 per unit.	☐ ______________________
☐ Cite an example in your workplace where a break-even analysis would be a valuable tool to apply.	☐ ______________________

Remember

- ✓ At the break-even point, an organization has not had any losses, but neither have they made any profit. They are equal and have "broken even."
- ✓ When looking at your break-even point, be sure to consider if it is realistic. For example, if you must serve 10,000 clients to break even in six months, is this even possible? If not, you may need to look at trimming costs somewhere or changing how things work to bring the break-even point down.

Enhance Your Learning

Watch the following 8-minute video to better understand break-even points.

TakingTheBiz. (2017). *A level Business Revision - Break Even.*	Available at: https://www.youtube.com/watch?v=578qM7sXo_M

Reinforce Your Learning

Calculate the Break-Even Point. Assume you started a business venture with fixed costs of $2000. The variable costs are $50 per unit sold. You plan on selling the units at $80 per unit. What would be the break-even point?

Hint: Break-even Point = Fixed Costs / (Unit Selling Price – Variable Costs)

Controlling Project Costs and Budgets #10.8

Consider the Present Value of Money

In investing and accounting, we often hear about the value of money over time. That is, a dollar today can be invested, earn interest, and be worth more than a dollar tomorrow. On the other hand, a dollar stuck inside a mattress today will have less value when pulled into the daylight ten years from now due to inflation. While that dollar could have bought a candy bar today, candy bars are likely to cost more in the future. Thus, the buying power will be less.

In this concept, we discuss how project managers consider the present value of money and the future value of that money as they plan and budget project expenditures.

The Value of Money Over Time

Given a choice, would you prefer to receive one dollar now or one dollar three years from now?

You should prefer to receive one dollar now because you could invest the money and earn interest for three years. As a result, the amount you would have after three years would be greater than one dollar.

Money is said to have a *time value,* which varies according to how near it is to, or how far away it is from, the present.

It is presumed that money will earn more interest the longer it is invested. Money invested today will have a higher value in a year, even higher in five years, and much higher in twenty years.

On the other hand, inflation is also a constant, and probably more dependable than interest. This means that $100 tucked in a pocket today may have the spending power of just $95 in a year or two.

Present Value

Present value is the idea that getting a specific amount of money today is more valuable than getting that same amount of money in the future.

This idea is the basis for advertisements that you see on television beckoning people to sell their structured settlement payments for a lump sum of cash now. These buyers convince the sellers that money in the hand now is worth more than the same money in the hand in the future. That is true, but what they often do not say is the annuity they are selling is earning interest and thus will be worth more in the future.

The concept of present value is useful because it helps a manager make better decisions about making loans, giving loans, giving customers credit, and other things that extend giving or receiving money into the future.

Future Value

Future value looks at how much a specific amount of money will be worth in the future.

It is a straightforward calculation that requires knowing how much money you nave now, the expected rate of return (or interest), and how long the period is into the future. However, careful managers also consider inflation when making decisions based on future value.

$$\textbf{Future Value} = \textbf{Present Value} \ (1 + \textbf{interest rate})^{\text{number of years}}$$

Compound Interest

Compound interest is a simple idea in investing. It means when money is invested and then earns interest, both the original sum and the interest earned now earn interest, and on and on.

When money "compounds," it builds on top of itself, over and over, in a cycle.

Net Present Value

Net Present Value is a calculation that looks at how much money going out now is worth compared to money coming in, in the future, and vice versa.

This concept is best understood by looking at an example. Assume your organization is considering the purchase of a competitor. You would want to look at the cost now and the estimated cost in the future based on what that money might earn. Figure out the time value of both amounts (costs), to determine if laying out the cash today will result in a net gain against the money earned in the future.

Present Value = Future Value/(1 + interest rate) number of years

The following table of **future** and **present values** demonstrates the impact interest rates (or risk rates) have on the decision-making process.

Future and Present Values

Value: $ 1,000	# Periods	Rate Per Period		
		3%	5%	10%
Future	2	$ 1,061	$ 1,103	$ 1,210
Present	2	$ 943	$ 907	$ 826
Future	5	$ 1,159	$ 1,276	$ 1,611
Present	5	$ 863	$ 784	$ 621
Future	10	$ 1,344	$ 1,629	$ 2,594
Present	10	$ 744	$ 614	$ 386

Future Value = Present Value (1 + interest rate) number of years.
Present Value = Future Value / (1 + interest rates) number of years.

What to do:	Other actions:
☐ Describe the difference between the present value of money and the future value of money.	☐ ____________
☐ Identify how the time value of money could apply to a project budget.	☐ ____________
☐ Describe how an increase in interest rates could impact a project.	☐ ____________

Remember

- ✓ A dollar today can be invested, earn interest, and thus be worth more than a dollar tomorrow.
- ✓ A dollar that is stuck inside a mattress today will have less value when pulled into the daylight in ten years due to inflation.
- ✓ In short, the time value of money depends on what would be done with that money.

Enhance Your Learning

Watch the following 6 ½-half-minute video to learn more about present value.

Subjectmoney. (2012). *Time Value of Money: Present Value & Future Value–Lesson–Formula.*	Available at: https://www.youtube.com/watch?v=XF_3Dt-8OPE

Reinforce Your Learning

Time Value of Money. Given a choice, would you prefer to receive $1000 now or $1000 three years from now?

Identify how the time value of money could apply to a project you are familiar with.

____

____

____

____

____

____

Describe how the project could be affected by an increase in interest rates.

____

____

____

____

____

____

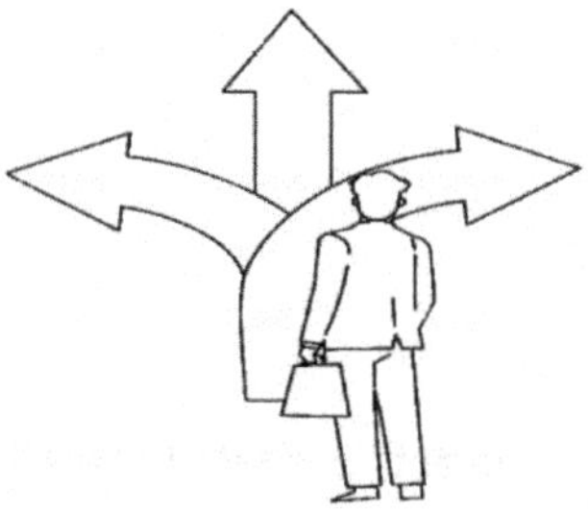

Controlling Project Costs and Budgets #10.9

Make Decisions Using Facts

Creating a budget is, in some ways, is similar to the process of a child deciding how to spend their allowance. Like the child, project managers cannot have everything they want, so they must determine what is most important and put resources there. Of course, project planning is more complicated than spending an allowance. That is why project managers use **alternative choice analysis**, a fancy way of saying that you must make the best decision based on costs, income, goals, and **return on investment**.

In this concept, we discuss how project managers use quantitative and qualitative information to make decisions and describe two useful tools: cost-benefit analysis and return on investment (ROI).

Project Managers Use Quantitative and Qualitative Information to Make Choices

Business and project leaders continuously choose among project alternatives. Alternatives can be evaluated and then ranked in terms of their benefits and costs.

Costs alone are not the only way to make a **quantitative** decision between multiple alternatives. You need to use your own experience and input from other knowledgeable people as qualitative input to evaluate your options.

Justification deliberations may include quantifiable and non-quantifiable elements. **Quantifiable elements** might be things such as anticipated profit, the future value of money, and competitive analysis. **Non-quantifiable factors** (qualitative) might include your own experience, your "gut feeling" for turns in the marketplace, possible litigation, and consumer surveys.

Making choices involves making both short-term and long-term decisions.

Short-term decisions are usually more evidence-based; that is, they rely on reliable, quantitative data. The further into the future you try to project, the less dependable your data become, and the more you need to rely on qualitative information.

Short-term decisions should support long-term goals. Long-term decisions should be a natural extension of your short-term goals and decisions.

Both short-term and long-term decisions/goals can have budgets created for them. But longer-term budgets may be less accurate and require revision as time goes on.

Cost-Benefit Analysis and Return on Investment (ROI)

A cost-benefit analysis lets you look at the costs versus the benefits of taking a specific action. There is no one way to do a cost-benefit analysis. All the concepts discussed in this course go into it.

One important consideration in making decisions is calculating the **return on investment** (**ROI**) of a project or initiative. That is, you want to look at how much profit will be realized, and consider things we have already discussed, such as the break-even point and present value.

Return on investment can be calculated using the following formula:

Return on Investment = (Benefit – Cost)/Cost

Or, as a percentage:

ROI = ((Project's Financial Benefit – Project's Cost)/Project Cost) x 100%

Also, look at the **payback period**, which is how long a new project or initiative will take to become profitable.

For example, installing solar panels on your roof might save you on energy bills in the long run, but it may cost tens of thousands of dollars today. Consider how long that would take to pay off and if it is worth it.

Payback Period = (Project Investment/Annual Cash Flow) x 12 Months

ROI and Payback Period

Return On Investment (ROI)	
Current Revenue	$300,000
Useful Life (Years)	3
Output Increase	5%
Total Benefit	$ 45,000
Cost	$ 22,000
ROI:	104.5%
Payback Period	
# Months	17.6

ROI = (Benefit – Cost) / Cost
Payback Period = (Project Investment / Annual Cash Flow) x 12 Months

What to do:	**Other actions:**
☐ Understand that project budgeting comes down to making better decisions based on facts.	☐ ______________
☐ Consider both quantifiable and non-quantifiable elements when choosing among alternative investment options.	☐ ______________
☐ Identify a project where calculating the ROI and payback period would be appropriate.	☐ ______________

Remember

✓ Alternative choice analysis is just a fancy way of saying that you have to make the best decision based on costs, income, goals, and return on investment.

Enhance Your Learning

Watch the following 5-minute video to learn more about the central questions, "What percent annual return do I get on my investment?" and "How long does it take to recover the initial investment?" in this video comparing alternatives using ROI and payback analysis.

The Finance Storyteller. (2018). *ROI vs payback method.*	Available at: https://www.youtube.com/watch?v=o4of1uNSRis

Reinforce Your Learning

Calculate ROI and Payback. Assume a process improvement project would cost $22,000 and has a useful life of three years. When implemented, the project would result in a five percent increase in the output of an operation that generates $300,000 net income per year.

What is the Return-On-Investment (ROI)?

What is the payback period?

Hint: Return on Investment = (Benefit – Cost)/Cost *Or, as a percentage*
ROI = ((Project's Financial Benefit – Project's Cost)/Project Cost) x 100%

Payback Period = (Project Investment/Annual Cash Flow) x 12 Months

Controlling Project Costs and Budgets #10.10

Use Budgets and Earned Value Analysis for Control

As we discussed, budgets can serve many organizational goals. Organizations use budgets for planning, motivating, facilitating communication, coordinating activities, allocating resources, and so forth. But the truth is that **budgets**, at their essence, **are about planning**, **monitoring,** and **controlling** business processes and opportunities.

Without a budget, it would be next to impossible to plan for success. This is true of a personal budget, where you decide how to spend your paycheck, how much to save, and so forth. Similarly, organizations must decide how to spend their money to meet their goals in the best way, grow the organization, serve more people, and satisfy shareholders.

In this concept, we discuss how project managers use budget variance tracking and earned value analysis to ensure project budgets are well planned and money is well spent.

Organizations Must Monitor and Control How Money is Spent

Budgeting methods can depend on who does the planning and spending of an organization's money. Milton Friedman, a Nobel Prize-winning economist known for being a staunch advocate of the free market economy, once wrote there are **four ways to spend money** (Friedman & Friedman, 1979):

- Spend your money on yourself
- Spend your money on someone else
- Spend someone else's money on yourself
- Spend someone else's money on someone else

As Friedman said, when **spending your money** on **yourself,** you are **cautious** about what you spend, and you look to get good value for your money. In short, when spending their money on themselves, a person tends to economize and look for value.

When you **spend your own money** on **another person,** you tend to **worry less about the value** that you get, yet you are still **careful** about how much money you spend. In short, when spending their own money on someone else, people tend to economize but do not look for the highest value.

Now, when **spending someone else's money** on **themselves,** people are usually **careful to get good value** for the money, but they **do not focus on how much they spend**. When spending someone else's money on themselves, people do not economize but do look to get the most value.

Finally, when a person **spends someone else's money** on **another person** or thing, they are generally the **least careful** because while people have an instinct to do good, they are simply **not as concerned** when it is someone else's money. When spending someone else's money, people do not try to economize and do not try to get the best value.

The point is that when managers spend an organization's money, the organization needs to be sure the managers are careful. Thus, **creating accurate budgets** and **watching** how closely they are adhered to is essential.

Budget Variances

Sometimes, as they say, the best-laid plans go awry, and a budget does not fit the needs of the organization or project once it is in action. When this occurs, the manager in charge may ask for a **budget variance** (Lotich, 2014). This is why budgets should be designed with a cushion of additional money set aside for emergencies.

Variances usually fall under the responsibility of project or department management, and it falls to them to ask questions when things do not come in on budget (Lotich, 2014). When a variance to a budget is needed, a report should be created that lists the old budget, the revisions, the benefits expected from the revisions, and a reason for the changes.

A **baseline plan** is a resource created and approved at the start of a project. It serves as a map of how you think the project will look. Specifically, it lists costs, timetables, and goals. You can adapt the baseline plan for monitoring and controlling purposes by periodically comparing the baseline *budgeted* spending to *actual* spending. Thus, you can see the variances, as illustrated below. Depending on the type of project, you can track variances daily, monthly, quarterly, or annually.

Baseline Plan

Talent Development	1st Quarter	2nd Quarter
Actual Spending:	$32,500	$24,000
Budgeted Spending:	$30,000	$30,000
Variance:	$2,500	($6,000)
Variance %	8.3%	-20.0%

Earned Value Management (EVM)

EVM is a project management technique for measuring project performance and progress. Using EVM, a project manager can combine measurements of the project scope, time, and costs.

Earned Value (EV), also known as Budgeted Cost of Work Performed (BCWP), looks at the portion of a task that is complete (technical performance).

When calculating earned value, managers consider the following factors (For Dummies, n.d.):

- Planned valued (PV): costs specified in the approved budget.
- Earned value (EV): approved budget for work completed up until that point in time.
- Actual cost (AC): the actual costs incurred up until that point in time.

As an example, assume you are going to build a house, and a contractor said it would take, depending on weather, nine to twelve months to complete at an estimated budgeted cost of $175,000. You planned and expected 50 percent of the work to be complete in the first six months.

If now, six months after the start of the project, the contractor would report that they have spent 50 percent of the budget, you might think the project is perfectly on plan. However, in reality, the provided information is not sufficient to come to such a conclusion as you do not know how much of the project is complete nor the actual cost of completed work. The project can spend 50 percent of the budget while finishing only 30 percent of the work, which would mean the project is not doing well. Or the project can spend 50 percent of the budget while completing 80 percent of the work, which would mean the project is doing better than planned.

EVM addresses such issues by including pre-defined methods of quantifying the accomplishment of work. At the end of each month, the contractor identifies every specific element of work completed, and sums the approved budget for each element, or Earned Value (EV), for each completed element. You and the contractor also track the actual cost (AC), which is also accumulated monthly for this case scenario.

The illustration below from Wikipedia shows the concept of Earned Value Management for the house building project. What might you conclude from reviewing the illustration?

Earned Value Management

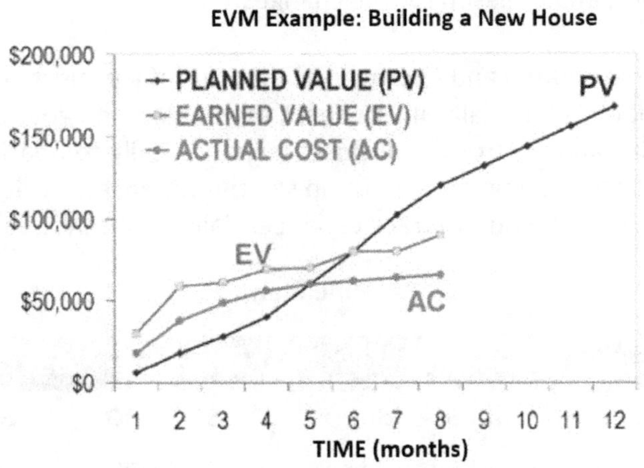

Contingency Planning

Contingency planning involves creating an alternative plan in case things do not go as initially planned. It requires considering what could go wrong and making plans for how to deal with the situation if it does go wrong. When it comes to budgeting, for example, contingency planning could include creating an alternative budget for Quarter 2 if sales in Quarter 1 are not sufficient to fund the Quarter 2 baseline budget.

Post Project Reviews

Post project reviews and evaluations are an essential step in the budgeting process. By reviewing how well projects met their goals and came in on budget, managers can learn more about the budgeting process and improve their ability to make accurate future budgets. In addition, the project manager can ensure that team members stick to the approved budget.

What to do:	Other actions:
☐ Recognize that people are less careful with an organization's money than they are with their own.	☐ ____________
☐ Identify where creating accurate budgets and watching them closely in your work unit is critical.	☐ ____________
☐ Explain why budgets are about planning, monitoring, and controlling, and that the devil is in the details!	☐ ____________

Remember

✓ People are less careful with an organization's money than they are with their own. This is one reason why managing budgets are an important part of leading people. Evaluating how well a budget was adhered to matters.

Enhance Your Learning

Watch the following two videos to learn the basics of Earned Value Management and get a few expert tips on project budgeting. Then review the Wikipedia website for a description of Earned Value Management.

Litten, D. (2022). *PMP Earned Value in 10 Minutes Flat.*	Available at: https://www.youtube.com/watch?v=OfwYr6Pw_x0
Stolovitsky, N. (2014). *Ten expert tips for project budgets.*	Available at: https://www.youtube.com/watch?v=wkmfizbhkY4
Wikipedia. (2022). *Earned Value Management.*	Available at: https://en.wikipedia.org/wiki/Earned_value_management

Reinforce Your Learning

Budgeting and Earned Value Management. Review each of the four charts from Wikipedia that follow for the Building a New House project previously described. Then assess how you think the project is going and why. Finally, consider situations where you either currently use, or might use, earned value analysis for project monitoring or control.

Chart 1. Your assessment: ______________________________

PV = Planned value (PV) or budgeted cost of work scheduled (BCWS)

AC = Actual cost

Viewing the Chart 1, how is the project doing?

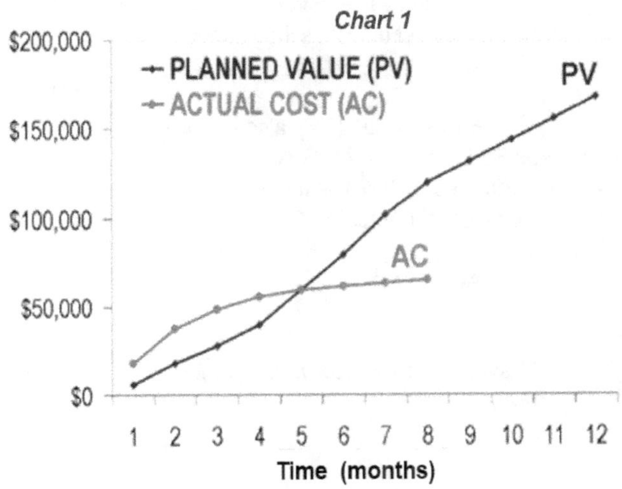

Chart 2. Your assessment: __________________________________

PV = Planned value (PV) or budgeted cost of work scheduled (BCWS)

EV = Earned value (EV), also known as Budgeted Cost of Work Performed (BCWP), is the Budget at Completion x Actual % Complete

(EV is calculated by multiplying % complete of each task (completed or in progress) by its planned value)

Viewing Chart 2, how is the project doing?

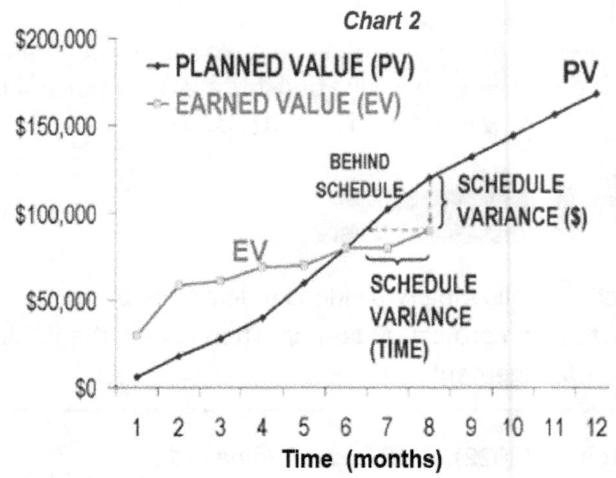

Chart 3. Your assessment: __________________________________

EV = Earned value (EV), also known as Budgeted Cost of Work Performed (BCWP), is the Budget at Completion x Actual % Complete

AC = Actual cost

Viewing Chart 3, how is the project doing?

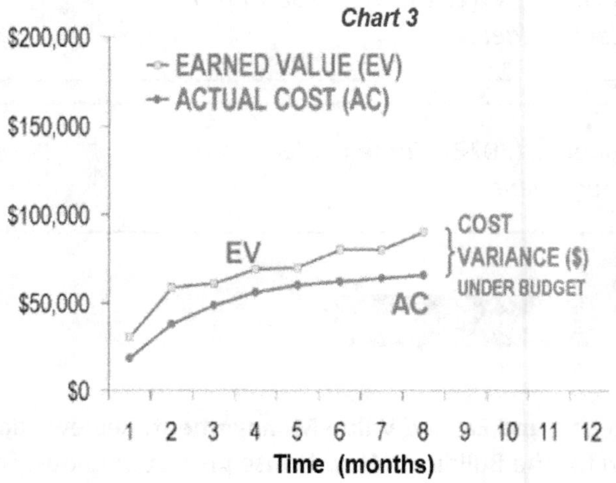

Chart 4. Your assessment: __________________________________

PV = Planned value (PV) or budgeted cost of work scheduled (BCWS)

EV = Earned value (EV), also known as Budgeted Cost of Work Performed (BCWP), is the Budget at Completion x Actual % Complete

AC = Actual cost

Viewing Chart 4, how is the project doing?

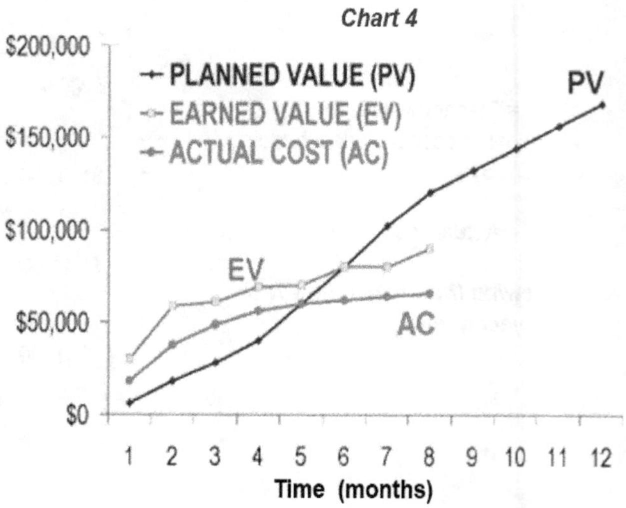

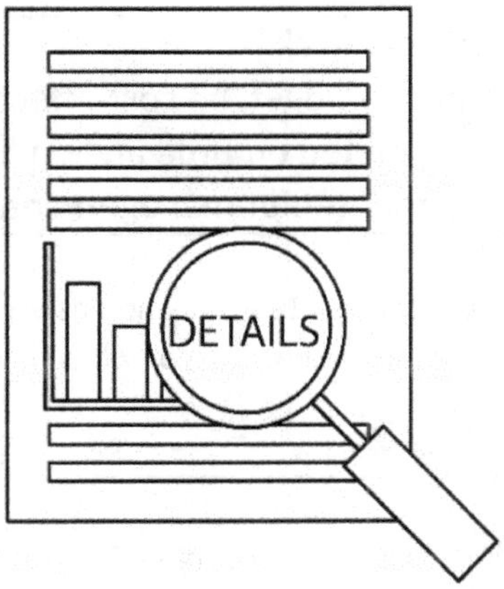

Summary

A Final Word: *Controlling Project Costs and Budgets*

Whether you are seeking approval from a client, sponsor, government agency, or any other stakeholder, you must know how to communicate the budgeting numbers, line items, and reasoning effectively and persuasively.

You have just completed a chapter on Controlling Project Costs and Budgets.

A few topics you encountered in this chapter include:

- Describing the typical types and uses of budgets.
- Explaining the general process of budgeting.
- Providing examples of typical costs and benefits.
- Discussing financial analysis tools and techniques.
- Describing how budgets and earned value analysis can help you control projects.

In the next Project Mastery chapter, we will focus on Managing Time and Project Meetings.

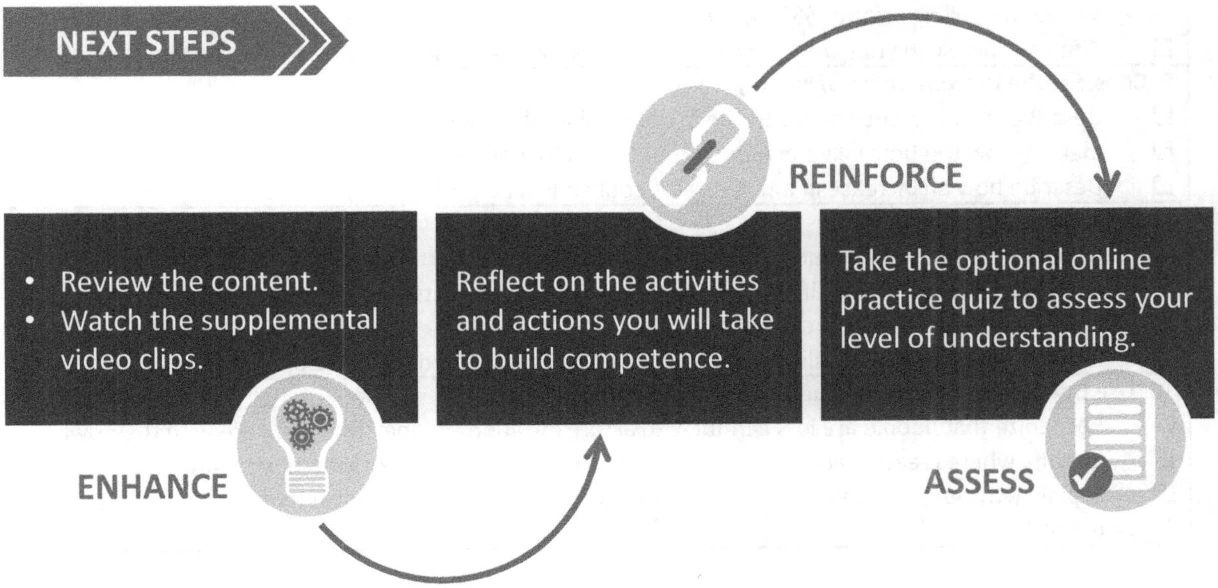

Controlling Project Costs and Budgets—Recap Checklist
Everything in business is bound by money. Every project must have a budget, be cost-effective, and justify expenses. However, **less than** a **third of all projects** are **completed on time** and **on budget,** and **33 percent of projects fail** because of a **lack of involvement** from **senior management**.
1. **Understand How Organizations Use Budgets** ☐ Recognize that managerial budgets include educated guesses about expectations, and answer questions such as how much money, from what sources, to whom, by when, and for what? ☐ Identify the types of budgets you use or will use. ☐ Describe the difference between a capital budget and an operating budget.
2. **Understand How Organizations Track and Report Money** ☐ Locate and review the Chart of Accounts for your organization or project or create one. ☐ Identify how daily transactions are recorded in your organization or for your project. ☐ Describe how General Ledger and Journal entries relate to a Balance Sheet.
3. **Appreciate How Budgeting Relates to Time** ☐ Describe the difference between cash basis accounting and accrual accounting. ☐ Cite where developing optimistic, pessimistic, and most likely budget scenarios would be appropriate. ☐ Define depreciation and cite an example.
4. **Understand Budget Forecasting** ☐ Explain how forecasting is making an educated guess and applying logic and common sense. ☐ Cite examples of where qualitative and quantitative budgeting approaches may be applied. ☐ Develop a pro forma or projected budget forecast for your project or one you are familiar with.
5. **Estimate Project Costs** ☐ Describe how estimating costs is part of accounting for expenditures. ☐ Describe the fixed and variable costs for your project or a project you are familiar with. ☐ Estimate the budget costs for a training program applicable to your job or position.
6. **Evaluate Project Benefits** ☐ Consider a project you are familiar with and identify the benefits to be derived from the project, such as employee safety, increased capacity, or decreased costs. ☐ Consider the benefits of conducting business safely and ethically. ☐ Estimate the anticipated benefits for a training program applicable to your job.
7. **Analyze the Break-Even Point** ☐ Recognize that the break-even point is total expenses matching total earnings. ☐ Calculate the break-even point for an initiative with fixed costs of $2000, variable costs of $50 per unit sold, and a selling price of $80 per unit. ☐ Cite an example in your workplace where a break-even analysis would be a valuable tool to apply.
8. **Consider the Present Value of Money** ☐ Describe the difference between the present value of money and the future value of money. ☐ Identify how the time value of money could apply to a project budget. ☐ Describe how an increase in interest rates could impact a project.
9. **Make Decisions Using Facts** ☐ Understand that project budgeting comes down to making better decisions based on facts. ☐ Recognize that both quantifiable and non-quantifiable elements should be considered when choosing among alternative investment options. ☐ Identify a project where calculating the ROI and payback period would be appropriate.
10. **Use Budgets and Earned Value Analysis for Control** ☐ Recognize that people are less careful with an organization's money than they are with their own. ☐ Identify where creating accurate budgets and watching them closely in your work area is critical. ☐ Explain why budgets are about planning, monitoring, and controlling, and that the devil is in the details!

Knowledge Review Test—Questions

Controlling Project Costs and Budgets

1. Which is about the present and details how daily operations of the project and will be handled and what the budget over the coming year will be?

   A. Operating budget.

   B. FISCAL.

   C. Capital plan.

   D. ROI.

2. Which is more long term, looking at the organization's plan for the future (expected revenue and growth), how investments will be made, what projects will be financed and so forth?

   A. Operating budget.

   B. FISCAL.

   C. Capital budget.

   D. ROI.

3. Which type of budget is easier to prepare?

   A. Fixed budget.

   B. Flexible budget.

   C. Project budget.

   D. All are easy to prepare.

4. During the allocation of project costs, where would a project manager typically allocate team training expenses?

   A. Direct costs.

   B. NPV.

   C. Indirect cost.

   D. Variable cost.

5. How close to actual costs should a budget estimate expect to be?

   A. -75% to +25%.

   B. -10% to +25%.

   C. +10% to -25%.

   D. -10% to +10%.

6. A contractor informs you that your activities have resulted in an increase in their costs. You should:

   A. Issue a change to the project costs.

   B. Ask the contractor for supporting documentation.

   C. Have a meeting with management.

   D. Ignore the request.

7. Cash based accounting means what?

   A. Everything must be paid for in cash.

   B. Nothing is paid for in cash.

   C. Expenses and income are reported when they happen.

   D. None of the above are true.

8. What word means: to predict a future condition or occurrence; to plan or arrange beforehand; foresight in planning.

A. ROI.
B. Forecast.
C. Budget.
D. Accounting.

9. The present value of US $500,000 received 3 years from now is:

A. US $550,000.
B. US $450,000.
C. US. 325,000.
D. Not enough information.

10. Which refers to a type of cost used in accrual accounting where the cost to create a product that will be sold is not listed on the balance sheet until the period in which the product is sold?

A. Period costs.
B. Product or service costs.
C. Materials costs.
D. Indirect costs.

11. Which are those necessary for the organization but cannot be directly tied back to an income stream?

A. Period costs.
B. Product or service costs.
C. Materials costs.
D. Indirect costs.

12. What is one way to manage variable costs as the budget manager determines a maximum amount that they are willing to spend on a specific line item and in the budget, overall.

A. A budget ceiling.
B. Variable costs.
C. Budgeting.
D. Calculating ROI.

13. Which of the following are true:

A. Efforts to improve employee morale and turnover are important benefits to consider when budgeting projects.
B. When employees have low morale, they are more likely to lose work time by goofing off on the job or using sick days.
C. It is more expensive to hire a new employee to replace one who leaves than it is to keep an employee happy.
D. All the answers are true.

14. An organization is said to "break even" when?

    A. On Black Friday.
    B. When they are "in the red."
    C. When total expenses match total earnings.
    D. All the answers are true.

15. This formula calculates which of the following: Fixed Costs / (Unit Selling Price – Variable Costs).

    A. ROI.
    B. Breakeven point.
    C. Overhead costs.
    D. Return on investment.

16. In a cost-plus fixed fee contract, the fee:

    A. Never changes.
    B. Changes if change orders are issued.
    C. Is small.
    D. Will always decrease.

17. An advantage of a fixed price contract for the buyer is:

    A. The buyer has the cost risk.
    B. The seller has the cost risk.
    C. There is little risk.
    D. The risk is divided equally.

18. The following formula is for what concept? (benefit – cost)/cost?

    A. ROI.
    B. Time value of money.
    C. Interest.
    D. Breakeven point.

19. ROI is expressed how?

    A. As a fraction.
    B. As a dollar value.
    C. As a percentage.
    D. As a fixed value.

20. With a fixed price contract, the profit is:

    A. Part of the negotiation.
    B. Unknown.
    C. Applied equally to both parties.
    D. Determined with the other party at the project's close.

Controlling Project Costs and Budgets — Answer Sheet

1.	_X_ A	___ B	___ C	___ D	___ E
2.	___ A	___ B	_X_ C	___ D	___ E
3.	_X_ A	___ B	___ C	___ D	___ E
4.	___ A	___ B	_X_ C	___ D	___ E
5.	___ A	___ B	___ C	_X_ D	___ E
6.	___ A	_X_ B	___ C	___ D	___ E
7.	___ A	___ B	_X_ C	___ D	___ E
8.	___ A	_X_ B	___ C	___ D	___ E
9.	___ A	___ B	___ C	_X_ D	___ E
10.	___ A	_X_ B	___ C	___ D	___ E
11.	___ A	___ B	___ C	_X_ D	___ E
12.	_X_ A	___ B	___ C	___ D	___ E
13.	___ A	___ B	___ C	_X_ D	___ E
14.	___ A	___ B	_X_ C	___ D	___ E
15.	___ A	_X_ B	___ C	___ D	___ E
16.	___ A	_X_ B	___ C	___ D	___ E
17.	___ A	_X_ B	___ C	___ D	___ E
18.	_X_ A	___ B	___ C	___ D	___ E
19.	___ A	___ B	_X_ C	___ D	___ E
20.	___ A	_X_ B	___ C	___ D	___ E

Chapter 11

Managing Time and Project Meetings

Overview

"Nothing is a waste of time if you use the experience wisely."

—Auguste Rodin

People and issues at work make many demands on your time, yet you only have 24 hours in a day and not that many hours at work. To be effective, you need to set priorities, develop daily and weekly plans to control your time, and make meetings productive.

This Chapter will help you understand how you currently use time, provides guidelines for focusing on priorities, reducing timewasters, and developing meeting agendas focused on project priorities.

After completing this Chapter, you will know how to:

- Identify your current time-use habits
- Sequence top-priority project tasks, responsibilities, and priorities
- Develop daily and weekly plans to control your time better
- Set ground rules for productive project meetings
- Establish structure and strategies for effective meetings

The competencies associated with this Chapter include:

Self-Direction	Resource Usage	Decisiveness

Take Your Temperature for Managing Time and Project Meetings

The following assessment will help you identify what you know about managing time and project meetings and the areas where you can improve. With you and your organization in mind, read each statement below, and consider how much you agree. Use the numbers from 1 to 10, where **1** means you **strongly disagree** and **10** means you **strongly agree**.

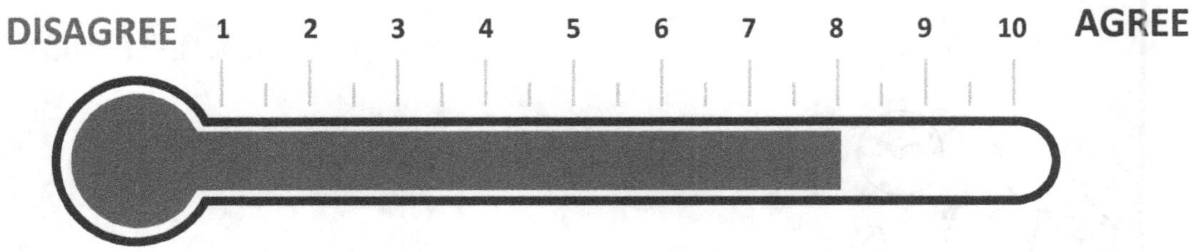

_____ 1. I am good at managing priorities in relation to the time allotted.

_____ 2. I can differentiate urgency from importance as a priority tool.

_____ 3. I use a system for setting and managing priorities.

_____ 4. I know how to plan for today and the future.

_____ 5. I do a good job of managing distractions.

_____ 6. I am good at delegation.

_____ 7. I know the benefits and pitfalls of meetings.

_____ 8. I am confident in my ability to prepare meeting agendas.

_____ 9. I have strategies for dealing with meeting participant problems.

_____ 10. I can develop guidelines for managing effective meetings.

_____ **Total**

Take a moment to reflect on the following before you move forward in the chapter.
In your work, what problems have you experienced or observed related to managing time and project meetings at work?

Take a few minutes to also reflect on your self-assessment. List two or three areas you want to improve.

As you progress through this chapter consider what actions you can take to demonstrate competence in the following three competency areas:

1. Self-Direction. Realistically assesses strengths and weaknesses; invests in self-development; demonstrates self-confidence; works persistently toward a goal; manages time effectively.

2. Resource Usage. Identifies, organizes, uses, and allocates resources.

3. Decisiveness. Decides and responds quickly; makes difficult decisions as necessary.

Managing Time and Project Meetings #11.1

View Time as a Resource

Time is not a renewable resource. Sure, "there is always tomorrow." This may be true; however, such thinking leads to procrastination, or at best, hope for a better future.

The truth is that while time can bring us money, time is more valuable than money. Time is the **only resource** given to everyone in **equal amounts**. We all have the same number of hours in a day. People use time differently according to their varying needs, values, attitudes, habits, and goals.

Think about it. Why do you work? Likely, you work to give your family security, food, shelter, and entertainment. If you are fortunate, you go beyond that and use your money to buy free time—time you spend with your family, friends, and loved ones doing what you enjoy. Time is the most critical resource we have (Mankins, 2014), and we need to spend it wisely.

In this concept, we discuss how increasing your awareness of time as a resource is the first step in learning to use time wisely.

Time is a Limited Resource

Time is the **most limited resource we have**. You can always buy more stock, hire more people, earn more money, and so forth, but you cannot go back in time, change the past, or extend a moment. With 365 days in a year, you have **8,760 annual hours to use wisely**.

How you spend your time is how you spend your life! Planning your time is a critical part of life, both in and out of work. When you are outside work, you do not want to waste precious family time thinking about work. When you are at work, you want to achieve as much as possible.

Some projects are inherently time-constrained. These projects must be completed within a specified period. A particular resource might limit other projects. Still, other projects might be financially constrained, limited by a budget. With time-constrained projects, you must ensure you are clear on what time management has allotted and before you commit, that you have ample time to complete the project.

On Budget, On Schedule, and High Quality?

There is a saying that you can get things done **on budget**, on **schedule,** or with **high quality**, but that you can only have two out of the three – one will always suffer. Experienced project managers know this tends to be true.

Whether time is the major constraint or not, it is always a factor. Projects without deadlines are rare, and time and money are so related that the two eventually influence each other. In addition, quality typically requires a time investment to plan and implement, again making time a critical aspect of any project.

Manage Your Time

To make use of time as a resource, you need to schedule your time, get the most out of each moment, avoid time-wasters, manage interruptions, and limit distractions, all concepts we discuss in detail in this course.

Managing your time can be difficult if you do not know how to do it. But once you develop specific strategies, time management will become an unconscious habit.

Mismanaging your time can lead to **stress**. For example, many people experience stress because their work takes up too much of their time and takes away from family life (WebMD, 2013). Or, without sufficient time management skills, you can lose focus on your goals and miss deadlines (Gaston, n.d.).

Overconfidence also contributes to poor time management. People often think they can get more done than they actually can. Less experienced project managers and those who have not learned how to manage time, sometimes assume strategies that have worked for small projects will work for large projects, but this is not necessarily true.

As a project manager, you must also help others manage their time, or at least help them develop their strategies for managing time.

What to do:	**Other actions:**
☐ Answer this question: What do you want most in life?	☐ ______________
☐ Identify how you currently manage your time daily.	☐ ______________
☐ Be mindful of what is most important to you.	☐ ______________

Remember

- ✓ How you spend your time is how you spend your life!
- ✓ The best way to improve how you use time is to commit to doing so. Tell yourself that how you use time is essential to you, your team, your organization, your family, and your friends. Commit to learning to use your time more efficiently, and you will become more productive and less stressed.
- ✓ Time is a limited and non-renewable resource. You only have as much as you have, and you cannot make more. Further, you cannot replace time with anything else, and you cannot buy more time.
- ✓ Time is the only resource given to everyone in equal amounts. We all have the same number of hours in a day. People use their time differently according to their needs, values, attitudes, habits, and goals.

Enhance Your Learning

Watch the following one-minute video to learn 10 tips for managing your time .

BeatYourDeadlines. (2013). *10 most important time management tips.*	Available at: https://www.youtube.com/watch?v=HFuCXvvXgQQ

Reinforce Your Learning

Current Preferences. To effectively control your time, you must understand your current preferences, patterns, and habits. The following questions will help with this analysis.

Know Your Job Responsibilities, Goals, and Abilities. To better understand your job responsibilities, goals, and the abilities you bring to the job, take a few minutes to consider these things and write your ideas below. Keep your answers in mind as you continue with the course.

1. What do you get paid to do? What are your most important tasks and responsibilities in a typical workday?

2. What tasks do you like doing at work? How much time should you spend doing them?

3. What tasks do you dislike doing at work? Do you procrastinate?

4. What tasks do you delegate? Do you tend to delegate tasks you do not like?

5. In what areas of your job do you feel less secure? Do you tend to put off doing work in those areas?

6. During what time of day do you work best—your *prime time*?

Managing Time and Project Meetings #11.2

Determine Your Priorities

You likely are pulled in many directions by various people throughout a typical workday. One of your time management goals must be setting priorities for what you will do and when. In other words, what is most important to you, your position, and the organization?

Many things may seem urgent, but when we consider them, we realize they are not significant. For example, your friend may call you at work to urgently ask if you want to go out to eat that night. While this may seem urgent to your friend, it is not important to your work life.

To be an effective project manager, you need insight into why you allocate your time as you do, and what factors affect your time-use behavior. Only by identifying and working with these factors can you change the time-use pattern of your job. What are your priorities today; over the next six months, and long term? Focusing on what is essential —prioritizing what you deal with— is critical in managing your time.

In this concept, we discuss how to identify your priorities. In the next concept, we will discuss strategies for managing your priorities.

What Matters Most?

Like most of us, you likely have many priorities. Some may even seem to compete with each other. Decisions can be difficult to make, and priorities are challenging to juggle. For example, you might have a responsibility to ensure good customer or client service and remember "the customer is always right." On the other hand, a coworker may complain that a customer has treated them rudely. You want to support your coworker, but you also must support the customer.

The number one priority of all employees is to do what is best for the organization. But doing that requires setting a variety of smaller priorities.

When it comes to your team, one business website offers an excellent list of priorities for you to consider. The website focus is on supervision, but the priorities apply to most leadership situations ("Now that I am," 2010):

- Focus 1: **Stabilization**. Review your team to ensure things are working well. Look for big problems. For example, if you work in a production facility, ensure supplies are on hand, safety warnings are posted where required, the facility is clean and safe, people know how to run the equipment and answer questions, the signage works, and the parking lot is safe.
- Focus 2: **Train new hires**. To ensure a good return on investment, training is a necessity.
- Focus 3: **Team development and communication**. Collaborate and communicate with team members to promote a cohesive team environment.
- Focus 4: **Team staffing**. Ensure your team is sufficiently staffed and that each person is doing the job for which they are the most qualified and the best fit.

- Focus 5: **One-on-one meetings with team members**. Invest time in your team. Get to know each person so they feel appreciated. Learn what they want for their development.
- Focus 6: **Personal productivity**. Devote time to your own development.

Depending on your organization and role, you may have a variety of duties. Consider all the things you do and prioritize what matters most.

Consider a **typical day on the job** and answer these questions:

- What did you accomplish?
- What were the results of your efforts?
- Compare what you accomplished with what you would like to have accomplished and record any conclusions or observations.
- If you accomplished what you set out to, how do you account for that?
- What helped you?
- If you did not accomplish what you set out to, how do you account for that?
- What hindered you?

What to do:	**Other actions:**
☐ List your daily work activities.	☐ ____________________
☐ Estimate the time spent on each activity.	
☐ Think about how an ideal workday might look.	☐ ____________________

Remember

✓ Knowing your priorities is an essential part of getting things done.
✓ What you do is integral to the successful functioning of your organization. But take care of yourself as well. You also need time to enjoy your family and friends and ensure your own physical and emotional health.
✓ You need insight as to why you allocate your time as you do, and what factors affect your time-use behavior. Only by analyzing these factors can you change the way you use your time.

Enhance Your Learning

Watch the following 3-minute video to learn more about prioritizing your work.

Ways To Grow. (2017). *How to Prioritize Tasks Effectively: GET THINGS DONE.*	Available at: https://www.youtube.com/watch?v=czh4rmk75jc

Reinforce Your Learning

Current and Ideal Workday. Consider a typical day at work. List your tasks and activities on a sheet of paper or even better, also start keeping track of each task and activity and time consumed.

Typical Daily Tasks and Activities	Time (%)

After you capture every task and activity, go back and estimate the percent of time you spent on each. This will give you an idea of what priority you placed on each one. Then, mark in the first pie chart below the estimated time you spent on each task during the workday. This will give you a snapshot of how you spend your time. Then reflect on how an ideal workday might look and mark those percentages in the second pie chart.

Current Workday | **Ideal Workday**

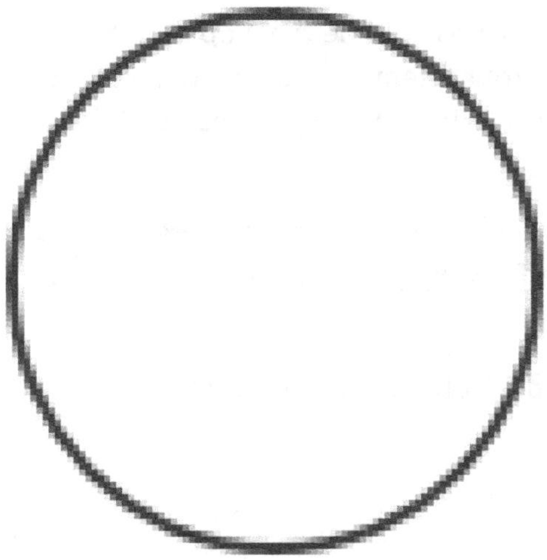

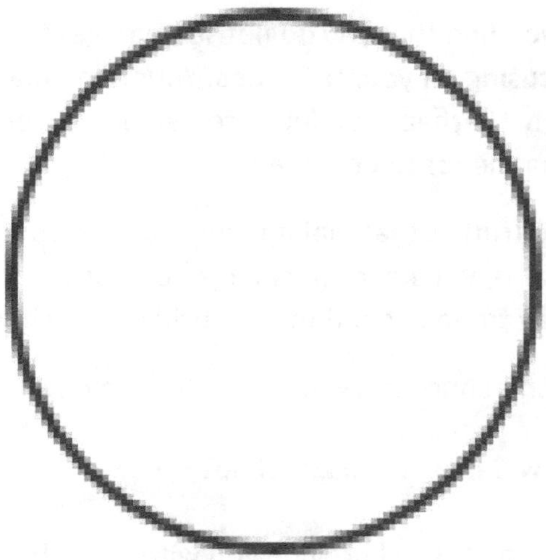

Top Priorities.

1. Based on what you have considered thus far, what are your top three or four priorities?

2. Review your task and activity list and pie charts. What tasks and activities directly contribute to your top priorities?

3. Are any of the tasks and activities that fill your days consuming too much time? Could any of these be better completed by someone else?

Manage Your Priorities

If you find that you do not seem to get things done, the problem might be that you are not **focusing on your priorities**. Your mind may be wandering. Your attention span might be short. Why? Perhaps you feel stressed out? Overwhelmed? Hungry? Tired? Simply incapable of doing what needs to be done?

The truth is that maintaining focus on your priorities **requires** a degree of **concentrated effort**. That is, you need to manage not just your time but yourself. Listing your tasks and activities in a daily **to-do list** and then prioritizing the items can help.

In this concept, we describe how to develop and use a to-do list to manage priorities.

How Can a To-Do List Help?

A to-do list will provide you with many benefits:

- Show you what needs to be done that day.
- Help you prioritize.
- Relieve your stress by eliminating the pressure of having to remember things.
- Make it easier for you to manage time, because you will see what needs to be done, and what tasks/activities are most important.
- Make it easier for you to know when to say "no" to additional responsibilities by showing you what you already must do.

Write Your To-Do List

Your to-do list **needs to be written down**. Just having it in your mind is not sufficient and trying to remember everything will cause stress.

Try different ways of creating a to-do list and find which works best for you; then make that method a habit. Research has found that behavior becomes a habit when you force yourself to carry it out for a period of about three weeks. After that, the behavior should come naturally.

Here are a few ways you might want to record your to-do list:

- **Cell phone notes**. Some people like to use a smartphone note tool.
- **Written notes**. Some people like to write the to-do list on paper and cross items off when they are accomplished.
- **Smartphone apps**. Many to-do list apps are available. Try a few to see which ones you like.
- **To-Do List Organizer**. Many paper organizers have to-do list formats. Or you can create one on your computer using Excel or Word and then print copies. You will also find a template in the Appendix, Part A.

Date ____________________

THINGS TO DO TODAY!

Priority A/B/C	
________	______________________________
________	______________________________
________	______________________________
________	______________________________
________	______________________________
________	______________________________
________	______________________________

How to Organize a To-Do List

You can organize a to-do list in a variety of ways, depending on what works best for you. Here are some ideas to help you get started.

Differentiate tasks that are **must do**, **should do**, and would **like to do**.

List items as a **high priority, middle priority** or **low priority**; or **A, B, C**.

Rank things by day. For example, list your entire week in advance, scheduling what will happen each day.

Organize by time. You might block out your time in hours, half hours, or fifteen-minute intervals, listing what you will do during each period.

You can **combine** these, listing what you will do each day and ranking tasks in priority order. For example, Monday: **Must do this**, would like to do this. The "Would like to do" items could get pushed to Tuesday (eventually, it might become a "Must do" if you push it out too far).

If you organize your to-do list by time, be realistic. Listing too many things to do within a period will leave you feeling rushed, stressed, and unproductive at the end of the day. This will set you up for failure. Make your time-bound to-do lists realistic and doable.

How to Use a To-Do List

How you manage your time is important but **managing yourself is crucial as well**. You can have a well-organized to-do list, but if you do not follow through—monitoring your behavior for **procrastination**, time-wasters, and so forth—you will not accomplish much, no matter how many to-do lists you write.

Whether you choose an electronic or a paper to-do list, do *not* erase or delete items after you have accomplished them. Instead, cross off items with a pen, a check button, or by changing the font to "strikethrough," so you see what you have done and feel a sense of accomplishment.

Everyone needs to take a break occasionally to refresh and refocus. At least once during the day, pause to contemplate achievements and refocus on your priorities. This might mean looking at your to-do list and feeling good about what you have accomplished, or it might mean reevaluating your priorities and adjusting your list.

What to do:	Other actions:
☐ Identify your top three or four priorities.	☐ ____________________
☐ Review your activity list and decide which activities directly contribute to your top priority outcomes.	☐ ____________________
☐ Consider whether any of the activities could be more efficiently completed by someone else.	☐ ____________________

Remember

✓ If you overload your to-do list, it can lose its effectiveness and instead create stress and make you feel behind schedule and unaccomplished. Be realistic, especially when setting time-bound lists.

✓ If you tend to overschedule and overwork, consider scheduling breaks into your day. Consider the 20-20-20 eye break (Mayo Clinic). Every 20 minutes, look up and off at a distance of 20 feet for at least 20 seconds. This important eye break can also be a mental break. So, set an alarm for 20 minutes and take a short break.

✓ How you spend your time is how you spend your life! Maintain focus on your priorities.

Enhance Your Learning

Watch the following 8-minute video to learn more about how to keep track of all your tasks.

Patreon. (2020). *The Beginner's Guide to Microsoft To Do.*	Available at: https://www.youtube.com/watch?v=6BfVtLD6iT0

Reinforce Your Learning

Plan to Delegate. With your to-do list as a starting point, use the template below to list project tasks or activities that you must perform yourself, and those you could delegate to others. Base your "delegate" list on the nature of the activity and not on the availability of people to whom you could delegate. Any activity you could legitimately give away should go in the "Activities I can delegate" column.

Activities I must perform	Activities I can delegate

Review the list of activities that could be delegated and select the one you would like to delegate. In the space below, list the activity and the name of the person to whom you plan to delegate responsibility.

Activity to be delegated: ____

To Whom: ____

Managing Time and Project Meetings #11.4

Plan Your Goals

You want to be a millionaire when you retire. We all do. That is a great long-term goal. But reaching that goal requires achieving many smaller goals in the meantime.

You might have even shorter-term steps, like to get through this holiday season on budget or to spend the next month without going out to eat. The point is that if you **genuinely want** to **accomplish anything,** you **need to set a goal**. Then you need to **plan** how you will achieve that goal.

Making a plan, writing it down, and committing to the plan emotionally are necessary steps to accomplishing any goal. Setting goals and planning can change your life now, and that will change your future.

In this concept, we discuss how to plan so that you do achieve your goals.

How Long Will It Take to Achieve Your Goals?

Not all goals take the same length of time to achieve. We recommend you think of goals as short-term, mid-range, and long-term.

Short-term planning is for goals you want to complete the **next day through the year**.

Long-term planning includes goals to be achieved **beyond three years.**

If applicable to your goals, you might also set **mid-range** goals for things you want to complete in **one to three years**.

Short-term goals and mid-range goals can stand alone, or they can be a subset of longer-term goals.

When we plan goals, we often think of long-term planning, such as building up our 401K or paying off a house before retiring. Indeed, these types of long-term plans do require many small steps that you must achieve over time.

But even **short-term goals** may **require multiple steps** and careful planning over time. For example, you may have a short-term goal to decrease workplace injuries by 90 percent within the next three months. But what smaller and shorter-term goals will be needed to accomplish that? Perhaps conducting a safety analysis or holding various training sessions would be beneficial in shorter-term steps.

What you plan to do today are short-term goals, whether they lead towards longer-term goals or not. Examples of immediate short-term goals might be:

- Complete personnel reports
- Clean out your cluttered email inbox
- File completed paperwork
- Improve customer service scores

Accomplishing your short-term goals requires you to develop **daily** and **weekly** plans to better control your use of time.

SMART Goals

All your goals need to be **SMART**, even your short-term goals. This does not mean you need to complete a full review of each goal against the acronym, but rather that you ensure each goal is **specific**, **measurable**, **achievable**, **relevant,** and **timely**.

As an example, consider a goal to clean out your cluttered email inbox. It is **specific** to say: Look at each message in my inbox and file or delete it. It is **measurable** to say: Get my inbox down to 10 messages or less. It is **achievable** to say: Go through the entire inbox and get it down to 10 messages or less by the end of the day. This goal is **relevant** because there might be things in your inbox that you have neglected to attend to because it is so cluttered. It is **time-bound** because you set today as the deadline.

One **short-term** time management strategy is just to **do it,** whatever "it" is. Another approach is to get smaller tasks done **quickly** and out of the way so you can focus on more significant tasks. Yet another method is to rank-order tasks by importance and accomplish the most important goals first.

No matter how you choose to get things done, evaluate your **short-term goals** to ensure they **align** with your **long-term goals** and that they are significant. If they are not, rethink whether they should be goals.

Planning Medium- and Long-Term Goals

As we discussed, **short-term goals** are intended to be achieved today **through the year. Medium-term goals** are intended to be completed between **one year and three years. Long-term goals** are expected to take **longer than three years to accomplish**.

How you plan for medium- and long-term goals is similar. Here are some examples.

- Short-term planning could include designing new packaging over the next three months or implementing a new software system this month.
- Medium-term planning might include increasing your customer or client base by 40 percent over the next two years or piloting a new product and having it ready for nation-wide distribution in three years.
- Long-term planning could include becoming an industry leader over the next ten years or taking your stock public in the next eight years.

Medium and long-term planning are critical for success both in projects and in business. Such planning gives you the ability to do research, evaluate risks, and properly think through plans before you take steps to achieve them. If you make decisions without thinking things through, you risk the future outcome of your project and organization. As we have said, short-term planning helps you take steps towards long-term goals.

The planning we are discussing here applies to both your business and your personal goals. One interesting thing to note is that research shows **when people look far into the future, their present priorities change** (Lebowitz, 2014). For example, if you ask a person today if they should save $50 in a 401K or indulge by spending $50 on a nice meal, most people will say 401K. However, if you prime people to think far into the future, say ten years, they might be more likely to vote for a nice meal now. This is because when we think far ahead, we tend to believe that current moments are what matter most, and we are more likely to indulge ourselves in the present.

This alerts you to the fact that you need to take an honest look at your goal-setting and find a way to balance your short-term, mid-term, and long-term goals.

Setting short-term goals can help you to plan for the future in ways that are more careful and less indulgent.

On the other hand, long-term goals can be good motivators. If you know that your goal is to retire in twenty years and buy a cottage on a lake, then you can set shorter-term goals to help you achieve that.

Long-term goals should be written down. They need to be **SMART**, which means they need to be specific, and time bound. Do not just say you want to increase enrollment numbers in your training program. Say you want to improve enrollment by 20 percent within 18 months. When you create long-term goals, take the time to set specific short or medium-term goals that will help you to work toward meeting the long-term goals. Improving your time management skills in the short-term can help you reach your long-term goals.

You Can Take Control

Here are a few things to remember (Lewis, 2014):

- With proper time management skills, you can take control of your life, operating on your schedule rather than letting work run you. When you plan your time effectively, you are in control of when you do things, rather than being pressured to do what might appear to be urgent.
- Managing your time effectively helps you to balance your priorities better. For example, it is easier to relax and take time out for a celebration or vacation if you have been managing your time well and meeting your short-term goals.

A goal-based planning approach to improve your application of time to your priorities is an effective strategy for improving the way you manage your work. Here are some questions that may help you plan your goals:

- What is currently the most important in your life? In your job/career?
- What can you do during the next week to move closer to one of your long-range goals or help you actualize what is most important in your life?
- Thinking specifically about your current job responsibilities, what are your goals or objectives for the next three to six months?
- If you were to select one of your goals or objectives to work on during the next week, what would it be, and what tasks or activities would you include in your weekly plan?

What to do:	Other actions:
☐ List priority tasks on your to-do list for today, as well as on your weekly schedule, and rate each task by its importance: A, B, or C.	☐ ____________________
☐ Document what short- and long-term goals are most important to you.	☐ ____________________
☐ Use SMART goal statement language.	☐ ____________________

Remember

- ✓ Creating a to-do list is the best short-term time management tool there is.
- ✓ Even short-term goals should be SMART.
- ✓ There are two aspects to time management: setting goals and planning the steps toward achieving those goals.
- ✓ Write down your goals—short-term and long-term—and the steps you need to achieve them. Use positive language such as, "I will pay off all my credit cards within two years."
- ✓ Use a goal-based planning approach to improve the application of time management activities.

Enhance Your Learning

Watch the following 2-minute video to learn more about long-term planning.

Howcast. (2011). *How to Achieve Long-Term Goals.*	Available at: https://www.youtube.com/watch?v=cEfeDxWRr8c

Reinforce Your Learning

Priorities Checklist. Do you focus on your priorities? Check the following that apply:

- ☐ I know what my priorities are.
- ☐ I list priority tasks on my to-do list for today, and on my weekly plan, and rate the importance of each task.
- ☐ I keep this idea in mind: A priority is something you do first, not last.
- ☐ When I am interrupted or become distracted, I review my priorities to regain focus.
- ☐ I recognize that when someone makes a demand on my time, it does not mean I must meet that demand; especially when it diverts time from my goals.
- ☐ I understand that priorities will change because of conditions beyond my control.
- ☐ I review progress on priorities regularly.
- ☐ Other:

Long-Term Goals. On the form that follows, list three of your long-term goals using the SMART goal format. Then, beneath each goal, write at least one short-term plan that will help you achieve your long-term goal.

GOAL #1:

Short term plans:

GOAL #2:

Short term plans:

GOAL #3:

Short term plans:

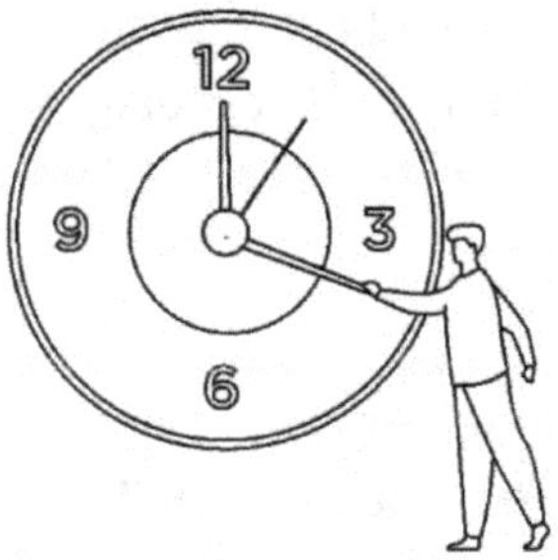

Managing Time and Project Meetings #11.5

Control Time Wasters and Distractions

The internet, while an excellent source of information, can be a time-waster. Besides surfing the net or looking at social media, gossiping about co-workers, and daydreaming also waste hours in the day.

It is also true that these activities can sometimes be useful as opposed to being a waste of time. The trick is to know the difference, and limit activities that do not help you complete your work. Conversely, some work activities might seem to be worthwhile, but upon analysis, are a waste of time.

Learning to pay attention to how you spend your time, to schedule yourself, and to look for and avoid time-wasters are necessary aspects of getting your work done and accomplishing your goals.

In this concept, we discuss how you can make better use of your time by planning and scheduling what you want to accomplish, and also by scheduling limited time for activities that otherwise could become time-wasters.

Recognize How You Use Time

Managing time well does not mean working every minute of every hour. It means *planning* your time, so you *consciously control* what you do, and when you take breaks.

This may sound simple, but it can be difficult because time-wasters are common in our modern world. It takes effort to become aware of what you are doing. The most critical aspect of avoiding time wasters is to recognize *your* biggest temptations.

Distractions such as chit-chat with co-workers, video games, cell phone apps, email, telephone calls, and surfing the net are time-wasters.

Other time-wasters are due to **inefficiencies**, such as the poor organization of files, failure to keep required paperwork, allowing problems to escalate due to procrastination, making uninformed decisions, and more.

Procrastination wastes time in a variety of ways:

- First, you do not accomplish what needs to get done.
- Second, the problem sometimes gets worse because you have put it off.
- Third, you stress out because you needed to accomplish a task, and now it is hanging over your head—and it is distracting you from achieving other goals.

Say you have thing X to do, and you are dreading doing it. Now you have a choice. You can put off X for days, even weeks. But in your mind, you will most likely be thinking about thing X while stressing over or fearing it, avoiding it, planning for it, or obsessing about it. Eventually, the deadline will approach, and you will have to rush to accomplish X.

Imagine, instead, that you must do thing X, which you dread. So, you just do it. Now, it is done. The stress is over, like pulling a Band-Aid off quickly!

Procrastination is a tough problem to overcome. To succeed, you must commit and exercise self-discipline to work through each task on your list today. Then do the same tomorrow. Eventually, you will break the procrastination habit.

Everyday Time Wasters and How to Avoid Them

Surfing the net. Today, you can surf online on your phone, tablet, and computer—maybe even on your watch! It is fun to read the news, check the weather, watch a funny video, and so forth. But these activities are time-wasters.

If surfing online is essential to you, set and allot a certain number of minutes for it. Decide you will do it for 10 minutes each morning when you wake up, or for 15 minutes at night before you go to bed, or an hour on the weekend.

Working from home and playing with your pet or housekeeping. If you work from home, a time-waster might be doing the dishes, playing with your pet, or doing some laundry. Again, you can do these things, but schedule them.

Being disorganized. If you are disorganized, you will waste time without understanding what you are doing. Imagine how much time you spend looking for your car keys, trying to find an essential piece of paperwork, or recreating a document you cannot find. By keeping your environment clean and organized, you will save time.

Not doing things right the first time. Another time-waster results from rushing through work and doing it wrong. Perhaps you misread the directions or glossed over a task that requires more concentration. Slow down, read carefully, and do things right the first time. It is more time-efficient to do a job right the first time than to have to redo it.

Anything that interferes with what you need to get done is a time-waster. Schedule time for activities that are important to you but focus on work-related tasks during the hours designated as work. The definition of time-waster varies from person to person, the situation, and the amount of time spent. Look at your time-wasters and figure out how you can avoid, reduce, or eliminate these issues.

Interruptions and Distractions

Distractions and interruptions happen to all of us. Besides taking our time, they also cause us to lose focus, and then we must regroup to get back "into the flow" of what we were doing. Even an important interruption can stop you from finishing what you were doing.

Learning to handle distractions and interruptions is crucial if you want to be an effective leader. You might think you have no control over **interruptions** and **distractions.** But in truth, you can manage these time wasters also.

Learn to Manage Interruptions and Distractions

Interruptions occur when you are working, getting things done, and someone or something stops your work, begging for your attention. Examples are a team member asking a question, your child running into the room while you are on a conference call, someone phoning you while you are writing a report, and so on.

Interruptions derail you by **taking time** you have **not scheduled,** and they cause you to **lose your train of thought** and slow your progress; not only by taking up time, but by making you start over, wasting time getting back into "the zone."

Besides desiring to be "kept in the loop" and to socialize, here are some everyday things you might be doing that make you prone to interruptions:

- Failure to delegate
- Tendency to "micro-manage"
- Fear of offending others
- Location and space layout invite interruption
- Inadequate screening of calls and visitors
- Confusion in priorities, or confusion of "urgency" vs. "importance"

Distractions are similar to interruptions in that they impact what you can accomplish in a day, often without your realization. Distractions are like interruptions in that they use up time and cause a lack of focus.

When it comes to interruptions and distractions, here are a few things you can do to minimize the effect they have on your work:

- Create a **to-do list** and stick to it.
- If your co-workers frequently interrupt you on unimportant issues, set some **limits**. Make office hours or appointments, create guidelines that tell people only to interrupt during certain times, or for emergencies, or consider setting open times where anyone is welcome for any reason.
- **Role-model** good behavior; do not interrupt others when they are working.
- Often, you can **stop interruptions** before they get far. If a person knocks on your door and you are busy, ask if it is an emergency. If it is not an emergency, ask them to come back at a specific time.
- When it comes to distractions, you must be **in charge** of your mind. Exercise self-discipline.
- Use technology to your advantage. **Turn off your cell phone** or email alerts when you need to work. Or, set your email to out-of-office and indicate you are consumed by a project today, and that you will respond to emails tomorrow, unless it is an emergency.

If someone interrupts you and you attend to their request, here are a few things you can do to get back to work after you handle the issue:

- When someone first asks for your attention, ask the person to give you five seconds. Then, jot down a couple of words that will refresh your memory about what you were doing.
- After you manage the problem, you might feel like jumping back into work— great! But if you lost your rhythm, take an extra five minutes to walk around, stretch, and relax before you go back to work.
- If you were writing something before the interruption, refocus your attention by rereading the last couple of paragraphs.

Interruptions and distractions will happen. You can try to minimize them, but some are going to get by you. One of the most important things you can do is to **schedule a little extra time into your day** to allow for such issues. Contractors who do things like floor tiles and drywall know that when you order supplies, you always factor in at least an extra 10 percent for waste. Time works the same way.

If you plan your workday down to the minute and do not accomplish every item on your list, you will feel stressed, frustrated, and unaccomplished. Instead, when you schedule your time, leave some wiggle room, some spare time to manage waste. For example, if your plan is for an eight-hour workday, schedule tasks you can complete in six or seven hours. That will give you one or two hours of wiggle room to handle unexpected issues, or to handle things that take longer than you anticipated, and without undue stress.

What to do:

- ☐ Identify your three or four biggest time-wasters and decide which ones are due to distractions and which ones are due to inefficiencies.
- ☐ Think of two or three distractions that regularly occur in your workday.
- ☐ Decide what you will do to better manage or prevent these distractions from happening.

Other actions:

- ☐ ______________________________
- ☐ ______________________________
- ☐ ______________________________

Remember

- ✓ Stress relievers and time-wasters are not the same. A time-waster is something you derive little benefit from, and it takes you away from your planned work.
- ✓ Some activities that seem useful are time-wasters in disguise. Say you are working on a project you do not enjoy, so you clean your desk, then review emails, then your files, and so on. Sure, these are important, but they are not your priority, so they waste time.
- ✓ Understand why you allow yourself to be interrupted or distracted to the point that you do not finish your priority tasks.
- ✓ Use phrases such as, "I'm busy now, but let's schedule a time to talk."
- ✓ Do not overschedule your time—leave "wiggle" room to allow for interruptions.

Enhance Your Learning

Watch the following 4-minute video to learn more about choosing the right projects; those that are not a waste of time.

Forleo, M. (2011). *How to Recognize Great Opportunities vs. Time Wasters*	Available at: https://www.youtube.com/watch?v=EM6saU7X4q8

Reinforce Your Learning

Time Wasters in the Workplace. Which of the following time-wasters do you experience? Check all that apply, and then circle the three biggest time-wasters.

- ☐ Telephone interruptions
- ☐ Drop-in visitors
- ☐ Poorly run or unnecessary meetings
- ☐ Cluttered desk/personal disorganization
- ☐ Lack of goals, objectives, or clear priorities
- ☐ Attempting too much at once
- ☐ Managing paperwork
- ☐ Responding to e-mails at random times throughout the day
- ☐ Lack of (or unclear) communication
- ☐ Inadequate, inaccurate, or delayed information
- ☐ Indecision
- ☐ Procrastination
- ☐ Ineffective delegation
- ☐ Confused responsibility and authority
- ☐ Inability to say no
- ☐ Constant distractions
- ☐ Lack of self-discipline
- ☐ Waiting for responses
- ☐ Others:

My Common Distractions. Think of two or three distractions that regularly occur during your workday. Then decide what mitigating actions you will take to better manage or prevent these distractions from happening.

1. Distraction: ______________________________

   Mitigating actions: ____

2. Distraction: ______________________________

   Mitigating actions: ____

3. Distraction: ______________________________

   Mitigating actions: ____

Managing Time and Project Meetings #11.6

Organize and Delegate Work

Sometimes, no matter how much you plan, how many distractions you avoid, and how hard you work, it may seem you do not have enough hours in the day to do everything. At least it feels that way. But perhaps you put too much on your plate. Maybe your plans are too lofty. Or, maybe you just need some guidance.

Organizing yourself, your work, and your environment will help you accomplish more in the time you have. This means that you take on an appropriate amount of work in an appropriate time frame, prioritize your tasks, organize your resources, and then get the job done.

Delegating some of your work is another way you can accomplish more. Remember, delegation is not "passing the buck." We are not talking about pushing difficult or tedious work off on others. We are talking about **making the best use** of your skills and your time and using **your team** effectively.

In this concept, we discuss how you can better organize your time and how to delegate appropriately.

Organize Yourself

As you work to manage your time more efficiently, be aware that everything begins with **organizing yourself** and your environment. You need a to-do list (organize your time). You need to organize items so you can find them (your environment). You need to organize your ideas, work tasks, and priorities, and organize how you handle interruptions and distractions (yourself).

Sometimes managing your time simply means saying "**No.**" Taking on more than can be reasonably accomplished in a certain period adds stress to your life and guarantees frustration. When you already have a full schedule, and someone asks you to do something additional, you must either say you cannot do it or before you accept the new work, remove something from your list of things to do.

Organize Your Workspace

How does your **workspace** look? Is it organized? Can you quickly find what you need? Here are some **tips** on space organization:

- Start with your **desk or work area**, and as with space organization, set priorities for what is most important.
- **Clean up** regularly and keep essential things in front of you**. Throw away** things that are not relevant or out-of-date.
- Keep your **files in** good order, so you can retrieve what you need quickly.
- Apply the above ideas to your **computer**, for example, organize your desktop around priorities.
- If you must carry **work home**, take **only important** items.
- Do not become **preoccupied** with **perfection.**

Set Priorities

Recognize when the things that compete for your time are **urgent** and when they are **important**.

- Some things are **urgent** in that they need to be attended to in a timely way, even if they are not that important.
- Other things are **important**, but they do not have to be done right now.

First, spend **most of your time** on things that are important and **urgent**. Then focus on things that are important and not urgent. Next, focus on things that are urgent but not important, and, if time permits, on things that are neither urgent nor important.

Consider these examples:

- Ordering office supplies before the 6 p.m. delivery cutoff might be urgent, but not important.
- Completing vendor evaluations is important but not urgent.
- Getting a haircut is neither important nor urgent.
- Calling a supervisor about a recent safety concern is both urgent and important.

There are many different theories of time management and prioritizing. However, what they all come down to is this: only you can manage your time. You need to set your priorities, determine what you can reasonably do in an allotted amount of time, and then fix your environment and your behavior so you can do these things. If you fail, try again.

Set Your Schedule

An article in *Entrepreneur* magazine offers a reminder that there are only **three ways to spend your time** (other than sleeping): **thinking, having conversations,** or **doing something** (Mathews, n.d.).

The following are a few more tips to getting the most out of time (Mathews, n.d.):

- **Schedule time with yourself**. When you have work to get done, block out time on your calendar to do it.
- **Spend most of your time on the thoughts, activities, and actions that get the most results**. This sounds logical, but if you stop and reflect upon yesterday, is this what you did? Or did you spend the most time chatting with someone about some future idea that is not relevant today?
- **Set office hours.** This is a time set aside for "planned interruptions." Any work you get done during that time is a bonus because you are planning time to be available to others.
- **Every time you make a phone call** or prepare to do a task, take a few moments beforehand to **determine what your goal is** and what you plan to **accomplish**. Then, follow through. Afterward, take a moment to evaluate how effective you were and what you could have done to make the task more effective.

In his classic book *How to Get Control of Your Time and Your Life,* Alan Lakein established the following six rules to gain control of your time and life:

1. List goals and set priorities.
2. Make daily to-do lists.
3. Start with A's, not C's.
4. Determine the best use of your time right now.
5. Handle each piece of paper once.
6. Do it now.

Learn How to Delegate

To delegate properly, you must find people with the skills to do some of the items on your to-do list. It might also mean training people so that in the future they can do some of the things you need done.

Never give a job to someone who lacks the training or ability to do the job. It will cost you more time than you save and may damage that person's confidence, which may make them less effective in the future.

Use the following guidelines to delegate successfully:

- Choose people you feel confident can get the job done and who agree to take it on.
- Be specific about what you want the person to do, to what standards, and the schedule.
- Check in during the process to see if the person needs help or asks questions.

Cultivate a Positive Attitude

Attitude is everything. If you have a **positive attitude**, if you believe that you can get done what you need to do, if you know that you can avoid distractions and follow your plan, then you are well on your way to success.

Success and accomplishment require **managing your stress**, avoiding **procrastination**, taking charge of situations, and taking charge of your destiny. You are the master of your destiny. You are the one who determines how much work you can do each day. You are the one who controls distractions and prevents interruptions. You are the one who plans your schedule, focuses on tasks, manages stress, and leaves work each day feeling accomplished.

What to do:	Other actions:
☐ Learn to say "no" when there is not enough of you to go around.	☐ ____________________
☐ Organize your work area regularly to keep essential items in front of you. Throw away things that have little relevance or are no longer important.	☐ ____________________
☐ Work toward accomplishing the items on your to-do list while avoiding interruptions and distractions.	☐ ____________________

Remember

- ✓ Work on managing yourself, your work, and your environment efficiently.
- ✓ Factor extra time into your schedule to allow for unexpected interruptions and distractions.
- ✓ The most effective way to take charge of each day is to write a to-do list.
- ✓ Having a positive attitude will help you accomplish more.
- ✓ Develop the discipline needed to use your system effectively.
- ✓ Create a to-do list for every day or week, whichever works better for you.
- ✓ Set long-term goals in your personal as well as your work life. Then, set short-term goals to help you achieve them.
- ✓ How you manage your time and priorities is how you live your life.

Enhance Your Learning

Watch the following 7-minute video to obtain tips to increase your productivity.

Tracy, B. (2012). *6 Time Management Tips to Increase Productivity.*	Available at: https://www.youtube.com/watch?v=3KafgR2WEgY

Reinforce Your Learning

Time Management Checklist. The guidelines below will help you use your time consistently in ways that support the achievement of your goals. Check the ones you think will help you the most right now.

- ☐ Start each day in a positive mood.
- ☐ Make a to-do list for the day and prioritize the tasks.
- ☐ Schedule only 60 percent of your time.
- ☐ Do not waste time getting to work.
- ☐ Coordinate your time plan with your team and others you work with.
- ☐ Start the most important task at the beginning of the day.
- ☐ Try to adjust deadlines if they interfere with your plan for the day.
- ☐ Decline new pressing tasks.
- ☐ Take breaks.
- ☐ Finish the most important tasks first, then move to the next most important.
- ☐ Use those extra bits of time to plan, prepare, or take care of routine activities.
- ☐ Do your best to try to set up and honor a "quiet hour" if possible.
- ☐ At the end of the day and week, analyze your results.
- ☐ Try to go home in a positive mood, leaving your work concerns behind.

Larry Late Simulation. Below is a day in the life of Larry Late, a team leader. Read what he did yesterday and check whether each activity was an effective use of his time.

Time Frame	Activity	Effective	
		Yes	No
8:20	Arrived at work 10 minutes early to get a cup of coffee.		
8:20-8:40	Shared vacation plans with two co-workers who were also getting coffee.		
8:40-8:45	Went to restroom for morning meditation.		
8:45-9:05	Told Ivan Intense he would hold a "Project Team" meeting today.		
9:05-9:15	Attended safety orientation for employees hired over the last three months.		
9:15-9:50	On way back to office, was asked to share opinion to settle conflict between two co-workers who were arguing about project work schedules.		
9:50-10:05	Learned the project manager had moved-up the scheduling meeting to 11:00		
10:05-10:30	Rushed to review material for 11 o'clock meeting. Answered two phone calls from the main office concerning some delays.		
10:30-11:00	Spoke with supervisor about delay problems and learned that problems were caused by material issues. Made note to call main office after lunch.		
11:00-11:45	Met with the scheduling manager to help plan schedule. Manager launched into a long story about how complex things have become and the value of leadership. He then discussed why geese flew in a "V" formation. Meeting ended.		
11:45-11:50	Reminded supervisor about attending lunch meeting with Business Group.		
11:50-1:30	Luncheon Meeting. Went well until he was introduced as a presenter. He forgot he agreed to talk about his PM Team actives. Winged it.		
1:30-1:45	Looked through emails. Found two that must be answered by tomorrow.		
1:45-2:05	Began to return phone calls left from yesterday about scheduling.		
2:05-2:20	Went to break room for coffee.		
2:20-3:00	Co-worker dropped-in to chat about progress of his Project Team.		
3:00-4:00	Participated in interviews of three candidates for a project scheduler position.		
4:00-4:20	One supplier called and wanted to know when they could deliver supplies.		
4:20-4:25	Ivan reminded him about Project Team meeting. He promised they would have it tomorrow.		
4:25-4:45	Tried to make a copy of a report due today for the project manager. Photocopier was out of toner. Tried to replace cartridge, only to spill toner on himself.		
4:45-5:00	Tried to make two important calls. Both parties gone for the day.		
5:00-5:10	Looked over stuff on his table and felt a migraine coming on. Said to himself: I'll get in early tomorrow and make a list to catch up. Went home.		

To-Do List for Tomorrow. Based on yesterday's tasks/activities, create a to-do list for tomorrow for Larry Late. Assign each task/activity a priority of A, B, or C.

Date ____________________

THINGS TO DO TODAY!

Priority A/B/C	
______	______________________________
______	______________________________
______	______________________________
______	______________________________
______	______________________________
______	______________________________
______	______________________________
______	______________________________
______	______________________________
______	______________________________
______	______________________________
______	______________________________
______	______________________________
______	______________________________
______	______________________________
______	______________________________
______	______________________________
______	______________________________

Time Management Systems and Role Models. Whether you use a notebook or computer, a planning and scheduling system keeps information associated with managing your time in one place. Such a system can help keep you organized and help you stay disciplined in managing your time and priorities.

What systems are available? Which ones might work for you?

List people you know who are good at managing their time and priorities.

What can you learn from these role models? What systems do they use?

Understand Meeting Benefits and Pitfalls

Many project managers think of meetings the same as they think about air. They believe they must have them. For example, we have Monday progress meetings, weekly team meetings, client meetings, staff meetings, stakeholder meetings, and the list goes on. But are all these meetings necessary? Why have meetings in the first place?

Effectively managed meetings are a significant source of communication, ideas, and innovations. Meetings facilitate information sharing with instant opportunities for feedback and clarification. In addition, the trend toward increased team and stakeholder involvement in problem-solving and decision-making suggests that we will have more meetings, not fewer, in the future.

So, the answer is yes, some meetings are necessary. But others are not. To realize the gains that can come from necessary meetings, project leaders must make meetings as useful as possible. Leaders must also eliminate unnecessary meetings. Alternatives to meetings, including one-on-one discussions, memos or bulletin postings, email, text or chat, phone and conference calls, video conferences, and internal social networks, also have their place. For example, it is typically not necessary to schedule a meeting if all you want to do is pass on straightforward information, or if there is nothing that a team needs to discuss.

In this concept, we discuss when it is advisable to hold a meeting, and common problems with meetings. In the next concept, we will discuss how to conduct productive project meetings.

When Do You Need a Meeting?

In general, hold a meeting when you want to:

- Share information
- Process information for problem-solving
- Plan actions and assign responsibilities
- Make decisions
- Generate new ideas

A meeting is an effective way to **kick-start a team**, launch a **project**, **celebrate success**, **gain consensus**, and **maintain team spirit**.

Meetings help people feel connected. This is especially the case when group members travel frequently, are based in separate locations, do not see each other regularly, or do not have other opportunities to connect.

Meetings offer some efficiencies over other means of exchanging information. For example, people often play phone tag, and especially when more than two people are involved. Setting up a meeting, even a brief one, establishes a definite time to connect. **Text** and **chat** are great ways to ask straightforward, factual questions and to get a quick response. But when people need to discuss an issue activities such as text, chat, and email can be unproductive.

Email is **useful in some situations**, but if you think about it, like most of us, you have probably spent time crafting an email only to realize that you could have resolved the issue in three minutes in person or by phone. Email is great for communicating long distances when time is not of the essence, for recapping discussions, and for reminding people of action items or **documenting decisions**. But email is often overused. If talking to a person or holding a meeting is more efficient, do it!

Many articles and websites recommend that you cancel regularly scheduled staff or team meetings if you have nothing to share. But canceling is not always the best choice. As a project manager, meeting with your staff each week can be beneficial. You will learn what they have been doing, and regular meetings are an excellent time to recognize what has been achieved and to ask if anyone needs assistance. You may choose to cancel a project team meeting, however, when there is nothing to report.

In essence, hold a meeting when you need to hold a meeting, and when it is the most efficient way to achieve something.

Meeting are useful when they:

- Facilitate information sharing and provide opportunities for clarification and feedback.
- Enhance communication of complex information, including verbal (words), vocalic (tone of voice), and nonverbal (facial expression and body language) communication.
- Provide opportunities for employees to participate in solving problems and making decisions.

When Not to Hold a Meeting

One widespread problem with meetings is that too many leaders hold meetings when getting together is not necessary. Having too many meetings, especially when the perception is that most are not useful, can cause participants to treat even important meetings with half-hearted interest. Participants find it hard to take preparation for any meeting seriously, which results in superficial discussions and meetings that fall short of their objectives.

Why are some meetings boring? Why does the word *meeting* strike dread in the hearts of busy employees? Why do meetings have such a bad reputation?

The reality is that holding a productive meeting is not as simple as it might seem. Many meeting problems arise from, among other things, insufficient planning, faulty communication, lack of proficiency in managing differences of opinion and conflict, and lack of closure and follow-through.

Meeting Problems to Avoid

As we discussed, **one problem** is having **too many** unnecessary **meetings.** Every meeting needs a clear purpose that is best served by specifically holding a meeting.

A related problem is when people are asked to attend who see little value or interest to them. Their input is not needed, or the agenda has no bearing on their work. If some people need to know about meeting decisions, you can distribute a summary after the meeting.

Wasted time, confusion, overly long meetings, and lack of conclusions are typical results when participants have not been adequately informed about why they are attending and what is expected of them. When people do not know why they have been asked to participate in a meeting, it can be challenging to accomplish meeting objectives.

Many meetings are **not well organized**. They may not include all the people with the most relevant information and interest in the meeting issues, and the meeting space, time, and equipment may be inadequate. Even when the leader provides an agenda, it may have been hastily prepared, not shared with all participants before the meeting, or not structured logically. Even worse, many meetings have no agenda.

Timing can also be a problem. Sometimes meetings are scheduled when important participants cannot attend, or at a time that is inconvenient for others.

Some meetings suffer from weak leadership or facilitation skills. Discussions may be superficial, rambling, or disjointed; issues may be raised but not resolved; the leader's focus may be on staying within a strict time limit rather than taking the time to deal with issues that require more discussion.

Also, without the appropriate skills, leaders can find it difficult to facilitate meetings when some people are long-winded and others quiet, and when some people are disruptive and others apathetic.

Meetings can be serious time-wasters! In one study, mid-level managers reported that only about **half the time** they spent **in meetings was effective** (Legacee, n.d.). Time is wasted when meetings start late, even only five or ten minutes can be time put to better use. Meeting time is also wasted in idle conversation, straying off-topic, or when people repeat the same points. A significant waste of time is when an issue under discussion is concluded with no resolution or decision and tabled until the next meeting. Even in well-run meetings, this may happen from time to time, but when it does, participants should be assigned to continue working the issue, so progress is made between meetings.

What to do:	**Other actions:**
☐ Hold a meeting when you need to and send a written agenda ahead of time to all attendees.	☐ ____________________
☐ Limit attendees by inviting the people who should be there.	☐ ____________________
☐ Allow enough time in the schedule to discuss topics and make decisions.	☐ ____________________

Remember

- ✓ Sometimes having a meeting is the most efficient way to discuss issues and make decisions.
- ✓ A quick fifteen-minute meeting can be more productive than a series of two emails, three missed phone calls, and a text. When you have things to discuss or need ideas from several people, have a meeting!
- ✓ On the other hand, do not have meetings just to have them. Sometimes an email or phone conversation is appropriate and productive.
- ✓ One problem with meetings is that people hold them when they are unnecessary. If a simple email, public posting, or phone call will do, choose one of those over a meeting.
- ✓ Conducting a productive meeting is not as simple as it may seem. Without attention to detail, time can be wasted, and people can become meeting intolerant.

Enhance Your Learning

Watch the following 3-minute video to learn more about effective meetings.

Tracy, B. (2013). *How to Conduct Effective Meetings.*	Available at: https://www.youtube.com/watch?v=wm0XtZ_K3kA

Reinforce Your Learning

Meeting Problems Checklist. Think back to recent meetings you have attended where you felt the meeting had problems. Check all the issues below that apply:

- ☐ Meeting was not necessary.
- ☐ No agenda, and the goals were not clear.
- ☐ Critical items (elephant in the room) were not discussed.
- ☐ Meeting too long and went over the allotted time.
- ☐ Discussion was sidetracked off the subject and moved to tangents.
- ☐ Discussion was rambling, redundant, or digressive.
- ☐ Leader did not distribute an agenda, and the attendees were not prepared.
- ☐ The right people were not in the meeting.
- ☐ I did not know everyone, and no one made introductions.
- ☐ Meeting leadership or facilitation was disorganized and ineffective.
- ☐ The meeting provided irrelevant information.
- ☐ My time was wasted.
- ☐ Started late.
- ☐ The meeting had no process for reaching decisions.
- ☐ The meeting had several interruptions, and there was a lack of control.
- ☐ One or two people dominated the discussion.
- ☐ A few participants were hostile to other participants.
- ☐ Conflicting views were raised but not resolved.
- ☐ The facilitator did not publish results or follow-up on actions.
- ☐ The meeting was scheduled at an inopportune time.
- ☐ Other:

Of those items checked, circle the two or three you most want to improve in future meetings.

Make Project Meetings Productive

As we discussed, **people meet to share information and ideas**, **make decisions**, and **get things done**. So, like it or not, you will need to plan and conduct meetings. It is up to you to make your meetings as effective and efficient as possible. Be aware that according to at least one survey (Forbes, 2014), many employees consider ineffective, nonproductive meetings as the top time killer.

In this concept, we discuss how to make your meetings productive for everyone involved.

What is an Effective Meeting?

The person in charge of a meeting must ensure that the meeting is productive, engaging, informative, and useful from the perspective of the participants. **Productive meetings** leave team **members feeling energized** and accomplished (Mindtools, n.d.).

Meetings are **productive** when they do not waste time, energy, or money. Ineffective meetings waste all these resources.

In estimating the costs of nonproductive meetings and inefficient meetings, consider the salaries of everyone attending, but also consider:

- Lost opportunity costs: what people are not doing because they are in a meeting.
- Quality costs: what is lost in decision-making when not all participants are prepared, when only one side of an issue is heard, and when people with creative ideas feel suppressed or at risk in speaking up.
- Satisfaction costs: when people leave a meeting confused, apathetic, or resentful, there are considerable costs in motivation, commitment, and productivity.
- Relationship costs: relationships are damaged if, for example, one or more participants are devalued, embarrassed, or attacked.

Productive Meetings Have a Goal and a Process that Works

Productivity is a **goal** for work-related meetings. That is, you hold a meeting to get something done, and to make the best decisions given the available information. Other useful goals are related to productivity; group or team maintenance, building work relationships among members so that productivity is not threatened, and commitment from the group to implement the decisions reached.

A productive meeting is one where (Mindtools, n.d.) the following happens:

- Objective accomplished
- Time used efficiently
- Participants think the process worked

Productive meetings have a **clear purpose** or **objective** (Forbes, 2014). They also have:

- The proper participants, people directly involved with the subject of the meeting and people with relevant skills, experience, and information.
- The right timing, with the meeting scheduled far enough in advance to give participants time to prepare, scheduled at a time convenient to participants, and sufficient time allocated to achieve the meeting objective.
- A meeting climate that supports an honest exchange of information.
- A process for making decisions, resolving conflicts, and gaining consensus or agreement.
- An effective leader or facilitator.
- Explicit ground rules for attendance, appropriate behavior and language, and use of technology.

Most of the time in a meeting should be spent in discussion among the participants. A meeting is a group event, and the participants are there because they have information and opinions to share and things to learn from each other.

As ideas are shared and decisions reached, it is essential that these not be lost. Designating a scribe is one way to capture ideas. Another way, and one that can also stimulate creative thinking, is *group memory*, or the public posting of ideas, concerns, and conclusions expressed during the meeting. A group memory can be created using a flip chart or whiteboard, or by projecting recorded items on an electronic device.

You may sometimes need to ban technology from your meetings (Forbes, 2014). This means that participants cannot use cell phones, computers, or tablets during the meeting. In other situations, such as when there is a need for information not held by anyone in the room, such devices may be useful. Or you may want participants to use electronic devices to record their notes.

Pre-Meeting Responsibilities

As a meeting leader, you must ensure that your meetings are productive. In order to do that, you also must plan the meeting and prep the attendees.

Before a meeting, you must perform the following tasks:

- Define the purpose of the meeting and the objectives.
- Understand who should attend and why.
- Choose the best format and leadership style for the meeting.
- Distribute an agenda, and, if desired, solicit additional items for the agenda.

Meeting Action Items

A productive meeting generates action items to be done after the meeting. If not, the meeting may have had little value other than an opportunity for socializing.

The end of the meeting is the beginning of following up on the action items generated during the meeting. Use a specific format for noting decisions and follow-up responsibilities. You can create a to-do list from the meeting notes or from a chart created at the end of the meeting that shows who is responsible for what, and when.

Types of Meetings and Formats

Meetings are held for a variety of reasons. For example, there are informational meetings, collaborative meetings, problem-solving and decision-making meetings, and training meetings. There are also regularly scheduled team or project-based meetings. Because these meetings have different objectives, they also have different formats. In planning a meeting, choose the format that supports your objective.

Kick-off meetings. A kick-off meeting is a great way to start a project. The meeting needs a stated purpose and an agenda with the ground rules you want participants to follow.

Keep in mind that the first meeting sets expectations for future meetings and, if it is their first meeting experience with you as the leader, the meeting will also set their expectations of you. If, for example, you start late, that can set an expectation that you will start late at every meeting, which sends a message that "time is not important, and it is okay to be late." Set your expectations at the first meeting, and you will have fewer problems later.

Brainstorming and planning meetings. These meetings provide a forum where participants can develop ideas to resolve issues or plan for change. The focus can be on anything, for example, a business process or current best practice, a resource reallocation, a market or customer shift, a restructuring, or how best to implement a change decision made at a higher level. Plan to record the ideas discussed during the meeting. It helps to have a dedicated note-taker who is not part of the idea-generating group. These types of meetings need a facilitator who can direct the brainstorming or planning process.

Team update meetings. Update meetings keep team members current with a project. They provide new information, allow people to share progress and status, and allow participants to raise questions and concerns. Status reports can be shared through email, but meetings foster creative collaboration. These meetings can be informal, but they should begin on time and end when everything is reported.

Issue management and problem-solving meetings. Hold this type of meeting to address a specific problem. For example, the purpose of the meeting might be to address something that is wrong, and that needs correction, or the meeting might address a potential problem. Even if called on short notice, the meeting should not be held until the people with the most relevant information can attend. As well as having a clearly defined purpose and agenda, the meeting needs a systematic process for problem-solving and decision making. That process may include brainstorming as one element. A facilitator may also be valuable in this type of meeting.

Committee meetings. Most organizations have standing and temporary committees to address various issues, and many issues cut across organizational units. There may be, for example, a safety committee, a social or celebration committee, a long-range planning committee, and committees focused on unit-specific issues. The committees may meet regularly or as needed, and they tend to use a standard agenda and format.

Shareholder or stakeholder meetings. Hold this type of meeting when people need to be updated on specific initiatives or outcomes. These meetings are often somewhat formal. If reports are given, the information should be provided in advance so that participants can prepare questions for the meeting.

Customer/client meetings. Many organizations hold meetings with their customers/clients to determine their needs, to get feedback about customer experiences with products or services, and to assess customer reactions to proposed changes. These meetings benefit from a clear purpose and agenda, and from a tight structure that still allows time for customers to have their say.

Personnel meetings. These are one-on-one meetings between a manager and employee, and the purpose is related to employee responsibilities and performance. Such meetings may address, for example, goal setting, coaching, disciplinary actions, and performance appraisals. These meetings should have a clear purpose and meet the criteria for a productive meeting.

Training or education meetings. These can be formal classroom meetings or informal on-the-job meetings.

Planning a Meeting

In real estate it is all about location, location, location. For meetings, it is all about planning, planning, planning. Productive meetings can add value to the workplace. They are a source of ideas, a place where teams come together, and an invaluable aid in creating shared responsibility among the members of a workgroup. Effective, productive meetings are, first and foremost, well-planned.

Some of the problems that arise from poor planning are the wrong people in attendance, participants not prepared in advance, lacking or disorganized agendas, and inadequacies in things such as the meeting space, time frame, or equipment.

When you convene a meeting, you are responsible for making decisions and taking actions **before** the **meeting** to increase the meeting's value. This means that you set the agenda, identify the participants, and take care of the physical arrangements.

As we have mentioned, meetings are more successful when participants understand why they were invited, and what they are expected to do before, during, and after the meeting.

Consider what information you want participants to have before attending the meeting. For example, consider the following questions:

- Is the meeting primarily informational, or are participants going to be involved in discussions, brainstorming, or solving problems?
- Are they expected to give a progress report during the meeting? If so, what format and time frame are they to follow?
- Do they need to prepare by reading any documents?
- Do they need to get information from individuals not included in the meeting and be prepared to share that information?
- What should participants bring with them: records, files, charts, materials to distribute, visual aids?

If it is the first meeting, or if some time has elapsed between meetings, do you need to share the ground rules?

For a meeting you are planning, review the following items to ensure success:

- ✓ Define the purpose and general objectives for the meeting.
- ✓ Establish the location, starting time, and duration of the meeting.
- ✓ Decide who should attend and for what reasons.
- ✓ Decide the best format and leadership style for the meeting.
- ✓ Decide who should chair or facilitate the meeting (highest-ranking member, project leader, another member, someone outside the team or group).
- ✓ Prepare and distribute the agenda and any prework. If different people will facilitate or present during the meeting, put their names on the agenda next to their topics.
- ✓ Solicit additional agenda items if desired.

Plan the Agenda

Prepare and distribute the agenda sufficiently in advance of the meeting to permit participants time to adjust their schedules, if needed, and to prepare.

Apportion agenda time according to the importance of each topic. Place the essential items first.

Sample Agenda

<table>
<tr><td colspan="2">To:</td><td>From:</td></tr>
<tr><td colspan="3">Subject / Purpose:</td></tr>
<tr><td colspan="2">Participants:</td><td>Place / Date / Time:</td></tr>
<tr><th>AGENDA ITEM</th><th>TIMING</th><th>BY</th></tr>
<tr><td>Warm-up Discussion
Review Past Meeting
Current Agenda
• Topic A
• Topic B
• Topic C
Round-the-Table
Summarize Assignments / Actions
Set Agenda for Next Meeting
Evaluate Meeting</td><td></td><td></td></tr>
</table>

Planning Large Meetings and Conferences

For most meetings, small conference rooms located close to participant worksites offer sufficient space and features so that setting up is not an issue. In planning for a large group, especially if the meeting is held at a hotel or conference center, more is involved.

Following are some of the things to consider, along with the usual guidelines (that is, being clear about the meeting purpose, knowing who should attend, and putting together a meaningful agenda):

- Collect specific information about the location and meeting space.
- Identify what pre-meeting materials are needed.
- Define what announcements, relevant points, or other information is to be communicated.
- Get an accurate count of who will attend.
- If the event location requires travel (driving more than an hour or air or rail travel), ensure that everyone has the information needed to make their plans and pay for, or be reimbursed for, travel expenses.
- Ensure that everyone has clear directions to the building and meeting room.
- If the meeting will last longer than a couple of hours, or if people will have traveled a long distance, arrange for snacks. At the least have cold water available. You may also coffee, juice, and tea. Snacks should be easy to eat, such as cheese, crackers, deli trays, granola bars, nuts, and cookies.
- Provide writing pads and pens for each person to take notes.
- Ensure that any audio-visual equipment is set up before people arrive. Also, familiarize yourself with things such as light and air conditioning controls.
- Set the room temperature cool enough for most people to be comfortable. It is a good idea to go a little cold rather than a little hot; when it is too warm, people become sleepy and are more likely to complain.

- Arrange tables and chairs appropriately. For a large group (more than 20) and a presentation style meeting, arrange chairs or tables facing the front. Seating people in an open U is effective for smaller groups if participants need to talk with each other. If you plan to have people work in small groups, you might use cluster seating, which is five to eight people per round or square table.
- For off-site meetings at locations unfamiliar to participants, spend the first few minutes on pleasantries to give stragglers a little extra time to find the place.
- Begin the meeting with a few basics, such as explaining where the exits and restrooms are, and telling people when the first break will be.
- When you schedule breaks stick to the times—start time and end time—exactly. This is respectful of everyone's time and will keep the momentum going.

What to do:	**Other actions:**
☐ Define the purpose of the meeting clearly and ensure participants understand it.	☐ ____________
☐ Use a checklist for planning meetings to ensure you are prepared well in advance of the meeting.	☐ ____________
☐ Develop an agenda that includes the meeting purpose, participants, start and stop times, and that lists agenda items with the most important items first.	☐ ____________

Remember

✓ A productive meeting follows a process to accomplish a specific purpose in a time-efficient manner.

✓ The meeting leader is responsible for ensuring the meeting is productive.

✓ In productive meetings, most of the time is spent in discussion among participants.

✓ Your first meeting with a group can be powerful in setting their expectations and behavior for subsequent meetings. Be sure to model the ground rules you want participants to follow.

✓ Before you hold a meeting, prepare for success. For every meeting, set the purpose and agenda, be clear about what outcomes are expected, and ensure participants are adequately informed.

✓ Take care in setting the agenda, so you do not schedule the meeting too tightly. Allow time for interaction without people feeling rushed. Be prepared to end early if you finish ahead of schedule.

Enhance Your Learning

Watch the following 6-minute video to learn more about running an effective project meeting and consider how you can apply this information in your work.

Whitt, J. (2012). *How to Run Team Meetings.*	Available at: https://www.youtube.com/watch?v=xW0CjH95K3Q

Reinforce Your Learning

Project Meeting Checklist. The following are meeting processes that help ensure a productive meeting. Check the items your team uses.

- ☐ Follow regular schedule
- ☐ Have a note taker
- ☐ Send agenda in advance and clarify what to bring
- ☐ Take attendance
- ☐ Review minutes from last meeting
- ☐ Solicit new agenda items
- ☐ Allow time for open discussion
- ☐ Review and recap commitments and document them
- ☐ Send minutes to stakeholders afterward—confirm date and time of next meeting
- ☐ Other:

Managing Time and Project Meetings #11.9

Handle Difficult Meeting Behaviors

Sometimes we attend meetings where everyone gets along well. Things go great. But when people from different backgrounds and with varying expectations get together, things can go wrong. They may even get heated. In situations like this the leader or facilitator, or both, must have the skills and confidence to put things right.

Achieving the goal of a productive meeting can be challenging when participants exhibit behaviors that are more extreme or inappropriate. Such behaviors can disrupt the meeting, derail the thought processes of other participants, shut down discussion, and leave people feeling uncomfortable—or worse, angry, alienated, or embarrassed. While you may find it hard to do, you must deal with extreme and inappropriate behavior swiftly and firmly.

In this concept, we discuss unproductive behaviors commonly seen in meetings and what you can do to get a meeting back on track.

Properly Plan and Prepare for Meetings

As we have discussed, ensuring that participants understand a meeting's goal and what is expected of them will help set the tone for positive exchanges. Sometimes, despite your best preparation and effort, people may seem to have trouble speaking up, while others may engage in counterproductive or self-serving behavior.

As issues arise, focus on the behavior, not the person. It is the behavior that is difficult or disruptive, not the person. For example, if you notice someone often has a particular problem in meetings, instead of saying, "You are disruptive," you might say, "I notice that when other people express an opinion, you sometimes speak over them. Are you aware of that?" The person may not be aware they are cutting people off, so your comment may help them adjust their behavior. Posing comments as observations, not judgments, is less likely to be seen as an attack.

Focusing on your observation of behavior rather than accusing the person is even more important when you address an individual in front of the group. Anything you say or do that is perceived as a put-down or attack will have a negative effect on the person you address and can suppress participation from the entire group.

Keep Ideas Flowing

Keeping ideas flowing can be challenging, but it is necessary because one of the main benefits of having a meeting is to share ideas. Several things can inhibit discussion.

For example, quiet participants need to be drawn out and reinforced for their contributions. Overly chatty people may need to be advised to allow others to speak. The entire group may have gotten off track and need to be redirected. Participants may not always be respectful in their communication. People may stop listening because of distractions or a personal preference not to listen.

As you facilitate or lead the discussion, watch for potential trouble spots. **Pay attention to the body language** of others and notice if some people seem to feel unheard or neglected while others feel overly empowered to express their opinions. Be prepared to **solicit feedback from those who are quiet**. Carefully turn conversations from those who are taking over to others who may have something to say. You may also be able to short-circuit a disagreement or argument between two participants by asking a third, who you know has relevant knowledge, to join the discussion.

You must handle all these fairly common behaviors if they occur, or else your meeting may not be productive. For these kinds of problems, reminding participants of group-developed ground rules is often sufficient.

If Things Get Heated

Take a break if things start to get heated. Or if a break is not feasible, another strategy is to recommend that the discussion be tabled for now and then move the conversation to something else. You can point out that little progress is being made, and that a fresh look later may be helpful.

The following are some typical **challenging behaviors** that you may encounter, along with some tips for handling them.

Dominating and monopolizing. People who dominate or monopolize the discussion attempt to do most, or all, of the talking. They may feel confident in their knowledge (of most every topic) and want everyone to listen. Or they may have low regard for the knowledge or potential contributions of other participants.

While it is nice to have people willing to speak, allowing one person to dominate can shut down the conversation and the group's creativity.

To handle this problem, first try gentle tactics of redirection, saying something like, "I appreciate your input, John. You obviously feel strongly about the topic. Now, let's hear what Mike has to say. Mike, any thoughts?" Usually, this will be the end of it. If the person continues to argue or dominate, try the tactic again with more force. If things get too bad, you may need to pull the person aside and give direct feedback about the behavior.

Quibbling and being argumentative. *Quibbling* is finding something to disagree with or argue about regardless of the issue. An individual may assume a contentious position for a variety of reasons—and they may just want to frustrate the leader. Or they may have solid, accurate information to share but not know a productive means of sharing.

Regardless, you must keep a cool head and not be drawn into arguing. Argumentative types often irritate others who prefer a more hospitable climate, and this can bring a discussion to a quick halt. If this happens, allowing a brief silence can be effective. Then, pick up the discussion with direct questions to other participants. At times, an argumentative person will recognize what has happened and not present further problems. If the person lacks this kind of sensitivity, you may need to discuss the behavior with the person directly and privately.

Negativity and quipping. Some people express snide comments or quips. These are negative tidbits that add no real content and are simply meant to annoy others or to demonstrate the speaker's annoyance. Negative jibes can turn a positive meeting into a negative one and can cause people to lose focus and become reluctant to voice their opinions.

You have two ways to handle this type of behavior. First, you could take a meeting break and ask to talk to the person in private. This is usually the best choice for someone who is usually not so negative or disruptive. You can point out the problem with the behavior, mention how you think it is affecting the group, and tell the person that you want their comments but without the negativity.

If that does not work, you may want to address the behavior in the meeting with a comment such as, "Marc, we all understand that you are unhappy about this situation. However, your comments are not getting us closer to a solution. What do you suggest we do as a positive next step?"

Joking and clowning. This behavior is similar to quipping in that the person tends to make brief (and sometimes loud) comments meant to be funny. And often, the comments are funny. But while jokes can help to break tension, too many of them can disrupt the pace of a discussion or offend people if the jokes are in poor taste.

You can handle a joker by pulling the person aside or making non-insulting comments in front of the group. Usually, pulling someone aside works well. State that you appreciate the injection of humor, but you need it to be less frequent and possibly more disciplined. Then provide one or two guidelines.

Skeptical. Most people are skeptical at times, but when skepticism becomes a frequent response to ideas or facts, it can be problematic. The best way to handle this behavior is to meet the skepticism with a direct challenge. If what has been suggested by someone else will not work, what does the person think will work?

Sometimes people use skepticism to play devil's advocate. That is, they bring up negatives so they can be discussed, and any potential problems can be resolved early. This behavior can have significant value. In fact, if you have a skeptic that you are having trouble getting through to, you might have a private conversation where you express your appreciation for their willingness to bring up potential problems, but you would like to hear the positive sides of things as well.

Wandering attention and being distracted. Some people tend to focus on something slightly different than what is being discussed. That is, if you are talking about a specific topic, they use that as a lead-in to bring up something else. Refocus these people, saying something like, "Martha, you are asking some good questions, and we will make a note to discuss that later. But first, I'd like to continue with our current discussion."

Sidebar discussions. A *sidebar* discussion is a private, whispered conversation between people that takes place during the meeting or group discussion. Quiet as they might try to be, their whispers and inattention are distracting to the group. During the meeting, gently remind these people to please give their attention to the meeting or to step outside if their private conversation is more important than the meeting.

What to do:	**Other actions:**
☐ Review the types of difficult behavior commonly seen in meetings and ensure you know how to respond should you encounter them.	☐ ________________
☐ Address the behavior. Do not attack the person.	☐ ________________
☐ Take a break if things start to get heated.	☐ ________________

Remember

✓ When addressing someone whose behavior is threatening the productivity or tone of your meeting, focus on the behavior. Avoid doing anything that attacks or embarrasses the individual.

✓ It takes both skill and confidence to manage unproductive behavior, and it is not something most of us look forward to doing. But allowing such behaviors to persist guarantees that both the group process and participant satisfaction will suffer.

✓ Take a break if things start to get heated.

Enhance Your Learning

Watch the following 4-minute video to learn more about passive-aggressive behavior.

Pentameter, M. (2017). *How to Deal with Passive Aggressive People.*		Available at: https://www.youtube.com/watch?v=xTxSP0BixA4

Reinforce Your Learning

Non-Team Behaviors. As a team leader, what is the most important thing you can do to ensure your team works together well? Consider the following non-team behaviors and prepare what you would say to a team member exhibiting the behavior.

- Misses important meetings:
- Comes late:
- Has frequent personality clashes:
- Rambles:
- Does not speak-up:
- Is pessimistic:
- Writes or reads texts in meetings:
- Laptop surfs in meetings:
- Others:

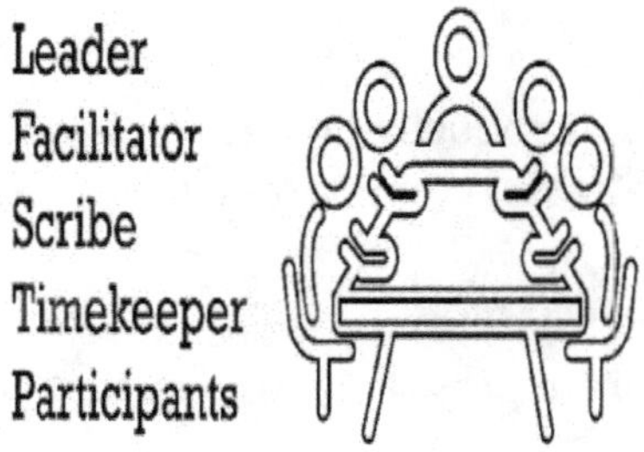

Managing Time and Project Meetings #11.10

Establish Structure and Strategies for Meetings

As we discussed, you must prepare for meetings by having a clear purpose, inviting the right people, and providing a written agenda before the meeting. These are vital to meeting success, but you will also need to provide structure. By this, we mean that you will need roles and responsibilities such as meeting planner, leader, facilitator, scribe, and participants. In some cases, one individual will perform more than one role. In other instances, each role is filled by a different person.

In this concept, we discuss the roles and associated responsibilities needed to structure successful meetings.

Meeting Roles and Responsibilities

Leader role. The *leader* provides structure and direction during the meeting.

The leader often performs the planning role, and the leader is typically responsible for:

- Opening the meeting.
- Focusing participants on the topics or issues being reviewed or discussed.
- Requesting input from participants.
- Clarifying, summarizing, and re-focusing discussions as needed.
- Obtaining action commitments from participants.
- Keeping the meeting on track and on schedule.
- Moving systematically through the agenda.
- Adjusting the meeting style as necessary to achieve the desired results.

Facilitator role. The *facilitator* is focused on process and can support meeting results in a variety of ways, often by assisting the leader in maintaining the focus and involvement of the participants.

The facilitator is responsible for:

- Functioning as an objective party with no stake in decisions.
- Observing the interactions among participants and intervening, if necessary, to improve the quality of the information exchange or decision-making process.
- Giving feedback to the leader about group process and recommending changes.
- If requested, planning the meeting in support of specialized activities or formats.
- If requested, creating the group memory, which is described under Scribe below.

Not every meeting has an individual designated as a facilitator, yet most meetings require facilitation. This includes stimulating the flow of ideas, ensuring that everyone is involved, and maintaining positive and respectful interactions.

Facilitation can be done by anyone; the leader, the named facilitator, or one or more participants.

Scribe role. The *scribe* is the meeting recorder or the person who takes notes. The person leading or facilitating a meeting should not take notes other than for themselves. Notes to summarize discussions and action items for the group and the official meeting minutes should be recorded by someone else. The scribe is a crucial role, and not everyone does it well.

The scribe is responsible for:

- Capturing participant ideas without editing or paraphrasing.
- Creating a group memory for major points and action items, although this may also be done by the leader, facilitator, or another participant.
- Taking notes that will become part of the meeting minutes.
- Summarizing at the end of the meeting when asked to do so by the leader or facilitator.
- Translating the notes into minutes and distributing them to participants or sharing them with the leader for distribution.

For regular team or staff meetings, the responsibilities for taking notes and writing the minutes are typically handled on a rotational basis, with the note taker being one of the participants. With standard formats and sufficient orientation to how notes should be taken, this can work, but it does divide the attention of the participant taking notes.

For more formal meetings, this is ill-advised because it is difficult to fulfill both sets of responsibilities equally. Either the person's contribution to the discussion or the quality of the notes will suffer. Even in regular team or staff meetings, if full involvement from all participants is needed, the notes should be taken by someone other than a participant.

Another option for recording what transpires in a meeting is to have it audio- or video-recorded. There are some concerns here, as well. First, everyone must be informed that the meeting will be recorded and agree to have their comments recorded. Second, there will be some, hopefully temporary, self-consciousness, or hesitation among participants because being recorded can alter the nature of an interaction. Some participants may remain guarded throughout the meeting. Third, processing the recording to condense it down to the key points and create a written summary takes time, even for only a one-hour meeting. Someone taking notes can condense the discussion as it takes place.

Some people suggest assigning note-taking to the last person arriving for the meeting, which makes it a type of punishment, but this may send the wrong message. While it might discourage people from being late, it is not treating the responsibility with the level of seriousness it deserves. And again, not everyone does it well.

Participant role. The quality of the meeting depends on the quality of the participants.

Participants are responsible for:

- Being on time and coming prepared to participate in the meeting.
- Listening to fellow participants and seeking to understand their input.
- Keeping an open mind.
- Observing established ground rules.
- Helping the leader, or the facilitator, manage the task, process, climate, or behavior of other participants.
- Offering constructive ideas, comments, and feedback.

Constructive Participation Guidelines

Giving participants guidelines on how to participate constructively in a meeting will benefit the overall productivity of the meeting. Sharing the guidelines with the participants *prior* to the meeting also sets the tone for how you expect participants to behave.

The following are guidelines that will assist participants in getting their points across effectively:

- **Organize your contribution.** Think before you speak. Make it as easy as possible for others to follow your thoughts.
- **Speak when your contribution is relevant.** Do not ask others to make mental leaps, or bounce around the agenda, to see the relevance of what you say.
- **Make one point at a time.** It is easier for others to understand and remember if they do not need to sort your ideas.
- **Speak clearly and forcefully.** Let the others know you believe in your ideas and want them to listen to you.
- **Monitor your nonverbal signals.** Guard against showing disdain, boredom, condescension, anger, or rejection in response to what someone says. Sit attentively. Assume the attitude of *looking for the benefit of someone else's idea.*
- **Listen actively to the discussion.** Work to understand each person's ideas.

Meeting Facilitation

Except for purely informational meetings, keeping the discussion productive is vital to meeting success. Productive discussions—discussions that retain the interest and involvement of participants and accomplish the objectives of the meeting—**do not happen without facilitation**. Whether you are a leader or a facilitator, you need to be engaged and focused on creating an atmosphere conducive to productive discussion. The success of any discussion depends upon authentic, constructive participation.

For the most part, the meetings that get the best results are those with the most productive discussions. If you are leading a meeting without the support of a facilitator, you are responsible for keeping the discussion focused and meaningful. But how do you ensure that discussions are productive? The weather could be discussed, but unless you are in the weather industry or planning an outdoor event, it would not move the meeting toward its purpose.

The first step is to ensure that everyone feels comfortable voicing their opinions. If participants are a new team or group, introductions will aid people in learning a bit about each other. You can ask people to introduce themselves, or you as the leader or facilitator can introduce and share information about each person's role. Or participants can pair up and introduce one another. No matter how you do the introductions, set guidelines for length and content to keep the introductions from being overly long.

Create a Productive Meeting Environment

The following is a discussion on strategies and guidelines for creating an environment that encourages discussion.

Start on time. This sends a message that the time set aside for the meeting is valuable and not to be wasted.

Stay with the agenda. Unless unavoidable, take the topics in the order planned and stay on schedule. This sends a message that the meeting is designed for specific objectives. If the group needs to spend more time than allocated on an important item, this should be permitted, but only with agreement from other participants that the agenda can be altered.

Build a supportive climate. Listen, encourage everyone's views, and accept criticism or disagreement with your views. Do not dominate, interrupt, or criticize. If divergent views exist within the group, allow for a fair hearing of all sides.

Establish criteria for discussion. Setting an agenda takes care of this, but you may want to share more information about what is expected. For example, you might mention that the first 20 minutes will be an open discussion on the topic. Then, you will brainstorm solutions for 10 minutes, with the last 30 minutes spent in choosing a solution and summarizing the decisions made.

Ask questions to stimulate discussion. Participants need to see that a mutual sharing of opinions and ideas is welcome, and the skilled use of questions helps create this atmosphere. Questions are also helpful in redirecting conversations that have gotten off-topic and in reminding participants of what the meeting needs to accomplish.

Ask safe questions when you start the meeting. If the participants do not know each other well, or if the group is quiet, asking safe questions can help start a discussion. *Safe questions* are ones that can be answered easily and where group agreement is likely. You might also do this if the group is struggling to reach a consensus. Get them to agree on simple things, even just the goal itself, and you will start building bridges toward an agreement.

Give people time to respond to the questions you ask. People sometimes prefer to wait to see if someone else speaks up first. So, if you ask a question and no one speaks up immediately, wait. Resist the urge to fill the silence with your voice. Wait a full minute, then repeat or paraphrase the question. If there is still no response, you can either call on someone or post the question on your group memory and tell them you will return to it after they have given the issue more thought. Use the same process for unanswered questions from a participant.

Use group memory as a discussion aid. Recording the important ideas, questions remaining to be answered, and decisions reached—in other words, recording the *group memory*—helps to keep participants focused and encourages them to share ideas and opinions. The scribe decides what is important by recording it in the notes. The facilitator can also make such decisions either by asking the scribe to add something to the notes or by writing important points on a whiteboard or large paper pad.

Are You an Effective Meeting Leader?

If participants develop a habit of deferring to you as the leader, the quality of discussion can suffer. You must support participants in taking ownership of the discussion and decisions. You may need to recast the role of leader, moving closer to the role of objective facilitator.

Ask questions such as, "What does the group think?" or "Based on your experience, what would you suggest?" Try to encourage participants to speak directly to each other rather than addressing everything toward you or looking to you for answers.

Meetings have the potential to be either highly engaging and productive or low energy and boring. And, while planning and adequate preparation are important, the difference in meeting quality often comes down to how effectively it is facilitated.

Participants must be engaged and challenged to think. The leader or facilitator must pay attention to visual cues that signal participant interest levels; facial expression, body language, apparent distractedness. Sometimes, the only correction needed is to take a break or to introduce an energizing activity. At other times, it is appropriate to ask questions to surface the causes for lack of interest.

How can you tell if participants are engaged? Regularly evaluating your meetings is one way. It will allow you to see how participants assess the quality and value of your meetings.

Presentation Skills

To improve your meeting leadership or facilitation skills, you may want to review your presentation skills. Do you express yourself clearly? Do you present your ideas with enthusiasm? Do you inspire others with your words?

To facilitate a productive discussion, you need to get people's attention, focus their attention on the topic under consideration, and persuade them to speak up and to listen to others. Developing your presentation skills by taking a course or by getting help from a coach or mentor may contribute significantly to the quality of your meetings.

Use a tool, such as brainstorming, to give everyone an equal voice. Brainstorming is often used to help groups generate ideas in meetings focused on solving problems, implementing changes, or coming up with innovations. Brainstorming is useful primarily because it separates the generation of ideas from their evaluation. Used correctly, it encourages creative thinking and objective evaluation.

If you are not familiar with the methods and rules for brainstorming, you may want to find out more, or use a facilitator before including it in a meeting. Depending on the problem or opportunity in focus, there are other helpful tools. You may want to check out *Nominal Group Technique* and the process of consensus decision making.

Always leave time at the end of the meeting to provide closure. Summarize the main points, highlight decisions and action items that were assigned, review any items discussed but not resolved, and talk about the next steps after the meeting. For an hour-long meeting, this should take only a few minutes.

Here is a summary of key ideas for generating participation and productive discussion:

- **Encourage participation**. Be prepared to get everyone involved by asking direct, specific questions.
- **Ask for feelings and opinions**. Be prepared to ask questions that will help people engage more deeply in the discussion.
- **Paraphrase and reflect back**. Use this active listening technique to help people stay attentive and to encourage both listening and speaking.
- **Ask for a summary**. Invite someone else to summarize a discussion.
- **Ask for clarification or examples**. Ask for clarification frequently, especially in highly detailed or emotion-laden discussions.
- **Test for consensus and surface differences**. Ask if everyone agrees.
- **Focus on actions and decisions**. Ask what decisions have been made and what actions have been discussed. Document responses on the group memory.

What to do:	Other actions:
☐ Ensure that meeting roles and responsibilities are assigned.	☐ ____________________
☐ Give participants guidelines for constructive participation.	☐ ____________________
☐ Ask questions to stimulate discussions and give people time to respond.	☐ ____________________

Remember

✓ As a meeting leader or facilitator, your job is to keep the meeting on track. This means asking questions and encouraging people to move along.

✓ The most effective meetings are those with the most productive discussions.

✓ For productive discussions, participants must feel comfortable and able to express their opinions.

✓ Tools such as brainstorming or nominal group technique can help make an idea-generating discussion productive. They can also be a welcome change from a straight question-and-answer format.

Enhance Your Learning

Watch the following 5-minute video to learn more about facilitating a virtual meeting.

Girdler, A. (2020). *How to Run a Virtual Meeting [BEST PRACTICES].*	Available at: https://www.youtube.com/watch?v=NPVTLroz2Ck

Reinforce Your Learning

Project Meeting Evaluation. Consider the last meeting you participated in or conducted. Evaluate the meeting with the intent of improving future meetings. Check all that apply.

- ☐ Did we start on time?
- ☐ Were the meeting roles established?
- ☐ Were the meeting purpose and agenda reviewed?
- ☐ Did we stay on track and focused?
- ☐ Was participation, for the most part, even?
- ☐ Was everyone prepared?
- ☐ Did we use good listening skills?
- ☐ Did we confront and resolve conflict?
- ☐ Did we use communication strategies such as brainstorming and consensus decision making effectively?
- ☐ Were any unproductive behaviors handled well?
- ☐ Were notes taken?
- ☐ Did we summarize decisions, assignments, and actions?
- ☐ Did we agree on the purpose or agenda for the next meeting?
- ☐ Did we end on time?
- ☐ Was sufficient time allotted to work through the agenda with meaningful discussion?
- ☐ Did we accomplish what we intended?
- ☐ Did we have fun?
- ☐ Overall, do you think the meeting time was time spent well?

Summary

A Final Word: *Managing Time and Project Meetings*

Time is the only resource given to everyone in equal amounts. You only have 24 hours in a day; use them efficiently. Meetings can be a productive use of time when conducted effectively because they allow people to communicate efficiently.

You have just completed a chapter on Managing Time and Project Meetings. A few important topics you have encountered include:

- Identifying your current time-use habits.
- Sequencing top-priority project tasks, responsibilities, and priorities.
- Developing daily and weekly plans to better control the use of time.
- Setting ground rules for productive project meetings.
- Establishing structure and strategies for effective meetings.

In the next Project Mastery chapter, we will focus on Completing and Closing Projects.

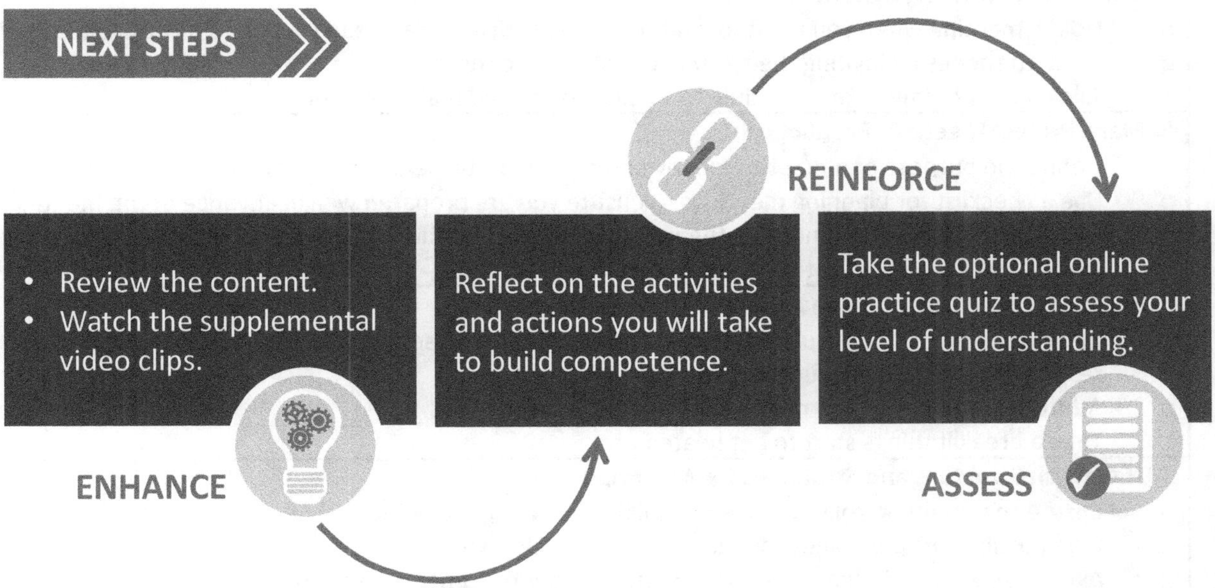

Managing Time and Project Meetings—Recap Checklist

The most important thing to know about time is that time is not a renewable resource. Time is the only resource given to everyone in equal amounts. You only have 24 hours in a day; use them efficiently. Meetings can be a productive use of time when conducted effectively because they allow people to communicate efficiently.

1. View Time as a Resource

- ☐ Answer this question: What do you want most in life?
- ☐ Identify how you currently manage your time daily.
- ☐ Be mindful of what is most important to you.

2. Determine Your Priorities

- ☐ List your daily work activities.
- ☐ Estimate the time spent on each activity.
- ☐ Think about how an ideal workday might look

3. Manage Your Priorities

- ☐ Identify your top three or four priorities.
- ☐ Review your activity list and decide which activities directly contribute to your top priority outcomes.
- ☐ Consider whether any of the activities listed could be more efficiently completed by someone else.

4. Plan Your Goals

- ☐ List priority tasks on your to-do list for today, as well as on your weekly schedule, and rate each task by its importance: A, B, or C.
- ☐ Document what short- and long-term goals are most important to you.
- ☐ Use positive SMART goal statement language.

5. Control Time Wasters and Distractions

- ☐ Identify your three or four biggest time-wasters and decide which ones are due to distractions and which ones are due to inefficiencies.
- ☐ Think of two or three distractions that regularly occur in your workday.
- ☐ Decide what you will do to better manage or prevent these distractions from happening.

6. Organize and Delegate Work

- ☐ Learn to say "no" when there is not enough of you to go around.
- ☐ Regularly organize to keep essential items in front of you. Throw away things that have little relevance or are no longer important.
- ☐ Work toward accomplishing the items on your to-do list, while avoiding interruptions and distractions.

7. Understand Meeting Benefits and Pitfalls

- ☐ Hold a meeting when you need to, and send a written agenda ahead of time to all attendees.
- ☐ Limit attendees by inviting the people who should be there.
- ☐ Allow enough time in the schedule to discuss topics and make decisions.

8. Make Project Meetings Productive

- ☐ Define the purpose of a meeting clearly and ensure participants understand it.
- ☐ Use a checklist for planning meetings to ensure you are prepared well in advance of the meeting.
- ☐ Develop an agenda that includes the meeting purpose, participants, start and stop times, and that lists agenda items with the most important items first.

9. Handle Difficult Meeting Behaviors

- ☐ Review the types of difficult behavior commonly seen in meetings and ensure you know how to respond should you encounter them.
- ☐ Address the behavior. Do not attack the person.
- ☐ Take a break if things start to get heated.

10. Establish Structure and Strategies for Meetings

- ☐ Ensure that meeting roles and responsibilities are assigned.
- ☐ Give participants guidelines for constructive participation.
- ☐ Ask questions to stimulate discussions and give people time to respond.

Knowledge Review Test—Questions

Managing Time and Project Meetings

1. Which of the following statements are true?
    - A. Time is not a renewable resource.
    - B. Time is the only resource given to everyone in equal amounts.
    - C. People use time differently according to their varying needs, values, attitudes, habits, and goals.
    - D. All statements are true.
2. Which of the following facts about project time are true?
    - A. Your time can be one of your primary resources, just as money.
    - B. Your ability to think of time use as an investment is critical to your plan.
    - C. Effective investment of time must be related to accomplishing goals.
    - D. All are true.
3. A good time management system should contain the following:
    - A. A daily/weekly plan.
    - B. A monthly plan.
    - C. Projects section.
    - D. Contact log/directory.
    - E. All are good elements of a time management system.
4. The number one priority of any project manager should always be what?
    - A. To do what is best for your organization.
    - B. To make the most money.
    - C. To reduce stress.
    - D. To make customers happy.
5. Delegation involves:
    - A. Assigning easy jobs to those less qualified.
    - B. Assigning jobs to those that volunteer.
    - C. Assigning jobs to people and giving them the necessary authority and responsibility to get the jobs done.
    - D. All the statements are correct.
6. What is the 20-20-20 rule?
    - A. Do 20 things on your list in 20 minutes so you feel good.
    - B. There are 20 timewasters in every day and 20 ways to stop them in just 20 minutes.
    - C. It's about the clock; 20 plus 20 plus 20 is 60.
    - D. It means if you are working on the computer every 20 minutes look 20 feet away for 20 seconds.
7. You can organize a to-do list in a variety of ways, depending on what works best for you, such as:
    - A. List items as a high priority, middle priority or low priority.
    - B. List items as A, B, or C.
    - C. List items as must do and like to do.
    - D. List items as 1, 2, or 3.
    - E. All are ways to organize a to-do list.

8. Which types of goals should be SMART?

   A. Short-term goals.
   B. Medium-term goals.
   C. Long-term goals.
   D. All goals.

9. Which of the following are NOT true of time management?

   A. Managing time well means working all the time.
   B. To manage your time, you need to conscious plan when you take breaks.
   C. Procrastination is perhaps the biggest time waster.
   D. All the answers are true.

10. Which of the following may be examples of time wasters?

   A. Writing a to-do list.
   B. Playing with your dog.
   C. Setting SMART goals.
   D. All the examples are time wasters.

11. When it comes to delegating tasks you should:

   A. Push people and give them more than they can handle so they grow..
   B. Give a job to someone you feel lacks the training or ability.
   C. Use the opportunity to develop your project team.
   D. None of the answers are correct.

12. You need to recognize when things that compete for your time are ___ and when they are ___.

   A. Urgent/important.
   B. Silly / frivolous.
   C. SMART/ DUMB.
   D. None of the answers are correct.

13. Planning is the best way that you can take charge of your day by:

   A. Creating a to-do list.
   B. Prioritize that list in a way that is meaningful.
   C. Work towards accomplishing the list while avoiding interruptions and distractions.
   D. All answers are correct.

14. Which of the following is not an important element of project meeting management?

   A. Assigning action commitments to team members.
   B. Clarifying the purpose of the meetings.
   C. Obtaining agreement about the agenda and objectives.
   D. Using disagreements to differentiate ideas from feelings.
   E. None of the actions are important.

15. A project team can increase its overall effectiveness by doing which of the following?

    A. Establishing ground rules.
    B. Prepare and Follow meeting agendas.
    C. Making decisions by consensus.
    D. All of the answers are correct.

16. Who should attend project meetings?

    A. Anyone who has direct responsibility/authority related to the topics of discussion.
    B. Anyone who has relevant information.
    C. Anyone who must implement decisions made.
    D. All the answers are correct.

17. In order to help team members prepare for a meeting, the project manager should:

    A. Establish the time frame.
    B. Clarify information needs.
    C. Clarify roles.
    D. Clarify participation.
    E. All the answers are correct.

18. What can be done effectively with a meeting participant who tends to "dominate" discussions?

    A. Confront the person and let them know that they have taken their allotted time.
    B. Ask other participants to answer a question.
    C. Have the "dominant" person serve as the meeting's recorder.
    D. All the answers are correct.

19. Which of the following is not a strategy that a project manager might use to manage a meeting?

    A. Make sure all team members understand their opinions are valued.
    B. Directly confront an overbearing team member, demanding team thinking.
    C. Ask a member to produce data for a strong opinion or position.
    D. Have all team members list possible roadblocks to project success.

20. All but one of the following are important in the Meeting Management process, which is not important?

    A. Start/end on time.
    B. Stick to the agenda.
    C. Summarize agreements.
    D. Take frequent breaks to refocus.
    E. Provide meeting closure.

Managing Time and Project Meetings — Answer Sheet

1. ___ A ___ B ___ C _X_ D ___ E
2. ___ A ___ B ___ C _X_ D ___ E
3. ___ A ___ B ___ C ___ D _X_ E
4. _X_ A ___ B ___ C ___ D ___ E
5. ___ A ___ B _X_ C ___ D ___ E
6. ___ A ___ B ___ C _X_ D ___ E
7. ___ A ___ B ___ C ___ D _X_ E
8. ___ A ___ B ___ C _X_ D ___ E
9. _X_ A ___ B ___ C ___ D ___ E
10. ___ A _X_ B ___ C ___ D ___ E
11. ___ A ___ B _X_ C ___ D ___ E
12. _X_ A ___ B ___ C ___ D ___ E
13. ___ A ___ B ___ C _X_ D ___ E
14. ___ A ___ B ___ C ___ D _X_ E
15. ___ A ___ B ___ C _X_ D ___ E
16. ___ A ___ B ___ C _X_ D ___ E
17. ___ A ___ B ___ C ___ D _X_ E
18. ___ A ___ B ___ C _X_ D ___ E
19. ___ A _X_ B ___ C ___ D ___ E
20. ___ A ___ B ___ C _X_ D ___ E

Chapter 12

Completing and Closing Projects

Overview

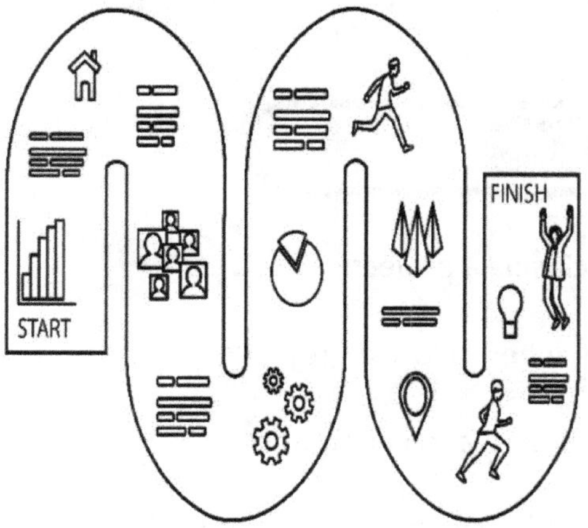

"A project is complete when it starts working for you, rather than you working for it."
—Scott Allen

You and your team work hard on a project, and then, before you know it, you are in the closing phase and preparing to perform the formal closing activities. Or, perhaps you received an unanticipated change in direction from management, and the project was terminated. Closing or termination will each affect your project differently, but they have processes in common.

This chapter describes how to manage the end of a project, whether you complete and close it, or management terminates it. The discussion includes how to leverage any plans for the closing phase that already exist.

After completing this chapter, you will know how to:

- Identify the critical elements to include in closing projects
- Describe project closing tools, techniques, and processes
- Communicate effectively about project changes and the change process
- Evaluate project outcomes
- Identify and document lessons learned

The competencies associated with this chapter include:

Planning and Evaluation	**Resource Management**	**Management Controls**

Take Your Temperature for Completing and Closing Projects

The following assessment will help you identify what you know about completing and closing projects and the areas where you can improve. With you and your organization in mind, read each statement below and think about how much you agree with it. Use the numbers from 1 to 10, where 1 means you **strongly disagree** and 10 means you **strongly agree**.

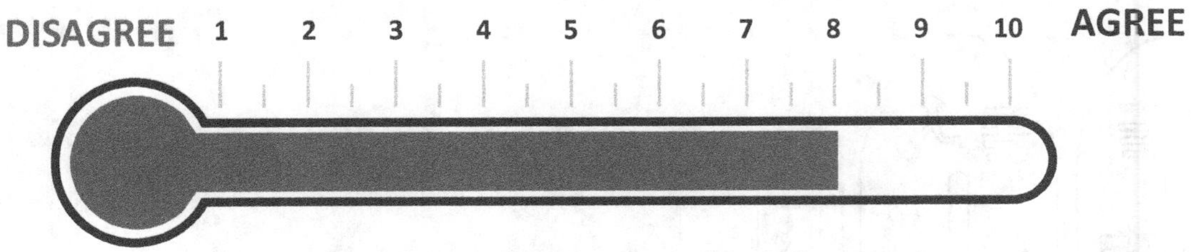

_____ 1. I can explain what it takes to prepare for closing a project.

_____ 2. I understand why laws and policies matter when I close a project.

_____ 3. I am skilled at evaluating resources, work issues, and conditions.

_____ 4. I am mindful of potential future property and casualty liabilities and risks.

_____ 5. I do a good job of identifying and using resources to be safe and legal.

_____ 6. I am effective at evaluating project outcomes: scope, schedule, and budget.

_____ 7. I am good at applying quality tools and techniques at project closing.

_____ 8. I know the importance of proper records retention.

_____ 9. I am good at identifying and documenting lessons learned.

_____ 10. I implement strategies to enhance stakeholder relationships.

_____ **Total**

Take a moment to reflect on the following before you move forward in the chapter.
In your work, what problems have you experienced or observed related to completing and closing projects at work?

Take a few minutes to also reflect on your self-assessment. List two or three areas you want to improve.

As you progress through this chapter consider what actions you can take to demonstrate competence in the following three competency areas:

1. Planning and Evaluation. Establishes policies, guidelines, plans, and priorities; plans and coordinates with others; aligns required resources; monitors progress and evaluates outcomes; improves organizational efficiency and effectiveness.

2. Resource Management. Demonstrates awareness of technical resources; knows how to apply resources to achieve desired outcomes.

3. Management Controls. Ensures the integrity of the organization's processes; promotes ethical and effective practices.

Completing and Closing Projects #12.1

Prepare for Closing or Terminating a Project

The project life cycle illustration below shows that we are in the **closing phase** of the project life cycle. The closing phase is when you will end a project and formally close it.

In this concept, we discuss how project managers prepare to end a project.

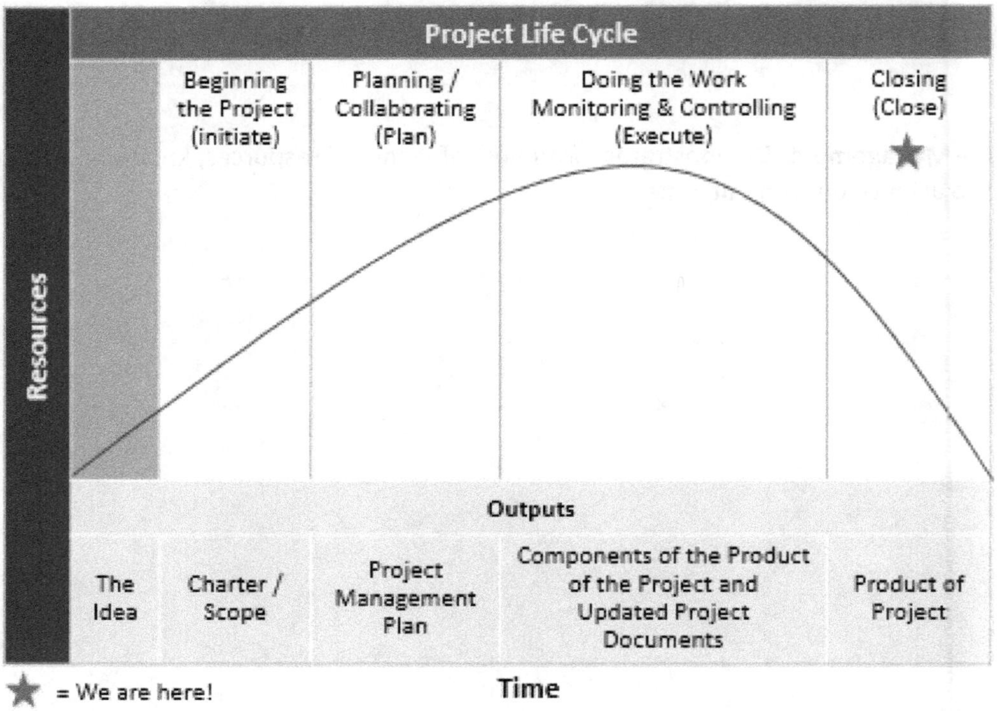

Completing a Project versus Terminating a Project

We previously said that you may arrive at the end of a project because you completed it, or management may terminate the project before you complete it. Either way, you will need to formally close the project.

Project differences dictate what closing activities to include in the closing phase, and completion activities can be different from terminating activities. However, in this course, we will consider completing and terminating activities to be mostly the same.

A project is typically **completed** when the product, service, or result of the project has been delivered and accepted by the client/stakeholders. However, for some projects post closure work issues, activities, and liabilities may still exist.

Management **terminates a project** when the benefits of the project are no longer worth the cost of completing the project.

Termination can occur for many reasons. For example, the organization's strategic objectives may change, or scope changes may have increased the cost beyond the benefits. Or a project could be technology-based, and the technology reached obsolescence more quickly than planned. (Guido, p. 307)

It can be difficult to terminate a project when the sponsoring organization has invested resources in it. However, if conditions turn the project into a losing venture, termination may be necessary. Termination can be perceived as unfavorable, but if a project's benefits will not be realized, there is no reason to continue.

The Closing Phase of the Project Life Cycle

The closing phase can sometimes seem like an independent project. This is because the closing phase has clearly defined tasks and responsibilities, schedules, budgets, and deliverables.

The illustration below shows how the project management plan enables integration among project knowledge areas. The closing phase includes updates and documentation in all these areas. The project manager must identify any loose ends and document them, determine a schedule and budget for each item, and assign someone to be responsible, along with a due date.

Project Management Knowledge Areas

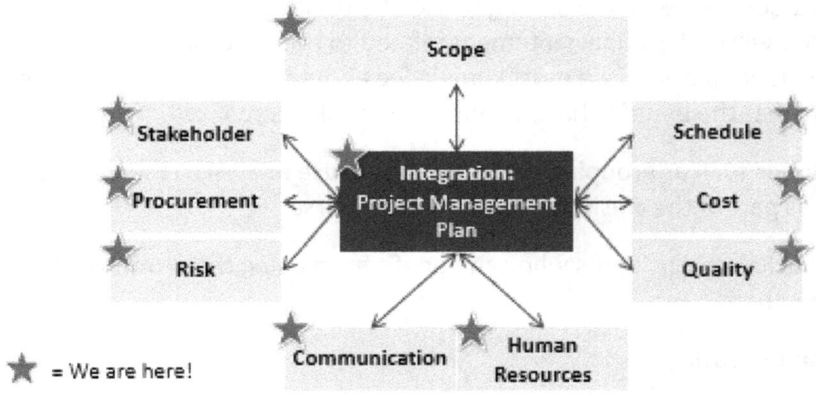

Information Needed to Start the Closing Phase

The following are the kinds of **inputs** you will use to start a project's closing phase:

- Project management plan
- Accepted deliverables
- Organizational process assets

By the time you reach the closing phase in a project, the **project management plan** will have been refined through **several iterations** or **modifications**. As the team works through the project, surprises and changes will occur, and the project management plan must be updated accordingly. We refer to this as a ***progressive elaboration***, which is the process of adding detail and updating changes to a plan. The closing process ensures that all project work is accounted for.

The list of **accepted deliverables** includes all original requirements of the project, as well as those added and changed along the way. Additions can come from surprises or change requests.

Ensuring that all deliverables have been completed and accepted by the client/stakeholders is an essential part of closing a project. You want to avoid having items you consider complete but that have not been accepted by the client/stakeholders. Failing to get stakeholder acceptance could create issues not only for your project but for future projects as well.

If the client/stakeholders discover **issues after work has been completed**, it may be easier for them to make minor adjustments and invoice a supplier for work done. This is referred to as a **back charge** and is common, especially when a project manager has a **good relationship** with the client/stakeholders. In the case of possible property and casualty liabilities, typically such issues are mitigated with assistance provided by an organization's insurance carrier.

Project managers use a list of **organizational assets** to identify and track anything from the organization that was used in the project. Examples are formal and informal plans, employee or project policy handbooks, documented methods, and knowledge databases.

Such a list documented at the beginning of the project might have been updated as the project progressed. The revised assets list must be used in the closing process. The updated list can also be used in future projects.

Project Closing Tools and Techniques

Project closing **tools** and **techniques** include:

- Expert judgment
- Analytical techniques
- Meetings

Expert judgment associated with the closing phase of a project includes input from all subject matter experts (SMEs). You will make a final statement related to the condition of the product, service, or result of the project. People with the most knowledge about the project's anticipated benefits must be the ones to say that the benefits have been, or soon will be, realized.

Analytical techniques such as grouping methods, **root cause analysis**, **reserve analysis,** and others will be discussed as part of the closing activities in this course.

Meetings should include participants who can benefit from being there or provide benefit to other attendees by being there.

Identify all Remaining Tasks

When you begin the closing phase of a project, identify all the tasks that still need to be completed to conclude the project. The list should include both the final project activities **and** all necessary closing activities/tasks.

Depending on your project, the necessary closing activities include the following kinds of things: writing the final progress report, informing the client/stakeholders that the project is coming to an end, communicating record retention requirements, and releasing project resources.

Stakeholders will need to know when the project will end, and any suppliers should be made aware of the project end date so they can prepare final bills. If the project team is composed of people from many divisions, their respective managers should also be informed.

Once the project is complete, you will need to release project resources, so you should identify these tasks as well. This involves returning borrowed or leased equipment, returning unused supplies for credit, and in the case of interdisciplinary teams, allowing the employees to return to their respective departments.

At the end of the project, you will also conduct a lessons learned review, and you should identify it as a closing task. The purpose of the review is to identify what went well in the project and what did not and to make this information available to future projects. In other words, the experience of previous projects can help future ones.

List all Closing Tasks Using a Project Closing Checklist

Managing the closing process can be facilitated by using a **project closing checklist**, as illustrated below. Listing all the remaining tasks will help you and the team focus during the final stretch of the project. The intent of a **closing checklist** is to ensure that all project work is accounted for.

Project Closing Checklist Example

Topic	Action Item	Comments	Owner	Due Date	Status
Integration	Update project plan.	Closing checklist not completed.	Joselyn Smay	3/3/yyyy	Complete
Scope	Conduct large impact scope change analysis.	Root causes to be included in lessons learned.	John Touke	4/2/yyyy	Late

The information you identify and track on a closing checklist is dictated by the project environment. In the example above, the first entries are related to integration and scope. Thus, for example, the "update project plan" action item on the checklist must be completed before the integration work can be closed.

In a comprehensive project closing checklist, all knowledge areas are accounted for by adding rows for schedule, costs, resources, quality, communication, risks, procurement, and stakeholders. In this way, you can account for all significant contributing factors to the project's success, or in the case of termination, lack of success.

Depending on the complexity of the project, a closing process can be somewhat involved; the checklist will help you stay organized.

Information to Update During the Closing Phase

The following are the kinds of **outputs** you will update in the closing phase:

- Product, service, or result of the project
- Project documentation
- Internal and external project information repositories

The product, service, or result of the project is why the project was started. Successful completion of the project must ensure that the product, service, or result has been delivered and also accepted by the client/stakeholders.

Several **project documents** are associated with closing a project, including the project scope, schedule, cost, resources, risks, communications, quality, procurement, and **stakeholder management plans**. Depending on your organization's requirements, supporting documents may also include an enterprise environmental factors list and an organizational process assets list.

You must also consider **internal and external project information repositories**. For example, these repositories could be knowledge databases that house historical data for all projects, networked file directories intended to consolidate project information, and any enterprise cloud-based project management software that can compare key metrics between projects.

What to do:	Other actions:
☐ Review the project closing checklist and consider its value for your project.	☐ ______________
☐ Are you aware of the closing plans for your project? Are those plans adequate?	☐ ______________
☐ For project tasks assigned, do you know the acceptance criteria for completion?	☐ ______________

Remember

- ✓ There are numerous reasons for project termination; not all are negative.
- ✓ The closing phase requires planning.
- ✓ Closing a project can be stressful for the project client/stakeholders.

Enhance Your Learning

Watch the following 3-minute video to learn more about killing a project.

Braun, R. and Dierking, C. (2013). *Feds: Killing Cincinnati Streetcar Project Unprecedented.*	Available at: https://www.youtube.com/watch?v=IK8Nf2GisvU

Reinforce Your Learning

Project Closing Checklist. Assume you are planning the closing phase of your current project or one you worked on, or might be familiar with, such as building a new house. Your goal in this exercise is to create a checklist to ensure that all project activities have been completed to prepare for closing the project.

Identify all the closing activities required for you to be confident that everything has been accounted for and that the client/stakeholders have accepted all project deliverables. Include all closing tasks, such as releasing project resources.

Using the Project Closing Checklist that follows, list 10 to 15 of your closing activities along with the appropriate information about each one. These activities should be associated with the key deliverables of the project.

Project Closing Checklist					
Topic	**Action Item**	**Comments**	**Owner**	**Due Date**	**Status**

Completing and Closing Projects #12.2

Understand Why Law and Policy Matter at Project Closing

As a project manager, you must **protect the interests** of your organization throughout the life cycle of a project and beyond. Just as you are responsible for ensuring that your team performs their work, you are also responsible for protecting your organization against lawsuits, or perhaps worse, losing its reputation due to a legal problem.

Keeping your organization free from litigation throughout a project and after closing the project is not only vital for the organization, but it will help ensure job security for you and the team.

Your Responsibilities as a Project Manager

Project managers face many situations, some of which are unique and many that cannot be anticipated. For these reasons, this course emphasizes the necessity for learning to **think through situations** and stresses the importance of using an **Employee Handbook** or **Project Policy Handbook.**

Being proactive is vital to preventing problems that can lead to litigation even after closing a project. Being a critical thinker is also essential. In other words, look at situations with a critical eye to determine what is going right and what could go wrong.

Recognize that safety comes first. Issues that occur throughout a project's life cycle should be handled immediately to prevent future problems and lingering litigation. Remember that your team and associates look to you as a role model. How you behave will influence how they behave.

Here are some examples:

If team members or associates see you walk past a puddle of water on the floor without attending to get it cleaned up, they too may ignore unsafe situations and use it as a defense should litigation occur.

If someone observes you being rude or failing to follow the rules, they may tend to do the same. They may also communicate such events to other people long after the project is completed and closed.

If your team sees you valuing safety, adhering to rules, double-checking that standards are being followed, and the like, they will be more apt to do the same and communicate such values to others.

Your Reputation and Career. Realize that your actions—or inactions—can affect your career. They may even hurt your reputation and damage your career. Taking the time to consider ethical, legal, and policy issues and taking proactive and appropriate action benefits everyone. Your inability to apply policies and laws, and to work with others to avoid litigation, may cause your organization to face liabilities and fines even on a closed project.

Wide-reaching legal implications lurk in what might seem to be the unlikeliest of places. Remember the McDonald's hot coffee case? Few people would have foreseen that, but it has changed the fast-food world and is the main reason behind "Caution: Hot!" stamps on many drink containers today. A spill on the floor, coffee that is too hot, a screw that is not tightened can all lead to litigation from employees, customers, or suppliers.

Proactively Handle Complaints. If someone on your team, a coworker, or a customer complains about a policy or legal issue, take it seriously. The complaint may also be an opportunity to improve and get better. There is no one size fits all answer. Keep the following guidelines in mind:

- Be **courteous** and choose **your words carefully**.
- You can apologize for how you feel about a particular action, but **don't apologize** for the action itself as this could be seen as an admission of guilt and open your organization to litigation. For example, if a customer says that your hot coffee burned them, you could apologize for them feeling injured, but do not admit that the coffee was too hot.
- **Document** (write a report) a questionable situation and discuss the situation with the appropriate person in your human resources department.
- **Never talk to media** about a situation unless you have been briefed by human resources or your organization's legal department, and you have their permission to speak on the topic.
- If legal charges are filed against the organization, get the proper human resources, legal, or leadership person involved immediately.

Document Employee Facts. Maintaining appropriate documentation throughout an employee's tenure on a project is an essential part of a project manager's job. This documentation includes observations and facts relating to recruitment and selection; orientation and onboarding; training and development; coaching and performance management; career planning and progression; retirement, separation, and termination; all infractions; and, in some cases, the need for progressive discipline.

Progressive discipline can include the following types of actions and will depend on your organization's policies and procedures:

- **Verbal conversation**. Documentation placed in a project manager's journal but usually does not go into the employee's file.
- **Verbal warning**. The next step can be to reprimand the employee for their behavior or lack of performance. This step is typically documented as a "verbal or oral warning" in an employee's file.
- **Written warning**. The next step is often a written warning given to the employee and a copy placed in their personnel file.
- **Suspension**. The next step might be a suspension for a set number of days or disciplinary layoff. This action may be with or without pay.
- **Discharge/termination**. If the problem continues, the individual may be discharged.

Laws and Policies Related to Employment

Project managers must know all laws and organizational policies and procedures that apply to employees throughout their **employment life cycle**.

The employment life cycle refers to everything from the time a person is recruited (this can apply even to job advertisements, job descriptions, and interviews) through all their time with the organization (including onboarding, training, orientation, career planning and development), through termination (which includes retiring, quitting, getting fired, and exit interviews).

The following describes a few of the most **common workplace laws** and **policies** and how they apply to the employment life cycle from pre-employment to post-employment. As you read through the list, consider how they might apply to your project or in your workplace:

At-Will Employment

This is a term in U.S. labor law that refers to a contractual relationship in which an employee can be dismissed by an employer for any reason. In theory, at-will employment exists, but in practice, it really does not. With the enactment of various civil rights laws and the establishment of protected classes, managers need documented reasons to terminate an employee.

Title VII of the Civil Rights Act

The act makes it illegal to refuse a service or a job to a person based upon gender, marital status, race, color, religion, or national origin.

Some states have laws that further protect additional groups, including gay, lesbian, bisexual, or transgender persons.

People with mental and physical disabilities are protected by a variety of laws, both state and national.

Always be aware of your specific state laws, as well as your organization's policies. Remember, the law that is the broadest (protects the most people) is the one that you must adhere to.

Protected Classes

The following are protected classes by federal law:

Race	Color	Religion	National origin
Sex	Pregnancy	Citizenship	Age (40 and over)
Disability	Veteran	Genetic	Familial Status

Project managers must treat everyone consistently and equally without regard to race, color, religion, gender, national origin, or any other characteristics protected under law.

People over 40 are a protected class according to the EEOC against age-related discrimination.

An employee may have grounds for filing an EEOC age harassment lawsuit if they believe they are being questioned in a certain way, treated unfairly, or there is some attempt to force or coax them into retirement.

A project manager must address performance concerns with a person without any reference to age or plans for retirement.

Sexual Harassment

There are two fundamental theories of sexual harassment: quid pro quo (employment benefits conditioned on sexual conduct) and a hostile work environment.

Under both theories, if the conduct is unwelcome and has a demonstrated effect on the employee's conditions of employment, it may constitute unlawful sexual harassment.

It is your responsibility as a project manager to take action if you know, or should have known, about questionable behavior in your workplace, and even if someone does not complain to you about sexual harassment.

Fair Labor Standards Act (FLSA)

This law established the 40-hour workweek and minimum wage and also prevents child labor. This is one of the most important laws to understand for any project manager involved in setting schedules or hiring.

Make sure you understand and follow this law, and also your organization's policies regarding work hours, breaks, lunches, vacation and sick time, and so forth.

Workers' Compensation

This law was designed to support employees in staying home while recovering from an injury. It specifies salary payments and benefits to employees who suffer an illness or injury at work and arising out of and in the course of employment.

The employee must promptly notify the employer of the injury.

Once someone advises you of an injury, or you become aware of an injury, it is your responsibility as a project manager to complete and file the WC form immediately with your workers' compensation coordinator or appropriate HR representative.

Family Medical Leave Act (FMLA)

FMLA is a law designed to allow employees to take time off work to deal with health issues in themselves or their immediate families without fear of losing their jobs.

The law allows for unpaid leave. Many organizations provide a certain amount of paid leave, so you must know the policy in your organization.

Employees who miss work or having trouble with work hours are encouraged or required to take personal time off (paid personal days or vacation) to cover their absences. Again, you will need to know the specific policies and procedures in your organization.

The rules for FMLA are complicated and you should defer to human resources (HR) for direction. Also, refer employees who may be eligible for FMLA to HR, where they can learn about their options.

If an employee is granted FMLA because of their health, the person must have a return-to-work authorization, signed by a doctor, before returning to work.

Never apply pressure to a person on FMLA to return to work or to forgo FMLA benefits.

Americans with Disabilities Act (ADA)

This is a federal law that prohibits discrimination against qualified individuals with disabilities in all conditions of employment.

A disability is defined as a physical or mental impairment that substantially limits one or more of the person's major life activities, has a history of such impairment, or is regarded as having such an impairment.

Prohibits an employer from retaliating against an employee that exercises their rights under ADA. The ADA is intended to "level the playing field" for employees with disabilities.

Qualified employees must all be afforded the same opportunities to succeed.

Reasonable accommodations must be made to allow individuals to participate equally unless it would impose an undue hardship on the institution.

Occupational Safety and Health Act (OSHA)

Project managers must provide a work environment that is free from recognized hazards that could cause physical harm, and they must comply with occupational safety and health standards.

Project managers may also be responsible for ensuring that people receive safety training,for assessing hazards in the work area, determining the type of protective equipment needed, investigating incidents and inspecting equipment, and reporting all accidents and injuries that employees have at work.

Avoid Management Pitfalls

Mistakes happen. Even experienced project managers are human. But there are two kinds of errors: ones you learn from and do not repeat, and ones you are unaware of and thus are doomed to repeat.

Obviously, any mistake you make should be a learning experience and not repeated.

If you take the time to **learn about the law** and consider how it may impact your projects and organization and **reflect upon what you know** and **do not know about the law** as well as **what you are doing right** and **not doing right**, you will be better prepared to protect yourself and your organization. Also, take the time to become familiar with all your organization's policies and procedures related to employment.

The following are key ideas to remember to avoid the pitfalls commonly encountered by managers, including project managers:

- Always file an injury report, in writing, the day an injury happens.
- Remember that employees who are out on Workers' Compensation or FMLA must have a return-to-work authorization before being allowed to return to work.
- Know that some employers do not require employees to produce a doctor's notice when they return to work following regular sick days or for personal injuries that happened outside of work, including short term disability. Talk to HR so that you know your organizational policies.
- Remember that an oral contract can be just as binding as a written one, so be careful what you promise employees and customers.
- Think through HR-related issues before you speak with an employee about their employment or conduct an interview. Ensure you are planning to ask legal and ethical questions and that you know your organization's policies and procedures.
- Know what not to ask. This includes questions about age race, national origin, sexual preference, and so forth. For example, even asking when a candidate graduated from high school or college can be considered an age question.
- Not paying overtime can result in legal problems, so be sure you are informed about overtime laws that apply to your situation.
- Terminating an employee can result in litigation. Make sure you have documented the cause for the termination. An employee should never be surprised. They should be made aware of each specific performance problem.
- Always have another manager in the room to witness a termination; however, be private about it with other employees. Avoid being apologetic during the termination process, or the employee might wonder if you have a valid cause for the action.
- Write a memo about the termination process to document it.

What to do:	Other actions:
☐ Identify the top two or three policy or legal issues your work unit could face.	☐ ______________________
☐ Formulate proactive measures you can take to prevent those issues from occurring.	☐ ______________________
☐ Know and avoid legal pitfalls often encountered by project managers.	☐ ______________________

Remember

- ✓ Remember that how you respond to any situation will impact you, your team and associates, and your organization.
- ✓ Be aware of and consider organizational policies and procedures throughout the employment life cycle.
- ✓ Laws and policies surrounding employees are there to protect both them and the organization.
- ✓ Always make a written report and inform HR and your manager when an injury occurs.
- ✓ Ensure that the injured party receives prompt medical attention.
- ✓ In most organizations, employees returning to work need a return authorization letter signed by a doctor. Any limitations to their work should be noted by the doctor and carefully followed.
- ✓ Be aware that your organization assumes liability if someone returns to work and injures themselves further or is assigned work that exceeds any medical limitations placed upon them by a physician.

Enhance Your Learning

Review the 10 most important laws that project managers should be aware of and their major responsibilities in ensuring compliance.

ERC Blog. (2013). *10 Employment Laws that Supervisors Need to Know.*	Available at: https://www.yourerc.com/blog/post/10-Employment-Laws-that-Supervisors-Need-to-Know

Reinforce Your Learning

Legal Scenarios. Review the scenarios that follow. Reflect on laws and policies applicable to these situations and what you should do to avoid litigation throughout the life cycle of your project, and after closing it. Then respond to the questions that follow.

Scenario A: Megan Hurts Her Back (workers' comp/safety)

While working on your project, an office worker, Megan, lifts a box of copier paper and hurts her back. She was out for 15 days on FMLA and returned to the office on Monday. She did not have any paperwork from her physician indicating her readiness to return to work. The project manager let her work her scheduled shift.

What are the potential future risks or liabilities of letting this worker return to work?

What references in your Employee Handbook or Project Policy Handbook (if available) might be applicable?

What actions should the project manager avoid?

What actions should the project manager take?

Scenario B: Greg, an Older Craftsman (hostile work environment/age discrimination)

Greg, who is nearing retirement age, performs his job duties satisfactorily, but his technical expertise is not keeping up with other employees, especially in newer technologies. Greg's boss is planning project budgets and staffing for the next year and is having difficulty because he does not know if or when Greg plans to retire. Needing insight for future planning, the manager has broached the retirement topic with Greg several times in the past few months.

What are the potential risks or liabilities in this situation?

Does your employee handbook or project policy handbook (if available) provide insight on this topic?

What actions should the project manager avoid?

What actions should the project manager take?

Scenario C: Krista and Aging Mother (personal time off/compensation law)

Krista is an hourly (non-exempt) project expediter assigned to your project who has been caring for her aging mother. This has affected her attendance record. With increasing frequency, she has arrived after her start time and left early to attend to her mother's needs. Krista has always made up the time.

John is a coworker who works as a project coordinator (exempt), and he is a single dad raising a middle school child. Noticing the flexibility Krista has been given, John asked his supervisor for permission to leave early every Tuesday to attend to his children after school events. John's request was denied. John has offered to make up his time, but he is upset and wonders why he is being treated unfairly.

Both Krista and John prefer not to use their personal time off for these absences.

What are the potential risks or liabilities in these situations?

Does your employee handbook or project policy handbook (if available) shed light on these issues?

hat actions should the project manager avoid?

What actions should the project manager take?

Scenario D: Lynda on Maternity Leave (Family Medical Leave Act and return to work)

Lynda has been employed as a client service representative on your project for three years. She consulted with HR regarding her FMLA benefits before going on maternity leave six weeks ago. Her plan was to return to work at the end of six weeks. However, yesterday she called and said she wanted to return to work early and forgo her remaining FMLA time. Your department really needs her back because of the workload, so you told her it would be great if she returned early. Today, she arrived to work to punch in without a medical release from her doctor and completed her workday.

What are the potential future risks or liabilities with Lynda returning to work?

Does your employee handbook or project policy handbook (if available) reference this issue?

What actions should the project manager avoid?

What actions should the project manager take?

Completing and Closing Projects #12.3

Evaluate Resources, Work Issues, and Conditions

Enterprise resources may include both human resources and operational process assets, which are often shared across projects. The timing of when a resource must be available is important because of relationships with other activities, such as activities on the critical path and successor activities. During the project closing phase, ensure that all resources are properly released, associated schedules are accurate, and work issues and conditions are documented.

In this concept, we discuss how to evaluate resources, work issues, and project conditions.

Information Needed for Closing Resources

The **inputs** for closing resources include:

- Resource calendars
- Project schedule
- Resource performance information
- Project close checklist

Resource calendars for people can be generated from the activities list within the project management software, if used. All assignments should be accounted for at 100 percent completion before a project is closed.

Enterprise resources, such as computing labs or specially equipped conference rooms, may have a **project schedule** established outside of the project documentation.

People's performance information is provided via the performance review and can be part of the project. It is most effective if included in the annual performance review conducted within the home department of the assigned individual. Input associated with performance should be gathered throughout the life of the project and recorded in a confidential setting. It is often difficult to remember how things actually happened a week or a month ago.

Feedback is typically provided by the **project manager** or a **team lead**. This information should be reviewed with the person as near the end of the project as possible. They need to know the good and bad (if any), so they can leverage that knowledge on the next project.

Enterprise resource performance information may come in the form of issues while trying to use the resource as planned. For example, a specially equipped conference room may have been overheated because of an air circulation issue, and the project team could not use the room. This can be disruptive to a team that has everything arranged in the first conference room and now must find an alternative. This is particularly problematic if the specialized equipment in this room is the only place to access it. This may cause a significant schedule impact, and you should confirm that these types of performance issues are recorded as part of closing that resource.

Another type of **enterprise resource performance review** could involve production equipment that produces a product through repetitive processes. This performance review will be included in the "quality closing" discussion.

Tools and Techniques Used for Closing Resources

Project managers use the following **tools** and **techniques** for closing resources:

- Expert judgment
- Meetings

Expert judgment comes from people. For people's performance closure, current performance is noted, along with any special mention associated with excellent or unacceptable behavior. It should also include observations concerning additional training that could be helpful for professional growth.

The opportunity to observe project team members during project work that is taxing and stressful can provide insight into their ability to handle pressure. It may reveal the need for additional core skill training or soft skill training. A skilled leader can make these observations and give appropriate recommendations.

The **meetings** for **people closure** should be **one-on-one** with **each team member**. They have contributed part of their career to the project and deserve time to discuss how things went, what might be different next time, and what could help them grow professionally. A little attention, particularly after a job is done well, is an investment in an employee.

Information You Will Update When Closing Resources

The **outputs** from closing resources include:

- Updated resource calendars
- Updated project documents

When a **problem occurs** that has potential legal ramifications, almost any evidence could be used in court. Witnesses will be called, pictures will be shown, and arguments will be made. The best defense is a good offense, as they say. In this litigious society, you must be proactive and document, document, document.

You also need to document to serve the needs of future project teams that may review your project materials.

Evaluating Resources and Documenting Work Issues and Conditions

Understand and apply employee privacy rights as outlined in your organization's policies and in law. Keep the following information in mind as well:

- **Privacy laws** can be **complicated**, and laws take precedence over organizational policies. Be aware of federal, state, county, and municipality laws.
- The **law does not protect an employee's right to privacy in the workplace in many situations**. This means that most employers can **read emails**, search through offices, listen in on business phone calls, and so forth (Find Law, 2015). However, under the Electronics Communication Privacy Act, an employer cannot listen in on personal phone calls or voice mails, even if they are made to or from the workplace (Find Law, 2015).
- **Video surveillance** is **legal**, but the organization must inform employees and customers that the grounds are video monitored (Find Law, 2015). Federal wiretap laws prohibit audio from being recorded (Find Law, 2015).
- Employers **can require drug testing** as a condition of accepting new employment. However, testing existing employees might be illegal; know the laws in your area. Drug testing existing employees may be allowed if it is objective (required of everyone) or if you have reasonable suspicion.

Privacy laws come into play during employee background checks and with employee record keeping. There are no federal laws to protect basic employee information unless you work in an educational institution where the Family Educational Rights and Privacy Act (FERPA) applies. For the most part, it is up to you and your personal code of ethics to determine what information to protect and how to protect it.

However, consider the following:

- Keeping information confidential makes it less likely that you will incur problems.
- Release personal information and employment records only if there is a court order or a compelling reason to do so.
- Background checks must be done, but you cannot perform a background check because someone is of a certain race, color, or another protected group. Your best course of action is to make background checks standard for all new hires.
- You must adhere to the **Fair Credit Reporting Act**, which requires written permission to pull a **background check**. Be familiar with this law if you consider doing background checks.
- Maintain records for at least one year after employees leave the organization.
- You must **dispose of any background reports in a secure manner**, such as shredding or burning them.

If legal action occurs, courts will view most paperwork and information as the property of your organization. This means that you control it, but also that it can be garnered as evidence. This includes such things as emails, voicemails, memos, intellectual property, files, and so forth.

Guidelines for Managing Harassment Situations

In 2011, 11,364 claims of sexual harassment were filed in the United States (Gaille, 2013).

Sexual harassment is a legal issue, and laws protect people. Your organization most likely also has policies and procedures that prohibit bullying, harassment, sexual harassment, and workplace violence.

Harassment can occur in any employee relationship. Know that **your organization can be held liable** for harassment in the workplace.

You are responsible for the behavior of your **staff** and also **non-employees** who visit the workplace, such as suppliers and customers.

As a project manager, if you become aware of questionable behavior, address the offending worker or employee directly. Your failure to address such a situation could result in a sexual harassment complaint.

If you have doubts about what to say, consult with your human resources department before you meet with the employee.

When you meet with the employee, either alone or together with someone from human resources, explain your organization's policies and the risks and liabilities concerning the behavior. Encourage the employee to use better judgment in the future.

Guidelines for Managing Complaints

Many situations in the workplace call for documentation, for example, from a minor accident, to an employee reporting to work in an inebriated state, to suspected theft. Anything can happen. But the most common issues in the workplace center on communication, harassment and bullying, and gossip (WiseGeek, n.d.). When does a complaint require further investigation?

When you are unsure, consult with your human resources department.

Use the following guidelines to help evaluate the situation and to help you determine an appropriate response:

- Even minor complaints warrant some level of investigation and documentation. Major complaints and issues, and particularly where safety or the organization's property or reputation is on the line, always require further investigation.
- Listen carefully to what the employee says and ask follow-up questions to ensure you fully understand the complaint.
- Assure the person reporting the issue that you understand and will take action if it is necessary.
- Document the conversation and whatever action you take, or the reasons for your decision not to act.
- If the situation turns out to be minor, it can end there. Otherwise, proceed to collect evidence and talk to others who are involved.
- Gather additional evidence. Observe the situation with your own eyes when possible.
- Consult HR or your legal department, or another supervisor, if necessary.
- Continue to document all that you learn and all actions that are taken.

Guidelines for Responding to Accidents

Accidents happen. When one occurs during work, take the following steps:

- Ensure that the employee is okay and receiving medical attention, if necessary. Make sure the work area is safe and that other employees are not in immediate danger.
- Make a timely report of the situation, including filing a workers' compensation claim.
- Investigate the incident that same day. Studies show that claims filed the same day as the event cost less in the end and are less likely to result in litigation (Mottern, n.d.). Take notes. Interview witnesses. Answer the usual who, what, where, why, when, and how (Mottern, n.d.).
- Document all details as you investigate.
- Secure a medical release from the injured party or other documentation as required by workers' compensation (Mottern, n.d.).
- Keep in touch with the injured worker, expressing concern and keeping them informed of the workers' compensation claim process (Mottern, n.d.).
- Stay in touch with the workers' compensation claim representative (Mottern, n.d.).
- Work with the employee to develop a return-to-work plan (Mottern, n.d.). Ensure the employee has a return-to-work authorization signed by a doctor before they return to work.

Guidelines for Discipling and Terminating Employees

Follow your organization's policies and procedures for **disciplining and terminating** employees. You also need to be aware of the law.

At a minimum, use a progressive discipline approach and consider the following guidelines:

- Before you issue any warning, document specifics about the problem behavior, including date, time, the behavior, and the negative impact it had.
- Issue a **verbal warning**. Explain to the employee why the behavior is problematic. For example, it may be because phones must be answered on time, and the person needs to set a positive example. Explain what behavior you expect to see in the future. Ask the employee to agree to that behavior.
- If the undesirable behavior continues, issue a second warning **in writing**. Again, be specific about the behavior, but do not attack the person. Focus on the specific behavior, sticking to one topic at a time. Describe specific actions. Do not give a litany of undesirable behaviors.
    - Ask the employee to sign the written warning.
    - Inform the employee of the repercussions if the behavior occurs again. Will they be fired? Hours cut? Seniority reduced? Receive another written warning?
- If the behavior does occur again, follow with the repercussions you promised.

Guidelines for Employee Record Keeping

Keeping records is an integral part of running an organization. **Certain records are required** under different laws, including FLSA, EPA, ERISA, COBRA, IRCA, HIPAA, FMLA, and more. In short, you must keep at least the following information for each employee (Nolo, 2015):

- Job description for the position
- Job application and resume
- Offer of employment
- IRS Form W-4 (employee's Withholding Allowance Certificate)
- Receipt or signed acknowledgment of employee handbook
- Performance evaluations
- Forms relating to employee benefits
- Forms providing next of kin and emergency contacts
- Complaints from customers and/or coworkers
- Awards or citations for excellent performance
- Records of attendance or completion of training programs
- Warnings and/or other disciplinary actions
- Notes on attendance or tardiness
- Any contract, written agreement, receipt, or acknowledgment between the employee and the employer, for example, non-compete agreement, an employment contract, or an agreement relating to a company car
- Documents related to the employee leaving the organization, such as the reason the employee left or was fired, unemployment documents, and benefits continuation forms

Wrongful termination is one of the most common lawsuits. A likely cause is an employee believing that their termination was caused by discrimination. To avoid this situation, remember:

- Always document behaviors. Document only what you know to be accurate and document as objectively as possible.
- Use progressive discipline and document every instance.
- Ensure you have clear, accurate job descriptions for every position as these will be vital in substantiating that the employee failed to perform their job.
- When firing an employee, do not over-explain the situation. Be firm and assert the cause.
- Never mention gender, age, religion, disability, or other protected situations.
- Consult your HR or legal department at the first sign of a lawsuit.

Guidelines Concerning Work Hours and Breaks

Ensure that you know the laws and your organization's policies and procedures concerning attendance, work hours, and breaks for all employees.

Federal and state labor laws require non-exempt, full-time, hourly personnel to receive a meal period if their workday exceeds six hours.

You must adhere to these laws and provide proper breaks and lunch periods. Failure to do so could result in fines and legal action.
You must not allow employees to forgo these breaks even if voluntarily.

Some laws, such as the Age Discrimination in Employment Act (**ADEA**), require that **records be kept** for **one year** following termination of employment. Others, such as the Americans with Disabilities Act (**ADA**), require records to be kept for **seven years**.

It is best to maintain all employee records for at least seven years. If uncertain, check with your HR department.

Do not allow employee files to become a drop zone for every single piece of paper related to the employee. Remember (Nolo, 2015):

- Review employee files every year to ensure they are complete.
- Do not keep medical records in an employee file. These need to be kept in a separate, confidential file.
- Do not put I-9 immigration forms in the personnel file; keep them separate.

What to do:	Other actions:
☐ Review your project's resource close information and ensure the most current resource calendars accurately reflect that information.	☐ ________________
☐ Review the most current project schedule to ensure all activities have been closed.	☐ ________________
☐ Recognize that when a problem occurs, almost any type of evidence could be used in court.	☐ ________________

Remember

- ✓ Releasing resources on time is essential to being an effective project manager.
- ✓ You must give timely feedback to people resources. They deserve it.
- ✓ Enterprise resource calendars may be managed outside of the project.
- ✓ If you are unsure whether a situation needs to be documented, then document it. It is better to have too much documentation than not enough.
- ✓ Your workplace should have no tolerance for harassment or bullying.
- ✓ If harassment is occurring in your workplace, you and your organization could be held liable.
- ✓ Remember that typical documentation such as employee files and emails can be garnered as evidence in a court case. Consider what information to record and always stick to facts. Avoid opinion, conjecture, accusations, and the like.

Enhance Your Learning

Workplace violence is an increasing problem in the world today. To learn more about preventing workplace violence, review the following comprehensive publication by the United States Office of Personnel Management.

United States Office of Personnel Management. (1998). *Dealing with workplace violence.*	Available at: https://www.opm.gov/policy-data-oversight/worklife/reference-materials/workplaceviolence.pdf

Reinforce Your Learning

Potential Liability Scenarios. Review the scenarios that follow. Reflect on laws and policies applicable to these situations and what you should do to avoid litigation throughout the life cycle of your project and after closing it. Then respond to the questions that follow.

Scenario A: Brian No Lunch (compensation laws)

Work is backed up in the drafting department, so everyone is working mandatory overtime to catch up. Feeling the pressure, one drafter, Brian, decides to voluntarily work through his lunches. The manager is aware of this and does not intervene because Brian is willingly working through his lunches, and besides, this is helping productivity.

What are the potential risks or liabilities associated with this situation?

Does your employee handbook or project policy handbook (if available) reference this issue?

What actions should the project manager avoid or have avoided?

What actions should the project manager take or have taken?

Scenario B: Chris Working in Pain (workers' comp and OSHA)

Chris complains that her typing is getting slower because she is having a lot of pain in her wrist and may have developed a repetitive motion injury. Chris asks you if she has surgery, will she be entitled to time off under the Family Medical Leave Act (FMLA), and will she be eligible for Workers' Compensation benefits?

What are the potential risks or liabilities associated with this situation?

Does your employee handbook or project policy handbook (if available) reference this issue?

What actions should the project manager avoid or have avoided?

What actions should the project manager take or have taken?

Completing and Closing Projects #12.4

Be Mindful of Property and Casualty Liabilities

Accidents happen and the workplace is particularly subject to their occurrence. Project managers must be alert to their surroundings and proactively respond to potentially hazardous situations to avoid negative consequences.

If you and your staff are aware, careful, alert, and trained to pay attention and proactively deal with issues, you can significantly minimize human injury, property damage, and litigation.

In this concept, we discuss what you need to know to protect you, your project, and your organization from litigation arising from accidents and property damage.

Property and Casualty Liability Insurance

Every organization must carry property and casualty liability insurance. This insurance helps offset the costs associated with most litigation that results from accidents and property damage.

However, do not think that because your organization has insurance that you do not need to be proactive in preventing accidents and property damage.

Here are the facts to keep in mind:

- Insurance is **expensive,** and rates typically go up after a lawsuit.
- Insurance has **limitations,** and some lawsuits may cost you more than the insurance will pay.
- **Lawsuits** not only **cost money**, but they can **damage** an organization's **reputation.** As much as 80 percent of the value of an organization comes from intangibles like brand equity and reputation, intellectual capital, and goodwill (Eccles, Newquist, & Schatz, 2007). All of these can be damaged in a lawsuit.

Employees versus Independent Contractors

Know what rights and obligations may vary according to whether a person is an employee or an independent contractor. Ensure you know the law in your locale and also your organizat on's policies and procedures concerning these categories of workers.

Here are some guidelines:

- An independent contractor is typically hired to complete a short-term task, one in which you do not specifically direct how that task is accomplished.
- A contract employee might be a temporary employee you supervise who is helping to accomplish a short-term task.
- There is a legal distinction between an employee and an independent contractor. Talk to your human resources or your legal department to ensure you accurately classify the people you manage.
- **All people—employees and contractors—must be treated fairly and equally.**

- Expect the same quality of work from contract workers as you do from employees. If you expect less of contractors because they are temporary, you risk setting expectations for low performance and inferior work. You also risk negative feelings from your employees.
- Set clear work product expectations with contract workers and what level of expertise you expect.
- Hold contractors accountable to the same behavior standards as your employees, and ensure they have the same resources (such as employee handbooks) to know what those standards are.
- Contractors may or may not be covered by workers' compensation. Consult with your human resources and legal departments before work begins.
- When working with contractors, remember the following (Javitch, 2006):
    - Provide the contractor with clear goals, expectations, milestones, and expected outcomes, as you would any employee.
    - Let the contractor know you value their work, as you would any employee. Help the person stay motivated to complete the job to your standards.
    - Tell the contractor immediately if the work they are doing is not meeting your expectations in terms of quality, timing, quantity, cost, etc.
    - Tell the contractor immediately if their behavior is not meeting organizational policies and procedures.
    - Check in with the contractor frequently to ensure that work is progressing and to see if any obstacles require your attention.

Work Equipment, Tools, and Premises

Your organization assumes liability whenever anyone uses its equipment, tools, or **premises** during **work hours** and **after hours**. If someone uses equipment after hours and they are injured, your organization will be liable. During work hours, this liability comes with the benefit of employee productivity; after hours, it does not.

Follow your organization's policies and procedures, but if in doubt, do not allow people to use an organization's equipment or property during non-business hours.

Clearly post policies regarding the use of equipment and property.

Safe Work Environments

It is your responsibility to provide a safe workplace. This is also an ethical imperative and makes good business sense.

OSHA provides a variety of standards that must be adhered to in the workplace, so be sure that you are aware of OSHA policies.

Consider the following guidelines:

- Be certain each employee receives proper OSHA training each year.
- Ensure that any OSHA safety equipment is readily available and that all employees know how to access it.
- Train your staff to be proactive about safety issues and preventing accidents.
- Document safety concerns promptly and take immediate action to resolve them.
- Know what your workers' compensation insurance requires you to do to promote safety in the workplace.

Employee Privacy

Employee privacy is a topic we have covered previously, and again, you will want to know the law and your organization's policies and procedures.

Keep the following guidelines in mind:

- Employees should not expect privacy in the workplace.
- That said, you should interfere with employee privacy only when absolutely necessary, such as the following situations:
  - You suspect that someone may be a danger to themselves or others.
  - You have reasonable suspicion of law violations.
  - Another employee has reported a situation that needs to be investigated further.
  - The suspected issue is impacting the employee's work or the workplace.

Your right as an employer to conduct a "search and seizure" includes:

- **Employers have a right**—and sometimes an **obligation**—to **investigate misconduct** in the workplace (Repa, 2015).
- In general, "**reasonable searches that have a legitimate purpose and are limited in scope do not violate the law**" (Repa, 2015).

While the U.S. Constitution provides the right to privacy, these rights do not extend to private workplaces (Repa, 2015). However, sometimes state laws provide additional privacy measures (Repa, 2015), so be sure to know the law in your state and consult your human resources or legal department before searching an employee's work area or personal items.

When the law is unclear or non-specific (which it usually is in these areas), the court is the deciding factor (Repa, 2015). This means that an employee can take you to court as a result of a search or seizure. No matter the outcome, such a case will cost time, money, and productivity, so be sure you have a good reason before conducting any search.

If you conduct a search, document everything that is done and found, and do your best to be respectful of the person and property and maintain privacy.
One law firm provides the following advice: "The law considers whether the employee had reasonable expectations of privacy. A worker who legitimately expects, based on the employer's policies, past practice, and common sense that the employer will not search certain areas has the strongest argument here. For example, a worker has a high expectation of privacy in the employee restroom or a changing area, particularly if the employer has not warned workers that these areas might be monitored" (Repa, 2015).

The more intrusive of a search you conduct, the better reason that you need to have (Repa, 2015). In other words, you may need little reason to look in an employee's desk drawer, but you had better have a good reason for searching their belongings.

Lower your risk of a lawsuit by ensuring that your policies clearly detail what level of privacy an employee can expect. Display these policies in places where all workers can see them. Every year have workers complete training on the policy and agree to it.

Remember that **you can never restrain employees against their will** (Repa, 2015). This would be false imprisonment and subject you to legal problems.

When things "look" wrong, pay attention to your instincts, and evaluate the situation as it could have consequences long after closing a project. An old adage says that everything that can go wrong will go wrong. But you can avoid many problems if you are aware, anticipate trouble, and take proactive steps throughout the life cycle of a project and beyond.

What to do:	Other actions:
☐ Recognize that it is your responsibility to provide a safe workplace.	☐ ______________________
☐ Do not allow workers to use your organization's equipment or property during non-business hours.	☐ ______________________
☐ Be certain that each worker receives proper OSHA training each year.	☐ ______________________

Remember

✓ While you may tell a child that it is okay when they spill milk because accidents happen, the attitude in the workplace should be to prevent all accidents.

✓ Safety is paramount because protecting human life is critical in every project and organization. Protecting property, preventing financial loss, and preventing litigation are next in importance.

Enhance Your Learning

Watch this 1-minute video for an overview of property and casualty insurance.

Generali Group. (2017), *Property and Casualty Insurance.*	Available at: https://www.youtube.com/watch?v=nVW3fQMM1As

Reinforce Your Learning

Potential Liability Scenarios. Review the scenarios that follow. Reflect on laws and policies applicable to these situations and what you should do to avoid litigation throughout the life cycle of your project and after closing it. Then respond to the questions that follow.

<u>Scenario A: Dwayne and the Personal Project (personal use of an organization's property)</u>

A shop employee, Dwayne, who has an excellent employment and safety record, asks if he can come into the shop on the weekend to use the company's drill press for about 10 minutes for a personal project. The project manager allows this to happen just this one time.

What are the potential risks or liabilities associated with this situation?

Does your employee handbook or project policy handbook (if available) reference this issue?

What actions should the project manager avoid or have avoided?

What actions should the project manager take or have taken?

<u>Scenario B: Eric Declines Gloves (work safety risks)</u>

Eric works in production in the hardware department. Yesterday, he was asked by the project manager Mark to move over to the crating and packaging department to help get out a large shipment to meet an important project deadline. The entire department was hustling to get the shipment into the FedEx trailer before it pulled out at 5 pm. Eric was given a utility knife and asked to help with the cardboard packaging. A coworker, Bill, asked Mark if he wanted to wear gloves while using the knife, but Eric declined, saying it would only slow him down. Bill knows the safety policy requires gloves to be worn, but no additional sets of gloves were available for Eric to use. Besides, the work area was cluttered, and neither Eric nor Bill could find any gloves. But this shipment did get out on time.

What are the potential risks or liabilities associated with this situation?

Does your employee handbook or project policy handbook (if available) reference this issue?

What actions should the project manager avoid or have avoided?

What actions should the project manager take or have taken?

Completing and Closing Projects #12.5

Identify and Use Resources to be Safe and Legal

It is impossible to know everything needed to handle all situations that may arise at work. For example, your job as a project manager is to deliver your project on time and within quality standards, but because you manage people, many safety and legal issues will also arise, and you will need to respond to them.

For example, you may have an idea that something improper is happening or has taken place because you have some facts. Maybe you **observed something** directly (this is best), or a reliable employee may have given you information you believe to be accurate. The situation could involve any of several improper or illegal behaviors, for example, theft, substance abuse, physical abuse, bullying, unwillingness to follow safety rules, and so on.

The issue of privacy is involved in many of these situations, as is the law in terms of what you can do. You might have questions. What is reasonable suspicion? How should you handle substance abuse? How serious does a problem need to be before you address it?

In this concept, we discuss reasonable suspicion and offer guidelines to assist you in determining your response to improper behaviors at work.

Reasonable Suspicion

Your awareness that a problem may exist often begins with what is known as "reasonable suspicion." Reasonable suspicion is a yardstick you can use to evaluate behavior.

In the **law,** to have **reasonable suspicion** means you have a **valid**, **quantifiable reason** for believing that a person has done or is doing something improper, and you think another person in your position would likely come to the same conclusion.

Here are some examples of reasonable suspicion:

- If a police officer sees a man with a bat climbing through a smashed window from a building whose burglar alarm is going off, this situation would provide reasonable suspicion for the officer to detain and question the man.
- If an officer sees a car driving erratically, pulls the car over, and notices that the driver is wearing a wrist band from a nightclub, he has reasonable suspicion to conduct sobriety tests.

Law enforcement uses the idea of reasonable suspicion in other situations as well, such as investigating child abuse cases or interfering with domestic quarrels or violence.
Reasonable suspicion is also used by organizations and is defined in the laws of some states when it comes to issues such as firing a person for **theft** or **substance abuse**.

Even if there is no specific reasonable suspicion law in your state, your organization may have a policy. You can also ask your human resources department for guidance.

In some situations, you may want to have a **reasonable suspicion checklist** to use in cases such as alcohol abuse in the workplace. This checklist can give you a way to review behaviors you notice, such as staggering or an uneven walk, unusual appearance, demeanor, speech, and so forth.

When you suspect that something inappropriate could be occurring in the workplace, the first thing to do is carefully observe and document the person or situation.

Look around. Gather the facts. Obviously, if the issue is related to safety, you need to act quickly. Otherwise, gather facts and critically think things through—**consult your human resources department if necessary**—and then take a well-planned action.

Always **document what you see,** whether you deem action appropriate at the time or not.

Guidelines for Managing Alcohol and Drug Abuse Problems at Work

Alcohol and drug abuse can lead to serious safety concerns in the workplace. If you suspect that someone is or has been using drugs or alcohol on the job, keep the following in mind:

- You may not fire someone for being an addict. However, you may fire them for specific behaviors that go against your organization's policy or job description.
- Document all behaviors you witness.
- Talking to a person that you suspect is an abuser can be difficult. Speak to a representative from your human resources department or a legal advisor first.
- Be sure to take the proper steps before you take disciplinary action. You must issue warnings and possibly offer assistance (appropriate resources) with the opportunity for improved performance. Doing so may protect you from a wrongful termination lawsuit.
- Always be sure that you are disciplining an employee based on actual violations of policy or safety protocol, not merely on their condition.
- Remember that laws surrounding alcohol abuse and employee privacy change rapidly, so stay current on the laws in your locale.
- Be careful to respect a person's privacy. Hold conversations in private, with another manager or a human resources or legal representative present. Do not talk about the incident within the earshot of other people. Do not gossip and do your best to keep gossip from occurring.

You might wonder if you can fire an employee for being drunk. Here are some facts:

- **Alcoholism** is **protected** under the **Americans with Disabilities Act**. Thus, you cannot fire a person because they are an alcoholic.
- You can fire an employee for breaking an organization's policies or endangering people. If a person shows up on the job drunk or their drinking causes them to miss work, be late, or make mistakes, you can fire them for those behaviors. Consult your human resources department for advice.
- Before confronting someone you suspect is drunk, complete a reasonable suspicion checklist.
- **Do not confront someone without another manager or human resources representative present.**
- Confront the person as quickly as possible if their behavior is putting anyone at risk of injury, including themselves, or if there is potential liability for the organization.
- **Never dismiss and send home a person who is under the influence of drugs or alcohol.** Provide transportation to make sure they arrive at their destination safely.

Be Proactive and Seek Helpful Resources

You may have noticed that the word "proactive" is a recurring theme in this course. Our job is to anticipate potential issues so that you are prepared to solve them quickly and before they become more significant issues.

Being proactive is one of the best ways to keep you and your organization out of court.

Consider, for example, if you or others have observed any **unusual behaviors** of a particular person. When you have a reasonable suspicion that a person is unfit for duty and you observe behavior that reinforces this, then you, as well as another manager acting as a witness, should document the observations.

This is also a good example of why it is essential to be aware of **what resources** are **available** to help you **handle various situations**.

A **resource**, for example, can be a **person who is more experienced** or **more knowledgeable,** an **employee** or **project policy manual** specific to your project or organization, or a **book** about your field. Resources can take a variety of forms. You must have access to many types of resources, and those you manage need to know what resources are available to them.

You also need to know where to locate the resources you need. For example, if you know your employee handbook or project handbook provides the information you need to answer FMLA questions, be sure you know precisely where the handbook is at all times and that it is current.

Your project team also needs to know where to find resources. Keep handbooks and manuals in a place where they can be easily located.

You should occasionally perform an information audit to ensure that necessary policy and procedure documents are **in place and up to date**.

Consider safety issues and resources in the workplace. Ensure you are aware of the resources or materials OSHA requires your organization to have and check for compliance regularly.

Ensure that all **policies** and handbooks are current and have **revision dates** and that you have the latest copy. Also, **create procedures** for making changes in documentation, considering how, and how often, to revise, edit, or replace existing policy documents and how to properly inform employees of the changes.

In many organizations, the employee handbook or project policy handbook is a well-used resource. But keep the following in mind when referencing a handbook:

- It is not a contract.
- It does establish appropriate and equitable expectations for behavior.
- It should cover necessary topics but should not be a policies and procedures manual.
- It should be reviewed and updated regularly.

Be aware of external **resources** that might be **available**. For example, these might include **wellness programs** (perhaps through your **insurance provider**), **employee assistance programs**, a **medical hotline**, the **human resources department**, and the **legal department**.

What to do:	**Other actions:**
☐ Perform an information audit and ensure that necessary procedures are in place.	☐ ____________
☐ Ensure that all employee handbooks and project policy handbooks are current and have revision dates.	☐ ____________
☐ Be aware of the resources available to you: experienced people, policy handbooks, wellness programs, insurance providers, employee assistance programs, a medical hotline, HR, legal department, and so on.	☐ ____________

Remember

- ✓ Resources include experienced people (a mentor or leader), specially trained people (the safety lead or technical person), policies and manuals (like an employee handbook or project policy handbook), industry sources (such as standards), issue related sources (such as law books or ethics texts), and much more.
- ✓ Think critically before taking action. But take immediate action when safety is at issue.
- ✓ You must strive to protect your employees, contractors, customers, yourself, and your organization from potential litigation. Preventing lawsuits is vital for every organization, particularly in today's highly litigious environment.

Enhance Your Learning

Watch the following video to learn more about safety as a workplace priority: *10 Commandments of Workplace Safety.*

SafetyVideoShop. (2009). *10 Commandments of Workplace Safety.*	Available at: https://www.youtube.com/watch?v=3C6js5JtClQ

Reinforce Your Learning

Potential Liability Scenario. Review the scenario below. Reflect on laws and policies applicable to these situations and what you should do to avoid litigation throughout the life cycle of your project and after closing it. Then respond to the questions that follow.

Scenario: Is Joey Drunk (reasonable suspicion/drug-free workplace)

Project manager John and team lead Mary observed employee Joey making frequent trips to the water cooler. His walk appeared unsteady, and he spilled water on the way back to his work area. John and Mary, out of concern for Joey, stopped by his work area and began a casual conversation. Joey's responses appeared to be detached and vague. His situational awareness seemed to be compromised. Joey was heard slurring his words and struggled in trying to complete his sentences. Mary detected what she thought was the smell of alcohol on Joey's breath. However, John did not distinguish the same odor.

What are the potential risks or liabilities associated with this situation?

Does your employee handbook or project policy handbook (if available) reference this issue?

What actions should the project manager avoid or have avoided?

What actions should the project manager take or have taken?

Completing and Closing Projects #12.6

Evaluate Project Outcomes: Scope, Schedule, and Budget

Scope, **schedule**, and **budget** (costs) are the triple constraints of project management. It is difficult to affect one constraint without affecting one or both of the others. For example, if you find a change to the scope, schedule, or budget, you will want to check the other constraints to ensure that any needed updates were made to them as well.

In this concept, we will review the processes for closing a project's scope, schedule, and budget.

Information Needed for Closing

Depending on the size and complexity of a project, the **inputs** for closing the **scope**, **schedule**, and **budget** may include:

- Project management plan
- Project statement of work (SOW), scope statement
- Change request log
- Scope management plan
- Schedule management plan
- Project management software, schedule output
- Resource calendars
- Costs management plan
- Project management software, costs output
- Agreements
- Procurement performance information
- Enterprise environmental factors
- Organizational process assets

Closing the Project Scope

To close the scope, you must have access to the up-to-date project management plan, scope statement, scope management plan, and change request log.

The project management plan provides insight into all aspects of the project. Because changes to this document may not have made it to other project documents, you must check that all documents are up to date with the project management plan.

The scope statement, or statement of work (SOW), identifies the project deliverables. When needed, you should also state when something is not included in a project. For example, if a deliverable was discussed and then determined to be outside the scope, you would document that this deliverable was not included and why. This way, a few months down the road, you will not have a situation where the client/stakeholders recollect discussing the deliverable but not recall the decision to exclude it.

You should have updated the scope statement throughout the project life cycle, so it accurately reflects approved change requests, as indicated in the **change request log**. Change requests are often scope changes, which require an addendum to the scope statement.

The scope management plan should include how to close the project scope. It should discuss **comparing** the **scope statement** with the **change request log** to ensure that the approved change requests are accurately reflected in the final version of the scope statement.

You should also have established a process for adding the change request log as an addendum to the scope statement.

If the scope management plan does not contain closing processes as discussed above, you should add the processes to the scope management plan template, so the processes will be included in future projects.

Closing the Schedule

To close the schedule, you must have access to the up-to-date project management plan, schedule management plan, project management software schedule output, and resource calendars.

The **schedule management plan** includes closing functions such as verification that all tasks have been completed and all resources accounted for. In other words, you must verify that all work in the schedule has been completed as planned and according to the agreed quality standards.

Schedule closing outputs are the updated project documents.

One benefit of using project management software is that it is typically updated throughout a project, and so can provide an instant status report. Only a few items will need to be reviewed to ensure you are ready to close the project.

However, project management software can be expensive and, even if it is not, you and the team will need time to learn how to use it. Either way, it costs the organization money. Each organization must determine the best tools for their environment.

Two types of **resource calendars** should be included in the schedule close process: people and enterprise resources.

People calendars are typically accounted for by using the work breakdown structure (WBS) in the project management software.

Enterprise resources (non-people) are things like conference rooms and specialized equipment. These are typically scheduled by people in the enterprise and outside the project. Review this calendar to ensure that any remaining resource reservations are removed once you no longer need them.

Budget Closure

To close the **budget (**project costs), you must have access to the up-to-date project management plan, cost management plan, actual expenses/costs, procurement management plan, and procurement performance information.

The **cost management plan** should include how to close costs, with tasks like **comparing** the **final cost baseline** with the **change request log** to ensure that approved change requests are accurately reflected in the final version of the cost baseline. There could be a pre-approved process of adding the change request log as an addendum to the final cost baseline.

If the cost management plan does not contain closing processes like the ones discussed above, you should add them to the cost management plan template, so the processes can be included in future projects.

To reduce the probability of cost overruns, contingency reserves are often included in project budgets.

Cost reporting at closure can help you review not only the final performance of the project but the progression of the project as well. Understandably, scope and schedule changes occurring throughout a project can result in cost baseline changes.

The illustration below shows the costs for baseline 1 as compared with baseline 3 and the cost variations between them.

Tasks 2 through 6 each cost more than initially projected. The root cause was the tasks took longer than planned, which affected both the schedule and cost.

Costs Baseline Comparison Example

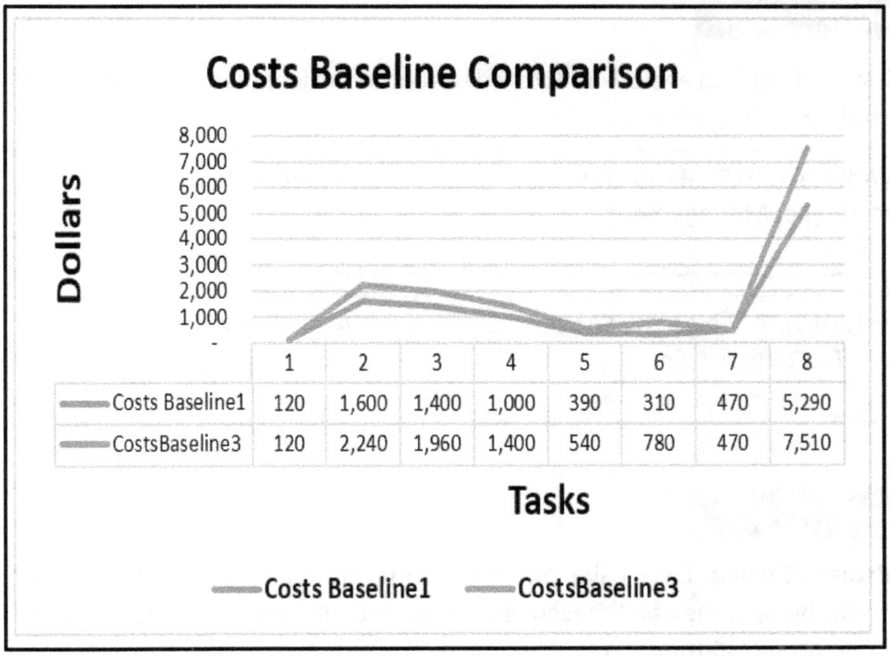

You can verify that the additional time was authorized correctly by using the resource management plan. This plan should outline what is necessary for additional resource allocations to be added to the project.

The procurement closing inputs include the procurement scorecard, which indicates the status of all contracts involved in the project. It will also report the overall performance of each supplier in terms of cost, quality, and timeliness.

What to do:	Other actions:
☐ Determine what type of project management software is in place or is needed.	☐ ______________________
☐ Access the project management software to learn what schedule and cost reporting is available.	☐ ______________________
☐ Consider if there is any schedule, cost, or scope information you need.	☐ ______________________

Remember

- ✓ The information reported from project management software is only as good as the data you and the team put into it.
- ✓ Compare information reported in various parts of the project management software to ensure it is accurate.
- ✓ You must consider software features and costs trade-offs in determining what software is best for your organization.
- ✓ Real-time reporting is only valuable if the data is up to date.
- ✓ System reporting can help ensure all aspects of the project have been accounted for and are ready to be closed.
- ✓ Project management software need not be expensive to be effective.

Enhance Your Learning

View the following 4-minute and 12-minute videos about project management software Basecamp Alternatives and Trello respectively.

TechnologyAdvice. (2015). Basecamp Alternatives: *Project Management Software.*	Available at: https://www.youtube.com/watch?v=pqFGpqQkwV8
Simpletivity. (2019). *7 Things You Should Do with Every Trello Board (Setup Guide).*	Available at: https://www.youtube.com/watch?v=HvkTamEjDXk

Reinforce Your Learning

Inadequate Project Closing. The closing phase can suffer from a lack of planning, not being completed, or not being done at all. Describe at least three things that could be negatively affected by an inadequate project closing.

1.

2.

3.

Completing and Closing Projects #12.7

Review Quality Tools at Project Closing

Even well-executed projects can fail during the closing phase of the project. Many project managers have been involved in projects that they could not completely finish and close.

We touched on the idea of the final 10 percent in terms of project monitoring and control, but project managers must also do their due diligence when closing a project. For example, at the end of the project, team members may want to make things perfect. Or they may run out of energy and enthusiasm. The project manager must ensure that such problems do not keep the project from being finished and properly closed.

In this concept, we discuss assessing how well the monitoring and control activities were performed by examining the effectiveness of those tools and identifying any loose ends. We will also consider internal and external audits.

Information Needed to Review Quality at Project Closing

The **inputs** for determining **quality** at project closing include the following:

- Contract review
- Project management plan
- Quality management plan
- Quality tools review
- Meeting minutes
- Progress reports
- Correspondence and emails
- Performance data
- Change request log
- Updated project documents

Contract review. Many project managers start the closing phase by reviewing the requirements outlined in the initial contract and subsequent change order log.

Project management and **quality management plans** should include quality processes for the closing phase. As an example, you might do an audit to review the critical aspects of quality and non-conformance from an internal and external perspective. Processes not included in the plan should be added to the quality management plan template, so they are included in future projects.

Details are required for the quality tools used to monitor and control the project, so the audit can be exhaustive. Tools we have discussed earlier include **checklists**, **check sheets**, **Pareto Diagrams**, **Cause-and-Effect Diagrams**, **Affinity Diagrams**, the **Requirements Traceability Matrix**, and the **Change Request Log**.

Checklists ensure that a particular set of activities are done, and sometimes that they are done in a specific sequence. Project managers use checklists because they know problems are more likely to occur if checklists are not used.

The next combination of tools is check sheets, Pareto diagrams, and cause-and-effect diagrams. With this combination, you can create a cause-and-effect diagram for what effect you are trying to address. Use a check sheet to get an identified cause to determine what categories of that cause may exist and a Pareto diagram to determine what categories of that root cause should be addressed first (possibly by frequency or impact). Then you can implement corrective action for the most important categories and determine if the anticipated impact on the effect has been realized. A quality audit can validate that this sequence of events was followed and whether the benefits were realized.

Minutes of meetings, progress reports, and correspondence. Review the information in these kinds of documents to determine that all requirements have been met, and to verify that all action items have been adequately addressed.

Performance data help to substantiate the effectiveness of corrective actions.

Cause and Effect Diagrams

Recall the Making Copies project discussed earlier and review the diagram below. If we follow through with the "Complicated to operate" cause on the diagram, we may surmise that it is one of the most frequent causes of copies taking longer than they should. An auditor can review the check sheet and Pareto Diagram to see why this cause was tracked.

Making Copies Project Cause and Effect Diagram Example

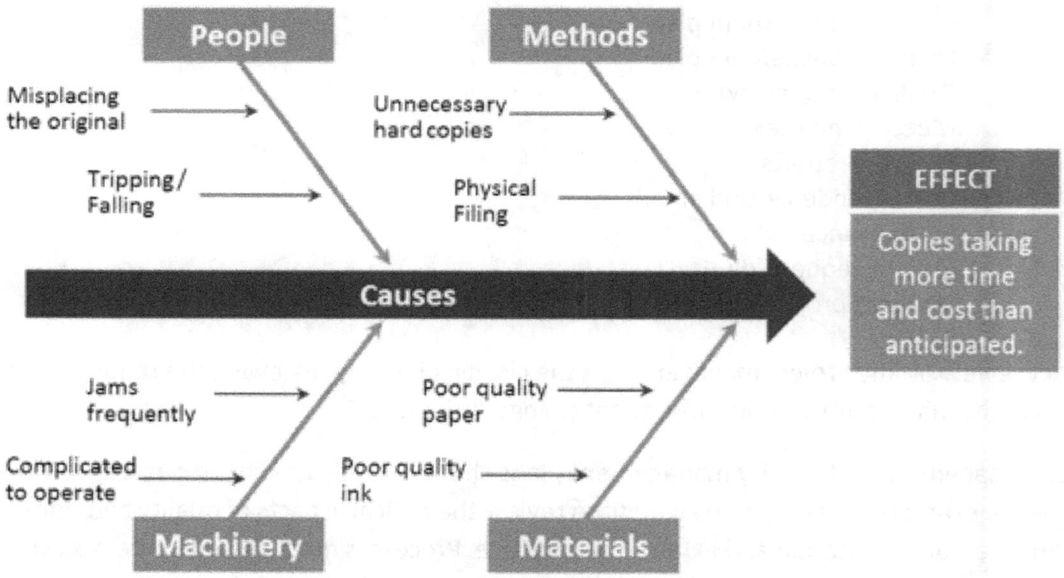

You could then inform the auditor of the corrective actions taken to help reduce the frequency and impact of the root cause. In the Making Copies project, for example, you would explain how user instructions were created that clearly illustrated the five functions most frequently identified as problematic, and short training sessions were held for all copier users. You would then explain that users continued to record the requested information on the check sheet near the copier, and over a few weeks, it reflected a significant decrease in time lost at the copier due to the "Complicated to operate" root cause.

In this example, the auditor could review the user instructions to ensure they are prominently displayed at the copier. The auditor could also talk with users at the copier to get their opinions on the improvement. Usually, when something makes people's lives better, they are happy to share their experiences.

Requirements Traceability Matrix

A Requirements Traceability Matrix is a tool to help you identify critical requirements and who is responsible for completion sign-off. At closing, all requirements must be accounted for and have a documented sign off.

The **Requirements Traceability Matrix** can also be used for quality auditing purposes. All requirements must be accounted for by fulfillment or cancellation. In either case, the closure must be documented on the Requirements Traceability Checklist.

Review the Webpage Creation project example below. An auditor could select an entry within the Requirements Traceability Matrix and determine if it was completed. There is enough information in the example matrix that the auditor could locate the requirements and make a validation.

Assume an auditor wants to review Requirement Identification Number 4, Website-next page 4. The auditor could reference the Requirements 1.0 version of the documentation for a description of the functionality. The auditor could also reference the test case used to test that the webpage functioned per the requirements. With these in place and signoff by a representative from the user community, the auditor should be in a position to sign off that this audit item was completed.

Webpage Creation Project—Requirements Traceability Matrix Example

Project Name:	Webiste Creation			Company:		Cars
Project Manager:	Jane Smith			Department:		Sales / Mktg.
Project Sponsor:	Cheryl Stratton					
Requirement Identification Number	Requirement Description	Requirement Document reference	Test Case Reference	User Signoff	Signoff Date	Comments
1	Website-homepage	Requirements 1.0	H0001	Jack Smith	2/13/yy	Job well done.
2	Website-next page 2	Requirements 1.0	H0002	Steph Nold	3/26/yy	
3	Website-next page 3	Requirements 1.0	H0003	Wes Gold	5/13/yy	
4	Website-next page 4	Requirements 1.0	H0004	Jane Ought	7/23/yy	

Use a **Change Request Log** to ensure that all change requests are accounted for by completion, cancellation, or future action. The illustration below shows a change request for an Age field to be added to one of the webpages, and all activities appear to be completed. An auditor can access the webpage and see if the Age field has been added and is functional.

Webpage Creation Project—Change Request Log Example

Request Date	Change Request Reference Number	Requestor Name	Requestor Contact Info.	Request Description	Why / Benefits	Est Cost Impact	Status (APP / REJ)	Status Date	Use Accept Name	User Accept Date
2/20/yy	CH001	Trudy Yates	tyates@org.com	"Age:" field added to web page 4	To provide targeted marketing data	$10,000	Approved	3/20/yy	John Sun	5/20/yy

In our example, the auditor finds everything in order and closes this audit item. Typically, you will have many change requests associated with a project. The auditor can randomly select a certain number of requests to audit and ensure completion.

Updating project documents can sometimes be a problem, particularly at the end of a project, when team members may want to get to their next assignment. This is where project management software can be invaluable because updating can be routine, which makes the closing activities seem less tedious.

During the closing phase, project managers must differentiate between **contract requirements** and **niceties**. Sometimes the closing documents contain both. In this event, the remaining work should be evaluated in terms of what will add value. In some cases, completing the niceties may cost a few extra nickels, which should be weighed against building client loyalty. However, the process of completing niceties may also be referred to by some as ***gold plating*** because they are outside of the original project scope and may add to the overall cost.

What to do:	Other actions:
☐ Identify critical requirements and who is responsible for completion sign-off.	☐ ______________
☐ Check to see if there are any project quality audits available for current or past projects.	☐ ______________
☐ Check all items on the Requirements Traceability Matrix you are involved with.	☐ ______________

Remember

✓ Project quality audits add value to a project.

✓ Project quality audits help ensure all quality items can be closed.

Enhance Your Learning

Watch the following four-minute video to learn more about quality tools typically used for efficient project management.

KnowledgeHut. (2018). *7 Basic Quality Control Tools for Efficient Project Management.*	Available at: https://www.youtube.com/watch?v=IFPF0TIFG-4

Reinforce Your Learning

Project Documents. Assume you are given the following project documents when a project's closing phase begins. What types of things can you do with what you have?

- Project Management Plan

- Schedule

- Change Request Log

- Traceability Responsibility Matrix

- Cause-and-Effect Diagram

Completing and Closing Projects #12.8

Appreciate the Importance of Records Retention

Your project records may seem to concern only your project, but they may be invaluable to your organization at a later time. For example, future events may be internal to the organization and planned, such as audits, or unplanned, such as a claim or lawsuit.

Identification and retention of appropriate records is a crucial part of a project's closing phase. If you do not retain all appropriate records, you and your organization may suffer consequences in the future.

In this concept, we discuss issues involved in record retention. You will need to perform an administrative closeout to ensure the proper records are kept available. You and your organization will also need to select appropriate hard copy and electronic formats for the records you retain.

Retention Policies and Procedures

Your goal is to identify and save project records that may provide documentation for audits; legal defense of the organization's products and services; administrative processes such as human resources and community relations; compliance with laws governing organizational record retention, and future projects.

Your organization should have retention policies and procedures in place to guide you in retaining records from your project. If not, you will need to research what records your organization needs to keep. To research and develop a project or an organization's document retention and destruction protocols, the project manager and other management representatives consult with competent legal counsel and tax and financial advisors.

In the event of litigation, the appropriate senior manager must issue a "litigation hold" to prevent the destruction of records.

Administrative Closing

A project's administrative closing is a process to resolve all contractual requirements for completed contracts.

The outputs from the administrative closeout may include:

- Project archives—a complete set of indexed records
- Project acceptance—confirmation that the project has met all client/stakeholder requirements and has been accepted
- Lessons learned—what did we do right; what did we do wrong, and what can we do better
- Contract file—a complete set of indexed records
- Formal acceptance and closure—official written notice to contractors and consultants that their work is complete

Clients often specify their record retention requirements in their contracts. Contract documents also often specify who owns products and services and who may use them for what purpose in the future.

You must review contract information as part of your closing process; most important, any clauses referring to:

- Ownership of documents
- Electronic files
- Confidentiality
- Copyrights
- Indemnities
- Prototype designs
- Unauthorized changes to plans

The final closeout is complete when all administrative and contractual actions have been accomplished, all disputes settled, and all final payments made.

How Long Should Records be Kept?
Generally speaking, an organization should retain relevant project files for the most extended practical time. Some organizations have policies and procedures specifying how long records must be retained. Examples are three years beyond the contract audit, 15 years after project completion, or permanently.

Aside from any legal requirements that may apply, or specific contract terms, the following are retention guidelines offered by the National Society of Professional Engineers (2016):

- Professional services agreements: minimum 3 years
- General correspondence: typically discarded at the end of the project
- Construction documents: minimum 3 years
- Studies and reports: usually 3 to 6 years
- Accounting and financial records: minimum 7 years
- Safety records: minimum 7 years
- Personnel records: preserve for 7 years past termination

Whether records are stored in hard copy or electronically, they should be organized and labeled carefully and logically for future access and stored in a safe environment.

Hard Copy Records
Hard copy (paper) document storage requires space, so purge unnecessary duplicate materials and distinguish between documents that should be archived and those that need to be available on a day-to-day basis as input to ongoing projects.

Do not archive multiple copies of marked up documents. Consolidate review comments on a single copy. Unless they are necessary, do not save editorial comments.

Reduce the bulk. Do not send things to storage in three-ring binders unless it is necessary.

Organize and label relevant materials carefully. Think in terms of retrieval as much as storage. Do not make others guess their way through your abbreviations. Keep in mind that t is easier to find something when the material is indexed.

Guidelines for managing the retention of hard-copy documents:

- Collect all information and file it in the project's filing system
- Label and date everything
- Discard intermediate products and duplicate materials
- Return library materials
- Keep copies of useful materials nearby for ongoing reference
- Package remaining materials for storage

Electronic Records

Archive organized electronic information on network servers or permanent media in read-only format.

Organize and label everything carefully. It could be a long time before something needs to be retrieved from an archive, and the person retrieving the information most likely will not have been involved in the project. Avoid abbreviations that people may not understand.

Use meaningful labels on computer files and date everything. Include the name of the person who created the electronic file on the copy being stored.

Guidelines for managing the retention of electronic documents:

- Store all project files (including email) on disks, hard drives, and network servers
- Carefully delete duplicates and outdated or superseded versions of project files
- Update CAD files to reflect final submittals

What to do:	Other actions:
☐ Consult your legal counsel and financial advisors to establish records retention protocols if they are not already in place.	☐ ______________________
☐ Be aware that clients often include clauses in their contracts specifying record retention requirements.	☐ ______________________
☐ Organize and label things carefully and do not use abbreviations.	☐ ______________________

Remember

✓ Review your organization's record retention policies and procedures.

✓ Avoid jargon and abbreviations when preparing records for storage.

✓ Index the information you retain because this will make it easier for people to locate what they need to know.

Enhance Your Learning

Watch the following 3-minute video to learn more about records retention.

Record Nations. (2017). *Records Retention—Requirements and Guidelines.*	Available at: https://www.youtube.com/watch?v=3YHnEeKiM34&t=1s

Reinforce Your Learning

Project Manager Closeout Self-Assessment. The following self-assessment lists tasks to do when closing a project. For a project you recently completed or observed, check all the statements that apply. Use this assessment in future projects.

- ☐ I confirmed with project team members that all changes in scope were resolved.
- ☐ I verified that all costs were included in the final invoice.
- ☐ I reviewed the project contract for document retention and ownership requirements.
- ☐ I purged duplicate and irrelevant materials before I stored the project records.
- ☐ I ensured the timely closing of all relevant contracts.
- ☐ I conducted a project closeout meeting with the team to discuss project performance.
- ☐ I included any overhead adjustments to a contract in the final invoice.
- ☐ I obtained the required releases and other documentation from consultants and contractors as part of the final payment process.
- ☐ I provided relevant project information such as photos and testimonials to appropriate business development and marketing personnel.

Identify and Document Lessons Learned

The PMI Pulse 2015 described lessons learned by referencing this quote: "We need to have people who capture the stories of what works and what doesn't . . . because people learn through content," said Dr. Ed Hoffman, Chief Knowledge Officer, NASA. "We need people who are skilled at capturing essential knowledge and putting it into a context that's appealing."

The idea behind the lessons learned process is to help projects **repeat what went well on a previous project** and **improve** on things that did **not go well** in future projects. Thus, pain from an earlier project can be a gain in a future project.

Project managers document **lessons learned** at the end of a project. However, be aware that many project managers, particularly those overseeing large or complex projects, also document lessons learned throughout their projects. This supports the team in focusing on problems and determining the best solutions to move forward. These ongoing lessons learned are incorporated into the final lessons learned at the end of the project.

Information you will need to prepare for the lessons learned process:

No matter when you capture and record lessons learned, the **inputs** for the process include:

- Project client/stakeholder experiences
- Project team experiences
- Project sponsor experiences
- Updated project documents

Lessons Learned Meetings

As you can see, the primary input for capturing lessons learned are the **experiences** of those involved in a project, the stakeholders and team members.

Project lessons are best reviewed in group discussions where everyone can contribute their expertise, brainstorm solutions, and in the end, reach agreement on what to do next time.

Keep these meetings to a manageable size; one where everyone who attends will have an opportunity to speak. One way to do this is to realize that not everyone from the same functional area or with the same skill sets must attend. For example, one member of a group of stakeholders who represent the same area can serve as the voice for that group. Likewise, if several people on a project have a specialized skill such as electrical engineering, one of them could represent that skill at the lessons learned meeting instead of having all of them attend.

You also want to ensure that the right people attend the meeting. For example, if you are at the stage in a project where you know you have a problem, for example, the requirements keep changing, but you do not yet have a solution, you will want to include people in the lessons learned meeting who do well at analyzing problems and who are not frustrated by lengthy discussion. Or, for example, if you already realize the cause of the problem is the requirements were not well documented before the project started, you will want to ensure that people skilled in writing requirements attend the meeting, so they can give ideas about how to proceed.

Capturing Lessons Learned

When closing a project, project managers should address lessons learned in each of the following project management knowledge areas:

- Scope
- Schedule
- Cost/Budget
- Quality
- Human Resources
- Communications
- Risk
- Procurement
- Stakeholders

When you are discussing lessons learned during the closing phase of a large project, you may want to schedule a series of meetings.

For smaller projects, a closeout meeting with the team can reinforce the things that made the project successful. Things that made the project less successful can be identified and solutions discussed and agreed on.

For large projects, it may be best to capture various lessons learned **iterations** along the way right after the lesson has been learned, or as part of a large **milestone review**, or as part of a **project lifecycle phase review**.

Regardless of when you discuss lessons learned, all participants must agree on the problem, the root cause(s), and the corrective actions.

Group Facilitation

No one likes criticism. When people feel criticized, they become defensive and do not participate productively in group discussions.

Every meeting must be facilitated by someone who can encourage participation and also keep the discussion from getting sidetracked by blame or criticism. The facilitator must be masterful at keeping the discussion productive and the tone collaborative. Everyone in the group must feel they are there to support each other in getting even better at what they already do well.

If you have a small project that is not complex, but you do not have a facilitator, you might be able to find a person in your organization who has facilitation skills. That person may be able to lead a lessons learned meeting.

Open-ended questions typically work best at generating group discussion. For example, a facilitator might lead the group by asking the following questions with a few words tailored to each project management knowledge area:

- What did we do right?
- What did we do wrong?
- What can we do better?
- So what?

However, the facilitator's job for a large or complex project can be challenging. If you do not have someone with facilitation skills, pick a few people you think may do well in the role and send them for facilitation training before you hold a session. You will most likely want to pick people who are familiar with the project because it could be a steep learning curve for people from outside the project.

Group Training

Before you conduct a lessons learned session, you will need to train everyone about how they are expected to participate. Ensure that everyone knows the session is not an opportunity to point fingers or assign blame for something that went wrong.

Ask participants to prepare for the meeting by considering what they think went well on the project and what could have been done better. Ask participants to think about themselves first—what they did well and where they could improve—and then to do the same for the project.

Just as you would for any successful meeting, send the participants an agenda before the meeting. With a lessons learned session, you will want to give enough detail so that the participants know the issue to be discussed. This will provide participants with the opportunity to review each item before the meeting, and some may want to do further research.

Implementing Solutions

The goal of a lessons learned meeting is to identify things done well and not so well in a project. When the discussion identifies things that did not go well, the facilitator will need to lead the group in defining the right problem, determining the cause, and formulating the best corrective actions.

Expert judgment helps determine the correct problem(s), the correct root cause(s), and appropriate corrective action(s). For example, when a schedule delay is identified as a problem, one team member may consider "bad duration estimation" to be the cause, while another may say "resources were not available."

For the first perceived root cause, we would want additional duration estimation training before the next estimates are put together. If the problem is the resources were not available, then we may want a scheduling tool to help with the early detection of a potential schedule delay. In other words, each perceived root cause will require its own solution.

The outputs from the lessons learned process include:

- Updated lessons learned log
- Updated knowledge database
- Updated project documents

Lessons Learned Log

One of the best ways to document and track the things you learn is to use a lessons learned log.

Another reason a log is useful to the project is it may help the team realize what other project documents are impacted by the lessons learned problem, cause, and corrective action. For example, if the discussion concluded that a problem was caused by three stakeholders being left off the stakeholder matrix, the team would realize they must update the stakeholder matrix—as well as the communications matrix, risk assessment matrix, and possibly other documents.

The **lessons learned log** should be updated throughout the lessons learned process.

The following example illustrates the kinds of information you might record on your log, and you can add other information as needed.

Lessons Learned Log Example

Ref #	Knowledge Area	Issue Description	So What?	Recommendation
LL001	Scope	No scope change thresholds established.	Numerous scope changes without approval.	Establish scope change thresholds requiring additional sign-off.
LL002	Schedule	The schedule changed without preserving the original schedule baseline.	It was not readily apparent in the project status reporting that the schedule slipped.	Include the use of schedule baselines. Include the baseline comparisons in the project status reports.
LL00n				

Your organization may have a project knowledge database that stores historical data associated with past projects. If so, you should also update this resource.

As another example of other sources to update, if the schedule was delayed because of a supplier that consistently missed delivery dates, you might want to introduce stiff penalties for late deliveries or drop the supplier altogether. Whatever the decision, it should be noted in the supplier information source so a future project can avoid the same kind of procurement performance issue.

What to do:

- ☐ Review your current project's lessons learned log. Create one if needed.
- ☐ Check the project knowledge database for entries of a similar project.
- ☐ Find a high-profile project's lessons learned log and read it.

Other actions:

- ☐ ______________________
- ☐ ______________________
- ☐ ______________________

Remember

- ✓ Not every project requires the same level of attention to each closing activity.
- ✓ Project managers are responsible for closing projects and documenting the lessons learned. For large projects, the project manager may also record lessons learned throughout the project.
- ✓ The project manager must update all project documents with information that comes to light from lessons learned.
- ✓ Project leaders who do not take the time to review the lessons learned from a similar project may be destined to repeat past mistakes and problems.
- ✓ Lessons learned are a way to help prevent or minimize the same negative things from happening again.
- ✓ The lessons learned log facilitates capturing lessons learned throughout a project.
- ✓ Lessons learned must be approached with an attitude of "let's help each other improve."

Enhance Your Learning

View the following 6-minute video by Project Management Videos about lessons learned.

Whitt, J. (2013). *How to Capture Lessons Learned at the End of a Project.*	Available at: https://www.youtube.com/watch?v=DBUqW_ek4hI&feature=youtu.be

Reinforce Your Learning

Lessons Learned Step 1: Brainstorming. This learning activity is most useful for a recent or current project. If neither applies, you could use this book as input for the exercise.

Start by brainstorming a list of lessons learned.

Category?	Lessons Learned
	1.
	2.
	3.
	4.
	5.
	6.
	7.
	8.
	9.
	10.
	11.
	12.
	13.
	14.
	15.

Lessons Learned Step 2: Lessons Learned Log. Using your brainstorming list, categorize the items according to project management knowledge areas, for example, scope, schedule, cost, and so on.

Using the Lessons Learned Log that follows, record your list by project management knowledge area, and complete the information across each row. Continue that process until you finish or run out of time.

Feel free to add additional columns for information that may apply to your project.

Lessons Learned Log				
Ref #	Knowledge Area	Issue Description	So What?	Recommendation

Completing and Closing Projects #12.10

Implement Strategies to Enhance Stakeholder Relationships

Every project needs some level of **stakeholder relationship building.** Communication, even at the most basic level, is necessary for every human interaction, and of course, in every aspect of a project. People like to work with those they trust.

Recognize that **stakeholders do not owe you, their loyalty. You must earn it—continually**. Too many project managers think it is enough to simply provide good service, but the truth is, you must work to build and maintain relationships with stakeholders. The same is true for your team.

Building and maintaining **stakeholder relationships** is a significant contributing factor in maintaining an environment that fosters a **successful** project. Doing so will require that you develop strategies for dealing with stakeholders.

In this concept, we discuss issues involved in building and maintaining relationships with stakeholders. The discussion includes the closing phase of the project.

Project Stakeholders

Stakeholders serve various roles or functions on projects. For example, a stakeholder could be a sponsor, a client, a representative from the operational area most affected by the project, or a representative from a support area, such as marketing, legal, or accounting.

A stakeholder's role or function on a project will affect how they engage with the project. For example, a client stakeholder will want to be up to date on all project activities, and they will make many project decisions. This is in contrast to a stakeholder who represents an operational department not directly involved in the project; they may only want to know limited project information.

Some stakeholders can be involved in more than one of your projects, and they can have different roles. For example, on one project a person could be a client stakeholder who you frequently communicate with; on another project, that person may only want to be copied on status reports.

Assess Your Confidence in Managing Stakeholder Relationships

The following questions can help you assess your confidence in managing the stakeholder relationships associated with your project:

1. Do you understand stakeholder requirements and expectations? How do you know?
2. Do you communicate your strategies for satisfying these requirements and expectations to the project team and to the stakeholders?
3. Do you behave to make it easy for people to do business with you, or do you think mostly in terms of your own convenience?
4. Do you know how your stakeholders feel about you and your project? Do your stakeholders give you repeat business?

Information Needed for Developing and Managing Stakeholder Relationships

Information you will need to build and maintain stakeholder relationships include the following inputs:

- Project management plan
- Stakeholder management plan
- Stakeholder register
- Stakeholder engagement matrix
- Stakeholder power/interest matrix

The **project and stakeholder management plans** will help you understand and manage stakeholder requirements and expectations. The **stakeholder register** and the **stakeholder matrices** will help you identify productive interactions with each type of stakeholder.

Managing Stakeholder Relationships During the Closing Phase of the Project

Tools and **techniques** used in managing stakeholder relationships and also in closing them include:

- Expert judgment
- Meetings

Expert judgment can be used to determine if the appropriate strategies were used to manage stakeholder relationships throughout the project. The same tools that were used to plan, execute, monitor, and control stakeholder relationships can also be used to close them.

For example, one part of the close process could be a review of the Stakeholder Engagement Grid. If the stakeholder relationship grid illustrated below were to be reported during a project as the status of the stakeholder engagement, we would notice issues.

In this example, Jim was thought to be in favor of the project and therefore designated under the "For" category as "P" (Planned level of engagement), which meant he was a proponent. However, when this current status was reported, Jim was not even aware of the project. Jack was planned to be "Neutral" but was reported as "Against" at the time of the status.

Both of these situations could be problematic and would need to be addressed.

Stakeholder Engagement Grid: Example During the Project

Stakeholder	Unaware	Against	Neutral	For	Promote
Jim	C			P	
Jack		C	P		
Susan				P, C	
Ralph					P, C
Legend	P = Planned level of engagement				
	C = Current level of engagement				

The same matrix reported as part of the closing activities for the project is illustrated below. It shows there has been some movement.

Stakeholder Engagement Grid: Example at Project Closing

Stakeholder	Unaware	Against	Neutral	For	Promote
Jim	C			P	
Jack			P, C		
Susan				P, C	
Ralph					P, C
Legend	P = Planned level of engagement				
	C = Current level of engagement				

Closing activities should include some explanation as to why differences between "planned" and "current" may still exist, as well as how other discrepancies were corrected. In this case, we would need to explain why Jim's status did not move. As it turns out, Jim's role changed in the early stages of the project. His new responsibilities no longer included an association with this project. That part of the grid should have been updated earlier in the project.

The change in Jack's status came from the project team inviting him to a meeting to discuss the project. Jack's perception of the project went negative when he heard that his personnel would have to change some of their processes. This would hurt the effectiveness of his team, reflect poorly on their performance ratings, and affect their annual salary increases.

In fact, the project activities would neither increase nor decrease his team's performance. What Jack heard was a miscommunication from someone outside the project team. Once the team showed Jack that the project would have no impact on his area, he was comfortable with his neutral status, and the project documentation was updated.

These types of situations can contribute to lessons learned. In Jim's case, the lesson is an emphasis on the importance of project documents being updated on time. In Jack's case, the lesson is the importance of preventing misinformation to avoid stakeholders unnecessarily alienating themselves from the project.

What to do:

- ☐ Review the stakeholder engagement matrix for the projects you are associated with.
- ☐ Participate in the stakeholder close process for a project you are involved with.
- ☐ Recognize that what is just one of many projects for you is likely a major deal for stakeholders.

Other actions:

- ☐ ______________________________
- ☐ ______________________________
- ☐ ______________________________

Remember

- ✓ The same person can play different roles on separate projects you are involved with.
- ✓ Determining stakeholder relationship strategies, and following them, is critical to managing successful projects.
- ✓ A stakeholder's role within a project may change during the project and should be assessed during the project for any changes to expectations.
- ✓ Managing your relationships with existing and past stakeholders is essential. It provides a good return on investment.

Enhance Your Learning

View the following 8-minute video by about stakeholder relationships.

Clayton, M. (2019). *Stakeholder Engagement Tips: 5 Tips For Project Managers.*	Available at: https://www.youtube.com/watch?v=APc9S_8v7YY

Reinforce Your Learning

Cultivating Stakeholder Relationships. List the best two or three examples of things you have done in the past year to ensure that stakeholders on your projects were satisfied:

1.

2.

3.

What reactions did you observe from others (such as other stakeholders, team members, customers, vendors) in these situations?

Summary

A Final Word: Completing and Closing Projects

You have just completed the *Completing and Closing Projects chapter*. A few important topics you have encountered include:

- Identifying the critical elements in closing projects
- Describing closing tools, techniques, and processes
- Communicating effectively about project changes and the change process
- Evaluating project outcomes
- Identifying and documenting lessons learned

This chapter emphasized the closing phase of the project life cycle. We have one final comment: Never forget to celebrate a successful project!

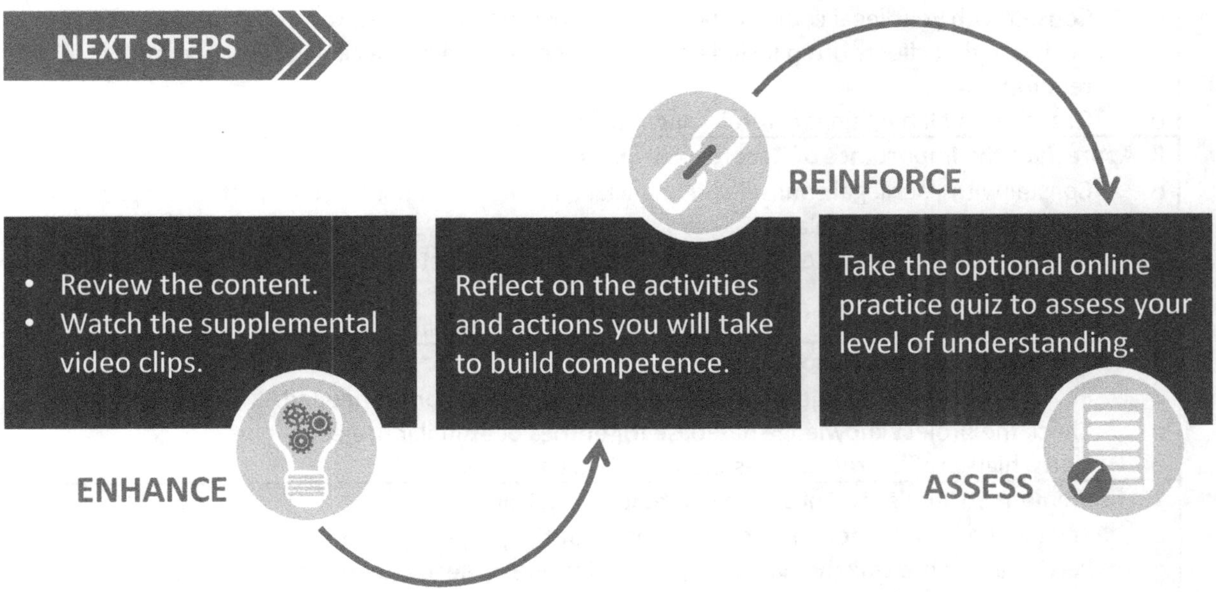

Completing and Closing Projects—Recap Checklist
The closing activities in a project are essential for project success. Many project managers spend too much time planning the other phases of the project life cycle and fall far short in preparing for project closing.
1. Prepare for Closing or Terminating a Project □ Review the project closing checklist and consider its value for your project. □ Are you aware of the closing plans for your project? Are those plans adequate? □ For project tasks assigned, do you know the acceptance criteria for completion?
2. Understand Why Law and Policy Matter at Project Closing □ Identify the top two or three policy or legal issues your work unit could face. □ Formulate proactive measures you can take to prevent those issues from occurring. □ Know and avoid legal pitfalls often encountered by project managers.
3. Evaluate Resources, Work Issues, and Conditions □ Review your project's resource close information and ensure the most current resource calendars accurately reflect that information. □ Review the most current project schedule to ensure all activities have been closed. □ Recognize that when a problem occurs, almost any type of evidence could be used in court.
4. Be Mindful of Property and Casualty Liabilities □ Recognize that it is your responsibility to provide a safe workplace. □ Do not allow people to use your organization's equipment or property during non-business hours. □ Be certain that everyone receives proper OSHA training each year.
5. Identify and Use Resources to be Safe and Legal □ Perform an information audit and ensure that necessary procedures are in place. □ Ensure that all employee handbooks and project policy handbooks are current and have revision dates. □ Be aware of the resources available to you: experienced people, policy handbooks, wellness programs, insurance providers, employee assistance programs, medical hotline, HR, legal department, and so on.
6. Evaluate Project Outcomes: Scope, Schedule, and Budget □ Determine what type of project management software is in place or needed. □ Review the project management software to learn what schedule and cost reporting is available. □ Consider if there is any schedule, cost, or scope information you need.
7. Review Quality at Project Closing □ Consult with your legal counsel and financial advisors to establish records retention protocols. □ Be aware that clients often include clauses in their contracts specifying record retention requirements. □ Organize and label things carefully and intelligently.
8. Appreciate the Importance of Records Retention □ Consult with your legal counsel and financial advisors to establish records retention protocols if they are not already in place. □ Be aware that clients often include clauses in their contracts specifying record retention requirements. □ Organize and label things carefully and do not use abbreviations.
9. Identify and Document Lessons Learned □ Review your current project's lessons learned log. Create one if needed. □ Check the project knowledge database for entries of a similar project. □ Find a high-profile project's lessons learned log and read it.
10. Implement Strategies to Enhance Stakeholder Relationships □ Review the stakeholder engagement matrix for the projects you are associated with. □ Participate in the stakeholder close process for one project you are involved with. □ Recognize that what is just one of many projects for you is likely a major deal for the stakeholders.

Knowledge Review Test—Questions

Completing and Closing Projects

1. All the following items normally take place as part of the final invoicing process, except:
   A. Confirmation that all scope changes have been resolved.
   B. Subcontractor releases.
   C. Applicable overhead adjustments.
   D. Notification to client that project is being closed.
   E. Work breakdown structure is finalized.

2. Appointing someone other than the project manager and his or her team to be responsible for the project closeout phase may be useful. However, a disadvantage of this approach is that the:
   A. Closeout manager probably lacks technical knowledge about the project.
   B. Closeout manager may get all the credit for the project.
   C. Project manager is the only person who can guarantee that no loose ends remain on the project.
   D. Project manager is the only person who will ensure that the closeout process is not dragged out unnecessarily.

3. During the contract close-out, the project manager needs to document the:
   A. Statement of work.
   B. Formal acceptance.
   C. Payment schedule.
   D. Change control process.

4. If someone on your team, a coworker, or a customer complains about a policy or legal issue, take it seriously. Which of the following should you NOT do?
   A. Be courteous and choose your words carefully.
   B. Apologize for how you feel about a particular action, but don't apologize for the action itself as this could be seen as an admission of guilt and open your organization to litigation.
   C. Document (write a report) a questionable situation and discuss the situation with the appropriate person in your human resources department.
   D. Talk to media about the situation.
   E. If legal charges are filed against the organization, get the proper human resources, legal, or leadership person involved immediately.

5. Due to funding cuts your project has been terminated. The scope verification process:
   A. Should be delayed until the project is completed.
   B. Should establish and document the level and extent of completion.
   C. Should determine the quality of the work completed.
   D. Should form the basis of a project audit.
   E. None of the answers are correct.

6. When accidents happen, a project manager should do all the following except:
   A. Ensure that the employee is okay and receiving medical attention, if necessary, that the work area is safe, and that other employees are not in immediate danger.
   B. Make a timely report of the situation, including filing a workers' compensation claim.
   C. Investigate the incident that same day.
   D. Document all details as you investigate.
   E. Encourage the employee to report-to-work ASAP.

7. A final key aspect of closing out a project is to:
   A. Transfer all project records to the owners.
   B. Disseminate information to formalize project completion.
   C. Disseminate progress reports and risk assessments.
   D. Review the specific project results to establish if they comply with the quality standards.

8. You are the project manager of a $2,000,000 project and your estimated calculations indicate that you will be done on time and 10% under budget. Based on your earned value calculations you estimate that your organization's profit will decrease by 50,000. At the completion you plan on documenting everything. The information may be valuable for future projects in all areas , except:
   A. Administering contracts
   B. Eliminating durations.
   C. Resolving conflicts.
   D. Allocating resources.

9. In the law, to have reasonable suspicion means all except:
   A. You have a valid, quantifiable reason for believing that a person has done or is doing something improper.
   B. You should ask another person in your position if they would likely come to the same conclusion.
   C. You should ignore the situation for fear that you may be liable.
   D. You should try and look at the situation objectively.

10. Alcohol and drug abuse can lead to serious safety concerns in the workplace. If you suspect that someone is or has been using drugs or alcohol on the job, consider all the following, except:
   A. Firing the person for being an addict.
   B. Talking to a person that you suspect is an abuser can be difficult.
   C. Be sure to take the proper steps before you take disciplinary action.
   D. Always be sure that you are disciplining an employee based on actual violations of policy or safety protocol, not merely on their condition.

11. During the project closeout phase, most conflict arises from:
   A. Schedule problems.
   B. Cost overruns.
   C. Technical problems.
   D. Working interfaces.

12. The following are the kinds of inputs you will use to start a project's closing phase, except:
   A. Project management plan.
   B. Accepted deliverables.
   C. Organizational process assets.
   D. Stakeholder resource assets.

13. A ______ is a tool to help you identify critical requirements and who is responsible for completion sign-off.
   A. Project Charter.
   B. Requirements Traceability Matrix.
   C. Cause and Effect Diagram.
   D. Lessons Learned Register.

14. At closing you should do what to all hard copy documents, except:
   A. Archive all.
   B. Consider the storage space requirements.
   C. Purge unnecessary duplicate materials.
   D. Identify documents that need to be available on a day-to-day basis as input to ongoing projects.

15. The PMBOK® approach to administrative closure includes:

   A. Confirmation that the project has met all customer requirements.
   B. Confirmation that the customer has formally accepted the final product.
   C. Appropriate project documents be retained.
   D. All the answers are correct.

16. Contract closeout is a process that involves:

   A. Customer satisfaction analysis and final payment.
   B. Administrative closeout and archiving records.
   C. Final subcontractor payment and lessons learned.
   D. Product verification and administrative closeout.

17. At the end of each project, the project team should prepare a lessons-learned summary that focuses on all the following except:

   A. Sharing best practices with other project teams in the organization.
   B. Warning others of potential problems on similar future projects.
   C. Suggesting methods to mitigate risk effectively to ensure success on future similar projects.
   D. Sharing only positive aspects of the project for future replication elsewhere in the organization.

18. The lessons learned inputs for the process include all the following, except:

   A. Project client/stakeholder experiences.
   B. Project team experiences.
   C. Project sponsor experiences.
   D. Updated project documents.
   E. None of the answers are correct.

19. Information you will need to build and maintain stakeholder relationships include the following inputs, except:

   A. Project management plan.
   B. Stakeholder management plan.
   C. Stakeholder rumor index.
   D. Stakeholder register.
   E. Stakeholder engagement matrix.

20. All the implement strategies to enhance stakeholder relationships are ones your you to consider, except:

   A. Review the stakeholder engagement matrix for the projects you are associated with.
   B. Participate in the stakeholder close process for one project you are involved with.
   C. Assume that stakeholders owe you their loyalty.
   D. Recognize that what is just one of many projects for you is likely a major deal for the stakeholders.

Completing and Closing Projects — Answer Sheet

1.	___A	___B	___C	___D	_X__E
2.	_X_A	___B	___C	___D	___E
3.	___A	_X_B	___C	___D	___E
4.	___A	___B	___C	_X_D	___E
5.	___A	_X_B	___C	___D	___E
6.	___A	___B	___C	___D	_X_E
7.	___A	_X_B	___C	___D	___E
8.	___A	___B	_X_C	___D	___E
9.	___A	___B	_X_C	___D	___E
10.	_X_A	___B	___C	___D	___E
11.	_X_A	___B	___C	___D	___E
12.	___A	___B	___C	_X_D	___E
13.	___A	_X_B	___C	___D	___E
14.	_X_A	___B	___C	___D	___E
15.	___A	___B	___C	_X_D	___E
16.	___A	___B	___C	_X_D	___E
17.	___A	___B	___C	_X_D	___E
18.	___A	___B	___C	___D	_X_E
19.	___A	___B	_X_C	___D	___E
20.	___A	___B	_X_C	___D	___E

Afterword

Reading is a valuable way to gain knowledge. This is why schools have textbooks, why online news sites and blogs are popular, and why business and self-help books are a billion-dollar industry. However, reading without reflecting and acting will not be productive.

As with any learning, simply reading without engaging will accomplish little. The level of effort you put forth will determine how much you improve and how close you come to realizing your full potential. You must reflect on what you learn and then apply that knowledge to your projects and career.

How do you do that? You do it by acting; by practicing what you have learned. This book is designed around the following three themes and chapters:

Processes

1. Contemporary Project Management
2. Defining Project Scope
3. Project Planning and Scheduling Tools
4. Monitoring and Controlling Projects

People

5. Enhancing Project Communication
6. Developing Interpersonal Project Relationships
7. Building High-Performance Project Teams
8. Leading and Managing Projects

Resources

9. Identifying and Mitigating Project Risk
10. Controlling Project Costs and Budgets
11. Managing Time and Project Meetings
12. Completing and Closing Projects

To become a more accomplished project manager or a more successful leader, you must engage with the material presented here. Question it, think about how it relates to your projects, and begin to internalize the concepts.

As Jim Lovell, Apollo 13 astronaut and famous author once said, "There are people who make things happen, there are people who watch things happen, and there are people who wonder what happened. To be successful, you need to be a person who makes things happen."

Action is the foundational key to all success.

—Pablo Picasso

References

Accenture Academy. (2016). *Spotlight on the Difference between Budgeting and Forecasting*. [Video]. YouTube. https://www.youtube.com/watch?v=FL2NcGYQ_Vc

Adams, S. (2013). How to communicate effectively at work. *Forbes*. http://www.forbes.com/sites/susanadams/2013/11/19/how-to-communicate-effectively-at-work-3/

Aguirre, D. (2014). *How to Lead Change Management*. [Video]. YouTube. https://youtu.be/PQ0doKfhecQ

Akbar, (2022). *Project Monitoring vs Controlling, Difference B/W Project Monitoring and Controlling.* [Video]. YouTube. https://www.youtube.com/watch?v=XTo99IsVS14&t=1s

Akers, M. & Porter, G. (2016). *What is emotional intelligence?* http://psychcentral.com/lib/what-is-emotional-intelligence-eq/

Akinwale, P. (1016). *Manage Communications/Stakeholder vs Control Communications/Stakeholder*. [Video]. YouTube. https://www.youtube.com/watch?v=ZKYmRKadujg

Andrade, R. (2013). *Affinity Diagrams*. [Video]. YouTube. https://www.youtube.com/watch?v=jvTSsJrDZec

Beare, S. & McMillan, M. (2007). *The power of teamwork*. [Video]. YouTube. https://www.youtube.com/watch?v=mR0_SK1K8xY&feature=related

BeatYourDeadlines. (2013). *10 most important time management tips*. [Video]. YouTube. https://www.youtube.com/watch?v=HFuCXvvXgQQ

Bender, M. (2013). *Project Risk Management — Simplified!* [Video]. YouTube. https://youtu.be/O6sVZ5ZwST4

Birkinshaw, J. (2013). *Two tips for developing good management skills*. [Video]. YouTube. https://www.youtube.com/watch?v=lLcgCEbsT08

Bizconsesh. (2018). *Risk vs. Uncertainty*. [Video]. YouTube. https://www.youtube.com/watch?v=aX6Cr6enozw

Blake, B. (2014). *The Johari Window Model*. [Video]. YouTube. https://www.youtube.com/watch?v=BWii4Tx3GJk

Bourne, L. (2009). *Stakeholder Relationship Management: A Maturity Model for Organizational Implementation.* Routledge.

Bradberry, T. (2014*).* Emotional intelligence – EQ. *Forbes.* http://www.forbes.com/sites/travisbradberry/2014/01/09/emotional-intelligence/#3beebe1a3ecb

Braun, R. and Dierking, C. (2013). *Feds: Killing Cincinnati Streetcar Project Unprecedented*. [Video]. YouTube. https://www.youtube.com/watch?v=lK8Nf2GisvU

Bridges, J. (2019). *How to Monitor Daily Progress as a Project Manager*. [Video]. YouTube. https://www.youtube.com/watch?v=E2Qahnv1j1g

Bridges, J. (2019). *How to Successfully Execute a Plan*. [Video]. YouTube. https://www.youtube.com/watch?v=802yQd8TNf8&t=1s

Bridges, J. (2017). *Top 5 Excel Project Templates - Project Management Training*. [Video]. YouTube. https://www.youtube.com/watch?v=BAY78mJr9cQ

Bridges, J. (2018). *What is Project Scheduling?* [Video]. YouTube. https://www.youtube.com/watch?v=WNWSQOynrl0

Bright Hub PM. (2022). *Project management legal issues: Watching out for Project Management Legal Issues*. https://www.brighthubpm.com/risk-management/63360-project-management-legal-issues-are-you-at-risk/

Bright Hub PM. (2022). *Project Risk Matrix Template — Probability and Impact.* http://www.brighthubpm.com/templates-forms/104889-using-a-probability-and-impact-matrix-for-project-risk-management/

Burke, C. (2021). *How to Create Effective Project Deliverables*. [Video]. YouTube. https://www.youtube.com/watch?v=7HADacRHSFs

Business Management Daily (BMD). (2012). *9 things employees want from their managers.* http://www.businessmanagementdaily.com/11630/9-things-employees-want-from-their-managers-and-5-things-they-dont

Can-Indian Channel. (2018). *MBA 101 Leadership, Situational Leadership Contingency Model.* [Video]. YouTube. https://www.youtube.com/watch?v=H1SiGs-SVdA

Centrestar. (2020). *Risk Analysis & Planning Audit (RAPA) – Part 1.* https://app.box.com/s/7um2qchru95j2y06v3z1apkmc7nz1t4y

Centrestar. (2020). *Risk Analysis & Planning Audit (RAPA) – Part 2.* https://app.box.com/file/686380339807?s=toqq4jur65zqpze3p7ifjciwxqum3ejj

Cholayil, S. (2013). *Budget Concept And Types*. [Video]. YouTube. https://www.youtube.com/watch?v=nsesYlcp2BE

Christianson, S. (2019). *Project Management Networks Part 2: Forward and Backward Pass*. [Video]. YouTube. https://www.youtube.com/watch?v=Momnar3Mr9Q

Clayton, M. (2022). *How to Create a Gantt Chart in 9 Easy Steps.* [Video]. YouTube. https://www.youtube.com/watch?v=FX70rWEE8eY

Clayton, M. (2019). *Project Management in Under 5: What is Project Integration?* [Video]. YouTube. https://www.youtube.com/watch?v=cHUP61KU3xE

Clayton, M. (2017). *What is a PERT Chart? Project Management in Under 5.* [Video]. YouTube. https://www.youtube.com/watch?v=i160aaBX7mE

Clayton, M. (2017). *What is Milestone Planning?* [Video]. YouTube. https://www.youtube.com/watch?v=aJcd1rQ0Z9w

Cobb, A. (2011). *Leading Project Teams: The Basics of Project Management and Team Leadership.* SAGE Publications, Inc.

Conservation Strategy Fund. (2014). *Intro to Cost-Benefit Analysis*. [Video]. YouTube. https://www.youtube.com/watch?v=7tdKkeNClPE&list=PLBfu1mD9hk64sgOIH_nUEsndUzACDe-4Y

Cooper, R. (2005). *Product Leadership: Pathways to Profitable Innovation*. Basic Books.

Craft, C. (2022). *Project Management QuickStart Guide: The Simplified Beginner's Guide to Precise Planning, Strategic Resource Management, and Delivering World Class Results*. ClydeBank Media LLC.

Creative Education. (2020). *How to draw a Project Network Diagram.* [Video]. YouTube. https://www.youtube.com/watch?v=pWkhR67bkXw

CREIO Newsletters. The Consortium for Research on Emotional Intelligence in Organizations (CREIO) regularly posts research studies and other information about emotional intelligence. http://www.eiconsortium.org.

Dartmouth University. (2014). *Five-Level Work Breakdown Structure.* [Video]. YouTube. https://www.youtube.com/watch?v=SqcuLsyFr-o

Deen, D. (2011). *Top 5 Project Tracking & Monitoring Tips for Project Managers*. [Video]. YouTube. https://www.youtube.com/watch?v=CTg_9K5K6hE

Denziel, G. (2004). "E-mail Use and Communication Perceptions of University of Vermont Extension Employees." *Journal of Extension* 42(3).

Dodd, J. (2014). *The Role of The Project Manager*. [Video]. YouTube. https://www.youtube.com/watch?v=dUhJuB69ZBo

Donahue, W. (2022). *Building Leadership Competence: A practical guidebook for professional development*. Second Edition. IngramSpark.

Donahue, W. (2020). *Small Projects Tool - EMPTY SPREADSHEET.* https://app.box.com/s/gq1t096g0cs51bbi4dwzi681xutzrh72

Donahue, W. (2021). *Unlocking Lean Six Sigma*. IngramSpark.

Dow, W. (2008). *Project Management Communications Bible*. Wiley.

Dynamic Flight. (2017). *Effective communication*. http://www.dynamicflight.com/avcfibook/communication/

ERC Blog. (2013). 10 *Employment Laws that Supervisors Need to Know*. https://www.yourerc.com/blog/post/10-Employment-Laws-that-Supervisors-Need-to-Know

ExpertVillage Leaf Group. (2008). *Non-Verbal Communication Tools: Non-Verbal Communication: Paralanguage*. [Video]. YouTube. https://www.youtube.com/watch?v=P5j0DFkbnE0

Faraaz, A. (2019). *Identify Stakeholders | Project Stakeholder Management.* [Video]. YouTube. https://www.youtube.com/watch?v=0xa2Q2XU2nM

Forleo, M. (2011). *How to Recognize Great Opportunities vs. Time Wasters*. [Video]. YouTube. https://www.youtube.com/watch?v=EM6saU7X4q8

Franklin Covey. (2018). *Weekly Planning- A Video from The 7 Habits of Highly Effective People*. [Video]. YouTube. https://www.youtube.com/watch?v=7yMh2QNRc_M

Fremouw, B. (2014). *Project Management Concept #2: Qualitative Risk Analysis vs. Quantitative Risk.* Analysis. [Video]. YouTube. https://www.youtube.com/watch?time_continue=3&v=hx-r0sdp2wk&feature=emb_logo

Generali Group. (2017). *Property and Casualty Insurance*. [Video]. YouTube. https://www.youtube.com/watch?v=nVW3fQMM1As

Gido, J., Clements, J. (2017). *Successful Project Management*. Cengage Learning.

Gilbert, T. (1996). *Human Competence: Engineering Worthy Performance.* ISPI.

Girdler, A. (2020). *Project Charter Guide [HOW TO WRITE A PROJECT MANAGEMENT CHARTER]*. [Video]. YouTube. https://www.youtube.com/watch?v=FlPeKqrvNCs

Goleman, D. (1995). *Emotional Intelligence.* New York: Bantam Books.

Goleman, D. (2019). *Emotional Intelligence: For a Better Life, success at work, and happier relationships.* Independently published.

Goleman, D. (2021). *Emotional Intelligence Takeaways*. [Video]. YouTube. https://www.youtube.com/watch?v=5_09ArEkFxY

Goleman, D. (2015). *What it takes to be a great leader*. [Video]. YouTube. https://www.youtube.com/watch?v=_QpUH7i7EhM

Gordan Training. (2010). *Leadership Training - Active Listening*. [Video]. YouTube. https://www.youtube.com/watch?v=ESujTCel6lM

Goulston, M. (2015). *Just Listen: Discover the Secret to Getting Through to Absolutely Anyone*. AMACOM.

Hall, H. (2022). *10 Ways to Engage Project Stakeholders*. https://projectriskcoach.com/engage-project-stakeholders/

Halloran, A. (2009). *Workplace Communication – Three Steps to Building Relationships*. [Video]. YouTube. https://www.youtube.com/watch?v=Hrd06zuiofI

Harvard Business Review. (2014). *Take Control of Your Nonverbal Communication.* https://hbr.org/video/3541641444001/take-control-of-your-nonverbal-communication

Harvey, J. *The Abilene Paradox*. Video Production.

Hasrian, F. (2013). *Wisdom of Geese*. [Video]. YouTube. https://www.youtube.com/watch?v=5rOg4WfNDfM

Haworth, S. (2018). *Start Your Projects Right: A Complete Guide to Project Initiation*. Digital Project Manager. https://thedigitalprojectmanager.com/project-initiation-guide-how-start-projects/

Heathfield, S. (2021). *10 Tips for Better Teamwork*. https://www.liveabout.com/tips-for-better-teamwork-1919225

Howcast. (2011). *How to Achieve Long-Term Goals*. [Video]. YouTube. https://www.youtube.com/watch?v=cEfeDxWRr8c

Improvement Pill. (2016). *Communication Skills - How to Improve Communication Skills - 7 Unique Tips!* [Video]. YouTube. https://www.youtube.com/watch?v=mPRUNGGORDo

James, J. (2011). *Jared James explains how Perception is Reality*. [Video]. YouTube. https://www.youtube.com/watch?v=maUICwTzK5k

Janis. (2020). *What is Level of Effort?* [Video]. YouTube. https://www.youtube.com/watch?v=u5JJOiSIaSI

Johnson, S. (2010). *Where Good Ideas Come From*. [Video]. YouTube. https://www.youtube.com/watch?v=NugRZGDbPFU

Kaufman, A. (2017). *How to Calculate Critical Path: Project Management Professional (PMP)® Exam Prep.* [Video]. YouTube. https://www.youtube.com/watch?v=0Lo4zsB-bjE

Kaufman, A. (2010). *Leading cause of project failure: Poor project planning*. [Video]. YouTube. http://www.youtube.com/watch?v=MwQ_5JfDIJs

Kaufman, A. (2015). *Sequence Activities Demonstration*. [Video]. YouTube. https://www.youtube.com/watch?v=H-1Ab30_rrM

Keirsey. (2022*). Keirsey Inventory*. https://profile.keirsey.com/#/b2c/assessment/start

Knowledgehut. (2019) *4 Stages of Project Life Cycle | Phases of Project Management Life Cycle*. [Video]. YouTube. https://www.youtube.com/watch?v=N3N9-RLSbvo

KnowledgeHut. (2018). *7 Basic Quality Control Tools for Efficient Project Management*. [Video]. YouTube. https://www.youtube.com/watch?v=IFPF0TIFG-4&feature=emb_logo

Kogon, K. (2015). *Project Management for the Unofficial Project Manager: A FranklinCovey Title*. BenBella Books

Kouzes, J.M. & Posner B.Z. (2017). *The Leadership Challenge*. San Francisco: Jossey-Bass, Inc.

Kumar, K. (2016). *Organization structure Influence project management | Functional Matrix Projectized organization*. [Video]. YouTube. https://www.youtube.com/watch?v=pbtqqZm8Ak0

Leduc, S. (2009). *Using Assessments to Learn about Yourself*. |Video]. YouTube. http://www.youtube.com/watch?v=4YwmaHiT7b8

Lencioni, P. (2019). *5 Dysfunctions of Teams*. [Video]. YouTube. https://www.youtube.com/watch?v=Ro0NBgHo_a8

Litten, D. (2022). *PMP Earned Value in 10 Minutes Flat*. [Video]. YouTube. https://www.youtube.com/watch?v=OfwYr6Pw_x0

List 25. (2013). *25 Biggest Corporate Scandals Ever.* [Video]. YouTube. https://www.youtube.com/watch?v=LrJaOOGJLW0

Madsen, S. (2015). *How to Build Trust - Project Leadership Training*. [Video]. YouTube. https://www.youtube.com/watch?v=afgzypHh8Oc

Magia3e. (2010). *Legitimate Power - Big Bang Theory style*. [Video]. YouTube. https://www.youtube.com/watch?v=v4LlADSF93c

Massey, K. (2011). *Effective email communication*. [Video]. YouTube. https://www.youtube.com/watch?v=HF79m5KjuJg

Marsh, F. (2022). *10 Great Leadership Skills of Project Management*. https://aboutleaders.com/10-great-leadership-skills-of-project-management/

Mercanix. (n.d.). *Employee likes and dislikes*. http://www.mercanix.com

MindToolsVideos. (2018). *The Blake Mouton Managerial Grid: Balancing People's Well-Being with Productivity.* [Video]. YouTube. https://www.youtube.com/watch?v=lSvVdEKq9io

Moore, S. (2009). *Strategic Project Portfolio Management: Enabling a Productive Organization.* Microsoft Executive Leadership Series.

Morphy, T. (2008). *20 Common Project Risks - Example Risk Register*. https://www.stakeholdermap.com/risk/register-common-project-risks.html

Mulcahy, R., (2020). *Rita Mulcahy's Hot Topics*. 10th edition. RMC Publications.

Mulcahy,R., (2022). *Rita Mulcahy's PMP Test Prep.* 10th edition. RMC Publications.

Muzio, E. (2011). *The Ladder of Inference Creates Bad Judgment*. [Video]. YouTube. http://www.youtube.com/watch?v=K9nFhs5W8o8

National Institutes of Health. (2022). *6 Safety Responsibilities for Supervisors*. https://www.ors.od.nih.gov/sr/dohs/HealthAndWellness/Pages/Safety-Responsibilities-for-Supervisors.aspx

Nutcache. (2022). *10 Attributes of an Effective Project Manager*. https://www.nutcache.com/blog/10-attributes-effective-project-manager/

O'Connor, D. (2009). *4 Magic Phrases You Can Use to Respond to ANYTHING | Power Phrases for Work*. [Video]. YouTube. https://www.youtube.com/watch?v=g5RknemM8Hw

People Rich. (2012). *Project Management 2 – Scope Management*. [Video]. YouTube. https://youtu.be/xjawa7wnd-w

Patreon. (2020). *The Beginner's Guide to Microsoft To Do.* [Video]. YouTube. https://www.youtube.com/watch?v=6BfVtLD6iT0

Pentameter, M. (2017). *How to Deal with Passive Aggressive People*. [Video]. YouTube. https://www.youtube.com/watch?v=xTxSP0BixA4

Peterson, D. (n.d.). *Learning styles inventory – Are you a grouper or a stringer?* http://adulted.about.com/od/applyingwhatyoulearn/a/Learning-Styles-Inventory-Are-You-A-Grouper-Or-A-Stringer.htm

Pink, D. (2010). *The surprising truth about what motivates us.* [Video]. YouTube. http://www.youtube.com/watch?v=u6XAPnuFjJc

Pritchard, C. (n.d.). *The Project Management Communications Toolkit*. Artech.

ProjectManagement.com. (2021). *Top 10 Reasons Why Projects Fail.* https://project-management.com/top-10-reasons-why-projects-fail/

Project Management Institute (PMI). (2015). *Pulse of the Profession: Capturing the Value of Project Management*. http://www.pmi.org/learning/pulse.aspx

Project Management Institute (PMI). (2017). *A Guide to the Project Management Body of Knowledge (PMBOK® Guide)*-Sixth Edition.

Project Management Institute (PMI). (2021). A Guide to the Project Management Body of Knowledge (PMBOK® Guide)-Seventh Edition.

Project Management Institute (PMI). (2017). *Q&As For The PMBOK® Guide Sixth Edition*.

Pritchard, C. (2013). *The Project Management Communications Toolkit*. Artech.

Ray, S. (2017). *12 Effective Communication Strategies at Work*. https://www.projectmanager.com/blog/12-effective-communication-strategies-work

Record Nations. (2017). *Records Retention—Requirements and Guidelines*. [Video]. YouTube. https://www.youtube.com/watch?v=3YHnEeKiM34&t=1s

Ridpath, R. (2012). *Managing stress in the workplace – The 4 A's*. [Video]. YouTube. https://www.youtube.com/watch?v=XeKfpCGwN8I

RiskDoctorVideo. (2015). *The Wasa – a true story of scope creep*. [Video]. YouTube. https://www.youtube.com/watch?v=kmJ59yyYza4

SafetyVideoShop. (2009). *10 Commandments of Workplace Safety*. [Video]. YouTube. https://www.youtube.com/watch?v=3C6js5JtCIQ

Scott, M. & Brydon, S. (1997). *Dimensions of Communications: An Introduction*. Mountain View: Mayfield Publishing Co.

Scott, S. (2021). *The Difference between Discrimination & Preference*. http://smallbusiness.chron.com/difference-between-discrimination-preference-10338.html

TechnologyAdvice. (2015). *Basecamp Alternatives: Project Management Software*. [Video]. YouTube. https://youtu.be/pqFGpqQkwV8

The Crowd Training. (2016). *Drawn Out Project Scope Management*. [Video]. YouTube. https://www.youtube.com/watch?v=KZn2ly_p8pY

The Finance Storyteller. (2018). *ROI vs payback method*. [Video]. YouTube. https://www.youtube.com/watch?v=o4of1uNSRis

TheFunkypineapple. (2012). *The Many, Many, Many Mispronunciations of Gloria from Modern Family*. [Video]. YouTube. https://www.youtube.com/watch?v=om9EhbBo1Yc

Thompson, J. (2011). Is nonverbal communication a numbers game? *Psychology Today*. http://www.psychologytoday.com/blog/beyond-words/201109/is-nonverbal-communication-numbers-game

Tracy, B. (2012). *6 Time Management Tips to Increase Productivity*. [Video]. YouTube. https://www.youtube.com/watch?v=3KafgR2WEgY

Tracy, B. (2013). *How to Conduct Effective Meetings*. [Video]. YouTube. https://www.youtube.com/watch?v=wm0XtZ_K3kA

Treasure, J. (2011). *5 ways to listen better*. TEDTalk. [Video]. YouTube. https://www.youtube.com/watch?v=cSohjlYQI2A

True Colors International. (2021). *What is True Colors*. [Video]. YouTube. https://www.youtube.com/watch?v=XsM1wZ5WT8Q&t=32s

Tuckman, B. (1965). Developmental sequence in small groups. *Psychological Bulletin*, Vol 63(6), Jun 1965, 384-399Whitt, J. (2013). *Conflict Resolution Training: How to Manage Team Conflict in Under 6 Minutes! [Video]. YouTube. https://youtu.be/PHJ8eybXJdw*

Ury, W. (2010). *The walk from "no" to "yes." TED*. [Video]. YouTube. https://www.youtube.com/watch?v=Hc6yi_FtoNo

United States Office of Personnel Management. (1998). *Dealing with workplace violence.* Office of Workforce Relations Publication. https://www.opm.gov/policy-data-oversight/worklife/reference-materials/workplaceviolence.pdf

Van Camp, E. (2015). *10 Emotional Intelligence facts that will blow you away!* LinkedIn. https://www.linkedin.com/pulse/10-astonishing-ei-facts-eric-van-camp

Ways To Grow. (2017). *How to Prioritize Tasks Effectively: GET THINGS DONE*. [Video]. YouTube. https://www.youtube.com/watch?v=czh4rmk75jc

Weatherburn, K. (2013). *Introduction to developing & managing performance management systems.* [Video]. YouTube. https://www.youtube.com/watch?v=LC7bWCEtyHc

Whitmore, J. (2020). *1974, TIME OF YOUR LIFE, narrated by JAMES WHITMORE.* [Video]. YouTube. https://www.youtube.com/watch?v=36-JaSqOQaw

Whitt, J. (2013). *Controlling Project Scope: Top 5 Project Scope Control Tips.* [Video]. YouTube. https://www.youtube.com/watch?v=D7Jx1ob1sPg

Whitt, J. (2013). How to Capture Lessons Learned at the End of a Project. [Video]. YouTube. https://www.youtube.com/watch?v=DBUqW_ek4hl&feature=youtu.be

Whitt, J. (2012). *How to Run Team Meetings*. [Video]. YouTube. https://www.youtube.com/watch?v=xW0CjH95K3Q

Whitt, J. (2012). *4 Project Financial Management Tips – Project Management Finance*. [Video]. YouTube. https://www.youtube.com/watch?v=5DybRTTvrmY

Whitt, J. (2014). *The project Management Lifecycle*. [Video]. YouTube. https://www.youtube.com/watch?v=POuGKD3xLqs&t=7s

Whitt, J. (2018). *Top 10 Project Management Responsibilities*. [Video]. YouTube. https://www.youtube.com/watch?v=vMWJyELLSQ4

Wonstolen, L. (2014). *Big Bang Expert Power*. [Video]. YouTube. https://www.youtube.com/watch?v=5thkrGlq-M0

Yap, C. (2014). *Qualities of an Effective Team Member*. [Video]. YouTube. https://www.youtube.com/watch?v=mf9FRXPHMUg

Zavvy. (2022). *What Is A Leadership Competency Model And How To Create One?* https://www.zavvy.io/blog/leadership-competency-model

Zink, T. (2017). *Project Management Tips: Working with Tasks and Milestones in Your Project Schedule*. [Video]. YouTube. https://www.youtube.com/watch?v=tP4wIlWjsqU

Index

Index

[Created with **TExtract** / www.TExtract.com]

A good plan violently executed now is better than a perfect plan executed next week.

—George S. Patton

About the Author

Wesley E. Donahue, PhD, PE, PLS, PMP®, Six Sigma Blackbelt

As a business owner, engineer, manager, and now educator, I have always thought of myself as being in the business of helping other people succeed. But like you, at each step of the way I had to learn, and I had to put that learning into action. This latest book, ***Project Mastery: Bridging Processes, People, and Resources,*** speaks from the voice of experience from being a project manager. It is a practical guidebook designed for individuals who manage or anticipate managing workplace projects.

As for my background, I am a professor of Organization Development at Penn State University. In this capacity, I am engaged in top-ranked graduate research and programming in learning and performance systems and lead a successful online graduate program in organization development and change. I am also president of Centrestar, Inc. a firm that offers a unique and straightforward approach for professionals to assess their leadership skills, develop personalized roadmaps for success, and access on-demand micro-learning courses and competency-based workbooks available on Amazon. View our resources at: https://www.centrestar.com

Prior to my current position at Penn State, I was Director of Penn State Management Development Programs and Services, where I led and provided education and training services to business and industry clients around the world. Before that, I had years of multi-faceted experience as a project manager, leader, and business owner. I was regional sales vice-president for a mid-sized plastics packaging manufacturer; co-founder and executive vice-president of a manufacturing company; corporate and international manager of technology for a *Fortune 200* multi-national company; and I also co-owned and operated a retail business.

I am a registered professional engineer and land surveyor with an MBA, Six Sigma black belt, certified project management professional (PMP®), author of ***Building Leadership Competence***, ***Unlocking Lean Six Sigma***, **Developing Effective Leadership in Organizations**, as well as a host of workbooks and and other education and training materials.

I would enjoy hearing from you and finding out how this book has helped you achieve your goals. Please contact me at wdonahue@centrestar.com

Best wishes for your continued success.

Knowing is not enough; We must apply.
Willing is not enough; We must do.

—Bruce Lee

www.ingramcontent.com/pod-product-compliance
Lightning Source LLC
LaVergne TN
LVHW062133200926
848484LV00007B/1811